with the financial support of:
Deutsche Forschungsgemeinschaft
Minister für Wissenschaft und Forschung
des Landes Nordrhein-Westfalen
National Science Foundation

sponsored by:
Deutsche Physikalische Gesellschaft
European Physical Society

PHYSICS OF SOLIDS
UNDER HIGH PRESSURE

PHYSICS OF SOLIDS
UNDER HIGH PRESSURE

Proceedings of the International Symposium on
the Physics of Solids under High Pressure
Bad Honnef, Germany, August 10-14, 1981

Edited by

James S. SCHILLING
Universität Bochum, Bochum, West Germany

Robert N. SHELTON
Iowa State University, Ames, Iowa, U.S.A.

NORTH-HOLLAND PUBLISHING COMPANY
AMSTERDAM · NEW YORK · OXFORD

ISBN: 0 444 86326 5

Publishers:

NORTH-HOLLAND PUBLISHING COMPANY
AMSTERDAM · NEW YORK · OXFORD

Sole distributors for the U.S.A. and Canada:

ELSEVIER NORTH-HOLLAND, INC.
52 VANDERBILT AVENUE
NEW YORK, N.Y. 10017

PRINTED IN THE NETHERLANDS

PREFACE

The use of high pressure as a parameter for the investigation of solid state phenomena has become increasingly important in recent years. Both the range and the quality of the high pressure techniques available to the experimenter have steadily improved, enhancing the intrinsic usefulness of this technique for creating new states of matter, as well as permitting an exacting comparison between experiment and theory. The true potential of the high pressure technique for studies of the properties of condensed matter is, however, only fully appreciated by a relatively small number of researchers. One of the main goals of the present international symposium in Bad Honnef was to enhance the significance of the high pressure method as a quantitative tool in solid state investigations by bringing high pressure specialists in contact with a group of leading theoreticians who had not necessarily been previously involved in high pressure research. A similar concept was behind the highly successful "First International Conference on the Physics of Solids at High Pressure" held in 1965 in Tucson, Arizona, U.S.A. The need, 16 years later, for a meeting of related character was underscored by the overwhelmingly positive response to the proposed symposium by leading authorities in the field. The highly active and stimulating level of participation during the conference itself provides perhaps the clearest measure of the correctness of the original concept.

The quality of the oral and poster contributions was very high, including many new results and innovative ideas. These papers contain ideas and suggestions for important future research and should be of considerable value to anyone actively involved in research on condensed matter, at ambient or high pressure. For this proceedings, the papers have been organized according to scientific content and are rearranged slightly from the conference program. For many readers the most valuable section of the book will be the final informal comments by a number of participants who summarized their views of high pressure research during the "Jam Session" at the end of the conference.

We would like to express our deep gratitude to the Deutsche Forschungsgemeinschaft, the Minister für Wissenschaft und Forschung des Landes Nordrhein-Westfalen, and the National Science Foundation for their generous financial support without which this conference could not have taken place. We also acknowledge the sponsorship of the symposium by the Deutsche Physikalische Gesellschaft and the European Physical Society. A contribution to the social program by the High Pressure Equipment Company, Inc. is also acknowledged. We would also like to express our deep indebtedness to J. Debrus, for his tireless assistance in planning the local arrangements, and to R. Offerzier and the staff of the Honnefer Haus for the superb culinary and room arrangements. The excellent preparation of the concert program by Elke Havenstein-Schilling was also appreciated.

The editors would also like to thank the staff of the North-Holland Publishing Company, in particular, W. Montgomery, Mary Carpenter and G. Andrew, for their superb handling of the publication of the Conference Proceedings. Particularly helpful for the speedy publication of the Proceedings was the presence of North-Holland representatives in Bad Honnef during the entire meeting.

The success or failure of a conference is determined, to a major extent, by the willingness of its participants to interact with each other and to engage in open discussions on matters of mutual interest. The participants to the present symposium formed a marvelously stimulating, questioning, and gesprächs-freudige group. Experiencing this comradeship of scientists was a rewarding experience for the conference organizers. Should the need for a conference of similar character evolve at a future date, we would be more than willing to expend the necessary effort to try it again.

JAMES S. SCHILLING
Bochum, W. Germany

ROBERT N. SHELTON
Ames, Iowa, U.S.A.

TABLE OF CONTENTS

PREFACE   v

LIST OF PARTICIPANTS   xi

Historical Perspective: Sixteen Years since Tucson
C.T. TOMIZUKA   1

High Pressure Optical Studies
H.G. DRICKAMER   3

Theory of Pressure-Dependent Spectra of Ions and Molecules in Crystals
F. WILLIAMS   15

Temperature Dependencies of Libron Frequency Shifts and Widths in
Solid Nitrogen and Carbon Dioxide
W.B. DANIELS, J. SCHMIDT and F. MEDINA   23

Solid O_2 near 298 K: Raman and Electronic Spectra of β and ε-O_2
K. SYASSEN and M. NICOL   33

Intramolecular Vibrational Mode Shifts with Pressure in Solid CO_2, N_2 and O_2
R.D. ETTERS and A.A. HELMY   39

Energy Dispersive X-Ray Diffraction Measurements at Simultaneously High Pressure
and Temperature using Synchrotron Radiation: Preliminary Data on P(T)
Calibration and Phase Transformations in MnF_2 and FeF_2
M.H. MANGHNANI, E.F. SKELTON, L.C. MING, J.C. JAMIESON, S. QADRI,
D. SCHIFERL and J. BALOGH   47

Electronic Structure and Properties of Nonmetals under Pressure
W.A. HARRISON   57

X-Ray Absorption Spectroscopy at High Pressures
R. INGALLS, J.M. TRANQUADA, J.E. WHITMORE and E.D. CROZIER   67

Structural Studies at High Pressure and Temperature using Synchrotron Radiation
I.L. SPAIN, S.B. QADRI, C.S. MENONI, A.W. WEBB and E.F. SKELTON   73

Energy Dispersive X-Ray Diffraction at High Pressure in Chess
A.L. RUOFF and M.A. BAUBLITZ, JR.   81

Recent Advances in the Study of Structural Phase Transitions at High Pressure
G.A. SAMARA   91

New ways of looking for Phase Transitions at Multi-Megabar Dynamic Pressures
J.W. SHANER   99

Studies on Phase Transitions in Liquid Crystals under High Pressure
R. SHASHIDHAR   109

Inter- and Intra-Molecular Disorder and Volume Contributions to the
Entropy of Melting
A. RUBČIĆ and J. BATURIĆ-RUBČIĆ   117

Optical Properties of Cesium and Iodine under Pressure
K. SYASSEN, K. TAKEMURA, H. TUPS and A. OTTO   125

Structure of Cesium and Iodine under Pressure
K-I. TAKEMURA, S. MINOMURA and O. SHIMOMURA   131

Interatomic Potentials for Solid Argon and Neon at High Pressures
G.T. ZOU, H.K. MAO, L.W. FINGER, P.M. BELL and R.M. HAZEN   137

Inversion of the Γ and X Conduction Bands in InP under High Pressure
T. KOBAYASHI, T. TEI, K. AOKI, K. YAMAMOTO and K. ABE   141

Absorption Spectra and Pressure Dependence of Pseudodirect Gaps in $ZnSiP_2$ Crystal
T. SHIRAKAWA and J. NAKAI   149

Ordered Ground States of Metallic Hydrogen and Deuterium
N.W. ASHCROFT   155

The Metallization of some simple Systems
M. ROSS and A.K. MCMAHAN   161

Shock Anomaly and s-d Transition in High Pressure Lanthanum
A.K. MCMAHAN, H.L. SKRIVER and B. JOHANSSON   169

Metallic Magnetism under High Pressure
E.P. WOHLFARTH   175

Magnetic Properties of Intermetallic Uranium Compounds under High Pressure
J.J.M. FRANSE, P.H. FRINGS, F.R. DE BOER and A. MENOVSKY   181

Electronic Structure of Bulk Solids and Artificial Materials at
Positive and Negative Pressures
A.J. FREEMAN   193

Organic Superconductors
M. WEGER   199

High Pressure Effects on Metal-Semiconductor Peierls Transition: $MEM-(TCNQ)_2$
D. BLOCH, J. VOIRON, J. KOMMANDEUR and C. VETTIER   203

Recent High Pressure Fermi Surface Studies on $(TMTSF)_2PF_6$ and ReO_3
J.E. SCHIRBER   207

Abnormal Compressibility Divergence in ReO_3 under High Pressure
B. BATLOGG, R.G. MAINES and M. GREENBLATT   215

The Evolution of the Charge Density Wave State of $2H-TaSe_2$
D.B. MCWHAN and R.M. FLEMING   219

High Pressure Studies of Graphite Intercalation Compounds
R. CLARKE   223

Some Recent Results in Metal Hydrogen Systems in the High Pressure Region
S. FILIPEK, B. BARANOWSKI and M. KLUKOWSKI   231

Pressure Dependence of the Electrical Resistivity of Some Metallic Glasses
G. FRITSCH, J. WILLER, A. WILDERMUTH and E. LÜSCHER   239

Electronic Structure of the Actinide Metals
B. JOHANSSON, H.L. SKRIVER and O.K. ANDERSEN   245

First Principles Calculation of the Electronic Pressure: Application to
Transition Metals, Semiconductors and Cerium
D. GLÖTZEL   263

5f-Delocalization in Neptunium Laves Phases under Pressure
J. MOSER, W. POTZEL, B.D. DUNLAP, G.M. KALVIUS, J. GAL, G. WORTMANN,
D.J. LAM, J.C. SPIRLET and I. NOWIK   271

Electronic Transitions in Praseodymium under Pressure
H.L. SKRIVER   279

Pressure-Induced Valence Instability in Praseodymium Metal and Dilute
La Pr and Y Pr Alloys: A Counterpart of the Cerium Case?
J. WITTIG — 283

Pressure-Induced Valence and Structure Change in some anti-Th_3P_4 Structure
Rare Earth Compounds
A. WERNER, H.D. HOCHHEIMER, A. JAYARAMAN and E. BUCHER — 295

Pressure Induced Semiconductor-Metal Transitions in $Tm_{1-x}Eu_xSe$
H. BOPPART and P. WACHTER — 301

High Pressure Diffraction Studies of YbH_2 up to 28 GPa
J. STAUN OLSEN, B. BURAS, L. GERWARD, B. JOHANSSON, B. LEBECH, H. SKRIVER
and S. STEENSTRUP — 305

Low-Temperature Electrical Resistivity of Unstable-Valent $CeIn_3$, $CePd_3$ and
YbCuA1 up to 225 kbar
J.M. MIGNOT and J. WITTIG — 311

L_{III} Edge proves Pressure-Induced Valence Transition in Yb Metal
K. SYASSEN, G. WORTMANN, J. FELDHAUS, K.H. FRANK and G. KAINDL — 319

Specific Heat Measurements on Intermediate-Valent YbCuA1 under High Pressure
A. BLECKWEDEL and A. EICHLER — 323

Effects of Pressure on the Mixed Valent Properties of Ce and its
Intermetallic Compounds
S.H. LIU — 327

Electronic Instabilities in Rare Earth Compounds: Spin Fluctuations in
Ce Compounds
M. CROFT, H.H. LEVINE and R. NEIFELD — 335

Electronic Instabilities in Rare Earth Compounds: "Recent Results" on
Cooperative Charge Fluctuations in $EuPd_2Si_2$
M. CROFT, J.A. HODGES, E. KEMLY, A. KRISHNAN, V. MURGAI, L. GUPTA
and R. PARKS — 341

Some Recent Results in Magnetism under High Pressure
J.S. SCHILLING — 345

Unusual Observation in some Unusual Rare-Earth Chevrel Ternary Compounds
C.W. CHU, M.K. WU, R.L. MENG, T.H. LIN, V. DIATSCHENKO and S.Z. HUANG — 357

High Pressure Study on a Linear Chain Compound TaS_3
D.L. ZHANG, M.K. WU, T.H. LIN, P.H. HOR, C.W. CHU and A.H. THOMPSON — 365

The Application of Thermodynamic Grüneisen Parameters to Volume Dependence Studies
T.F. SMITH — 369

Nonlinear Pressure Effects in Superconducting Rare Earth – Iron – Silicides
C.U. SEGRE and H.F. BRAUN — 381

High Pressure Studies of the Superconductivity of La-Chalcogenides
A. EILING, J.S. SCHILLING and H. BACH — 385

Superconducting Transition under Pressure in Some Eutectic Alloys
F.S. RAZAVI and J.S. SCHILLING — 397

Electrical Behavior of Ca, Sr, Ba, and Eu at very High Pressures and
Low Temperatures
F.P. BUNDY and K.J. DUNN — 401

Electrical and Magnetic Properties of Pressure Quenched CdS
C.G. HOMAN, D. KENDALL and R.K. MACCRONE — 407

CONCLUDING REMARKS — 413

AUTHOR INDEX — 419

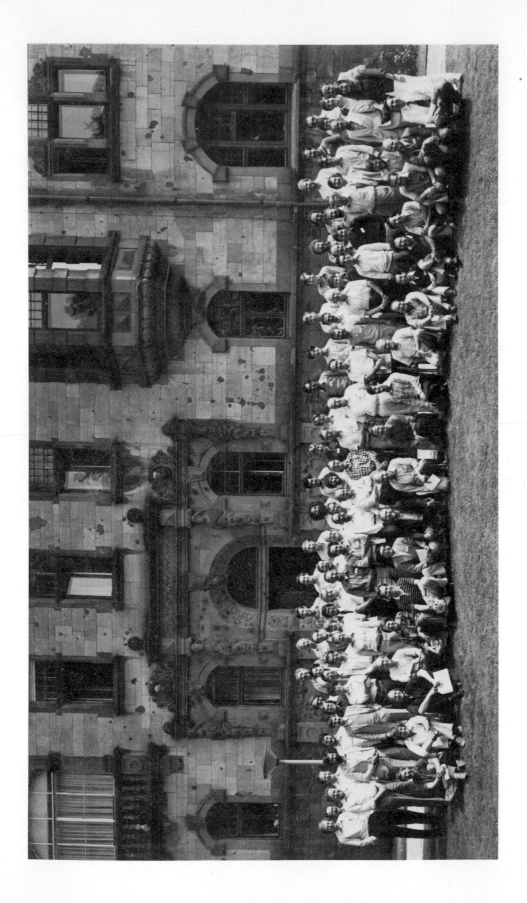

LIST OF PARTICIPANTS

M. Abd-Elmeguid
Institut für Experimentalphysik IV
Ruhr-Uni Bochum
4630 Bochum
WEST GERMANY

K. Aoki
National Chemical Laboratory
Tsukuba Research Centre
Yatabe
Ibaraki 305
JAPAN

N.W. Ashcroft
Laboratory of Atomic and Solid State Physics
Clark Hall, Cornell University
Ithaca, N.Y.14853
U.S.A.

B. Batlogg
Bell Laboratories, 1D-237
Murray Hill, N.J.07974
U.S.A.

U. Benedict
Europäisches Institut für Transurane
Postfach 2266
7500 Karlsruhe 1
WEST GERMANY

D. Bloch
Lab. L. Néel, CNRS
Avenue des Martyrs
B.P. 166 centre de Tri
38042 Grenoble Cedex
FRANCE

H. Boppart
Lab. für Festkörperphysik
ETH Zürich
8093 Zürich
SWITZERLAND

H.F. Braun
Dép. de Physique
Université de Genève
32 Bd. d'Yvoy
Geneva
SWITZERLAND

F.P. Bundy
General Electric Corporate
Research & Development Laboratory
P.O. Box 8
Schenectady, N.Y.12301
U.S.A.

C.W. Chu
Department of Physics
University of Houston
Houston, TX 77004
U.S.A.

R. Clarke
Department of Physics
University of Michigan
Ann Arbor, MI 48109
U.S.A.

M. Croft
Physics Department
Rutgers University
Piscataway, N.J. 08854
U.S.A.

W.B. Daniels
Physics Department
University of Delaware
Newark, DE 19711
U.S.A.

M. Dietrich
Kernforschungszentrum
7500 Karlsruhe 1
WEST GERMANY

H.G. Drickamer
105 R. Adams Laboratory
University of Illinois
1209 W. California Street
Urbana, IL 61801
U.S.A.

A. Eichler
Institut Technische Physik
TV Braunschweig
Mendelssohn Strasse 1B
3300 Braunschweig
WEST GERMANY

A. Eiling
Ruhr Universität Bochum
Experimentalphysik IV
Universitätsstrasse 150
4630 Bochum
WEST GERMANY

R.D. Etters
Physics Department
Colorado State University
Fort Collins
Colorado 80523
U.S.A.

List of Participants

S. Filipek
Institute of Physical Chemistry
Polish Academy of Science
Kasprzaka 44
Warsaw
POLAND

K. Frank
FU Berlin
Institut für Atom- und Festkörperphysik
Boltzmannstrasse 20
1000 Berlin 30
WEST GERMANY

J.M.M. Franse
Natuurkundig Laboratorium
Valckenierstraat 65
1018 XE Amsterdam
THE NETHERLANDS

A.J. Freeman
Physics Department
Northwestern University
Evanston, IL 60201
U.S.A.

P.H. Frings
Universiteit van Amsterdam
Natuurkundig Laboratorium
Valckenierstraat 65
1018 XE Amsterdam
THE NETHERLANDS

W. Gey
Institut für Technische Physik
TV Braunschweig
Mendelssohn Strasse 1B
3300 Braunschweig
WEST GERMANY

D. Glötzel
Max-Planck-Institut für Festkörperforschung
Heisenberg Strasse 1
7000 Stuttgart 80
WEST GERMANY

T. Gregorian
GHS Paderborn
Hochdruckphysik
Postfach 1621
4790 Paderborn
WEST GERMANY

R.A. de Groot
Physics Laboratory
Universiteit van Nijmegen
Toernooiveld
6525 ED Nijmegen
THE NETHERLANDS

B.N. Harmon
Department of Physics
Iowa State University
Ames, Iowa 50011
U.S.A.

W.A. Harrison
Applied Physics Department
Stanford University
Stanford, CA 94305
U.S.A.

G. Hilscher
Institut für Experimentalphysik
Technische Universität Wien
1040 Wien
Karlsplatz 13
AUSTRIA

H.D. Hochheimer
Max-Planck-Institut für Festkörperforschung
Heisenbergstrasse 1
7000 Stuttgart 80
WEST GERMANY

W.B. Holzapfel
Uni-GH Paderborn
Postfach 1621
4790 Paderborn
WEST GERMANY

R.L. Ingalls
Physics Department
University of Washington
Seattle, WA 98112
U.S.A.

A. Jayaraman
Bell Laboratories
Murray Hill, N.J.07974
U.S.A.

J.R. Jevais
Centre d'etude de Limeil
B.P. 27
94194 Villeneuve St. Georges
FRANCE

B. Johansson
Physics Department
University of Aarhus
8000 Aarhus
DENMARK

T. Kobayashi
Department of Electrical Engineering
Kobe University
Rokkodai, Nada
Kobe 657
JAPAN

J. Lauterjung
Minerol.-Petrol. Institut
Poppelsdorfer Schloss
5300 Bonn 1
WEST GERMANY

S.H. Liu
Oak Ridge National Laboratory
Oak Ridge, TN 37830
U.S.A.

R.K. MacCrone
Department of Physics
Rensselaer Polytechnic Institute
Troy, N.Y.12181
U.S.A.

M.H. Manghnani
University of Hawaii
Institute of Geophysics
2525 Correa Road
Honolulu, Hawaii 96822
U.S.A.

A.K. McMahan
Lawrence Livermore Laboratory
P.O. Box 808
Livermore, CA 94550
U.S.A.

D.B. McWhan
Bell Laboratories, 1D-234
Murray Hill, N.J.07974
U.S.A.

M. Methfessel
Physics Laboratory
Universiteit Nijmegen
Toernooiveld
6525 ED Nijmegen
THE NETHERLANDS

J.M. Mignot
CRTBT/CNRS
B.P. 166 X
38042 Grenoble
FRANCE

J. Moser
Physik Department E 15 der
Technische Universität München
James-Franck-Strasse
8046 Garching
WEST GERMANY

F.M. Mueller
Physics Laboratory
Universiteit Nijmegen
Toernooiveld
6525 ED Nijmegen
THE NETHERLANDS

H. Myron
Physics Laboratory
Universiteit Nijmegen
Toernooiveld
6525 ED Nijmegen
THE NETHERLANDS

M. Nicol
Department of Chemistry
University of California
Los Angeles, CA 90024
U.S.A.

J.S. Olsen
Physics Laboratory II
H.C. Ørsted Institute
Universitetsparken 5
2100 København
DENMARK

M. Penicaud
Centre d'Etudes de Limeil
B.P. 27
94190 Villeneuve St. Georges
FRANCE

D. Rainer
Physikalisches Institut
Universität Bayreuth
8580 Bayreuth
WEST GERMANY

F. Razavi
Experimentalphysik IV
Ruhr Universität
4630 Bochum
WEST GERMANY

M. Ross
L-355 Lawrence Livermore National Laboratory
Livermore, CA 94550
U.S.A.

A. Rubcic
Faculty of Science
University of Zagreb
Bijenicka Cesta 46
41001 Zagreb
YUGOSLAVIA

A.L. Ruoff
Department of Materials, Science & Engineering
Cornell University
Ithaca, N.Y.14853
U.S.A.

G.A. Samara
Department 5130
Sandia National Laboratories
Albuquerque, NM 87185
U.S.A.

J.S. Schilling
Experimentalphysik IV
Universität Bochum
Postfach 2148
4630 Bochum
WEST GERMANY

J.E. Schirber
Department 5150
Sandia Laboratories
Albuquerque, NM 87185
U.S.A.

J.W. Shaner
MS 970
Los Alamos Scientific Laboratory
Los Alamos, NM 87545
U.S.A.

R. Shashidar
Lehrstuhl für Physikalische Chemie II
Ruhr Universität
4630 Bochum 1
WEST GERMANY

R.N. Shelton
Department of Physics
Iowa State University
Ames, Iowa 50011
U.S.A.

T. Shirakawa
Department of Electronic Engineering
Osaka University
Osaka 565
JAPAN

H.L. Skriver
Risö National Laboratory
4000 Roskilde
DENMARK

T.F. Smith
Physics Department
Monash University
Clayton, Victoria 3149
AUSTRALIA

I.L. Spain
Department of Physics
Colorado State University
Fort Collins, Colorado 80523
U.S.A.

F. Steglich
Institut für Festkörperphysik
Fachbereich 5
6100 Darmstadt
Hochschulestrasse 2
WEST GERMANY

K. Syassen
Physik. Institut IV
Universität Düsseldorf
Universitätsstrasse 1
4000 Düsseldorf 1
WEST GERMANY

K. Takemura
Physikalisches Institut
Universität Düsseldorf
4000 Düsseldorf 1
WEST GERMANY

C.T. Tomizuka
Physics Department
University of Arizona
Tucson, Arizona 85721
U.S.A.

D. Vigren
Institut für Theoretische Physik
Universität des Saarlandes
Bau 38
Saarbrücken
WEST GERMANY

Y.K. Vohra
Universität Gesamthochschule
Experimentalphysik FB 6
4790 Paderborn
WEST GERMANY

D. Wagner
Institut Theoretische Physik
Ruhr Universität Bochum
4630 Bochum
WEST GERMANY

U. Walter
Universität Köln
II Physikalisches Institut
Zülpicher Strasse 77
5000 Köln
WEST GERMANY

M. Weger
Hebrew University
Physics Department
Jerusalem
ISRAEL

A. Werner
Max-Planck-Institut für Festkörperforschung
Heisenbergstrasse 1
7000 Stuttgart 80
WEST GERMANY

J. Willer
Physik Department
Technische Universität München
8046 Garching
WEST GERMANY

F. Williams
Physics Department
University of Delaware
Newark, Delaware 19711
U.S.A.

J.A. Wilson
H.M. Wills Physics Laboratory
University of Bristol
Bristol, RS8 1TL
U.K.

J. Wittig
IFF KFA Jülich
Postfach 1913
5170 Jülich
WEST GERMANY

E.P. Wohlfarth
Department of Mathematics
Imperial College
London SW7 2BZ
U.K.

G. Wortmann
Institut für Atom- und Festkörperphysik
Freie Universität Berlin
Boltzmannstrasse 20
1000 Berlin 30
WEST GERMANY

M.K. Wu
University of Houston
Physics Department
Houston, TX 77004
U.S.A.

G. Zou
Geophysical Laboratory
Carnegie Institute of Washington
Washington, DC 20008
U.S.A.

PHYSICS OF SOLIDS UNDER HIGH PRESSURE
J.S. Schilling, R.N. Shelton (editors)
© *North-Holland Publishing Company, 1981*

HISTORICAL PERSPECTIVE: SIXTEEN YEARS SINCE TUCSON

C. T. TOMIZUKA

Department of Physics, University of Arizona
Tucson, Arizona, USA

Some of us who gathered in Tucson in the fall of 1964 to plan the First International Conference on the Physics of Solids at High Pressures had to face the fact that not many solid state physicists had been actively employing pressure as a parameter in their study of solids. Very few theoretical physicists had been seriously investigating the effect of pressure on the properties of solids.

In fact, in some cases, we felt very strongly that an important purpose of the Conference was to invite those solid state physicists who are not engaged in high pressure research but whose specialization fell into the area where high pressure might potentially be a powerful tool. The exchange of ideas by mixing various solid state physicists, especially those involved in high pressure work with those who were not, for three days was expected to contribute in the long run to a better understanding of solids.

I am delighted to find a large number of active participants of sixteen years ago are again active participants today at this conference along with those who have since joined our ranks.

In a sense, this is the Second International Conference of its kind devoted to the physics of solids as the central theme and pressure as its tool, and not the other way around. I do not need to remind those who are here of the spectacular progress made both in solid state physics and its parameter, high pressure and its techniques since our First Conference. This progress has been made by standing on the shoulders of the past grant achievements in both areas. Unfortunately, some of those who made vital contributions in the field have since the Tucson Conference been lost from among us. Leonid Vereschagin and Andrew W. Lawson are very much missed by all of us. Andy Lawson was on the planning committee for the First Conference. He had also been my mentor and played a same role for a large number of physicists of my generation.

Let us hope that the Third International Conference takes place in less than 16 years from now and that all of us will be present without loss to observe another significant progress in the physics of solids at high pressure.

PHYSICS OF SOLIDS UNDER HIGH PRESSURE
J.S. Schilling, R.N. Shelton (editors)
© *North-Holland Publishing Company, 1981*

HIGH PRESSURE OPTICAL STUDIES

H. G. Drickamer

School of Chemical Sciences and Materials Research Laboratory
University of Illinois
Urbana, Illinois 61801

High pressure experimentation may concern intrinsically high pressure phenomena, or
it may be used to gain a better understanding of states or processes at one atmosphere. The latter application is probably more prevalent in condensed matter
physics. Under this second rubric one may either use high pressure to perturb
various electronic energy levels and from this "pressure tuning" characterize states
or processes, or one can use pressure to change a macroscopic parameter in a controlled way, then measure the effect on some molecular property.
In this paper we emphasize the pressure tuning aspect, with a lesser discussion
of macroscopic - molecular relationships.
In rare earth chelates the efficiency of 4f-4f emission of the rare earth is controlled by the feeding from the singlet and triplet levels of the organic ligand.
These ligand levels can be strongly shifted by pressure. A study of the effect of
pressure on the emission efficiency permits one to understand the effect of ligand
modification at one atmosphere.
Photochromic crystals change color upon irradiation due to occupation of a
metastable ground state. In thermochromic crystals, raising the temperature
accomplishes the same results. We demonstrate that, for a group of molecular
crystals (anils) at high pressure, the metastable state can be occupied at room
temperature. The relative displacement of the energy levels at high pressure also
inhibits the optical process. Effects on luminescence intensity are shown to be
consistent.
In the area of microscopic - molecular relationships, we demonstrate the effect of
viscosity and dielectric properties on rates of non-radiative (thermal) and
radiative emission, and on peak energy for luminescence. For systems which can
emit from either of two excited states depending on the interaction with the
environment, we demonstrate the effect of rigidity of the medium on the rate of
rearrangement of the excited state.

1. INTRODUCTION

It is, in general, useful to consider high pressure research in two categories: the investigation of phenomena which occur only or primarily at high pressure, and the use of pressure to gain a better understanding and
characterization of states of matter or processes which occur at one atmosphere, or to
test theories concerning those properties or
processes. Both of these objectives are
important; the first contains many novel events,
but the second accounts for the wide application
of pressure in so many aspects of modern
science.

Both of these classes of experimentation have
had an impact in condensed phase physics, but
in this presentation I shall concern myself
with the second category of studies. Electronic
investigations of this type are undertaken
either to characterize electronic states or
excitations or to test theories concerning them,
or for a combination of these reasons. Theories
which have been verified by high pressure
studies include: ligand field theory, Mulliken's
theory of EDA complexes, van Vleck's theory of
spin state changes, the Förster-Dexter theory

of energy transfer in phosphors, various
theories of luminescence efficiency, and
theories concerning insulator-metal transitions
in various oxides and chalcogenides. This list
is, of course, representative rather than
exhaustive.

Within the bounds of this category of experimentation we may consider a further subdivision
into two classes. A general effect of pressure is to perturb the energy levels of the
outer electrons on atoms or molecules. Since
different levels are perturbed in different
degrees, one can use this "tuning" effect
directly in testing of theories, in characterizing states, or in bringing about transitions to new electronic configurations. In a
second type of experiment, the emphasis is on
the use of pressure to change a macroscopic
variable such as viscosity or dielectric
constant without change of temperature or
composition of the medium. From measured
changes in a molecular property such as luminescence rate or efficiency, rotational motion,
or reactivity, one can establish more definitive
tests of the relationship between the macroscopic and the molecular.

2.   PRESSURE TUNING OF ELECTRONIC STATES

We take up at some length two examples of pressure tuning of energy levels and then discuss more briefly the relationship between the macroscopic and the molecular as the latter might be considered a more typically "chemical" approach to the understanding of phenomena.

The pressure tuning of energy levels is especially well illustrated in molecular crystals of organic materials as the perturbations induced by pressure are large. The ideas can frequently be carried over to apply to simpler inorganic crystals.

2.1   Rare Earth Chelates

A study of the emission characteristics of rare earth chelates is an example where pressure experiments have permitted us to choose among alternative descriptions in a rather complex energy transfer process. The results are relevant to the design of rare earth lasers and lighting systems. A high pressure study has been made of a series of tris chelates of various rare earths with derivatives of acetylacetonate.[1] The rare earth ion is surrounded by six oxygens whose electronic character is modified by the substituents on the terminal carbons of the acetylacetonate. These modifications establish the energies of the excited singlet and triplet states of the ligands. In this presentation we shall restrict ourselves to one rare earth, Eu^{+3}. We shall largely be concerned with the ligand dibenzilmethide (DBM) with two phenyl substituents, but will briefly mention results for a complex with thenoyltrifluoroacetylacetonate (TTF).

The usual excitation is to first excited singlet state (S_1) of the ligand, while emission is observed from the lowest excited 4f level (labeled ^5D_o) of the rare earth.

The question under review is the nature of the path from S_1 to ^5D_o and the effect of various paths on the emission intensity. If the lowest triplet state of the ligand lies high in energy above the $^5D_1$ and ^5D_o states of the Eu^{+3} ion it is not involved in the energy transfer of e.g. $^5D_1 \rightarrow {}^5D_o$. If it lies between the $^5D_1$ and ^5D_o levels or lower, it can be profoundly involved in the process.

The location of the ligand S_1 and T_1 levels varies with type of substituents on the ligand and with the medium, crystalline or polymeric. With increasing pressure the S_1 and T_1 levels shift to lower energy vis a vis the Eu^{+3} levels so that from pressure studies we can test various hypothesis concerning the effect of changing the character of the ligands or the medium on the luminescence efficiency.[1]

In Figure 1 we show the energy of the ^5D_o and $^5D_1$ levels of Eu^{+3} on the right.

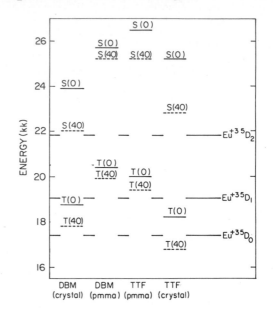

Figure 1 : Excited state energy levels for Eu^{+3} and ligands

On the left are the S_1 and T_1 levels of DBM in the crystalline state and dissolved in PMMA (polymethyl methacrylate). The solid lines represent the energies at one atmosphere pressure while the dashed lines represent the energies at 40 kilobars. (By 90 kilobars the S_1 and T_1 levels shift again as far to lower energy.) As one can see, the ligand levels lie at higher energy and shift less in the PMMA than in the crystalline state. As a result, there should be no important change in the ligand-metal energy transfer with pressure in PMMA. In the crystalline state we see two possible paths for transfer of excitation from $^5D_1$ to ^5D_o. The direct transfer occurs only slowly. There is, however, an alternative possibility. The path from $^5D_1$ to T_1 and back to ^5D_o is allowed. As the energy difference between T_1 and ^5D_o decreases the transfer rate will increase. However when T_1 comes within a few k_BT of ^5D_o back transfer can become significant, and as it comes below ^5D_o this quenching mechanism will dominate. In addition as T_1 moves to lower energy the rate of quenching via thermal crossing to the ground state will increase.

If transfer via T_1 is the most important process, one would anticipate that for DBM in PMMA one would observe a modest increase in intensity with pressure as $T_1 \rightarrow {}^5D_1$ transfer improves, and possibly a slight decrease at very high pressure due to back transfer.

In the crystal we would predict a distinct increase at lower pressures followed by a
dramatic drop as quenching takes over. It
should be mentioned that, superimposed on these
effects is a modest increase in emission intensity (a factor of 1 1/2 to 2 in ~ 60 kilobars)
because of increase in the radiative rate due
to increased spin-orbit coupling and better
d-f orbit mixing at high pressure. In Figure 2
we see the relative emission efficiency

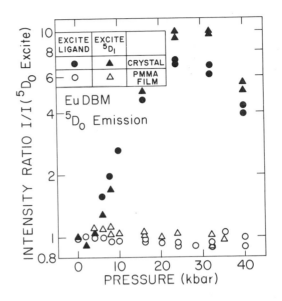

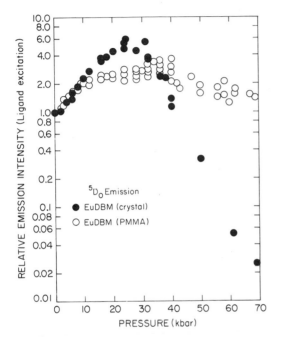

Figure 2 : Relative emission efficiency vs
pressure-DBM chelate, crystalline and in PMMA

Figure 3 : Emission efficiency vs pressure for
excitation in ligand and in $^5D_1$ relative to
excitation in $^5D_0$ - EuDBM

from $^5D_0$ vs pressure. The results are exactly
as predicted above. Evidently, transfer via
T_1 is dominant.

There are a couple of ways to check this conclusion. From Figure 1 we see three possible
methods to stimulate emission from $^5D_0$. If one
excites directly in $^5D_0$ there should be only
a modest change in intensity due to the change
in radiative rate mentioned above. If one
excites in the ligand, for the crystal, one
would expect, relative to what one observes
by exciting in $^5D_0$, a sharp increase at low
pressure and a drop beyond 30 kilobars. When
one excites in $^5D_1$, if the major path from
$^5D_1$ to $^5D_0$ is via the triplet state, the relative effect should be the same as for excitation in the ligand. On the other hand, for
the PMMA solution there should be no effect
of the type of excitation. In Figure 3 we
observe that the relative intensity effects are
exactly as predicted.

Another check involves the rate of back donation
from $^5D_0$ to T_1 which is limited by an energy
barrier E. One can extract E from the

temperature coefficient either of the emission
intensity or lifetime. (Both have been used.)
E should decrease as T_1 shifts to lower
energy. The shift of T_1 can be established
from the shift of phosphorescence from the
gadolinum derivative. In Figure 4 we show
the change in E plotted against the shift of
the phosphorescence. The dotted line has a
slope of 45°, precisely confirming the
hypothesis.

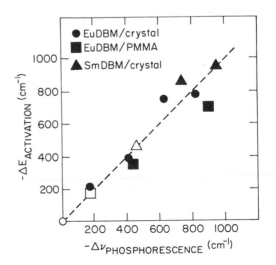

Figure 4 : Change in energy barrier for back
donation vs phosphorescence shift

Finally, from Figure 1, we deduce the behavior
of the TTF complex with Eu^{+3}. The ligand
levels lie slightly lower than for DBM. In the
plastic this should have no important effect,
but in the crystal T_1 already lies near the
optimum for ligand to $^5D_0$ transfer. We thus
anticipate little if any increase at low pressure and a dramatic decrease at high pressure.
These predictions are confirmed in Figure 5.

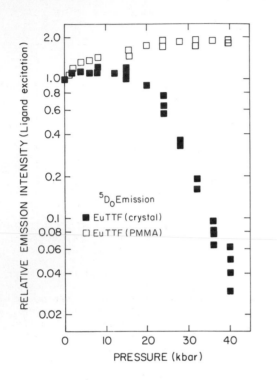

Figure 5 : Relative emission efficiency vs
pressure-TTF complex, crystalline and in PMMA

This study illustrates clearly how one can use
high pressure 'tuning' of energy levels to gain
a better understanding of the effects of ligand
substitutions at one atmosphere.

2.2   Thermochromic and Photochromic Anils

There is a broad class of materials in which an
electron can be transferred either by heating or
by optical excitation to a metastable state
with a different absorption spectrum. These
compounds are called thermochromic if the transfer mechanism is thermal, and photochromic if
it is optical. The anils constitute a group
of compounds where different derivatives or
crystal structure can exhibit either of these
phenomena.[2] In general, derivatives or
structures which are thermochromic are not
photochromic, and vice versa. We discuss here
salicylideneaniline (SA) which is photochromic
and the 5-bromo derivative (5BrSA) which is
thermochromic. The various possible structures
appear in Figure 6. The thermochromic yield is
generally only a few percent even at elevated

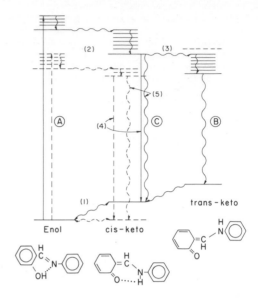

Figure 6 : Structures and energy levels for
thermochromic and photochromic pressures in anils

temperatures, but photochromic yields may be
50% or more at low temperature.

As shown in Figure 6, there are three possible
molecular arrangements. The normal ground
state is the enol form (A). The cis-keto
form (C) is the thermochromic product and the
structure which fluoresces. The trans-keto
form (B) is the photochromic product. The
solid horizontal lines represent energy levels
at one atmosphere. The straight vertical
lines represent optical transitions, while
the wavy lines represent thermal processes.
The dotted lines describe the situation at
high pressure.

From Figure 7 we see that absorption peaks
correspond to the enol (A) and cis-keto (C)
structures, as well as the fluorescence peak,
shift to lower energy by 2500 to 3200 cm^{-1}
(0.3 - 0.4 V) in 100 kilobars. So the enol (A)
and cis-keto (C) excited states are stabilized significantly with respect to their
ground states. The photochromic (trans-keto)
peak (B) shows little or no shift with
pressure.

We consider first the behavior of the thermochromic compound (5BrSA). At room temperature
and one atmosphere there is at most a trace
of thermochromic product, but with increasing
pressure at 296K the intensity of the thermochromic peak increases rapidly and it
constitutes 80% of the spectrum at 80 kilobars as shown in Figure 8, so the cis-keto
ground state is stabilized vis a vis the
enol ground state with pressure. We call

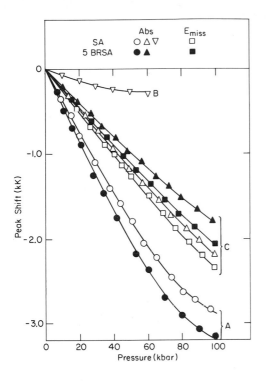

Figure 7 : Shift of absorption and emission peaks in anils

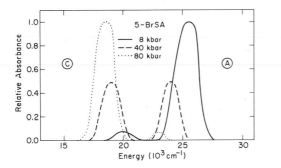

Figure 8 : Absorption bands A and C for 5-BrSA

this process piezochromism. From the areas under the peaks at steady state we extract the equilibrium constant (Figure 9). From the relationship

$$\left(\frac{\partial \ln K}{\partial P}\right)_T = \frac{\Delta V^o}{RT}$$

we establish that the volume change in transforming from the enol to the cis-keto form is 0.7% of the one atmosphere enol volume.

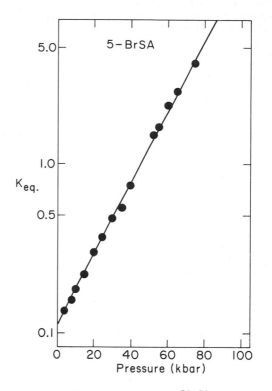

Figure 9 : E_{eq} vs pressure – 5BrSA

A series of isobars (Figure 10) permit us to establish ΔH^o as a line function of pressure. The results are shown in Figure 11. The line represents a plot of $p\Delta V^o$ normalized at one atmosphere. The agreement between the temperature and pressure data is excellent.

The emission intensity from 5BrSA decreases by a factor of 50-60 in 100 kilobars. This is associated with the large red shift of the emission peak. It has been clearly demonstrated[3,4] that a shift of the emission peak to lower energy is accompanied by an increase in k_{nr}, the non-radiative rate to the ground state, because of increased overlap between excited and ground state vibrational levels.

We turn now to the photochromic compounds (SA). At one atmosphere there is no sign of the thermochromic peak even at the most elevated temperatures. However, with increasing pressure at room temperature a modest but quite measurable amount of the thermochromic (piezochromic) peak appears - about 10% yield at 90 kilobars (see Figure 12). Evidently the cis-keto ground state is stabilized significantly with pressure even in the SA structure.

In Figure 13 we exhibit the photochromic yield after one hour's irradiation as a function of

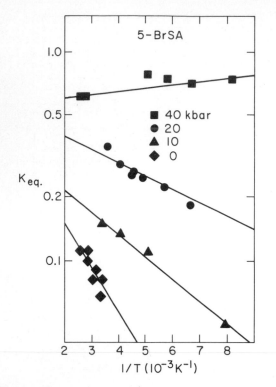

Figure 10 : Ln K_{eq} vs 1/T for 5BrSA

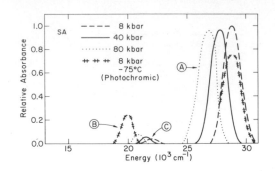

Figure 12 : Absorption spectra - SA

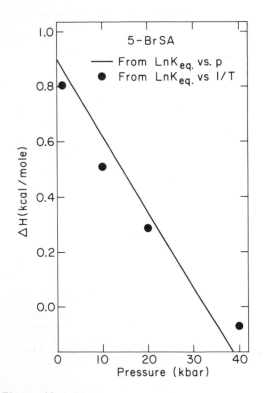

Figure 11 : ΔH vs pressure -5BrSA

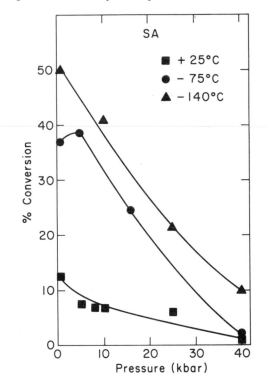

Figure 13 : Photochromic yield vs pressure (SA)

pressure at three temperatures. It should be emphasized that these results represent rates, not equilibrium yields. The rate of formation of the photochromic state decreases with increasing pressure at all temperatures. From Figure 6 we see that at high pressure the excited state of the cis-keto form lies below the excited state of the trans-keto form. Evidentally the mechanism of formation of the photochromic state is:

enol(gd) ⟶ enol(ex) ⟶
cis-keto(ex) ⟶ trans-keto(ex)⟶
trans-keto(gd).

Pressure inhibits the rate of the second last step.

Finally, in Figure 14 we exhibit the fluorescence yield for SA as a function of pressure.

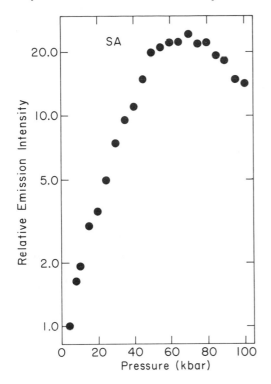

Figure 14 : Fluorescence emission yield vs pressure (SA)

It should be kept in mind that the emission occurs from the excited cis-keto states (C). The inhibition of the cis-keto (ex) → trans-keto (ex) step by pressure permits the increase in fluorescence yield by a factor of over twenty in seventy kilobars. At the highest pressures the increase in k_{nr} discussed above more than compensates, so that there is a drop in emission efficiency.

These results illustrate rather well the use of pressure tuning to sort out the steps in a group of interacting processes.

3. MACROSCOPIC-MOLECULAR CORRELATIONS

In relating bulk properties to molecular properties it is most convenient to consider liquid solutions, as large changes in bulk properties are induced by moderate pressures. In justification of discussing liquids at a "solid state" meeting one should remember that the former Solid State Division of the American Physical Society is now the Condensed Phase Division.

We discuss three molecular properties: k_{nr}, the rate of thermal dissipation of electronic excitation energy, k_r, the radiative rate of emission of excitation, and the energy (peak location) of the emission. The macroscopic variables include viscosity, dielectric constant (or square of the refractive index for nonpolar molecules) and the polarizability, which varies with the refractive index.

We study three types of molecules selected for their biological or technical importance. The porphyrins (Figure 15) with various metal

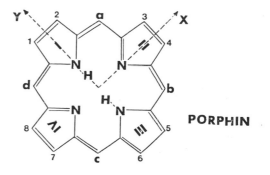

Figure 15 : Structure of porphin

ions in the center and various substituents are involved in photosynthesis, in oxygen transport in the blood, and as constituents of energy recovery systems. The important factor here is the effect of macroscopic variables on the distribution between radiative and non-radiative dissipation of excitation.[5]

Indole (Figure 16) is the chromophore in tryptophan, an amino acid which occurs in many proteins. As one of the few luminescent amino acids it is a valuable probe of protein environment and conformation, in that the wave length of the emission as well as k_r and k_{nr} depend on whether it is buried in the protein in an environment like a hydrocarbon or heavy alcohol, or exposed on the surface to water. It can be used as a marker for the steps in the denaturation process. 5-methoxy indole is useful to compare with indole since its lowest excited state is well established and is the same in all solvents. With indole and tryptophan there is such a wide difference in peak energy and k_r in different solvents that there are many speculations that the character of the excited state must be entirely different in different solvents. We show that a single macroscopic parameter can account for the changes in $h\nu$ and k_r observed.[6] This is

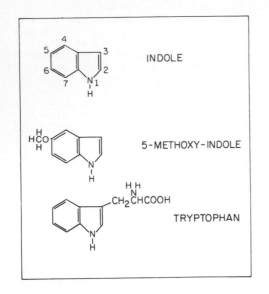

Figure 16 : Structure of indoles

important in differentiating between a triptophan
exposed to the solvent (H₂0) and one in an essentially hydrocarbon environment inside the
coiled protein. We present also a correlation
between peak energy and polarizability for a
series of polyenes which are prototypes for
rhodopsin, a molecule essential in vision.[7]

The solvents used include, hydrocarbons and
fluorocarbons, alcohols, chloroform and water.

The thermal dissipation of electronic excitation in liquid media should take place largely
by collisional effects. In the experiment
mentioned here, care has been taken to eliminate
impurity and concentration effects so that the
solute-solvent collisions are dominant. The
relevant macroscopic parameter is the viscosity.
In Figure 17 we present a plot of k_{nr} vs log η

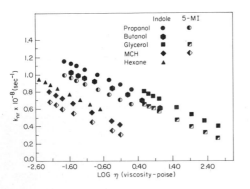

Figure 17: k_{nr} vs log η (viscosity) for indoles
in various solvents

for indoles in a series of solvents.[6]
Clearly the marcoscopic viscosity is the
controlling feature in thermal dissipation
of excitation. A similar correlation has been
obtained for metalloporphyrins.[5] Another
illustration of this correlation is shown in
Figure 18 where we plot k_{nr} vs pressure for

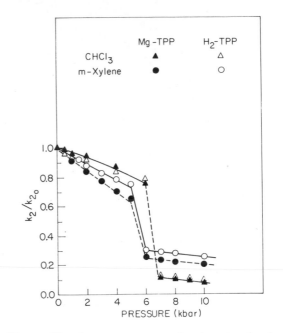

Figure 18 : k_{nr} vs pressure for two porphyrins
in m-xylene and CHCl₃

two prophyrins in m-xylene and chloroform
which freezes at ~ 4.5 and 5.5 kilobars
respectively. The sharp drop (by a factor
of 5-7) in k_{nr} at the freezing point clearly
shows the effect of the rigidity of the medium
on this parameter.

We turn now to the effect of the medium on the
radiative rate. In Figure 19 we show k_r vs
pressure for two metalloporphyrins in a
variety of solvents, mostly with small dipole
moments. We see that for these cases n^2
(the square of the refractive index) correlates
closely with k_r. This is quite reasonable
since n^2 measures the strength of the dispersion forces, and in non-polar media k_r
should depend on these forces. Note that
there is little or no discontinuity in k_r
at the freezing point in m-xylene or chloroform, in contrast to the effect of freezing
on k_{nr}. In Figure 20 we present the effect
of the dielectric constant (ε) on k_r for
several indoles in solvents from hexane and
MCH through a series of alcohols to water.
ε varies from 2 to 110. For the hydrocarbons and lower alcohols ε varies little
with pressure, but for water it increases
from ~ 80 to ~ 110 in ten kilobars. In
spite of secondary variations in some alcohols
it is clear that ε is the controlling para-

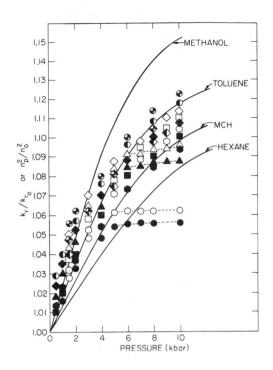

Figure 19 : k_r vs pressure - porphrins in various solvents (see Ref. 5 for meaning of symbols)

meter over this very large range of solvents. A group of studies in which pH and ε were varied separately at one atmosphere in aqueous solutions confirmed this conclusion.

A third set of studies involves the effect of the medium on peak energy. In Figure 21 we show the correlation of peak energy with a

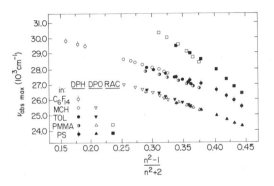
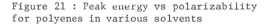

Figure 21 : Peak energy vs polarizability for polyenes in various solvents

parameter which measures polarizability for a series of polyenes.[7] The continuity is remarkable for solvents with as widely varying properties as two polymers, toluene and perfluoroheptane. In Figure 22 we show

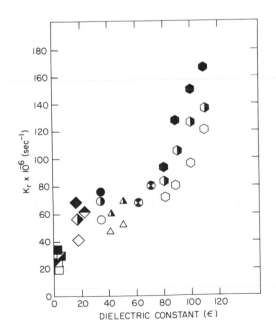

Figure 20 : k_r vs dielectric constant - indoles in various solvents (see Ref. 6 for meaning of symbols)

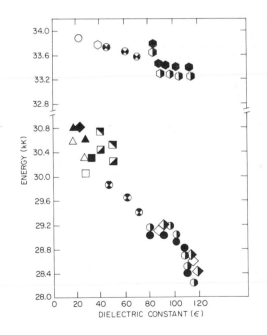

Figure 22 : Peak energy vs dielectric constant - indoles in various solvents (see Ref. 6 for meaning of symbols)

the correlation of peak energy with dielectric constant [6] for 5-methoxy-indole (5MI) and indole (IN). 5MI has an excited state of 1_{Lb} symmetry under all conditions. This is known to be a state of relatively low polarizability. There is considerable debate about the excited state of indole. In alcohols it probably has the relatively polarizable 1_{La} configuration, but it is frequently considered to be entirely different in water than in less polar solvents such as alcohols. We see, however, that the energy varies continuously with ε so that this macroscopic parameter can account for all changes of the emission energy.

A variety of molecules and complexes can emit from two different excited states depending on the rigidity of the medium. Based on the effects of temperature and solvent there have been speculations that the effect of increasing viscosity is to increase the energy barrier to transformation along an electronic coordinate associated with the emission. One would then expect that the rate of rearrangement should be dependent on temperature at constant viscosity; in fact the temperature coefficient should increase with increasing viscosity.

We have studied a series of complexes of tetracyanobenzene (TCNB) with aromatic hydrocarbons, especially xylenes.[8,9] These EDA complexes exhibit emission from a geometry like the ground state (the FC state) in rigid media and from quite a different geometric arrangement (the EQ state) on fluid media. A series of paraffin hydrocarbon solvents were used: methylcyclohexane (MCH), heptamethylnonane (HMN) tetramethyl pentadecane (TMPD) and a mixed solvent. These have in common about the same polarizability but have very different pressure coefficients of viscosity so that a range of $\sim 10^5$ in viscosity can be covered at constant temperature.

At both 25°C and 0°C we have measured the distribution of emission at steady state between the two states and the time dependent emission (decay), as a function of pressure. From these results we were able to extract the rate of rearrangement k_{RE} from the FC to EQ state. In Figure 23 we plot log k_{RE} vs log η. Clearly k_{RE} is independent of solvent, donor, or temperature except in so far as these effect viscosity. The slope at quite low viscosity is ~ -1 but above three poise it is constant at -0.39 independent of temperature. A more direct check of the slope is obtained from the ratio

$$x = \left(\frac{\log \dfrac{k_{RE}(25)}{k_{RE}(0)}}{\log \dfrac{\eta(25)}{\eta(0)}} \right)_{P, \text{ solvent}}$$

In Figure 24 we exhibit x vs viscosity. It changes from −1 at low viscosity to an average value of −0.42 above 2-3 poise, in excellent agreement with the results of obtained above from Figure 23. The effect of viscosity is clearly a diffusional one, inhibiting geometrical rearrangement of the complex.

From the Debye equation for molecular relaxation as a function of viscosity one can write

$$k_{RE} = C \frac{T}{\eta}$$

This predicts a relaxation rate inversely proportional to viscosity with a small temperature dependence. Exactly this condition obtains in the low viscosity region. The Debye equation was derived for rigid spheres. MCH, the solvent used for the lower viscosity data is a reasonable approximation to this condition.

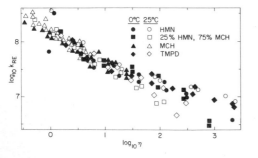

Figure 23 : log k_{RE} vs log η at 0° and 25°C.

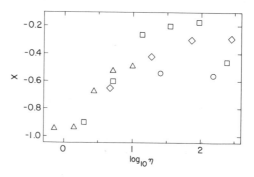

Figure 24 : x vs log η (see text for definition of x)

The other solvents are long flexible molecules and segmental motion or rotation around the long axis could permit rearrangement without contributing significantly to viscous flow.

These results illustrate how pressure data can establish the essential features of the relationship between bulk viscosity, local rigidity and the rearrangement of excited states of complexes.

4. SUMMARY

While many other examples could be presented, these results illustrate the power and versatility of high pressure optical studies as a means of resolving electronic problems in condensed systems.

5. ACKNOWLEDGEMENT

It is a pleasure to acknowledge financial support from the Department of Energy under contract DE-AC02-76ER01198.

6. REFERENCES

[1] Hayes, A. and Drickamer, H. G., High Pressure Luminescence Studies of Energy Transfer in Rare Earth Chelates, J. Chem. Phys. (in press).

[2] Hockert, E. N. and Drickamer, H. G., High Pressure Study of Crystalline Anils, J. Chem. Phys. 67 (1977) 5178-88.

[3] Mitchell, D. J., Drickamer, H. G., and Schuster, G. B. Energy Dependence on the Radiationless Deactivation Rate for Azulenes Probed by High Pressure, J. Am. Chem. Soc. 99 (1977) 7489-95.

[4] Tyner, C. E. and Drickamer, H. G. Studies of Luminescence Efficiencies of Tungstate and Molybdate Phosphors as a Function of Temperature and High Pressure, J. Chem. Phys. 67 (1977) 4103-15.

[5] Politis, T. G. and Drickamer, H. G. High Pressure Luminescence of Metalloporphyrins in Liquid Solution, J. Chem. Phys. (in press).

[6] Politis, T. G. and Drickamer, H. G. High Pressure Luminescence of Indole and Substituted Indoles in Liquid Solution, J. Chem. Phys. (in press).

[7] Brey, L. A., Schuster, G. B., and Drickamer, H. G. High Pressure Fluorescence Studies of Radiative and Nonradiative Processes in Polyenes, J. Chem. Phys. 71 (1979) 2765-72.

[8] Thomas, M. M. and Drickamer, H. G. High Pressure Study of Viscosity Effects on the Luminescence of Tetracyanobenzene-EDA Complexes, J. Chem. Phys. 74 (1981) 3198-3204.

[9] Thomas, M. M. Ph.D. Thesis, University of Illinois, Urbana (1981).

PHYSICS OF SOLIDS UNDER HIGH PRESSURE
J.S. Schilling, R.N. Shelton (editors)
© North-Holland Publishing Company, 1981

THEORY OF PRESSURE-DEPENDENT SPECTRA OF IONS AND MOLECULES IN CRYSTALS*

FERD WILLIAMS

Institut für Festkörperphysik II, Technische Universität, 1000 Berlin 12, Germany**

and

Physics Department, University of Delaware, Newark, DE 19711, U.S.A.***

After a brief description of the experimental situation we review recent theoretical studies on the effects of hydrostatic pressure on radiative transitions of ions and molecules in crystals. The coupling between pressure and electronic states is analysed in a double adiabatic approximation. A procedure is established for calculating transition energies and probabilities as a function of pressure, both for broad band and vibronic spectra, including effects of anharmonicity. Limitations of these and related analyses are clarified, and opportunities for further studies are emphasized. The analysis is generalized to include the effects of uniaxial stress. Experiments are proposed to test the theories and to obtain information on electron-phonon interaction and on radiative transitions of ions and molecules in solids.

1. INTRODUCTION

The experimental situation on the effects of hydrostatic pressure on optical absorption and luminescent spectra of ions and molecules in crystal divides as follows:

a) Broad band spectra from point defects with strong electron-phonon interaction, for example, alkali halides, doped with Tl^+, extensively investigated by Drotning and Drickamer (1) - the shifts in peak transition energies are of the order of 3 meV/kbar.

b) Broad band spectra arising in part from inhomogeneous broadening and in part from electron-phonon interaction, for example unresolved donor-acceptor pair spectra in ZnS investigated by Koda et al (2) - the shifts are of the order of 6 meV/kbar, somewhat less than that of the band gap.

c) Vibronic spectra arising from molecules or molecular ions such as O_2^- at halide sites in alkali halides and S_2^- in scapolite, respectively investigated by Laisaar (3) and Curie et al (4) - the shifts of the individual vibronic transitions are approximately 1 meV/kbar, however, the relative intensities change with pressure resulting in a greater displacement of the envelop of the spectrum.

d) Zero-phonon lines from deep centers such as the R-lines of ruby investigated by Forman et al (5) and transitions within the 4f levels of rare earth ions investigated by Huber et al (6) - the shifts are of the order of 0.1 meV/kbar.

We shall be mainly concerned in the following with the groups a) and c). These are related in that both involve strong electronic-vibrational interaction, with c) providing the more information, because of the vibronic structure. Group b) will certainly get serious theoretical attention once the inhomogeneous broadening has been eliminated, as will be described later.

Group d) has smaller pressure dependences which we shall show can be attributed to effects of anharmonicity.

The interpretation of group a) spectra has evolved as follows: Johnson and Williams (7) in explaining their early data on the pressure-dependence of the absorption spectrum of KCl:Tl made the hypothesis based on the configuration coordinate model that main effect of pressure is to change the occupational probability of any arbitrary configuration of the initial electronic state; Alers and Dolecek (8) included the effects of pressure on the final state; Drickamer et al (9), Lin (10) and Kelley (11) generalized the analysis but did not include the effects of pressure on the final state, arguing that the only pressure effect is on the probability distribution of the initial state with the transition energy at an arbitrary configuration being determined by crystal forces alone; and Curie et al (12) have analyzed group a) spectra taking account of the effect of pressure both on the probability of initial configuration and on the transition energy and also analyzed group c) spectra by calculating the effects of pressure on vibronic transition energies and on their oscillator strengths. These concepts have been critically reviewed (13, 14), from which the Curie et al analysis was shown to be based on a double adiabatic approximation with the inertia of the pressure apparatus assumed to be large; the Drickamer et al analysis based on an inertialess pressure apparatus.

2. ADIABATIC APPROXIMATIONS AND CONFIGURATION COORDINATE MODEL

In the following we shall first review the usual adiabatic approximation, then review the configuration coordinate model for radiative transitions, and finally present a double adiabatic approximation for the crystal plus pressure apparatus.

2.1. Adiabatic Approximation

The usual adiabatic approximation to the Born-
Oppenheimer expansion of the many-particle wave-
function $\psi(\underline{r},\underline{R})$, when $\underline{r}$ specifies the electronic
coordinates and $\underline{R}$ the nuclear coordinates, is
based on the orbital time for electron motion
τ_e being short compared to the oscillation time
for nuclear motion τ_N. The electrons are thus to
a good approximation in stationary states de-
scribed by the many-electron wavefunction $\phi(\underline{r};\underline{R})$
with eigenvalues $E_e(\underline{R})$, both with parametric dependence on $\underline{R}$; nuclear motion is in accordance
with $\chi(\underline{R})$ which are eigenfunctions of the effective potential - the adiabatic potential:

$$V_a(\underline{R}) = V_N(\underline{R}) + E_e(\underline{R}), \qquad (1)$$

where $V_N(\underline{R})$ is the potential energy for the nuclei interacting with each other. The eigenfunctions are related as follows:

$$\psi(\underline{r},\underline{R}) = \phi(\underline{r};\underline{R})\chi(\underline{R}), \qquad (2)$$

certain terms identified as electron-phonon interaction having been neglected in the separation of the many-body problem of electrons and
nuclei into a many-electron problem and a many-
nuclei problem.

2.2 Configuration Coordinate Model And Radia-tive Transitions

The configuration coordinate model is based on
the adiabatic approximation: $V_a(\underline{R})$ for the different electronic states, each with characteristic
vibrational levels and eigenfunctions, as shown
in Fig. 1. We emphasize that the $V_a(\underline{R})$ are not
pure electronic energies, as is evident from
Eq.1. With small displacements in R from equilibrium, $V_a(\underline{R})$ varies quadratically; $E_e(R)$,
linearly. Radiative transitions occur in accordance with the Franck-Condon principle: classi-
cally with fixed $\underline{R}$ between $V_a(\underline{R})$ for different
electronic states; quantum mechanically, between
individual vibrational levels of one electronic
state to individual levels of the other, as illustrated in Fig. 1 for the zero-phonon (ZP)
transition, $h\nu_0$. We note that $h\nu_0$ depends on
minima for two $V_a(\underline{R})$, each at different R, and
thus there is a change in $V_N(\underline{R})$, as well as in
the two $E_e(\underline{R})$, and $h\nu_0$ is not a pure electronic
transition. In addition, the differences in two
zero point energies are included in $h\nu_0$, and
thus for different force constants K for the two
electronic states isotope effects exist for $h\nu_0$
and for vibronic transitions.

2.3 Double Adiabatic Approximation

This approximation is applied to systems with
three classes of modes: fast, intermediate and
slow, sufficiently well separated in characteristic frequencies so that quasi-stationary
states of the fast modes are determined with parametric dependence on the intermediate and slow
modes, and then quasi-stationary states of the
intermediate modes are determined with parametric
dependence on the slow modes. Examples of previous applications are to electronic, intramolecular vibrational and intermolecular crystal

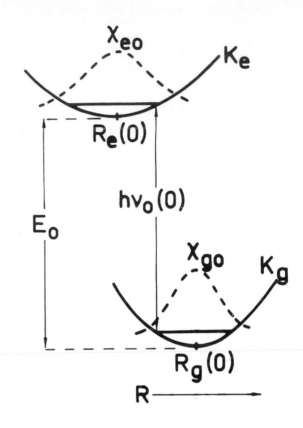

Fig. 1: Configuration coordinate model including
zero point levels and wave-functions,
and the zero-phonon transition.

modes as the fast, intermediate and slow modes,
respectively.

In our application of the double adiabatic approximation we divide the electronic and vibrational modes of the crystal as the fast and
intermediate modes, respectively, and the modes
of the apparatus involved in the application of
pressure, as the slow modes. Thus the many-particle wavefunction for the system including the
pressure apparatus is:

$$\Psi(\underline{r},\underline{R},P) = \Phi(\underline{r};\underline{R},P)\Lambda(\underline{R};P)\mu(P) \qquad (3)$$

Both the electronic and vibrational wavefunctions have in general parametric dependences on
the applied pressure, and therefore their eigenvalues will also. This leads, as we shall show
in detail in the next section, to a configuration
coordinate surface for each electronic state
which is different for different applied pressures.

3. DEPENDENCES OF RADIATIVE TRANSITIONS ON PRESSURE

We first consider application of hydrostatic pressure to a cubic crystal in the harmonic approximation, develop a procedure for determining the new equilibrium for each electronic state, and then determine the new adiabatic potential curves including the effect of pressure. The dependence of the peak transition energies of group a) broad band spectra on pressure is determined by classical application of the Franck-Condon principle, generalized to include $\Delta P=0$, as well as $\Delta R=0$ and $\Delta \dot{R}=0$; the dependences of zero-phonon and vibronic transition energies and probabilities are then determined quantum mechanically for group c) spectra. Finally, effects of anharmonicity will be discussed.

3.1 Pressure Dependence of Broad Bands in the Harmonic Approximation

The zero pressure adiabatic potentials are as shown in Fig. 1 with equilibria for the ground and excited states at O, $R_g(0)$ and E_o, $R_e(0)$, respectively. The pressure P is applied reversibly and new equilibria are established. We assume linear coupling such that the applied force F_a on the microscopic system is PA, where A can be interpreted as the effective area of the center but is in general the linear coupling constant. Equilibrium is thus determined by the total force F being zero:

$$F=F_c+F_a=0=K(R-R(0))+PA\Big|_{R=R(P)} \quad (4)$$

where F_c is the crystal force, the derivative of the zero pressure adiabatic potential. Incidentally, the existence of a deformation and thus of a new equilibrium means that there must exist a new force - a non-crystalline force - on the microscopic system. Reversible application of F_a results in the energy change $P^2A^2/2K$, consistent with the position change - PA/K, for each state. The new equilibria are shown in Fig. 2 and are $P^2A_g^2/2K_g$, $(R_g(0)-PA_g/K_g)$ and $(E_o+P^2A_e^2/2K_e)$, $(R_e(0)-PA_e/K_e)$, respectively, for the ground and excited states.

With only harmonic crystalline forces the force constant K remains unchanged with deformation. Thus the adiabatic potentials, including the effects of the hydrostatic pressure are:

$$V_a(P,R)=\frac{1}{2} K\left[(R-R(P))\right]^2 + P^2A^2/2K+E(0) \quad (5)$$

for each state, as shown in Fig. 2.

The shift in peak position with pressure mainly arises from unequal displacement of minima for the two states, i.e. $A_e/K_e \neq A_g/K_g$. For emission the shift in peak is as follows (12):

$$\frac{dh\nu_m}{dP} = (A_e \frac{K_g}{K_e} -A_g) (R_e(0)-R_g(0))$$
$$+ \left[(A_e^2/K_e-A_g^2/K_g) - K_g(A_o/K_e-A_g/K_g)\right]P \quad (6)$$

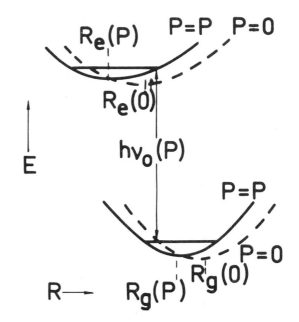

Fig. 2: Generalized configuration coordinate model including effect of pressure. Displacements from zero pressure adiabatic potentials are shown.

The corresponding expression for excitation has e and g permuted.

3.2 Pressure Dependence of Vibronic Spectra

In the harmonic approximation the vibrational levels for each electronic state are shifted uniformly, and quadratically, with pressure. For the transition between the n-level of the excited state and the m-level of the ground state, we have

$$h\nu_{nm}(P) = \frac{1}{2} (\frac{A_e^2}{K_e} - \frac{A_g^2}{K_g}) P^2 + h\nu_{nm}(0). \quad (7)$$

However, the transition probability has been shown to contain linear terms in a complex pressure dependence, arising from the change in overlap of the harmonic oscillation wavefunctions $\Lambda(R)$ (12).

Because the group a) broad band spectrum constitutes an envelop function of a vibronic spectrum groups a) and c) can be reconciled. The a) broad band spectrum is the limit where the c) vibronic structure is unresolvable. Their linear pressure shifts have a common origin in the change with P of $(R_e(P) - R_g(P))$, although the peak is shifted for the broad and the relative intensities changed for the vibronic spectrum - these are equivalent in the limit of unresolved vibronic structure.

3.3 Effects of Anharmonicity

To include anharmonicity we choose the zero pressure adiabatic potentials of the form:

$$V_a(O,\underline{R}) = \frac{K}{2}(R-R(O))^2 + \beta(R-R(O))^3 + \alpha(R-R(O))^4 + E(O) \tag{8}$$

for each electronic state. The portic term insures stability for large displacements. The coefficients α and β are inter-related so that $V_a(O,R)$ has a single minimum.

The pressure dependences for zero-phonon and vibronic transitions are found to be of the following form, including the effects of anharmonicity in the change of the adiabatic potentials with pressure to first order in β_g and β_e (12):

$$h\nu_{nm}(P) = 3\hbar\left[(m+\tfrac{1}{2})(\frac{K_g}{M_g})^{1/2}\frac{A_g}{K_g^2}\beta_g\right.$$

$$\left. -(n+\tfrac{1}{2})(\frac{K_e}{M_e})^{1/2}\frac{A_e}{K_e^2}\beta_e\right]P + (\frac{A_e^2}{K_e} - \frac{A_g^2}{K_g})\frac{P^2}{2} + h\nu_{nm}(0) \tag{9}$$

where M_g and M_e are reduced masses for the K_g and K_e vibrations. Thus, two effects are evident from Eq.(9), which originate from anharmonicity: 1) linear dependence on pressure, and 2) dependences which vary with vibronic quantum numbers n and m. Both effects are observed experimentally, for example, the first for the R-lines of ruby (5) and the second for O_2^- in alkali halides (3). The experimental displacements with pressure which are dependent on the specific vibronic transition can be used to determine anharmonicity.

4. EFFECTS OF UNIAXIAL STRESS

Experiments with uniaxial stress can supplement those with hydrostatic pressure in obtaining the basic interactions of ions and molecules in crystals;with non-cubic crystals the deformations with hydrostatic pressure are anisotropic and thus are related to those with uniaxial stress on cubic crystals. We are interested in a general formulation which applies to the effects on optical spectra of either hydrostatic or unixial stress.

The experimental, as well as the theoretical, studies on effects of hydrostatic and of uniaxial stress have developed quite undependently: the hydrostatic being at high stresses $\sim$ 100 kbar and mainly at room temperature; the uniaxial, at low stresses $\sim$ 1 kbar and helium temperatures.

In the following we shall first consider the effects of uniaxial stress on non-degenerate electronic states, and second, effects on degenerate electronic states.

4.1 Effects on Non-Orbitally-Degenerate Electronic States

For non-orbitally-degenerate electronic states of ions or molecules in solids the theoretical ana-

lysis of effects of uniaxial stress is essentially the same as for hydrostatic pressure. The deformation is divided spatially among orthogonal components and the analysis proceeds as in Section 3.

An example of experiments with uniaxial stress on non-orbitally-degenerate states, which supplement experiments with hydrostatic pressure, are those of Rolfe (15) on O_2^- in alkali halides. From the combination of separate optical experiments with hydrostatic and uniaxial stresses, force constants and anhamonicities can be determined (4). Both absorption or excitation spectra and luminescent emission spectra are necessary to obtain K_g, K_e, β_g and β_e. The determinations of A_g, A_e and $(R_e(0) - R_g(0))$ require improved and different experiments, as will be discussed in Section 6.

4.1 Effects on Orbitally-Degenerate Electronic States

For orbitally-degenerate electronic states of ions and molecules in cubic crystals the degeneracy is not removed by hydrostatic pressure but may be by uniaxial stress. However, the degeneracy may be reduced or removed by a spontaneous deformation without the application of any stress at all - this is the well-known Jahn-Teller effect. The effects of uniaxial stress on ions and molecules which show the J-T effect are large even for small stresses and therefore these have been extensively investigated experimentally. The sensitivity of Jahn-Teller systems to uniaxial stress is evident from the following extension of the procedure described in Section 3 for hydrostatic pressure. The formal basis for the extension was recently reported by Berry and Williams (16).

The zero-stress adiabatic potential is shown in Fig. 3 for a Jahn-Teller excited state and for a non-degenerate ground state. The double minima adiabatic potential for the excited state results from two coupled diabatic potentials, for example p_x and p_y orbitals in a square lattice which spontaneously deforms to different rectangles on occupancy of one or the other orbital. Thus the coordinate in Fig. 3 is of course different from that in Fig. 1 and 2 where a symmetric, "breathing" mode was the coordinate for hydrostatic deformation of a cubic crystal. We shall return in Section 5 to the question of the correct modes to describe effects of a specific applied stress and those to describe spectral characteristics such as Stokes shifts and band widths. The procedure to describe the effects of uniaxial stress on a Jahn-Teller system is divided into two parts, similiar to what was done in Section 3 for hydrostatic pressure: First, the stress is applied reversibly so that the applied force just balances the crystal force. In this way zero work is done moving the system from the outer well to the inner well of the excited state of Fig. 3, and then work is done on the system moving the inner minimum to the new equilibrium: $W_1 = P^2A^2/2K$, where K characterizes small oscillations in the inner

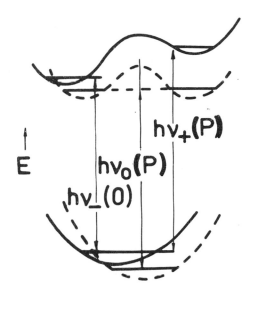

$hv_+(P)$

$hv_0(P)$

$hv_-(0)$

E

X ⟶

Fig. 3: Configuration coordinate model for Jahn-Teller system with and without applied uniaxial stress. Zero-phonon transitions with and without stress are also shown.

well and A is again the linear coupling constant. Second, dynamics at constant stress are analyzed: in addition to work done against crystal forces there is now work done against constant stress, specifically in going from the inner to outer well the work done is:

$$W_2 = PA(X_2 - X_1) \qquad (10)$$

therefore a linear splitting of zero-phonon and phonon-replicas occurs and the magnitude is large because $(X_2 - X_1) \gg PA/K$, where the smaller is the deformation involved in W_1, with attainable uniaxial stresses. The zero-phonon transitions are shown in Fig. 3. Implicit in the procedure just given for obtaining the stress-adiabatic potential from a multi-minima zero-stress adiabatic potential is the following rule for selecting the proper minimum on which to apply the procedure: the minimum is selected so that for all values of the stress and all configuration the stress-adiabatic potential is greater than or equal to the absolute minimum of the zero-stress adiabatic potential. This is evident from the fact that the potential energy of the system must increase on application of the stress.

5. LIMITATIONS OF PRESENT THEORY

The limitations of the present theory of the effects of pressure on the spectra of ions and molecules divide into several catagories. Among these are : 1) coupling between the macroscopic stress and the microscopic system; 2) the modes responsible for the spectral changes with stress compared to those responsible for zero-stress spectral properties; and 3) the single mode approximation in the usual configuration coordinate model. These and others have to some extent been discussed previously (13,14). In the following we shall consider them in sequence.

5.1 Coupling Between the Macroscopic Stress and the Microscopic System

In the analyses of Sections 3 and 4 the basic assumption is made that the inertia of the pressure apparatus is large compared to that of any component of the crystal within which the radiative transitions occur. In some sense the pressure apparatus extends right down to the ion and its ligands or to the molecule where the transitions to a good approximation occur. For those materials with radiative centers which have stiff local modes compared to the lattice modes which transmit the pressure, the basic assumption is surely valid. This should be case for molecular crystals and for molecular ions in ionic crystals. On the other hand, unusual cases with soft local modes may occur for which the double adiabatic approximation as applied in Sections 3 and 4 may not be valid.

As a matter of fact there are cases for which the ordinary adiabatic approximation does not apply. These include shallow dopant states in elemental semiconductors and in some III-V semiconductors where $\tau_e \gtrsim \tau_N$. These shallow effective mass states and radiative transitions between them are outside the range of validity of the analyses herein. On the other hand, for charged dopants and their pairs in GaP and in most II-VI semiconductors $\tau_e < \tau_N$(17), and the present theory with small modifications appears to be valid (4). Opportunities for experimental high pressure studies on radiative transitions of donor-acceptor pairs in some of these materials will be discussed in Section 6. Linear coupling between the applied pressure and the microscopic center has been assumed. A more general pressure-dependent coupling constant of the following form:

$$A(P) = \sum_{k=o}^{\infty} A_k P^k \qquad (11)$$

has been proposed and found to leave the essentials of the present theory unaffected, for realistic systems, within the order to which the theory is taken (18).

5.2 Modes Involved With Applied Stress and Modes for Radiative Transitions

A basic assumption of all existing analyses is that the modes involved with applied stress are the same or are simply related to the modes responsible for the zero-stress spectral properties.

F. Williams

The latter include Stokes shifts of broad band emission compared to broad band absorption and band widths of both. This assumption is most clearly valid for point defects with deep, non-orbitally degenerate electronic states in cubic crystals where the symmetric "breathing" mode dominates both hydrostatic and zero-stress optical properties. At the other extreme are aromatic molecular crystals for which the zero-stress optical absorption and luminescent emission involve molecular electronic states coupled to intramolecular vibrational modes whereas applied stress involves intermolecular crystal modes. This difference and the absence of any simple relation between the intra- and intermolecular modes appears to be partly responsible for the difficulties in explaining the pressure-dependent spectra of aromatic crystals with present theories (19).

5.3 Single Mode Approximation

The single mode configuration coordinate model has been generally used to explain stress-dependent spectral properties and other optical properties of ions and molecules in crystals. We have recently shown that it is possible to transform a multi-coordinate system to a single coordinate system in explaining the pressure-dependence of spectra of ions and molecules in crystals, however, the single coordinate model may be complicated. For example, a two mode system, each mode of which is harmonic, on transformation becomes a one mode system with a pressure-dependent force constant. Thus, a linear pressure dependence of zero-phonon and vibronic transition energies is obtained, although the real modes are harmonic. The effect of multi-modes is similar to that due to anharmonicity. Experimentally there are difficulties in distinguishing between the multi-mode and anharmonic single mode explanations of linear pressure dependences of zero-phonon and vibronic transition energies.

6. PROPOSED EXPERIMENTS TO EVALUATE THEORIES

Experimental studies which are needed to evaluate the different theoretical models and to give detailed information on the structure and states of deep centers and molecules in solids are of two types:

1) more quantitative measurements over larger ranges of pressure on materials with well-resolved optical spectra, and 2) qualitatively new experiments involving hydrostatic pressures. Representative examples of each will be discussed.

As noted earlier, K_e, K_g, β_e and β_g can be estimated from existing moderate pressure data, i.e. ~6 kbars, on a few materials exhibiting vibronic spectra. However, the data are neither sufficiently complete, nor sufficiently quantitative, to obtain precise values of these quantities or even to obtain the magnitudes of A_e and A_g. Pressure-dependences of both absorption or excitation and

luminescent emission spectra are needed in order to obtain the parameters.

The selection of materials and experimental conditions are important. As noted earlier the group b) broad band luminescent spectra of II-VI semiconductors originate in part from inhomogeneous broadening of donor-acceptor pair transitions. On the other hand, GaP and ZnSe doped with shallow acceptors and donors - the depths are still sufficient for the adiabatic approximation to remain valid - give well-resolved zero-phonon lines and phonon-replicas at low temperature T<20K, from pairs with fixed donor-acceptor distances. The dependences of the transition energies and intensities of these lines on hydrostatic pressure is being measured. Interpretation of the results will depend on having quantitative information on the pressure dependences of the band gap, electron and hole effective masses, and static and optical dielectric constants.

The experimental determinations of $(R_e(0)-R_g(0))$, and especially of $(R_e(P)-R_g(P))$, depend on qualitatively new experiments. The first requires a measurement of dimensional changes of the crystal with excitation; the second, on the pressure-dependence of this dimensional change. These measurements will put the configuration coordinate model on an experimental foundation and also contribute to the resolution of current controversies in the theory. The experiments are in progress.

7. CONCLUSIONS

The pressure-dependence of the optical excitation of ions and molecules in crystals is interpreted as a collective excitation with the pressure apparatus included in the total system (20). Separations are made with the double adiabatic approximation. Materials with vibronic structure are found to give detailed information on luminescent centers, from spectral measurements with hydrostatic pressure. Data from uniaxial stress measurements supplement those with hydrostatic pressure. General theoretical formulation applicable to all systems awaits further studies which remove some of the limitations of present theories. More quantitative and imaginative experimental studies are needed to keep pace with recent theoretical advances.

FOOTNOTES AND REFERENCES

  * Supported in part by a grant from the U.S. Army Research Office and in part by a Humboldt Prize (Senior U.S. Scientist Award).
 ** Present address
*** Permanent address

(1) W.D. Drotning and H.G. Drickamer, Phys. Rev. 13B (1976) 4568.
(2) T. Koda, S. Shionoya, M. Ichikawa and S. Minomura, J. Phys. Chem. Solids 27 (1966) 1577

(3) A.I. Laisaar, Proc. Oolloq. Int. CNRS 188, Paris (1970), 291.

(4) D. Curie, B. Canny, P. Jaszczyn-Kopec, H.K. Liu, D. Berry, F. Williams,Proc. Int. Conf. Luminescence 1981, J. Luminescence (in press).

(5) R.A. Forman, B.A. Weinstein, G.Piermarini, Proc. Colloq. Int. CNRS 255, Paris (1976)51.

(6) G. Huber, K. Syassen, W.B. Holzapfel, Phys. Rev. 15B (1977) 5123.

(7) P.D. Johnson and F. Williams, Phys. Rev. 95 (1954) 69.

(8) P.B. Alers and R.L. Dolecek, J. Chem. Phys. 38 (1963) 1046.

(9) H.G. Drickamer, C.W. Frank and C.P. Slichter, Proc. Nat. Acad. Sci.U.S.A. 69 (1972) 933.

(10) S.H. Lin, J. Chem. Phys. 59 (1973) 4458.

(11) C.S. Kelley, Phys. Rev. 12B (1975) 594.

(12) D. Curie, D.E. Berry and F. Williams, Phys. Rev. 20B (1979) 2323.

(13) C.P. Slichter and H.G. Drickamer, Phys. Rev. 22B (1980) 4097.

(14) D. Curie, D.E. Berry and F. Williams, Phys. Rev. 22B (1980) 4109.

(15) J. Rolfe, 5. Chem. Phys. 70 (1979) 2463.

(16) D.E. Berry and F. Williams, Proc. Int. Conf. Luminescence 1981, J. Luminescence (in press).

(17) E. Kartheuser, R. Evrard and F. Williams, Phys. Rev. 21B (1980) 648.

(18) D. Curie, D.E. Berry and F. Williams, J. Luminescence 18/19 (1979) 823.

(19) D. Berry, R.C. Tompkins, and F. Williams, J. Chem. Phys. (submitted)

(20) F. Williams "Collective Excitations in Solids" Edited by Di Bartolo (Plenum, N.Y. - in press).

PHYSICS OF SOLIDS UNDER HIGH PRESSURE
J.S. Schilling, R.N. Shelton (editors)
© *North-Holland Publishing Company, 1981*

TEMPERATURE DEPENDENCIES OF LIBRON FREQUENCY SHIFTS AND WIDTHS IN SOLID
NITROGEN AND CARBON DIOXIDE

William B. Daniels, James Schmidt*
Physics Department, University of Delaware
Newark, DE 19711, USA

Fernando Medina
Physics Department, Florida Atlantic University
Boca Raton, Florida 33431, USA

Anharmonic spectral shifts and widths have been measured for the Raman
active librations in solids Nitrogen and Carbon Dioxide. A useful experimental method is described. An analysis is given of the temperature
dependencies of the shifts and widths which yields insights into the
detailed anharmonic couplings and decay processes which contribute to
each.

1. INTRODUCTION

Lattice dynamics in the harmonic approximation is a relatively well developed
field at the present time. Experimentally such powerful techniques as inelastic neutron scattering and laser
Raman scattering have been major contributors to this state of affairs, while
on the theory side a significant role
has been played by large scale facilities for computation as well as more
sophisticated understanding of intermolecular forces and new techniques
such as self consistent phonons.
Numerous measurements of phonon dispersion relations exist, as do harmonic
(or quasi-harmonic densities of states
derived from them.) The exchange between these measurements and calculations has contributed enormously to insight and understanding of intermolecular forces and of the detailed dynamics
of molecular motions in crystals.
Further, connections between these quasi
harmonic spectra and thermodynamic
properties explicitly heat capacities
have been made yielding almost always
at least qualitative agreement between
"spectroscopic" and experimental values
and their temperature dependencies.
However, considering the state of knowledge in what might be called an exper-
imentallists anharmonic lattice dynamics
which in my view could reasonably consist of representative sets of phonon
mode gammas $[\gamma_G \equiv \frac{\partial \ln \omega_0}{\partial \ln V}]$ or shifts
arising from volume changes, anharmonic
temperature shifts [$\Delta_T \equiv$ change in phonon energy with temperature in a crys-
tal confined at constant volume] and
phonon lifetimes τ_j or energy widths
Γ_j. There exist relatively few data on
mode gammas at low phonon frequencies
derived from pressure dependence of
elastic constants, even less data for
phonons throughout the Brillouin zone

derived from inelastic neutron
scattering or second order Raman scattering, and very few indeed of quantitative measurements of shifts and widths.
On the other hand, it is these shifts
and lifetimes which enter into microscopic decompositions of contributions
to such phenomena as thermal conductivity. Thus it seems worthwhile to examine the possibilities for augmenting
the experimental base of values of anharmonic shifts and widths to seek the
insights which may be generated.

In general, anharmonic effects in simple
insulating crystals seem to be discuss-
able on three distinct but interacting
levels: (1) Macroscopic, involving such
observables as thermal expansion, a
"linear term" in the specific heat,
thermal conductivity, acoustic damping..
(2) The anharmonic Hamiltonian
$H = H_0 + H_2 + H_3 + H_4 \ldots .$ in which
the terms of higher than second order
in an expansion in powers of atom displacements are responsible for anharmonic effects. c.f. Ziman,[1] Wallace,[2]
Choquard[3] for standard treatments.
(3) A quasi-particle approach in which
phonons are treated as objects characterized by both energies and lifetimes
each of which depend in general on
pressure (or volume) and temperature.

Significant progress has been made in
relating various of these levels of
understanding. There now exist a substantial number of theoretical calculations of anharmonic properties in
simple systems based on a variety of
intermolecular potentials, some of which
are usefully related to the general
questions associated with currently
available data and (perhaps more importantly) suggest particularly fruitful
areas for experimental investigation.
Some correlations of properties i.e. of
phonon energy shifts with thermodynamic

data have been quite successful. One might note in this regard the work of Weinstein and Piermarini[4] and of Eckert et. al..[5] The former used an elegant second order Raman scattering experiment at high pressures to determine mode gammas at critical points of the phonon spectra in Si and GaP from which they synthesized "spectroscopic" values of the temperature dependent thermal expansion in these materials which were then compared with experimental values. The latter experiment used inelastic neutron scattering on crystals of neon at high pressures and obtained essentially a complete mode gamma dispersion plot $\gamma_j(\bar{q})$ which was in turn used to generate a spectroscopic Gruneisen parameter. (The latter may be thought of as a kind of dimensionless thermal expension coefficient and is related to other thermodynamic quantities by $\Gamma_G = \frac{\alpha \; B_T V}{C_v}$ where α is the volume coefficient of thermal expansion, B_T the isothermal bulk modulus and Cv/V the heat capacity per unit volume. The quasi harmonic model which limits mode frequencies to a volume dependence only predicts $\gamma_G = \frac{\sum \gamma_j \; C_{vj}}{\sum C_{vj}}$ where the mode weighting factors Cvj, are the Einstein heat capacity of the j th mode, the summation is over all modes of the crystal and other terms have already been defined.) These suffice to suggest at least approximate validity of the simple quasi harmonic oscillator model used in their interpretation and to support the identification of mode gammas as the microscopic seats of thermal expansion.

Going beyond thermal expansion, data on thermal conductivity in insulators is interpreted in terms of contributions to the lifetime (singular) of phonons transporting the heat current as a microscopic parameter. There have been as far as we know no opportunities to attempt to correlate microscopic phonon lifetime measurements with the thermal conductivity mean lifetimes. (Indeed the task would be a difficult one. The necessity of isolating separately the crystal momentum conserving and non-conserving contributions to the τ_j would be a formidable one!) The idea of a phonon τ_j (or Γ_j) dispersion plots in a simple system such a solid neon to be used together with thermal conductivity data in the Umklapp region as detailed test of the theory relating them remains as a difficult experimental task. Short of this, it would still be desirable to measure some phonon shifts and widths in almost any system or set of similar systems to attempt to gain insights in-

to the general behavior of these elusive quantities.

Solid Nitrogen and Carbon Dioxide in their Pa3 phase qualify reasonably well and a systematic study has been made of Raman response in these to compare and contrast the results, especially respecting lifetimes and shifts, and the processes contributing to each.

Experimental-Helmholtz Representation Experimentation

The experiments were done using a sample preparation and manipulation technique which is well described by the term "Helmholtz representation experiments". In brief, the sample crystals are grown completely filling the interior of an appropriate pressure cell at points at elevated pressures on the melting curve. Provided precautions are taken in the design to ensure that the connecting pressure tubing will not freeze off and block before the isobaric growth of the crystal is complete, one can prepare a crystal at a pressure, temperature and molar volume corresponding to the point at which it was grown. The tubing is then allowed to block by reducing the power to appropriate heaters, and the crystal is left clamped within its envelope. Since the thermal expansion coefficients of technical materials from which the pressure cell is constructed are small compared with weakly bound solids such as N_2 or CO_2, and corresponding elastic moduli are very large compared with the sample materials, the sample remains substantially at constant volume as the temperature of the sample-cell combination are cooled from the melting curve. That is, the sample travels along an isochore of its surface of state. The pressure in the sample decreases during such a process, and if the sample molar volume is less than that of a free standing crystal at zero pressure and temperature it will cross the sublimation line if cooled sufficiently. Finally the small parasitic volume changes which occur because of the thermal contraction of the cell in cooling are usually not large, and are treated as a correction. Considering a particular phonon (or libron) frequency to be a function of volume and temperature, we may write

$$d\omega = (\frac{\partial \omega}{\partial T})_V dT + (\frac{\partial \omega}{\partial V})_T \; dV \qquad (1)$$

in which $d\omega$ is a general shift in phonon energy and other terms have their conventional definitions. Crystals prepared in the manner just described permit direct measurement of the "delicate" (i.e. small) shifts with temperature at constant volumes because the second term vanishes along an isochore ($dV \equiv 0$). The coefficient of the second term multiplied by V/ω; is a mode gamma. It is determined by measuring ω_o in crystals grown at several different molar volumes (i.e. at several different points along the melting curve.) The separation into thermal and volume dependent contributions is of value conceptually and practically. In brief, one can conceive the frequencies to depend relatively strongly and in a relatively well understood way on the mean distances between atoms, but more weakly and more interestingly (?) upon the state of motion of the molecules, that is upon the occupation numbers $\{n_j\}$ of all phonon states of the crystal. It is the latter dependence which is reflected in the isochoric temperature dependence of ω in which the mean distances between atoms are constant and only occupation of excited states is changing. Finally, to obtain the $(\frac{\partial \omega}{\partial T})_V$ from the more usual "Gibbsian" experiments measuring ω (P,T), the problem of small differences appears. That is, by standard thermodynamic manipulation, $(\frac{\partial \omega}{\partial T})_V = (\frac{\partial \omega}{\partial T})_P + (\frac{\partial \omega}{\partial P})_T \alpha B_T$. In most examples the desired term on the left hand side of the equation appears as a small difference of two much larger terms on the right - a procedure which is not conducive to producing accurate measurement of the former, and which is avoided in the isochoric experiments.[6]

Solid Nitrogen and Carbon Dioxide.

These substances together with methane probably form the second most simple class of materials, following the solidified noble gases in their approach to a realization of the simplest relevant lattice dynamical models. N_2 and CO_2 may be thought of as "dumbell" shaped molecules. Table I contains information related to N_2 and CO_2, specifically the molecular length and diameter and the ratio of the two from the .002 electron density contour as calculated by Kihara and Sakai (7), the electrostatic quadrupole constants, the root mean square librational amplitudes at low and high temperatures and for each substance a ratio of a libron frequency to a translational mode frequency. (The term libron refers to orientational oscillations of the molecular axis.) These parameters give some idea of the relative strengths of the orientational vs translational forces and through the root mean square angular displacement $< \theta^2 >^{1/2}$ some measure of a 'priori relative importance of anharmonic effects in the near

quantum crystal N_2 and the more classical CO_2.

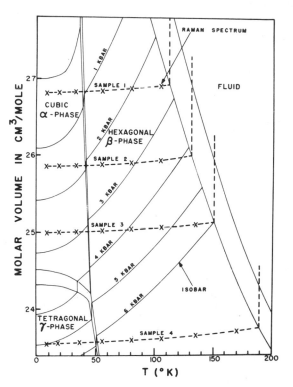

FIG. 1 Nitrogen Phase Diagram

TABLE I Parameters related to N_2 and CO_2.

	CO_2	N_2
Quadrupole Constant	.854	.31
Molecular Length, ℓ	5.36Å	4.13Å
Molecular Diameter, d	3.23Å	3.24Å
ℓ/d	1.7	1.3
Nearest neighbor distance	3.93Å	4.00Å
ω Libron/ω Tranlation	1.1	.67
$<\theta^2>^{1/2}$	3.6° (T=0)	14° (T=0)
	7.8° (T=212^k)	18° (T=34)

Solid carbon dioxide crystallizes in the cubic Pa3 structure and to date no other crystalline phases have been reported. The Pa 3 structure has four molecules per primitive cubic unit cell with the molecular centers arranged on a face centered cubic lattice. Each molecular axis has an equilibrium orientation along one of the four cube diagonals. Solid nitrogen on the other hand has a relatively rich field of structures as shown in Fig. (1). The high temperature β phase is hexagonal close packed with nearly the ideal hard sphere c/a ratio, the high pressure γ phase is body cnetered tetragonal. The α phase structure of N_2 has been a matter of extensive discussions. Nominally the α phase is in the Pa 3 structure similarly to CO_2. However a number of studies supported the $P2_13$ structure which is not centrosymmetric in that each molecule is displaced a small distance along its long axis in (111) type directions. The excellent review article by Scott [8] discusses the question and provides relevant references. The question has in a sense been abandoned at this time, and present day calculations are based almost entirely on the assumption that the centrosymmetric Pa3 is the correct structure. The discussion in this note is dedicated to anharmonic shifts and widths of the low frequency Raman active lattice modes in the Pa3 phase of N_2 and CO_2. Additional discussion and representation of the data as well as experimental details may be found in papers by Medina and Daniels [9] and Schmidt and Daniels [10].

In the Pa3 structure, two stretching modes of symmetry Ag and Tg and three librational modes of symmetry Eg, Tg and Tg are expected in the first order Raman spectrum. Two translational modes of symmetry Tu are infrared active. Since the Pa3 structure is centrosymmetric coincidences between Raman and infrared frequencies are not allowed. Fig. (2) displays three spectra in CO_2 showing qualitatively thermal and compression effects on the spectrum in the

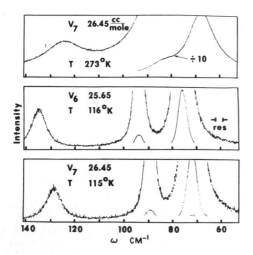

FIG. 2 Raman spectrum in Carbon Dioxide showing the volume and temperature dependencies for the three Raman active Libron modes.

librational region. Fig. (3) shows a spectrum taken in N_2 at low temperature. The features near 33.3 cm^{-1}, 37.5 cm^{-1} and 61.3 cm^{-1} are identified as the Eg, Tg$^-$ and Tg$^+$ libron modes respectively while the broad structure beyond 70 cm^{-1} is due to multiple phonon-libron combinations. N.B. the relatively sharp features at approximately 72 cm^{-1} and 95 cm^{-1} and their shifts with compression are consistent with Eg + Tg$^-$ and Eg + Tg$^+$ combinations respectively.

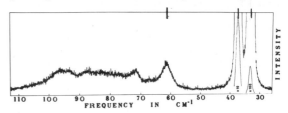

FIG. 3. Raman Spectrum in N_2. The vertical lines indicate the first order libron lines. The continuous spectrum beyond about 70 cm^{-1} is identified as multi-phonon scattering.

The feature at 71.8 cm^{-1} lies at a frequency about 1 cm^{-1} higher than the Eg + Tg$^-$ mode frequencies and possibly forms a two libron bound state in the sense of Ruwalds and Zawadowski [11]. The second order region in CO_2 has insubstantial intensity. Our interest lies in the temperature shifts and widths of the first order lines. Unfortunately, in N_2 the closeness of the lines, and the widths were such that only the Eg line could be studied quantitatively. Its temperature dependent shifts and widths at constant volume are shown in Fig. (4) and (5). Corresponding data in CO_2 are shown in Figs. (6), (7) and (8). Several qualitative points may be made about these data. (1) The scatter from the fitted curves is small, especially in the N_2 data. (2) The temperature ranges differ by almost a factor of 10 between the two materials. Even considering the characteristic temperatures in CO_2 are higher than those in N_2, the anharmonic effects are much weaker in CO_2. (3) The shape of the temperature dependence of the Eg linewidth is quite different in N_2 and CO_2. (4) Some mode linewidths become very small as T → 0, others do not.

Frequently one uses measurements of temperature dependencies of accurately measured quantities to gain insight into the spectral contributions to these quantities. Thus for example the measurement of heat capacities of Germanium and Silicon by Morrison et. al. [12] hinted at the existence of a very low energy phonon branch in Ge and Si even before the direct measurements by neutron scattering [13] found the anomalously low energy TA branch. Likewise, interpretation of thermal expansion data [14] indicated that these branches had negative mode gammas. We shall attempt in what follows to make an analysis of the temperature dependent frequency shifts and widths represented above in an attempt as far as possible to gain insight into the detailed processes contributing to each.

Let us examine the frequency shift results represented in Fig. (4) in some detail. In circumstances where translations and librational amplitudes are in some sense small, a perturbation treatment of anharmonic shifts and widths may be carried out. For the librons, Harris and Coll [15] take as a measure of librational amplitude $<\theta^2>^{1/2}/\pi$ which is about .1 in α phase N_2 and a factor two to four smaller in CO_2. They have shown the amplitude to be small enough to warrant a perturbation treatment and have calculated the shift in libron frequency due to the zero point amplitudes through cubic and quartic anharmonic corrections. Assuming that the sole contribution to the orientational part of the potential arose from quadrupole-quadrupole interactions they calculated a 12 ± 3% reductions of libron energies. If the high temperature approximately linear part of the curve of Fig. (4) is "athermally" [16] extrapolated to zero temperature, one obtains a shift of about 15% from the observed low temperature limit in remarkably close agreement with the Harris and Coll result. The agreement is rather surprising because the measured librational mode Grüneisen parameters of about 2 are inconsistent with purely quadrupolar interaction for which the value $\gamma = 5/6$ would follow. Other calculations of the anharmonic frequency shift of libron frequencies in N_2 using more realistic intermolecular potentials have been less successful. Raich, Gillis and Anderson [17] in a self consistent phonon calculation using a 6-12 atom-atom potential obtained too small a shift. Kobashi [18] using perturbation theory and an atom atom potential generated a shift of opposite sign from the observed values. Raich and Gillis [19] using an effective single particle anharmonic potential obtain a zero point shift near the observed value. On the other hand it may be noted that their procedure yields an incorrect sign of the shift when applied to the N_2 γ phase librons raising some questions about the method.

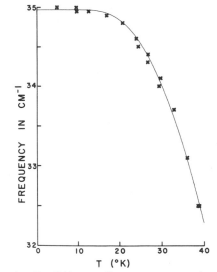

FIG. 4 Eg libron frequency vs temperature in a Pa3 Nitrogen crystal clamped at constant volume.

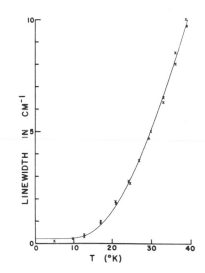

FIG. 5 True linewidth of Eg libron vs temperature in N_2.

In the absence of detailed calculations, it is of interest to test as far as possible the form of $\omega(T)$ against the general predictions resulting from application of perturbation theory to the cubic and quartic terms in an expansion of the crystal potential in a conventional way. In this we follow Wallace [2] and note that to this order, the frequency shift has the form

$$\Delta(\bar{q}) = C(\bar{q}) + \Sigma_{q'} C(\bar{q},\bar{q}') M(\bar{q}') \qquad (3)$$

where C(q) and C($\bar{q}$ $\bar{q}$¹) are temperature dependent terms which depend only on the strengths of the cubic and quartic terms in the crystal potential, and on geometrical factors related to the crystal structure. Thus all of the temperature dependence is contained in the set of occupation numbers n($\bar{q}$¹) given by the usual Bose-Einstein function:

$$n(\bar{q}^l) : \{\exp[h\omega_0(\bar{q}^l)/kT]-1\}^{-1} \quad (4)$$

Of course the fact that the sum in equation (3) is over an entire spectrum suggests that it will be difficult in general to identify individual terms i.e. individual modes to which the object excitation is coupled to generate an anharmonic shift. In the case of the Eg libron in N_2, consideration of a single term fit from the sum in Eg. 3 generates the solid line shown in Fig. (4) The best fit frequency ω_0 is 83 cm⁻¹ which lies close to the frequency of the Tu^+ optic mode frequency at this density (77 cm⁻¹). The quality of the fit suggests that the coupling of the Eg libron to the Tu^+ phonon through quartic terms in the potential dominates the Eg mode thermal shift. Unfortunately, the frequency shift data for other modes in N_2 are not sufficiently accurately determined to warrant such an analysis at this time, and in CO_2 the uncertainty of a correction due to expansion of the nearly clamped crystal reduces the quantitative quality of the shift data in Fig. (6).

In the same spirit however one may analyse the temperature dependence of libron lifetimes in N_2 and CO_2. Kobashi [18] has calculated values of Γ for the Eg mode in N_2 using perturbation theory but the values are substantially less than the observed values. Formally, from Wallace [2] the imaginary part of the self energy which is related to the linewidth Γ of a Raman line has a contribution from cubic terms given by

$$\Gamma(\bar{q}) = \frac{18\pi}{\hbar^2} \sum |\phi(\bar{q}\,\bar{q}'\,\bar{q}'')|^2 \times$$
$$\{[n(\bar{q}')+n(\bar{q}'')+1]\,\delta[\omega_0-\omega_0'-\omega_0''] +$$
$$[n(\bar{q})-n(\bar{q}')][\delta(\omega_0+\omega_0'-\omega_0'')-\delta(\omega_0-\omega_0'+\omega_0'')]\} \quad (5)$$

In this expression, $\phi(\bar{q},\bar{q}',\bar{q}'')$ is related to the cubic anharmonic term in the potential and other symbols have their conventional meanings. Contributions to Γ from quartic terms have been expressed by Wallis, Ipatova and Maradudin [20] A simplified approximation to their results has

been made by Gervais, Piriou and Cabannes [21] who assumed that the object phonon decays into two or three phonons having frequencies near an average of one half or one third of the object phonon energy for three and four phonon processes respectively. Their result yields the relation

$$\Gamma = \Gamma_3[n(\tfrac{\omega}{2})+\tfrac{1}{2}] + \Gamma_4\{[n(\tfrac{\omega}{3})+\tfrac{1}{2}]^2 + \tfrac{1}{12}\} \quad (6)$$

which was then applied to interpretation of phonon lifetimes in several silicates. One may note that in these expressions as in the frequency shift relation, the

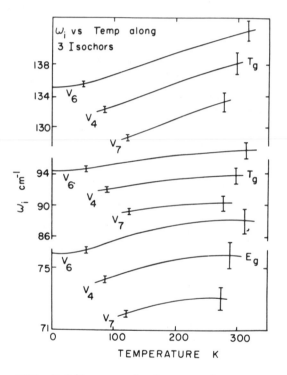

FIG. 6 Libron mode frequencies vs Temperature along isochores in Carbon Dioxide.

temperature dependencies occur solely as occupation numbers of phonon states involved in the scattering process. The fit of eg. (6) to the data of Fig. (5) was inadequate, largely because the observed linewidths were very small at low temperatures. The small linewidth at T = 0, followed by a strong increase turning on relatively abruptly at ∿ 15K suggested that a combination process probably dominates the decay of the Eg libron in N_2. Examination of the second set of terms in equation (5) indicates that these will have the proper "form", involving differences of occupation numbers. The simplest assumption one can make is that a single pair of states is involved in the combination-decay yielding a form:

$$\Gamma = \Gamma_0 + \Gamma_3 \left[n(\omega') - n(\omega + \omega') \right] \quad (7)$$

in which the constant term Γ_0 is taken to represent a residual contribution due to less important processes, and the second term refers to a process in which the Eg libron combines through a cubic anharmonic coupling with a second phonon to create a third, conserving energy and crystal momentum. A fit of the linewidth of the Eg libron in N_2 to this function yielded the solid curve shown in Fig. (6) which represents the "true" linewidth obtained assuming both the line and the instrument response are given by Lorentzian curves. Parameters resulting from the fit are: $\Gamma_0 = 0.2 \text{cm}^{-1}$, $\Gamma_3 = 140 \text{ cm}^{-1}$, $\omega' = 63 \text{ cm}^{-1}$, $\omega'' = 98 \text{ cm}^{-1}$. Error bars on these values are difficult to evaluate, and are probably large but they support an interpretation that the decay process occurs by combination of the object Eg libron with a Tg^- libron ($\omega_{Tg} \simeq 39.7 \text{ cm}^{-1}$) to create a Tu^+ phonon ($\omega_{Tu} \cong 77 \text{ cm}^{-1}$). The reverse process, i.e. decay of the Tu^+ phonon into Eg and Tg librons has been proposed as an explanation of the anomalous width of that phonon by Harris and Coll [15]. The accidental near degeneracy of the zone center combination of Eg and Tg energies with the Tu^+ phonon energy is an probably important factor in the large linewidths observed. This will be more clearly seen by comparison with the corresponding Eg libron linewidth in CO_2 below. It is not understood why the fitted frequencies ω' and ω'' are substantially larger than the directly observed Tg^- and $Tg^- + Eg$ frequencies and further study is presently being directed at that question.

In CO_2, it was possible to make quantitative measurements of the temperature dependent linewidths on each of the Eg, Tg^- and Tg^+ librons. Let us consider first the Eg libron to compare and contrast its behavior with that of the corresponding response in N_2. Referring to Figs. (7) and (8) we see that the behavior is qualitatively different in CO_2 and N_2. A fit of the data to eg. (7) which was quite satisfactory in N_2 gave poor results in CO_2. The quantitative reason for this seems clear. In N_2 as already mentioned there is an accidental near energy degeneracy of the combination Eg + Tg^- with the Tu^+ mode. In CO_2 the same $\omega_{Tg} + W_{Tg} \simeq 170 \text{cm}^{-1}$ while the Tu^+ phonon lies near 114 cm^{-1}. (22) An observation that the function $\Gamma(T)$ was

was nearly pure quadratic in T suggested a fit to the equation (6) which is shown as the solid line on Fig. (7)

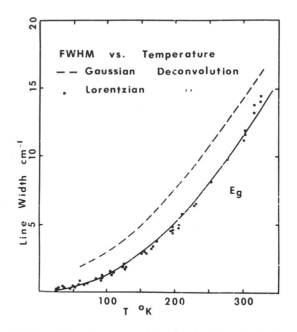

FIG. 7 Eg libron linewidth vs Temperature in Carbon Dioxide.

The result suggests that the seat of this linewidth in CO_2 lies in a process in which the Eg libron decays into three phonons dispersed around a frequency of 26 cm^{-1}. A final point of contrast may be made, i.e. that the linewidths of its Eg mode at corresponding temperatures are much smaller in CO_2 than in N_2. Once again it seems probable that the large density of final states for Eg + $Tg^- \rightarrow Tu^+$ is responsible for the extra linewidth in N_2.

In CO_2 the Tg^- and Tg^+ libron modes show qualitative differences from one another. The linewidth of the lower frequency Tg^- mode goes to approximately zero at low temperatures while the upper Tg^+ mode retains a substantial residual linewidth. The lower solid line in Fig. (8) corresponds to a fit of the Tg^- mode lifetimes based on the three phonon combination process represented by Eq (7) yielding $\Gamma_0 = 0$, $\Gamma_3 = 0.60 \text{ cm}^{-1}$, $\omega' = 25 \pm 15 \text{ cm}^{-1}$ and $\omega + \omega' = 116 \pm 15 \text{ cm}^{-1}$. The 116 cm^{-1} final phonon is remarkably close in energy to the Tu^+ optic mode at 114 cm^{-1} and the 25 cm^{-1} excitation would be among the acoustic modes of the spectrum. The upper solid line of Fig. (8) corresponds to a fit of the

Tg+ mode lifetimes based on a three phonon decay in which the object mode is annihilated and two other excitations are created. This may be represented by the first set of terms in eg. (5) which simplify to the form:

$$\Gamma = \Gamma_3 \left[m(\omega') + m(\omega'') + 1 \right] \qquad (8)$$

In this fit, the results

$$\Gamma_3 = 1.14 \text{ cm}^{-1}, \omega^1 = 26 \pm 15 \text{ cm}^{-1}, \omega'' = 110 \pm 15 \text{ cm}^{-1}$$

are obtained. N.B. once more that the value 110 cm^{-1} is remarkably close to the Tu$^+$ phonon frequency supporting a proposal that the Tg$^+$ libron decays into a Tu$^+$ phonon and a lower energy excitation, probably on acoustic phonon near 26 cm^{-1}.

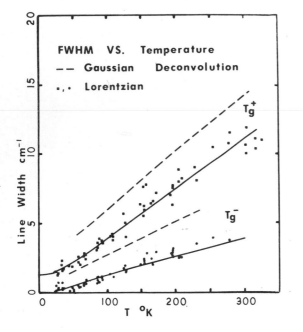

FIG. 8 Tg$^-$ and Tg$^+$ libron linewidths vs temperature in Carbon Dioxide.

It is amusing but probably an artifact of the form fit to the Eg mode lifetime, that a frequency near 26cm^{-1} occurs in the fitted descriptions for each of the three modes in CO_2.

In general, the results above support surprisingly simply anharmonic couplings as decay channels for the modes studied. That is, all results can be well fitted by considering essentially one spectral term of the general anharmonic shift and lifetime relations, and that in most cases the fit appears to have a more than merely formal physical significance because the energies

identified in the fits correspond to energies of independently identified features mode structure. Although the uniqueness of the assignments is untested, and flaws are potentially present (related principally to extraction of "true" lifetimes from instrumentally broadened observations) the results indicate that carefully measured temperature dependencies of shifts and widths may constitute a kind of internal spectroscopy of anharmonic effects.

The authors would like to thank the National Science Foundation for support under Grants DMR8100660 and DMR7801307. One of us (W.B.D.) would like to thank the Unidel Foundation for salary support during these studies.

*National Bureau of Standards, Washington, DC   20234

REFERENCES

[1] J. M. Ziman, "Electrons and Phonons" (Oxford, Clarendon Press, 1960).

[2] D. C. Wallace, "Thermodynamics of Crystals" (Wiley, New York, 1972).

[3] P. F. Choquard, "The Anharmonic Crystal" (W.A. Benjamin Inc., New York, Amsterdam, 1967).

[4] B. A. Weinstein and G. J. Piermarini, Phys. Rev. B12, 1172 (1975).

[5] J. Eckert, W. B. Daniels and J. Axe, Phys. Rev. B14, 3649 (1976).

[6] W. B. Daniels, in Proceedings of the VIII Int. AIRAPT Conference ed. B. Vodar (Pergamon Press, 1981).

[7] T. Kihara and K. Kobashi, Chem. Phys. Lett 65, 12 (1979).

[8] T. A. Scott, Physics Reports 27, 89-157 (1976).

[9] F. D. Medina and W. B. Daniels, J. Chem. Phys. 64, 150 (1976).

[10] J. W. Schmidt and W. B. Daniels, J. Chem. Phys. 73, 4848 (1980).

[11] J. Ruvalds and A. Zawadowsky, Solid State Coomun. 9, 129 (1971).

[12] P. Flubacher, A. J. Leadbetter and J. A. Morrison, Phil. Mag. Ser. 8, 4, 273 (1959).

[13] B. Brockhouse and P. K. Iyengar, Phys. Rev. 108, 894 (1957).

[14] W. B. Daniels, Report on the Int.
Conf. on the Physics of Semiconductors at Exeter, 1962. Pub.
by the Institute of Physics.

[15] A. B. Harris and C. F. Coll III,
Solid State Commun. 10, 1029
(1972).

[16] See for example: G. Leibfried
and W. Ludwig, Solid State Physics
12, 276 (Academic Press, New York,
1961).

[17] J. C. Raich, N. S. Gillis and
A. B. Anderson, Jl. Chem. Phys.
61, 1399 (1974).

[18] K. Kobashi, Molecular Physics 36,
225 (1978).

[19] J. C. Raich and N. S. Gillis,
Jl. Chem. Phys. 66, 846 (1977).

[20] R. F. Wallis, I. P. Ipatova and
A. A. Maradudin, Sov. Phys. -
Solid State 8, 850 (1966).

[21] F. Gervais, B. Piriou and F.
Cabannes, J. Phys. Chem. Solids,
34, 1785 (1973).

[22] A. Ron and O. Schnepp, Jl. Chem.
Phys. 46, 3991 (1967).

PHYSICS OF SOLIDS UNDER HIGH PRESSURE
J.S. Schilling, R.N. Shelton (editors)
© North-Holland Publishing Company, 1981

SOLID O_2 NEAR 298 K: RAMAN AND ELECTRONIC SPECTRA OF β AND ε-O_2 [§]

Karl Syassen[‡] and Malcolm Nicol[†]

‡ Physikalisches Institut der Universität Düsseldorf, Düsseldorf, West Germany
† Department of Chemistry, University of California, Los Angeles, California, USA

Raman and electronic (350–900 nm) spectra of O_2 near 298 K have been studied at pressures to 40 GPa. Phase transitions occur near 5.9 (fluid to β), 9.6 (β to α), and 9.9 GPa (α to a new ε-O_2 phase whose structure is not yet known). β-O_2 is dichroic; its visible spectrum is dominated by the $2\,{}^1\Delta_g \leftarrow 2\,{}^3\Sigma_g^-$ transition which is polarized normal to $\vec{c}$. The transition shifts to higher energies with increasing pressure. ε-O_2 also appears to be optically uniaxial, and its spectra have complex pressure dependences. For one axis, ε-O_2 crystals absorb strongly throughout the visible and have a strong near uv reflection band at pressures above 35 GPa. These spectra are interpreted in terms of transitions between states of $\pi^4\pi^{*2}$ and $\pi^3\pi^{*3}$ configurations.

The crystallization of oxygen near room temperature and two solid-solid phase transitions were first described by Nicol, Hirsch, and Holzapfel.[1] Visual observations and Raman spectra of the O_2 internal mode located the freezing pressure (5.9 ± .2 GPa near 298 K) and other first-order transitions near 9.6 and 9.9 GPa. Some Raman spectra near the coexistence points are reproduced in Figure 1a–f. In that first report, the low pressure solid phase was described as colorless; and the very high pressure phase was reported to absorb very strongly throughout much of the visible, with an absorption edge that shifted rapidly toward long wavelengths with increasing pressure.

These solids were identified by d'Amour, Holzapfel, and Nicol [2] from x-ray patterns of single crystals frozen from the fluid and the temperature dependences of the phase boundaries. The solid that freezes near room temperature was identified as β-O_2. Visual observations of these single crystals demonstrated that they were not colorless but had a faint magenta hue in certain settings. The solid formed near 9.6 GPa was tentatively identifies as α-O_2, while the temperature dependence of the higher pressure phase line demonstrated that the phase stable above 9.9 GPa at 298 K was new. It has been designated ε-O_2. Some previously unpublished spectra obtained near the melting line at lower temperatures (Figure 1g–i) suggest that γ-O_2 may have a narrow range of stability between the fluid and β-O_2 at temperatures as high as 273 K; but the identification of this phase is incomplete. The partial phase diagram for O_2 based on these previous studies is reproduced in Figure 2.

This report describes new studies of the optical spectra of β and ε-O_2 near room temperature and extensions of the Raman measurements to 40 GPa. Polarized absorption spectra (350–900 nm) of single crystals of β-O_2 are shown to be dominated ba the $2\,{}^1\Delta_g \leftarrow 2\,{}^3\Sigma_g^-$ transition which shifts to higher energy with increasing

pressure. Two characteristic settings of ε-O_2 could be identified from absorption and reflection spectra measured at pressures from 10 to 40 GPa. For one setting, the strong visible absorption features described by Nicol, Hirsch, and Holzapfel [1] were apparent, while the principal feature of the visible spectrum of the other setting could again be identified as the $2\,{}^1\Delta_g \leftarrow 2\,{}^3\Sigma_g^-$ transition. The strong absorption bands of ε-O_2 may be interpreted in terms of closing the gap between states with $\pi^4\pi^{*2}$ and $\pi^3\pi^{*3}$ configurations.

EXPERIMENTAL

Oxygen samples were loaded from liquid of nominal 99.9% purity by the previously described method.[1] Single crystals of β-O_2 were grown by first preparing a polycrystalline sample of β-O_2 and compressing it to about 9 GPa which also reduced the gasket to a convenient thickness, then reducing the pressure to the melting line until all but one seed crystal fused, and finally growing that seed by slowly increasing the pressure on the sample. This cycle typically produces a crystal of about 0.1 mm thickness. Recompression of this crystal to 9.5 GPa typically reduces the sample and gasket thickness by another 0.02 mm; however, precise pathlengths were not measured. Thus, relative absorptivities are not known with better that 20% precision. The sample thickness is reduced still further (to about 0.03 mm) upon transformation to ε-O_2 and compression to 40 GPa.

Polycrystalline ε-O_2 samples grown by compression of β crystals through the β–α and α–ε transformations occur in two characteristic settings, which suggest that the structure is uniaxial or only weakly biaxial. These settings have been characterized as "dark" and "light" by transmission spectra. Although systematic studies were not performed, no correlation between the occurrence of the two settings and the orientation of the mother β crystal have been noticed. At the same pressure, the Raman

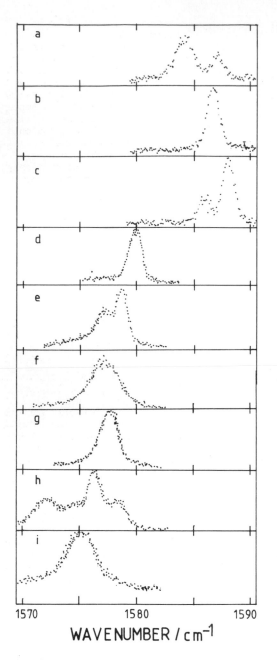

Figure 1 : Raman spectra of the internal mode of O_2 near several phase boundaries: a-c, ε-α, α, and α-β spectra at 298 K near 9.9, 9.6, and 9.4 GPa respectively; d-f, β, β-fluid, and fluid spectra at 298 K near 6.8, 5.9, and 5.6 GPa respectively; g-i, spectra at 273 K above (β-O_2, 5.7 GPa), at, and below (fluid, 5.2 GPa) the melting pressure. The additional features in spectrum h also occur near the melting line at lower temperatures. These may be the sidebands characteristic of the Raman spectrum of γ-O_2.[3]

spectra of the two settings occur within 2 cm^{-1} (typical of the reproducibility of these data), although the wavenumbers of the "dark" sample are more frequently lower than those of the "light" sample. This may result from heating of the darker samples by absorbed light. The intensities of the Raman spectra of the dark samples also are reduced by competition with absorption.

Raman and luminescence spectra were measured in conventional manners.[1] Temperatures were measured with a Pt-resistance thermometer mounted on the cell near the diamonds. This measurement cannot take into account heating of the sample that may result from absorption of light. The methods used to obtain absorption and reflection spectra are described elsewhere in this volume.[4] These spectra were collected from a fine focus of about 0.03-mm diameter that could be positioned throughout the 0.2-mm diameter sample. The small focal area was especailly useful for obtaining spectra of the two characterixtic types of ε-O_2.

RAMAN SPECTRA OF ε-O_2

The pressure dependence of the wavenumber of the internal mode of O_2 is reproduced in Figure 3. Earlier observations below 15 GPa [1] are represented in this figure by curves fitted to those data. Neither the rate of shift nor the linewidth of the Raman band appeared to be discontinuous between 10 and 39 GPa. Furthermore, visual observations of the sample during compression provided no evidence for another phase transition. Thus, the curve representing the lower pressure data for ε-O_2 has been extended through the new observations.

Except at the phase transition and between 10 and 15 GPa, the wavenumber of this band increases with pressure at 3.7 $\pm$.1 cm^{-1}/GPa for the fluid and 2.6 $\pm$.3 cm^{-1}/GPa for the solid phases. The increase from 1556 cm^{-1} for the isolated molecule to 1639 cm^{-1} for ε-O_2 at 39 GPa is equivalent to 25% of the difference between the vibrations of doubly-bonded O_2 and the O_2^+ ion whose formal bond order is 2.5. The rates at which this vibration shifts are similar to those of hydrocarbon skeletal vibrations [5] and the internal mode of D_2 to 30 GPa.[7] However, the O_2 vibrational frequency changes almost twice as rapidly as that of N_2.[8] Differences between these rates reflect to some degree the ease with which valence electrons are redistributed by compression. Thus, the difference between O_2 and N_2 may result from the ease with which π^* electrons are delocalized by intermolecular bonding; such delocalizaztion also is implied by effects of compression on the spectra of ε-O_2.

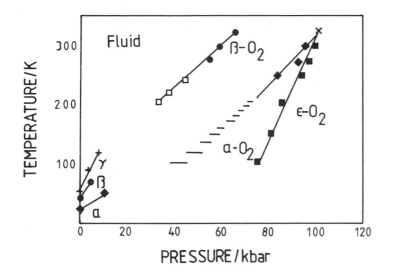

Figure 2 : Partial phase diagram of O_2. Low temperature data from Ref. 6. Points where coexistence of two phases has been observed are indicated by • (f-β), ◆ (β-α), ■ (α-ε), and x (β-ε); points designated by □ appear to be on the f-β or β-γ line. The horizontal bars indicate the extension of the β-α line to low temperatures where it cannot be precisely located by Raman spectra.

The origin of the anamolous pressure dependence of the vibrational frequency between 10 and 15 GPa is unknown. A plausible explanation is that, near 10 GPa, the molecular orientations in the ε structure are relatively unrestricted. Thus, initial compression of this structure occurs more by reorienting molecular axes than by compressing molecular bonds. This situation might occur if, for example, the structure of ε-O_2 were similar to those of the halogens, which differ from α-O_2 by relative orientations of the molecular and crystal axes. Diffraction data are needed to resolve this question.

ELECTRONIC SPECTRA (350–900 nm) OF β-O_2

Figure 4 compares transmission spectra of fluid O_2 at 5.4 GPa, a large β crystal with $\vec{c}$ parallel to the diamond culets at 6.7 and 9.0 GPa, and a β crystal with $\vec{c}$ normal to the culets at 7.6 GPa. These spectra have been corrected for all but the reflection losses and resulting optical interferences between reflections at the four parallel diamond surfaces. In the fluid phase spectra, the origin and one vibronic component of the collision-induced $2\,^1\Delta_g \leftarrow 2\,^3\Sigma_g^-$ transition are apparent near 610 and 570 nm. At

Figure 3 : Pressure dependence of the wavenumber of the Raman spectrum of O_2. New data for ε-O_2 are represented by open circles. Data below 15 GPa from Ref. 1 are represented by solid lines.

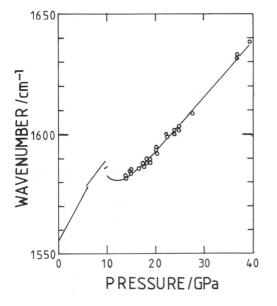

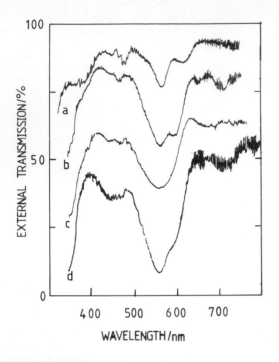

Figure 4 : Room temperature transmission spectra through the diamonds and O_2 sample at four pressures: a, fluid O_2 at 5.4 GPa ($l \simeq 0.1$ mm); b, β-O_2 with $\bar{c}$ parallel to the diamond culets at 6.7 GPa ($l \simeq 0.1$ mm); c, the same crystal as b compressed with some cracking to 9.0 GPa ($l \simeq 0.08$ mm); d, β-O_2 with $\bar{c}$ normal to the diamond culets at 7.6 GPa. Spectra a, b, and d have been displaced by +20, +10, and -10% for clarity. Transmission interference created by reflection at the four parallel diamond surfaces appear as noise and distort weaker absorption bands.

low pressures, the intensities of these two-molecule transitions are known to vary with the square of the oxygen density. Thus, it is of interest to compare the intensity of the 5.4-GPa spectrum ($\ln_{570}I_0/I \simeq .35$, $l = 10^{-4}$ m) with that reported by Blickensderfer and Ewing [9] at low pressures ($\ln_{570}I_0/I \simeq .07$, $l = 168$ m, $d \simeq 3$ mg/cm³). Were the dependence on the square of the density to extend to 5.4 GPa, these data imply that the density of the compressed fluid exceeds 8 g/cm³ which is more than four times the density of β-O_2 at 5.8 GPa. [2] Apparently, many-body effects or non-random orientations enhance the intensity of this transition at high densities.

Upon crystallization, the intensity of the $2^1\Delta_g \leftarrow 2\ ^3\Sigma_g^-$ spectra increase several-fold, and both bands move to higher energies by about 0.02 ev/GPa. Observation of the crystal used for x-ray diffraction showed that these bands are polarized normal to the $\bar{c}$ axis of the β-O_2 structure; thus, they appear weaker for the

crystal with $\bar{c}$ parallel to the culets. Further compression of the β-O_2 shifts the origin to shorter wavelengths and broadens the spectrum so that the vibronic structure cannot be resolved at 9 GPa. However, for a given crystal, the integrated intensity of this transition does not vary significantly over the range of pressures in which β-O_2 is stable at 298 K because the nearest-neighbor separation changes by at most 4%. Weak features in the fluid spectrum near 480 and 390 nm may be associated with the collision-induced transitions to the $^1\Sigma_g^+ + ^1\Delta_g$ and $^1\Sigma_g^+ + ^1\Sigma_g^+$ states, and weak features in the crystal spectra near 700 and 450 nm may be the $^1\Sigma_g^+ \leftarrow ^3\Sigma_g^-$ and $^1\Sigma_g^+ + ^1\Delta_g \leftarrow 2\ ^3\Sigma_g^-$ transitions. However, these weak features are not clearly discriminated from the interferences caused by reflections at the diamond surfaces.

The pressure dependence of the energies of these transitions can easily be understood. The $^1\Sigma_g^+$, $^1\Delta_g$, and $^3\Sigma_g^-$ states derive from $\pi^4\pi^{*2}$ orbital configurations that are degenerate in the absence of electron-electron repulsion. These states differ, however, in the average distance between the two π^* electrons. Thus, the electron-electron repulsion lifts the degeneracy. To the extent that compression of the molecular volume more tightly confines these valence electrons, the electron-electron repulsion and splittings among these states will increase.

The intensity and polarization of the $2^1\Delta_g \leftarrow 2^3\Sigma_g^-$ transition is more difficult to explain. Tsai and Robinson [10] have performed perhaps the most detailed analysis of the intensities of the collision-induced spectra and estimated transition probabilities for several O_2-O_2 orientations appropriate to γ-O_2. For the orientation most appropriate to the average β-O_2 structure, that is for a pair of O_2 molecules whose axes are parallel to each other but perpendicular to the vector between their centers, the transition probability induced by mixing with the allowed $^3\Sigma_u^- \leftarrow ^3\Sigma_g^-$ or $^1\Delta_u \leftarrow ^1\Delta_g$ transitions was shown to vanish! However, as Cahill and Leroi argue,[3] the instantaneous orientations of the O_2 bonds are not parallel to $\bar{c}$ but are tilted by 10-to-15 degrees from $\bar{c}$, around which the molecules librate. Interaction between librating molecules might explain the observed spectra; however, detailed calculations have not been performed for many of the orientations appropriate to a pair of librating molecules.

ELECTRONIC SPECTRA OF ε-O_2

Transmission spectra of the light and dark samples of ε-O_2 at several pressures are shown in Figure 5 which demonstrates that, at all wavelengths and pressures, the transmission of the dark sample is at least an order of magnitude lower than that of the light sample. Two features are apparent in the spectra of the light sample: an intense near uv absorption edge and a weak structureless band near 500 nm. Movement of the edge appears to be responsible

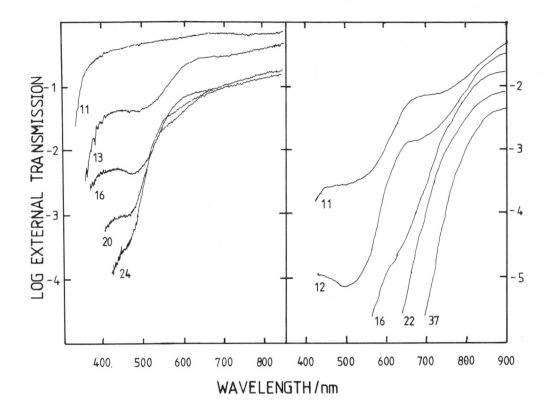

Figure 5 : Room temperature transmission spectra of the light (left panel) and dark (right panel) settings of ε-O_2 at several pressures. Note the 1 OD change of intensity scales between the panels.

for the generally lower transmission of this sample at high pressures. The origin of this band is not known although it appears to be associated with the strong uv absorption that is responsible for the high uv reflectivity of the dark setting which is attributed to the $^3\Sigma_u^- \leftarrow {}^3\Sigma_g^-$ transition.

The intensity, position, shape, and pressure dependence of the absorption band near 500 nm at pressures above 13 GPa are characteristic of the 2 $^1\Delta_g \leftarrow 2\ ^3\Sigma_g^-$ transition observed in compressed β-O_2. However, this band is noticably absent from spectra obtained between 10 and 12 GPa. Apparently, the relative orientations of O_2 molecules in the ε-O_2 structure at pressures just above the α-ε transformation are such that the transition probability for this two-molecule transition vanishes; but compression-induced reorientation, that also was invoked to interpret the Raman spectra, yields a configuration in which this two-molecule transition is as probable as in the

β-O_2 structure.

The transmission spectra of the dark ε-O_2 at 11 and 12 GPa consist of weak structureless shoulders near 700 and 500 nm on a rising absorption edge that extends into the uv. These weak bands are assigned as the $^1\Sigma_g^+ \leftarrow {}^3\Sigma_g^-$ and 2 $^1\Delta_g \leftarrow 2\ ^3\Sigma_g^-$ transitions, both of which appear to move toward shorter wavelengths and may gain intensity with increasing pressure. However, by 16 GPa, these features are swamped by the rapid shift of the strong absorption to longer wavelengths. Survey visible reflection spectra of ε-O_2 samples identified a weak feature whose pressure dependence corresponds to this edge; this suggests that the edge corresponds to a spin-allowed but orbitally-forbidden transition such as the $^3\Sigma_u^+ \leftarrow {}^3\Sigma_g^-$ or $^3\Delta_u \leftarrow {}^3\Sigma_g^-$ transitions. Near uv reflection spectra of dark ε-O_2 at high pressures, such as that reproduced in Figure 6, suggest that a strongly allowed transition, such as $^3\Sigma_u^- \leftarrow {}^3\Sigma_g^-$, is located in the middle uv at pressures of the order of 40 GPa.

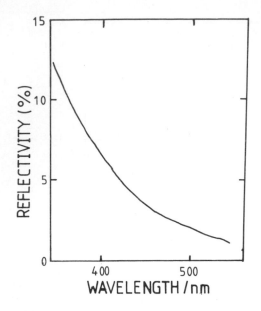

Figure 6 : Reflection spectrum of a dark sample of ε-O_2 at 37 GPa. The reflectivity of the O_2-diamond surface is calibrated by assuming that the reflectivity of the diamond-gasket surface is 50%.

Transitions from $^3\Sigma_g^-$ to the $^3\Sigma_u^+$, $^3\Delta_u$, and $^3\Sigma_u^-$ states involve promotion from $\pi^4\pi^{*2}$ to $\pi^3\pi^{*3}$ configurations, decrease the formal bond order from 2 to 1, and increase the equilibrium bond length from 1.2 to 1.6 A. For the free molecule, the minima of the $^3\Sigma_u^+$ and $^3\Delta_u$ states lie about 4 ev above the $^3\Sigma_g^-$ state; and $^3\Sigma_u^-$ is about 6 ev above $^3\Sigma_g^-$.[11] The change in equilibrium geometry and vertical nature of optical excitations place the corresponding absorption spectra (the Herzberg and Schumann-Runge bands) in the mid and vacuum uv. Molecules excited to states with $\pi^3\pi^{*3}$ configurations may be more compressible that ground state molecules; furthermore, delocalization of π^* electrons by intermolecular bonding in compressed O_2 may lower the equilibrium energies of $\pi^3\pi^{*3}$ states relative to $\pi^4\pi^{*2}$ states. A combination of these factors, which are not completely independent, could shift the energies of the Herzberg and Schumann-Runge transitions by the amount required to account for the visible and mid-uv spectra of ε-O_2.

REFERENCES AND NOTES

§   This work was supported in part by NSF DMR 77-27428 and 80-25620 and NATO 1952.

[1]   Nicol, M., Hirsch, K.R., and Holzapfel, W. B., Oxygen Phase Equilibria Near 298 K, Chem. Phys. Letters 68 (1979) 49-52.

[2]   d'Amour, H., Holzapfel, W.B., and Nicol, M., Solid O_2 Near 298 K. The Structure of β-O_2 and Identification of a New ε Phase, J. Phys. Chem. 85 (1981) 130-1.

[3]   Cahill, J.E., and Leroi, G.E., Raman Studies of Molecular Motion in Condensed Oxygen, J. Chem. Phys. 51 (1969) 97-104.

[4]   Syassen, K., and Takemuro, K., Optical and Structural Studies on Cs and I Under High Pressure, in: Schilling, J.S., and Shelton, R.N. (eds.), Proc. Int. Symp. Physics of Solids Under High Pressure (North-Holland, Amsterdam, 1981).

[5]   See, for examples, Ellenson, W.D., and Nicol, M., Raman Spectra of Solid Benzene Under High Pressure, J. Chem. Phys. 61 (1974) 1380-9.

[6]   Stewart, J.W., Phase Transitions and Compressions of Solid CH_4, CD_4, and O_2, J. Phys. Chem. Solids 12 (1959) 122-9.

[7]   Sharma, S.K., Mao, H.K., and Bell, P.M., Raman Measurements of Deuterium in the Pressure Range 8-537 Kbar at Room Temperature, Carnegie Inst. Wash. Year Book 79 (1980) 358-64.

[8]   LeSar, R., Ekberg, S.D., Jones, L.H., Mills, R.L., Schwalbe, L.A., and Schiferl, D., Raman Spectroscopy of Solid Nitrogen up to 374 Kbar, Solid State Comm. 32 (1979) 131-4.

[9]   Blickensderfer, R.P., and Ewing, G.E., Collision-Induced Absorption Spectra of Gaseous Oxygen at Low Temperatures and Pressures. II. The Simultaneous Transitions $^1\Delta_g + {}^1\Delta_g \leftarrow {}^3\Sigma_g^- + {}^3\Sigma_g^-$ and $^1\Delta_g + {}^1\Sigma_g^+ \leftarrow {}^3\Sigma_g^- + {}^3\Sigma_g^-$, J. Chem. Phys. 51 (1969) 5284-9.

[10] Tsai, S.C., and Robinson, G.W., Why is Condensed Oxygen Blue?, J. Chem. Phys. 51 (1969) 3559-68.

[11] Krupenie, P.H., The Spectrum of Molecular Oxygen, J. Phys. Chem. Ref. Data 1 (1972) 423-519.

PHYSICS OF SOLIDS UNDER HIGH PRESSURE
J.S. Schilling, R.N. Shelton (editors)
© *North-Holland Publishing Company, 1981*

INTRAMOLECULAR VIBRATIONAL MODE SHIFTS WITH PRESSURE IN SOLID CO_2, N_2, and O_2^+

R. D. Etters and A. A. Helmy

Physics Department, Colorado State University
Fort Collins, Colorado 80523

The pressure dependence of intramolecular mode frequencies in solid CO_2, N_2 and O_2 has been calculated using a theoretical description that depends on the details of the inter and intramolecular potentials. Results are in good agreement with experiment. A discussion of the predicted high pressure breakdown of this model is presented.

I. Introduction

In concert with the development of diamond cell technology [1], there has been a great advance in the utilization of Raman scattering and infrared absorption to study the pressure dependence of the intramolecular vibrational modes in molecular crystals [2]. The pressure dependence of the stretching mode in solid H_2 at room temperature has been measured to P = 150 kbars by infrared absorption [3] and by Raman scattering [4] to 650 kbars. The frequency of this mode increases at a decreasing rate until approximately 330 kbars, beyond which it decreases. Similar experiments [5] to 120 kbars in CO_2 show that the pressure dependence of the breathing mode frequency $\nu_1(P)$ is qualitatively similar to H_2 although $d\nu_1/dP$ is everywhere positive. In addition to the breathing mode, CO_2 supports a bending mode, ν_2, and an antisymmetric stretching mode, ν_3.

Raman scattering experiments [6] on the stretching mode in solid N_2 at room temperature have been performed to 374 kbars. There is a splitting of this mode in the region $54 \leq P \leq 118$ kbars because the crystal structure has two inequivalent sites [7]. Again $d\nu/dP$ is a decreasing positive quantity in the region $39 \leq P \leq 374$ kbars. The stretching mode in O_2 has also been measured [8] for $0 \lesssim P \lesssim 160$ kbars at room temperature.

The purpose of this work is to show that the pressure dependence of the intramolecular vibrational modes in simple molecular crystals can be calculated from the inter and intra-molecular potentials on the basis of the following model. At relatively low pressures, where the compressional energy is small compared to the electronic energies, the intramolecular interaction can be described by the properties of an isolated molecule. The crystal field of the solid acts as a perturbation and exerts a Van der Waal's force on the molecules which alters internal natural frequencies. At very high pressures it is certain that compressional forces are large

enough to significantly disturb the molecular electron charge distribution. Under such circumstances charge transfer is expected and the intramolecular interaction is drastically different than in the gas phase. A dramatic example of this must occur near the molecular to atomic phase transition predicted for all molecular solids at sufficiently high pressures. This article presents results for solid CO_2, N_2, and O_2, in the relatively low pressure regime, where the crystal field is assumed to act as a small perturbation on the internal modes of the molecule.

II. Method

The pressure dependence of intramolecular vibrational frequencies has not been calculated for molecular crystals other than CO_2. Hanson and Bachman [5] determined the axial force on a CO_2 molecule using the relation

$$F_{axis} = AP/(3)^{1/2}n$$

where A is the area of the cubic unit cell face and n is the number of molecules per plane in the unit cell. F_{axis} causes a change in the intramolecular separation and, using the intramolecular force constants of Suzuki [9], the pressure dependence of the normal mode frequencies were determined. As will be shown, the agreement of these results with experiment is fairly good considering the simplicity of the analysis.

Instead of this method we utilize a microscopic model which depends in detail on the inter and intramolecular interactions. The crystal potential energy is of the atom-atom form

$$U_c = \sum_{i<j} U_{AA}(R_{ij}) , \tag{1}$$

where the sum extends over all pairwise interactions between atoms on different molecules, and R_{ij} is the interatomic distance. The intramolecular potential can be described as an expansion in powers of the atomic displacements from equilibrium.

$$U_M = \frac{1}{2} k (r-r_o)^2 + g(r-r_o)^3 + \dots \qquad (2)$$

The constants k and g are determined by fitting to spectroscopic data or the actual ab initio potential, r_o is the equilibrium separation between atoms in an isolated diatomic molecule, and r is its instantaneous value. Keeping only terms through cubic in Eq. (2), the total potential energy per molecule in the solid is

$$U = \frac{1}{2} k (r-r_o)^2 + g(r-r_o)^3 + \frac{1}{N} U_c(r) \qquad (3)$$

where N is the total number of molecules in the system. Similarly, U can be expressed as an expansion about the equilibrium intramolecular atomic separation in the solid, r_c.

$$U = \frac{1}{2} k'(r-r_c)^2 + g'(r-r_c)^3 + \dots \qquad (4)$$

where k' and g' are the quadratic and cubic force constants, respectively. All constant terms in the expansions, Eqs. (2) and (4), are of no consequence to the analysis and are thus ignored. Taking second derivatives of Eqs. (3) and (4) with respect to r and equating them at $r = r_c$, we get

$$k' = k + 6g(r_c-r_o) + \frac{1}{N} [\partial^2 U_c/\partial r^2]_{r=r_c} \qquad (5)$$

for the quadratic intramolecular force constant in the solid. Utilizing the binomial expansion the normal mode frequency change in the solid is

$$\delta\nu(P) = \nu(P)-\nu_o =$$

$$A[r_c(P)-r_o] + B[\partial^2 U_c/\partial r^2]_{r_c(P)} \qquad (6)$$

where

$$A = 3g/(4\pi^2\mu\nu_o)$$

$$B = (8\pi^2\mu\nu_o N)^{-1}$$

where μ is the molecular reduced mass and ν_o is the isolated molecule normal mode frequency. The contribution to the gas to solid frequency shift from the crystal field perturbation is

clearly $\delta\nu(o)$. In addition to this term, the physical shift also includes an expression involving the change in polarizibility between the two phases.

$$\Delta\nu(P) = \nu(P)-\nu(o) = A[r_c(P)-r_c(o)] +$$

$$B[(\partial^2 U_c/\partial r^2)_{r_c(P)} -$$

$$(\partial^2 U_c/\partial r^2)_{r_c(o)}] \qquad (7)$$

is the frequency shift in the solid between zero and the pressure P. The second derivative terms in Eq. (7) while small compared to the first term, are not negligible at high pressures. Clearly, the calculation of $r_c(P)$ is a most important element in the analysis. The calculation proceeds by evaluating the total potential energy per molecule, Eq. (3), for the appropriate crystal structure. Lattice sums are taken over the first 10 shells. The intramolecular separation r is then varied in Eq. (3) until U is minimized at $r = r_c$. Note that U_c is a function of r as is the intramolecular potential because the force centers in the atom-atom potential are separated by r. This procedure is repeated at each molar volume until a mapping of r_c versus molar volume is established. The pressure-volume relation is determined by taking the derivative $P = -\partial U_c/\partial V$ of the energy U_c, which has been fitted to power series in the volume V.

From the calculated zero temperature equation of state, an approximate high-temperature expression can be estimated using the quasi-harmonic approximation

$$P(T) = P(o) + 3N_o kT\rho\gamma/M \qquad (8)$$

where N_o is Avogadro's number, T is the temperature, k is Boltzman's constant, ρ is the density, M is the molecular weight, and γ is Gruneisen's parameter.

III.   The Potentials

The intermolecular interactions used in this work were assumed to be pairwise additive and of an atom-atom form. They are semiphenomenological in character with parameters that were deduced using gas and solid data. Because of the uncertain quality of these expressions, several different potentials were investigated for CO_2 and N_2 so the dependence of our results on the specific form of the potential could be determined. A brief description of the potentials used in this work follows.

A. CO_2

The exponential-6 expression of Kitaigorodskii [10]

$$\sum_{i<j} U_{AA}(R_{ij}) = \sum_{i<j} [A_{ij}\exp(-\alpha_{ij}R_{ij}) - B_{ij}R_{ij}^{-6}] + U_{EQQ} \quad (9)$$

has been formulated by determining the parameters A_{ij}, B_{ij}, and α_{ij} from solid CO_2 data. Here the sum extends over all pairwise interactions between carbon and oxygen atoms on different molecules. Thus, between each pair of molecules there are nine terms plus the electric quadrupole-quadrupole interaction U_{EQQ}. The zero pressure quadrupole moment is given by $Q = -4.2 \times 10^{-26}$ esu and its pressure dependence is fitted to the theoretical results of Morrison and Hay [11]. Both the asymptotic and point charge model were used to evaluate U_{EQQ}, with little difference between them.

A Lennard Jones 6-12 potential form

$$4\varepsilon \sum_{i<j} [(\sigma/R_{ij})^{12} - (\sigma/R_{ij})^6] + U_{EQQ}$$

was also used. The potential by Murthy, et al. [12], MSKM, is employed where $\varepsilon = 142.1$ K, $\sigma = 3.005$ Å, and $Q = 3.54 \times 10^{-26}$ esu., and $r_0 = 2.3$ Å. These parameters are those which give the best comparison between experiment and calculated lattice energy and lattice spacing in the solid, and second virial coefficients in the gas. Other potentials of the 6-12 form, but with different parameters, were also studied.

The intramolecular force constants were taken from Suzuki [9].

B. N_2

An atom-atom representation of the ab initio potential of Berns and Van der Avoird [13] is given by

$$U_{AA}(R_{ij}) = q_i q_j R_{ij}^{-1} - C_{ij}R_{ij}^{-6} + A_{ij}\exp(-B_{ij}R_{ij})$$

where q, A, B, and C were determined to give the best fit to the ab initio results. With this potential, good agreement between calculation and experiment has been obtained for the gas phase and the α and γ phases of solid N_2 [13,14].

Another potential used [15] is of the form

$$U_{AA}(R_{ij}) = A\exp(-\alpha R_{ij}) - BR_{ij}^{-6}$$

with A = 6.4155×10^4 kcal/mole, B = 4.6268 kcal/mole (Å)^6, α = 3.635 Å^{-1}, and a distance between force centers of 0.9012 Å. These parameters were chosen to give the best fit between calculated results and experimental data for solid and gas properties.

The 6-12 potential of Zunger and Huler [16] utilize the parameters ε = 0.31 kcal/mole, σ = 3.3 Å, and r_0 = 1.098 Å. The 6-12 potential A of Thiery and Chandresakharan [17] also includes an electric quadrupole-quadrupole interaction term. Again, the parameters of these potentials were chosen to fit various gas and solid phase data.

The intramolecular force constants were taken from Herzberg [18].

C. O_2

For O_2 the 6-12 potential of Laufer and Leroi [19] and the 6-12 + U_{EQQ} expression of English and Venables [20] were used. The intramolecular force constants were taken from Herzberg [18].

IV. Results

A. CO_2

Figure (1) shows the pressure dependence of the breathing and bending mode frequencies, ν_1 and ν_2, respectively. The squares and circles represent the room temperature experimental results of Hanson and Bachman [5] and the dashed lines are their calculated results for these modes. The solid lines represent our results for the Pa3 cubic structure of CO_2 based upon the exp-6 potential of Kitaigorodskii [10] and the dot-dashed lines are based upon the MSKM [12] potential. The room temperature, experimental equation of state of Olinger and Cady [21] was used to convert $\Delta\nu$ from a function of V to P. It is represented by the circles on Fig. (2). The solid line shows the zero temperature P-V relation calculated using the exp-6 potential [10] and the dashed line is an approximate room temperature isotherm determined by adding to the zero temperature isotherm (solid line) a thermal pressure given by Eq. (8), where $\gamma = 3$. This value was chosen on the basis of the data compiled by Gibbons and Klein [22]. The triangles represent the experimental data of Stevenson [23]. The quantity $N^{-1}(\partial^2 U_c/\partial r^2)$ in Eq. (5) contributes about 8 cm^{-1} to $\Delta\nu$ at 100 kbars and about 1 cm^{-1} at very low pressures. Figure (3) shows the change in the antisymmetric stretching mode frequency, ν_3, with pressure. The squares show the experimental

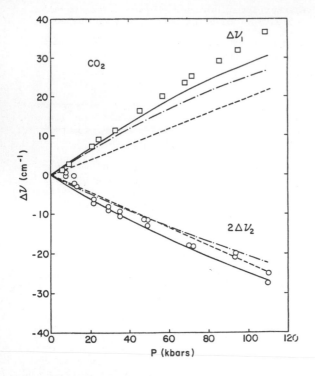

Fig. (1). Frequency change versus pressure for
the breathing mode $\Delta\nu_1$ and the
bending mode $\Delta\nu_2$ in solid CO_2.
Squares and circles represent experimental data [5] and the dashed
curve gives their calculated results.
The solid and dashed-dot curves are
our calculations on α-CO_2 based upon
the Kitaigorodskii [10] and MSKM [12]
potentials, respectively.

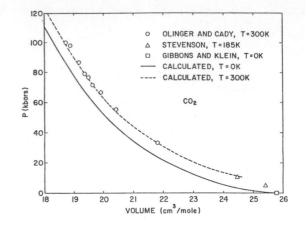

Fig. (2). Pressure versus molar volume for
solid CO_2. The circles and triangles represent experimental data
and the solid line is calculated,
based upon the Kitaigorodskii [10]
potential. The dashed line is the
same calculation corrected for
thermal pressures.

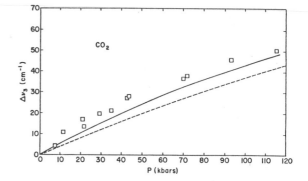

Fig. (3). Antisymmetric stretching mode frequency change versus pressure for
CO_2. The squares represent the
experimental data of Hanson and
Jones [5] and the dashed line is
their calculated result. The solid
line is our calculation based on the
Kitaigorodskii [10] potential.

data [5] and the dashed line shows the calculated results of Hanson and Jones [5]. The
solid line represents our calculation based
upon the exp-6 potential [10]. Also displayed
on Table I for this potential is the zero
temperature volume dependence of r_c. From Eq.
(6) $\delta\nu(o)$ has been calculated for ν_1, $2\nu_2$,
and ν_3 using the Kitaigorodskii potential
[10]. The results are 5.12, -5.62, and 9.45
cm^{-1}, respectively.

B. N_2

On Fig. (4), the squares represent the room
temperature experimental data of LeSar, et al.
[6] and the solid line represents our zero
temperature calculation of $\Delta\nu$ for γ-N_2, based
upon the potential of Berns and Van der Avoird
[13]. The values of r_c are listed on Table I.
The results of exp-6 potential [15] are shown
by the dot-dashed line and dashed line shows
results based upon the 6-12 potential of
Thiery and Chandrasekharan [17]. Figure (5)
shows the zero temperature P-V relations based
upon these potentials. The quantity $\partial^2 U_c/\partial r^2$

in $\Delta\nu$ contributes about 8 cm^{-1} to $\Delta\nu$ at 400
kbars down to approximately 0.6 cm^{-1} at very
low pressures. The volume dependence of r_c
is tabulated on Table I for the Berns-Van der
Avoird potential [13]. With this potential
$\delta\nu(o) = 1.43$ cm^{-1} for γ-N_2 at T = 0 K.

C. O_2

Figure (6) shows $\Delta\nu = \nu(P)-\nu(59.1$ kbars) for
β-O_2. The solid line shows the calculation
using the crystal potential of Laufer and

Table I. Values of r_c that minimize the total energy versus volume. For γ-N_2, c/a = 1.291 for all molar volumes. For β-O_2, different c/a values minimize the total energy at each volume.

CO$_2$		γ-N$_2$		β-O$_2$		
V(cm^3/mole)	r_c(Å)	V(cm^3/mole)	r_c(Å)	V(cm^3/mole)	r_c(Å)	c/a
25.19	2.29790	27.19	1.09377	20.99	1.2071	3.462
22.67	2.29604	25.82	1.09367	19.45	1.2067	3.482
21.41	2.29461	23.96	1.09346	18.78	1.2064	3.489
20.90	2.29392	21.33	1.09295	17.12	1.2055	3.53
19.74	2.29184	18.14	1.09159	15.98	1.2045	3.558
18.82	2.28974	15.28	1.08876	14.89	1.2030	3.592
18.37	2.28854	13.97	1.08640	14.05	1.2013	3.618
17.71	2.28648	12.74	1.08309	13.41	1.1996	3.642

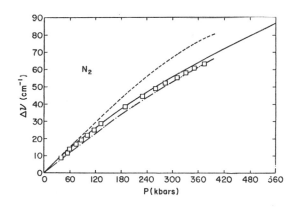

Fig. (4). Frequency change versus pressure for solid N$_2$. The squares represent the data points of LeSar, et al [6]. The solid line is calculated for γ-N$_2$, based upon the Berns-Van der Avoird potential [13]. The dot-dashed line is based upon the Murthy, et al potential [12], and the dashed line is from a 6-12 potential [17].

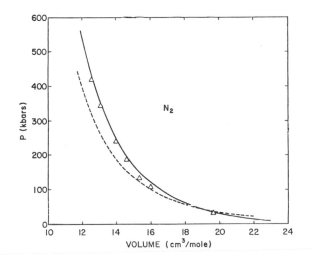

Fig. (5). Calculated pressure versus molar volume for γ-N$_2$. The solid line is based on the Berns-Van der Avoird potential [13], and the dashed line is based on the Murthy [12] potential. The triangles are based upon a 6-12 potential [17].

Leroi [19] and the dashed line is calculated from the potential of English and Venables [20]. The squares are taken from the experimental work of Nicol, et al. [8]. The values of r_c and the c/a crystal axes ratios that minimize the free energy at each volume, calculated from potential [20], are tabulated on Table 1.

V. Discussion and Conclusions

The quality of the results for $\Delta\nu$ depend on the quality of the intermolecular potentials that characterize the solid. Of the four potentials tested for CO$_2$, the exp-6 expression [10] gives superior agreement with experiment for $\Delta\nu$. This is not surprising since the other three have unphysical R^{-12}

repulsive cores which cannot adequately represent the important short ranged part of the actual energy surface. The results shown in Fig. (1) for the MSKM potential [12] are typical of the 6-12 expressions tested. The large EQQ potential energy in CO$_2$, U_{EQQ}, does not contribute much to $\Delta\nu$ because its $V^{-5/3}$ volume dependence is weak compared to the repulsive overlap interaction, diminishing its importance at high pressures. As with CO$_2$, the most satisfactory forms of the potential for N$_2$ were those with exponential repulsive cores [13,15]. We favor the expression of Berns and van der Avoird [13] because it was carefully derived using recent information and then tested by being used to calculate a wide variety of physical properties [13,14]. Nevertheless, the exp-6 form of Kuan, et al. [15] also gave very good

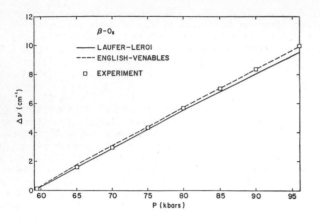

Fig. (6).   The pressure dependence of
             $\nu(P)-\nu(59$ kbars) for $\beta-O_2$. The
             squares represent the experimental
             data [8] and the solid and dashed
             lines are calculations based upon
             the Laufer-Leroi [19] and English-
             Venables [20] potentials, respec-
             tively.

To test the importance of crystal structure, calculations on α and $\gamma-N_2$ were performed up to 400 kbars and the results agreed with one another to within 5%. Similar results were obtained by comparing $\Delta\nu$ for two different structures of CO_2. On the basis of this we conclude that different crystal structures do not seriously change the pressure dependence of the intramolecular vibrational modes. A likely exception to this is for partially orientationally disordered phases with inequiv- alent sites, which occur in room temperature O_2 and N_2 at high pressures [24,25].

The results for O_2, shown on Fig. (6), can be compared with the experimental work of Nicol, et al. [8] only in the β phase regime, $59 \le P \le 96$ kbars. Below 59 kbars at room temperature O_2 is fluid and above 96 kbars, O_2 forms a structure that is partially orienta- tionally disordered. Our calculations have been limited to only the β phase, although work is in progress on the room temperature phase above 96 kbars and on low temperature $\alpha-O_2$. We have extended our β-phase calcula- tions to zero pressure and find that $\Delta\nu$ is in reasonably good agreement with experimental fluid results [8].

agreement with experiment for $\Delta\nu$. As shown by the dashed line in Fig. (4), the 6-12 potential of Thiery and Chandrasekharan [17] gives very poor results. This is not surprising since its parameters have not been carefully selected for general applicability and because the R^{-12} term becomes seriously unrealistic at the very high N_2 pressures investigated.

Our results have been calculated at zero temperature using the appropriate $T = 0$ K crystal structures. In order to compare these results with the $T = 300$ K experimental data, a study was made of the temperature dependence of $\Delta\nu$. For CO_2, the pressure dependence of $\Delta\nu$ was determined using our calculated zero temperature equation of state and using the experimental, room temperature equation of state of Olinger and Cady [21]. It was found that $\Delta\nu$ is only weakly dependent on tempera- ture, particularly at high pressure where the compressional energy dominates the thermal energy. To test for consistency, a thermal correction was added to our zero temperature P-V relation assuming the Grüneisen parameter at $T = 300$ K is only weakly pressure dependent. The results agreed well with experiment [21]. Using a similar technique for N_2, thermal contributions to $\Delta\nu$ were found to be negli- gible.

Another problem in comparing $T = 0$ K to $T = 300$ K results is that the high pressure crystal structures are different, except for CO_2.

The results of this work indicate that the low pressure model of compressed molecular solids, discussed earlier, is correct. The good agreement with experiment implies that no substantial charge transfer is occurring at the pressures so far reached. It is tempting to argue that charge transfer is responsible for $\partial\nu(P)/\partial P$ becoming negative in solid H_2 at 330 kbars. However, one would not expect major charge rearrangements at pressures so low compared to the predicted insulator-metal transition at about 2 megabars [26]. It is possible that higher order terms in the expansion of the intramolecular potential are responsible for this turnover in $\Delta\nu(P)$. Hydrogen, being somewhat unique among the molecular solids, may require special con- sideration.

VI.   Acknowledgements

We wish to thank Dr. V. Chandrasekharan and Tom Koehler for valuable comments and sugges- tions.

VII.   References

1.   Vodar, B. and Marteau, Ph., High Pressure
     Science and Technology, Vols. 1 and 2,
     Pergamon Press (1980).
2.   See Reference 1.
3.   Coulon, A. and Jean-Louis, M.,
     Proceedings of the 7th International
     AIRAPT Conference, Ed. Pergamon Press,
     Paris 1980, p. 598.
4.   Sharma, S. K., Mao, K. K., and Bell, P.
     M., Phys. Rev. Lett. 4, 886 (1980).

5. Hanson, R. C. and Bachman, K., Chem. Phys. Lett. 73, 338 (1980); Hanson, R. C. and Jones, L. H., Preprint (1981).

6. LeSar, R., Ekberg, S. A., Jones, L. H., Mills, R. L., Schwalbe, L. A., and Schifer, D., Solid State Commun. 32, 131 (1979).

7. Cromer, D. T., Mills, R. L., Schiferl, D., and Schwalbe, L. A., Acta. Cryst. B37, 8 (1981).

8. Nicol, M., Hirsch, K. R., and Holzapfel, W. B., Chem. Phys. Lett. 68, 49 (1979).

9. Suzuki, I., J. Mol. Spectry. 25, 479 (1968).

10. Kitaigorodskii, A. I., Soviet Phys. Cryst. 14, 769 (1970).

11. Morrison, M. A. and Hay, P. J., J. Chem. Phys. 70, 4034 (1979).

12. Murthy, C. S., Singer, K., Klein, M. L., and McDonald, I. R., Mol. Phys. 40, 1517 (1980).

13. **Berns, R. M. and van der Avoird, A., J. Chem. Phys. 72, 6107 (1980).**

14. Luty, T., van der Avoird, A., and Berns, R. M., J. Chem. Phys. 73, 5305 (1980).

15. Kuan, T. S., Warshel, A., and Schnepp, O., J. Chem. Phys. 52, 3012 (1970).

16. Zunger, A. and Huler, E., J. Chem. Phys. 62, 3010 (1975).

17. Thiery, M. M. and Chandrasekharan, V., J. Chem. Phys. 67, 96 (1977).

18. Herzberg, G., Spectra of Diatomic Molecules, Van Nostrand Reinhold Co., New York (1950).

19. Laufer, J. C. and Leroi, G. E., J. Chem. Phys. 55, 993 (1971).

20. C. A. English and J. A. Venables, Proc. Roy. Soc. London, Ser. A 340, 57 (1974).

21. Olinger, B. and Cady, H., Proceedings of the 6th Symposium on Detonation, ONR, Dept. of Navy ACR-221 (1976), p. 700.

22. Gibbons, T. G. and Klein, M. L., J. Chem. Phys. 60, 112 (1974).

23. Stevenson, R., J. Chem. Phys. 27, 656 (1957); 27, 673 (1957).

24. Schiferl, D., Cromer, D. T., and Mills, R. L., Acta Cryst. B., Submitted (1980).

25. See Reference 7.

26. Etters, R. D., Danilowicz, R., and England, W., Phys. Rev. A. 12, 2199 (1975).

†This work supported by NASA Grant #NGR-06-002-159 and NATO Grant #RG132.80.

PHYSICS OF SOLIDS UNDER HIGH PRESSURE
J.S. Schilling, R.N. Shelton (editors)
© North-Holland Publishing Company, 1981

ENERGY DISPERSIVE X-RAY DIFFRACTION MEASUREMENTS AT SIMULTANEOUSLY HIGH
PRESSURE AND TEMPERATURE USING SYNCHROTRON RADIATION: PRELIMINARY DATA
ON P(T) CALIBRATION AND PHASE TRANSFORMATIONS IN MnF_2 AND FeF_2

M. H. Manghnani,[1] E. F. Skelton,[2] L. C. Ming,[1] J. C. Jamieson,[3]
S. Qadri,[4] D. Schiferl,[5] and J. Balogh[1]

A feasibility study has been conducted to carry out the energy dispersive x-ray
diffraction measurements at high pressure (to 30 GPa) and temperature (to 250°C) with
three types of diamond-anvil pressure cells at the Stanford Synchrotron Radiation
Laboratory (SSRL).
In view of a recent suggestion that gold (Au) can serve as an excellent pressure
calibrant at high temperatures, we have obtained cross calibration data on the internal
standards NaCl and Au by measuring scattered synchrotron-produced x-ray photons with a
Si(Li) detector. Data were collected from an internal resistively heated diamond-anvil
pressure cell to pressures of 10 GPa and temperatures of up to 250°C. The shifts in
the NaCl and Au calibrant lines with pressure were cross calibrated against the shift
in the ruby fluorescence lines, and a good agreement was found in the pressure values
deduced from the two methods. The preliminary measurements made on Au + NaCl under
simultaneously high pressure and high temperature show promise of success in future
endeavors in establishing the high P-T equation of state of Au.
Preliminary energy-dispersive measurements made to study the phase transformations in
MnF_2 and FeF_2 are also reported.

1. INTRODUCTION

The feasibility of using energy dispersive x-ray
diffraction (EDXRD) was first demonstrated by
Buras *et al.* (1) and Giessen and Gordon (2). The
technique involves illuminating a sample with
polychromatic x-ray source and analyzing the
diffracted beam, in terms of its energy content,
under conditions of fixed geometry. Although
numerous papers employing this techinque have
appeared in the last 13 years or so (for a
recent review, see Ref. (3)), only a few have
utilized the technique for high pressure
studies. The earliest high-pressure EDXRD
studies involved a split belt type pressure cell
(4) and a piston-cylinder device (5). Later
studies utilized a cubic press (6-8) and a
diamond-anvil pressure cell (9-11). Both types
of pressure apparati have potential use in *in
situ* high-pressure and high-temperature EDXRD
studies. In the case of a diamond-anvil cell

---

[1] Hawaii Institute of Geophysics, University of
Hawaii, Honolulu, HI.

[2] Naval Research Lab, Washington, D.C.

[3] University of Chicago, Chicago, IL.

[4] Colorado State University, Fort Collins, CO.

[5] Los Alamos Scientific Laboratory, Los Alamos,
NM.

the attainable pressures are much higher (>50
GPa); however, the size of the sample is much
smaller, so much longer radiation exposure time
is required.

Until 1977, most of the reported high-pressure
EDXRD studies utilized conventional x-ray
generators (∿1 kW). In such cases the exposure
times were of the order of ∿60 min. This time
could be greatly reduced (to one-third to one-
fifth) with the use of rotary-anode x-ray generators (∿12 kW). The exposure time can be
reduced much more appreciably by employing
intense synchrotron radiation. Although speed
per se is not the main advantage in using synchrotron radiation, it can be a means to accom-
plishing certain types of experiments that would
otherwise be impossible (e.g., kinetics of a
transition at simultaneously high pressure and
temperature).

The synchrotron radiation (SR) source has unique
features that are unavailable with conventional
x-ray sources: extreme spectral brilliance,
well-defined polarization and highly collimated
beam. These characteristics make synchrotron
radiation an ideal x-ray source for EDXRD measurements under the extreme environments of
pressure and temperature.

The feasibility of interfacing diamond-anvil
cell technology with the EDXRD techniques,
utilizing synchrotron radiation facilities, has
only recently been exploited. It has been
employed by Buras and his associates (12) and

Holzapfel, Will and associates (personal communication, 1981) at Deutsches Elektronen-Synchrotron (DESY) and Skelton, Spain and associates at the Stanford Synchrotron Radiation Laboratory (SSRL) and, most recently, by Ruoff and Blaublitz at Cornell High Energy Synchrotron Source (CHESS) (13).

One of the purposes of this paper is to present preliminary data on the EDXRD measurements carried out by using the SR source at SSRL, in order to investigate the compressibility and phase transformations in two rutile-structured difluorides (MnF_2 and FeF_2). By using three types of diamond-anvil cells for these measurements, we have been able to evaluate the suitability of each of the cells for future work. The other purpose of this paper is to test the feasibility of using an internally heated diamond-anvil pressure cell for carrying out *in situ* equation-of-state measurements simultaneously at high pressures and temperatures. For this part of the study, we chose gold (Au), a potential pressure calibrant in the pressure-temperature range beyond the region in which Decker's equation for NaCl holds. It has been suggested recently (14) that Au can serve as an excellent pressure calibrant at high temperatures (e.g., >500°C), chiefly because of its

chemical inertness, the availability of experimental volume data in the wide range of pressure and temperature, and its simple x-ray diffraction pattern. We limited our feasibility study runs to 100 kbar and 250°C, although the pressure and temperature capabilities of our diamond-anvil cell were higher.

2. EXPERIMENTAL METHODS

2.1 Synchrotron Radiation Source and Allied EDXRD Facilities

The experimental setup designed for high pressure x-ray diffraction study at SSRL (Fig. 1) has been described in detail elsewhere (15-16); a brief summary is given here. A diamond-anvil pressure cell was mounted on an x-y-z stage inside the radiation hutch. The position of the cell was adjusted so that the synchrotron radiation impinged on the sample area between the anvils of the pressure cell. Photon energy resulting from the x-ray diffraction as well as the x-ray fluorescence of the sample (± gasket material) was collected simultaneously at a fixed 2θ angle by a Si(Li) solid state detector, with a resolution of 149 ± 3 eV at 5,894 eV (the weighted energy of $MnK\alpha_1$ and $MnK\alpha_2$ peaks) (16).

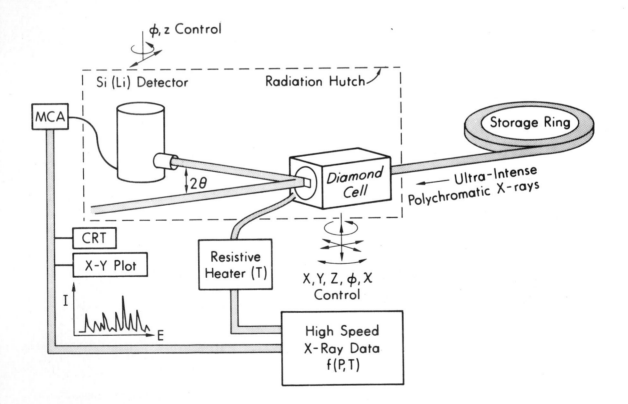

Figure 1. The experimental setup used for interfacing dispersive x-ray diffraction (EDXRD) system and diamond-anvil pressure cell with the synchrotron radiation source at SSRL.

The detector was located ∿60 cm from the pressure cell inside the hutch. The collected photon energy was stored in the 1024-channel analyzer (MCA), which was calibrated against 18 characteristic x-ray peaks. When the operation of the storage ring was maintained at 3 GeV and 60 mA, the exposure time for each of the pressure runs was 200 sec. As only the diffraction peaks shift when the angle of the detector (2θ) is varied, the 2θ angle desired for each sample assemblage was chosen to separate, as markedly as possible, the diffraction peaks from the stationary fluorescence peaks from the sample assemblage. After the 2θ angle was set approximately at a desired value, its true value was determined from the energies of the NaCl at ambient pressure and temperature. The true value, in turn, was used to calculate the interplanar spacings (d-spacings) of the sample assemblage at different pressures from the measured photon energy and the Bragg equation:

$$d = \frac{6.199}{E \sin\theta} \qquad (1)$$

where d is the d-spacing in Å, E is the photon energy of the diffraction peak in keV, and θ is one-half of the angle between the detector and the direct beam (2θ). The accuracy of d-spacings thus obtained has been estimated to 0.1% for 200 sec exposure (16).

2.2 Diamond-Anvil High-Pressure Cells

Three different types of diamond-anvil pressure cells were used in this study. The piston-screw type cell (17-18) and the modified level-spring type megabar cell (19) were used for high-pressure studies to ∿30 GPa at ambient temperature. Both of these cells have been described in detail previously. The third pressure cell employed by us for making the runs at simultaneously high pressure *and* high temperature was the lever type modified by Sung (20) for studying the kinetics of the olivine-spinel transition. Although this cell also has been fully described elsewhere (20), a short description, including some modifications, is in order here. Whereas the original upper piston of Sung's cell was a single piece containing an anvil seat with a straight hole, the modified piston consists of three separate parts: a piston body, a rocker, and an anvil seat where each one of the parts has a slot for allowing the x-ray diffraction up to 2θ = 30°. The rocker and the anvil seat, both made of tungsten carbide (WC), were glued together by Sauereisen (P1) high-temperature cement. After the diamond anvils were aligned, a micro-heater and a gasket were properly positioned on the lower piston and on the anvil of the upper piston, respectively, and then secured with the high-temperature cement.

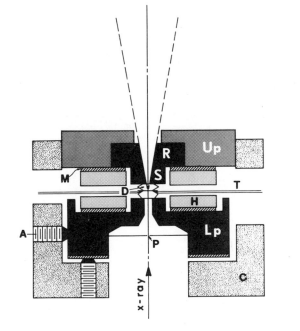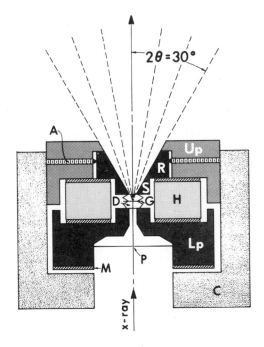

Figure 2. Cross sections of the modified piston and cylinder assemblage including a micro-heater in the high pressure-high temperature cell where A is adjusting screw; C, cylinder; D, diamond anvil; G, gasket; H, micro-heater; Lp, lower piston; M, mica plate; P, pinhole in the lead plate; R, WC rocker; S, WC diamond-anvil seat; Up, upper piston body.

Cross sections of the modified piston-cylinder assembly of the pressure cell, including a micro-heater, are shown in Figure 2. The original glass collimator was removed from the pressure cell, and no other collimator was introduced. Because the synchrotron radiation is highly collimated, the pinhole instead of the original glass capillary collimator served the purpose of directing the beam. In the first two types of pressure cells used, the WC-rocker with a 0.3-mm hole was sufficient to direct the radiation to the sample area. In Sung's high-temperature cell, instead of the original glass collimator, a lead plate (∿1 mm thick) with a pinhole of 0.4 mm was used to guide the radiation. A difficulty was experienced in using the lead pinhole in some of the runs in that the lead pinhole deformed at some high temperature and blocked the radiation from reaching the sample. Use of some other high-temperature material (e.g., tungsten carbide, high-temperature steels, and alloys) instead of lead would remove this difficulty with the pinhole.

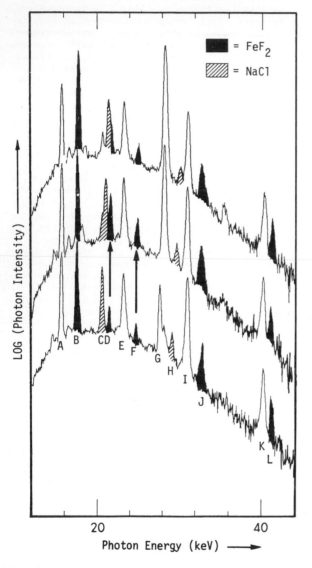

Figure 3. X-ray photon energy spectra of FeF_2 + NaCl assemblage recorded at pressures of 1 atm (bottom), 1.6 GPa (middle) and 3.1 GPa (top); each spectrum was recorded for a period of 200 sec. SPEAR conditions were 3.0 GeV and 35 mA. Diffraction peaks from FeF_2 (rutile) are solid and identified as B for (110), D for (101), F for (200) and (111), J for (211) and L for (301). Diffraction peaks from NaCl are hatched and identified as C for (200) and H for (220). Diffraction peaks from the gasket are marked by G, I, and K. Fluorescence peaks (unidentified) are marked as A and E.

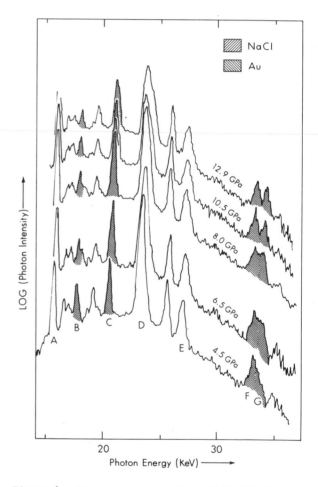

Figure 4. X-ray energy spectra of Au + NaCl assemblage at various pressures (pressure increases from bottom to top). X-ray diffraction peaks are hatched and labeled as C for Au (111), F for Au (220), B for NaCl (200), G for NaCl (400), and D and E for gasket (111) and (200), respectively. X-ray fluorescence peak from Zr (a thin shim underneath the diamond anvils) is identified at A.

3. EXPERIMENTAL DATA AND DISCUSSION

3.1 Isothermal Compression Studies at High Pressure and Ambient Temperature

Sample assemblages of MnF_2 (rutile) + NaCl + ruby + alcohol, FeF_2 (rutile) + NaCl + ruby and Au + NaCl + ruby were subjected to pressure of up to 3.4, 5 and 30 GPa at room temperature, respectively. The x-ray diffraction data were obtained during both the increasing and decreasing pressure cycles and were corrected on the basis of 0.1 MPa run results. Figures 3 and 4 are examples of the energy spectra for the sample assemblages of FeF_2 (rutile) + NaCl, and Au + NaCl, respectively, at increasing pressures. As can be seen, the diffraction peaks shift with pressure, whereas the position of the fluorescence peaks is invariant. The shift of the diffraction peak toward higher energy with increasing pressure is expected because at constant θ the photon energy (E) of the diffraction peak is inversely proportional to the d-spacing. The x-ray diffraction lines observed for each of the phases involved at different conditions are shown in Table 1. The d-spacings for each of the diffraction lines were calculated from the energy peaks observed and equation (1). By using the least-squares fit to the d-spacings, the lattice parameters for MnF_2, FeF_2, and Au at different pressures were calculated.

The pressure of the sample was determined optically by means of the ruby fluorescence method

(18, 21–22) and by the internal pressure calibrant (NaCl) using Decker's equation of state for NaCl (23). The two pressure values were found to be consistent within experimental errors (± 1%).

The effect of pressure on the molar volume V/V_o for MnF_2 (rutile), FeF_2 (rutile) and Au are plotted in Figures 5, 6 and 7, respectively. Isothermal bulk modulus at zero pressure (K_o) can be evaluated by using the least squares fit of the experimental data to the Birch-Murnaghan equation:

$$P = \frac{3}{2} K_o \left\{ \left[\frac{V}{V_o}\right]^{-7/3} - \left[\frac{V}{V_o}\right]^{-5/3} \right\}$$

$$(1 - \xi) \left\{ \left[\frac{V}{V_o}\right]^{-2/3} - 1 \right\} \qquad (2)$$

where V_o anv V are the molar volume at zero pressure and pressure P, respectively, $K_o' = (\partial K/\partial P)_T$ evaluated at P = 0, $\xi = 4/3 \,(4 - K_o')$. The present P-V data are not sufficiently accurate to determine unambiguous values of K_o and K_o' from equation (2). By assuming the value of K_o' to be 4.2 (Ref. 24) for MnF_2, 4.65 for FeF_2 (25), and 6.29 for Au (26), we calculate K_o from equation (2) to be 85 ± 7, 126 ± 7 and 220 ± 8 GPa, where the uncertainty includes the errors from both V/V_o and pressure determinations. Comparison of these K_o values with those previously reported (Table 2), shows that in the case of MnF_2, the values of K_o

TABLE 1

Typical Diffraction Lines Observed in MnF_2,

FeF_2, NaCl and Au Under Various Pressure-Temperature Conditions

Phase	(hkℓ)	Conditions
MnF_2 (rutile phase)	(110), (101), (111), (211) (301)	0.1 MPa–1.3 GPa (RT)
MnF_2 (α-PbO_2 phase)	(111)	1.3–1.8 GPa (RT)
MnF_2 (tetragonal phase)	(111), (220)	2–3.4 GPa (RT)
FeF_2 (rutile phase)	(110), (101), (200), (111) (211), (301)	0.1 MPa–5 GPa (RT)
Au	(111) ± (220)	0.1 MPa–30 GPa (RT –250°C)
NaCl	(111) ± (220)	0.1 MPa–30 GPa (RT – 250°C)

Note: RT stands for room temperature condition.

± means "with or without."

are in fairly good agreement, whereas in the
case of FeF_2 and Au, the K_o values from the
present study are about 20-25% higher than those
previously reported. These discrepancies may be
explained by the stress inhomogeneity caused by
nonhydrostatic environments (27) in our experiments. In spite of this discrepancy, the success in obtaining fairly good spectra for FeF_2
to ∿5 GPa using Bassett's type cell, and for Au
to ∿30 GPa using the Mao and Bell type cell has

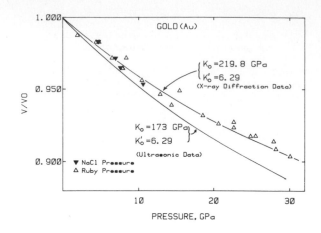

Figure 7. The effect of pressure using the
synchrotron pressure on the molar volume of Au
under nonhydrostatic pressure conditions and at
room temperature. All the data points are
based on measurements using the synchrotron
radiation.

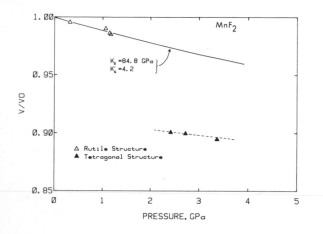

Figure 5. The effect of pressure on the molar
volume (V/V_o) of MnF_2 (rutile str.) and MnF_2
(tetragonal str.) under truly hydrostatic pressures at room temperature. All the data points
are based on measurements using the synchrotron
radiation.

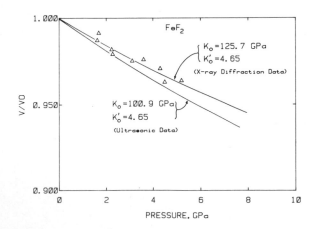

Figure 6. The effect of pressure on the molar
volume of FeF_2 under nonhydrostatic pressures at
room temperature. All the data points are based
on the measurements using the synchrotron
radiation.

fulfilled the main purpose (see Figs. 3 and 4)
of this feasibility study. Further equation-of-
state studies on different types of materials
under hydrostatic pressure, using the SR source,
are now being planned.

3.2 Phase Transformation Studies

3.2.1 MnF_2. In light of phase transformation
studies of MnF_2 (28-30, among others) it is
believed that the phase transformation sequence
in MnF_2 is:

$$MnF_2 \text{ (rutile str.) } \xrightleftharpoons{\sim 1.3 \text{ GPa}} MnF_2(\alpha\text{-}PbO_2 \text{ str.})$$

$$\xrightleftharpoons[]{\begin{array}{c}1.8\text{-}\\2.0 \text{ GPa}\end{array}} MnF_2 \text{ (tetrag. str.)}$$

The present *in situ* x-ray diffraction data for
MnF_2 confirm the transformation in MnF_2 from
MnF_2 (rutile str.) into MnF_2 (α-PbO_2 str.) at
∿1.3 GPa, and from MnF_2 (α-PbO_2 str.) into MnF_2
(tetragonal phase) at ∿2.0 GPa. Both these
transformation sequences and the transition
pressures are in good agreement with the previous studies (27-29). Because only one line was
observed for the α-PbO_2 phase, its molar volume
could not be evaluated. The molar volume change
from rutile to tetragonal phases is estimated to
be -8% at 2.0 GPa. This result is also in good
agreement with those of previous studies (e.g.,
28). The presence of only one or two line(s)
observed for the high-pressure phases of MnF_2,
such as MnF_2 (α-PbO_2 str.) and MnF_2 (tetragonal
str.) is probably due to the short durations of
data acquisition.

TABLE 2

Comparison of K_o Values Based on Ultrasonic,
Shock Wave, and Static Compression Data

Material	K_o (GPa)	K_o'	Method	Reference
MnF_2	89	4.2	ultrasonic	(24)
	94		x-ray (single-crystal)	(30)
	85	4.2*	x-ray	This study
FeF_2	101	4.65	ultrasonic	(25)
	126	4.65	x-ray	This study
Au	173	6.29	ultrasonic	
	220	6.29*	x-ray	This study
	180	6.39	shock	(32)

*Assumed value.

3.2.2 FeF_2. FeF_2 (rutile) has been found to transform into FeF_2 ("distorted fluorite") at ∿4.5 GPa and room temperature (31). In the present study up to 5.0 GPa, however, no high-pressure phase was observed. This discrepancy may be because the data acquisition period was too short to reveal the presence of the high-pressure phase, and that the maximum pressure attained was in the proximity of the phase transition pressure (i.e., 4.5 GPa). Further studies involving longer durations are now being planned.

3.2.3 Compression studies at high pressure and high temperature. The sample assemblage of Au + NaCl was subjected to pressures up to ∿10 GPa and was heated to 250°C by using the modified high-pressure and high-temperature cell (20). A series of spectra at various pressures and temperatures is shown in Figure 8. Diffraction peaks of NaCl (200) and Au (111) were observed. As can be expected, both diffraction peaks for NaCl and Au shift toward higher energy with increasing pressure at constant temperature, and shift toward lower energy with increasing temperature at constant pressure. Although such limited data do not permit reliable deduction of the equation of state at high temperature, this study was successful in obtaining good spectra of Au + NaCl at high pressure (∿10 GPa) and high temperature (250°C) using Sung's high pressure-high temperature cell. The present study has demonstrated realistically the feasibility of using SR source to make high-quality EDXRD measurements at simultaneously high pressures and high temperatures.

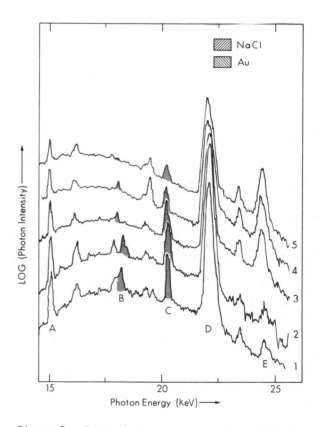

Figure 8. X-ray photon energy spectra of Au + NaCl assemblage recorded: (1) at ambient P and T; (2) at P = 0.7 GPa and T = room temperature; (3) P = 0.7 GPa and T = 255°C; (4) P = 1.0 GPa and T = 255°C; and (5) P = 4.7 GPa and T = 255°C. X-ray fluorescence from Zr is identified as A. Diffraction peak(s) (hatched) are identified as B for NaCl (200), C for Au (111), and D and E for gasket (111) and (200), respectively.

SUMMARY

1. We have demonstrated, for the first time, that the diamond-anvil pressure cell, when interfaced with the synchrotron radiation and energy dispersive x-ray diffraction (EDXRD) system, allows rapid collection of the diffraction data related to equation-of-state and phase transformation studies at simultaneously high pressure and high temperature. Such measurements would be extremely useful in several studies such as those related to the kinetics of the structural changes in materials.

2. All three types of diamond-anvil cells employed (Bassett type, Mao-Bell type, and modified Sung type) have yielded good energy dispersive spectra at high pressures and ambient temperature; however, only the modified Sung design is suitable for measurements at both high pressures and high temperatures.

3. The compression data for MnF_2 are in fairly good agreement with the previous data. The phase transformation from MnF_2 (rutile str.) $\rightleftharpoons$ MnF_2 (α-PbO_2 str.) $\rightleftharpoons$ MnF_2 (tetragonal str.) reported previously using the film method has also been revealed in the present EDXRD measurements.

Acknowledgments. We thank the staff of the Stanford Synchrotron Radiation Laboratory (SSRL) especially Prof. A. Bienenstock, Katherine Cantwell and Claire Burke for facilitating our research work. Work done at SSRL is supported by the NSF through the Division of Materials Research and by the NIH through the Biotechnology Resource Program in the Division of Research Resources (in cooperation with the Department of Energy). The financial support for our research was provided by the National Science Foundation (Grants EAR 79-11906 and EAR 80-20471 to the University of Hawaii; Grant EAR 79-22044 to the University of Chicago), Office of Naval Research, and the Department of Energy (Los Alamos National Laboratory). Thanks are due R. Rhodes for drafting and R. Pujalet for editorial help.

Hawaii Institute of Geophysics Contribution No. 1191.

REFERENCES

(1) B. Buras, J. Chwaszczewska, S. Szarras, and Z. Szmid, Inst. Nucl. Res. (Warsaw), Rep. No. 894/II/PS (1968).

(2) B. C. Giessen and G. E. Gordon, Science 159, 973-75 (1968).

(3) E. Laine and I. Lähteenmäki, J. Mat. Sci. 15, 269-78 (1980).

(4) P. J. Freud and P. M. La Mori, Trans. Am. Cryst. Asso. 5, 155-62 (1969).

(5) L. M. Albritton and Q. L. Margrave, High temp.-High pressure 4, 13-19 (1972).

(6) K. Inoue, Ph.D. Dissertation, Univ. of Tokyo (1975).

(7) S. Akimoto, N. Hamaya, and T. Yagi, in High Pressure Science and Technology, p. 194, B. Vodar and Ph. Marteau (eds.), Pergamon Press (1980).

(8) T. Yagi and S. Akimoto, J. Geophys. Res. 85(B10), 9661-95 (1980).

(9) E. F. Skelton, in Rept. NRL Prog., 31-33 (1972).

(10) K. Syassen and W. B. Holzapel, Europhys. Conf. Abstr. 1A, 75-76 (1975).

(11) E. F. Skelton, C. Y. Liu, and I. L. Spain, High temp-High pressure 9, 19-26 (1977).

(12) B. Buras, J. Staun Olsen, L. Gerward, G. Will, and E. Hanze, J. Appl. Cryst. 10, 431-38 (1977).

(13) A. Ruoff and M. A. Baublitz, Jr., Energy dispersive x-ray diffraction at high pressure in CHESS (this volume).

(14) J. C. Jamieson, J. N. Fritz, and M. H. Manghnani, in High Pressure Research in Geophysics, S. Akimoto and M. H. Manghnani (eds.), Center for Academic Publications, Japan (1981) (in press).

(15) E. F. Skelton, in High Pressure Research in Geophysics, S. Akimoto and M. H. Manghnani (eds.), Center for Academic Publications, Japan (1981) (in press).

(16) E. F. Skelton, J. Kirkland, and S. B. Qadri, to be published in High Pressure Science and Technology (1981).

(17) W. A. Bassett, T. Takahashi, and P. Stook, Rev. Sci. Instrum. 38, 37-43 (1967).

(18) L. C. Ming and M. H. Manghnani, J. Appl. Phys. 49(1), 208-12 (1978).

(19) H. K. Mao and P. W. Bell, Carnegie Inst. Washington Yearb. 74, 403-05 (1975).

(20) C. M. Sung, Rev. Sci. Instr. 47, 1343046 (1976).

(21) G. J. Piermarini, S. Block, and J. D. Barnett, J. Appl. Phys. 44, 5377-82 (1973).

(22) G. J. Piermarini, S. Block, J. D. Barnett, and R. A. Forman, J. Appl. Phys. 46, 274-80 (1975)1

(23) D. L. Decker, J. Appl. Phys. 42(8), 3239-44 (1971).

(24) M. H. Manghnani, T. Matsui, and J. C. Jamieson, EOS, Trans. 60(18), 387 (1979).

(25) M. H. Manghnani, L. C. Ming, and T. Matsui, in High Pressure Science and Technology, B. Vodar and Ph. Marteau (eds.), Pergamon Press, Oxford (1980).

(26) M. H. Guinan and D. Steinberg, J. Phys. Chem. Solids 35, 1501-12 (1974).

(27) R. D. Wilburn and W. A. Bassett, High temp.-High pressure 9, 35-39 (1977).

(28) T. Yagi, J. C. Jamieson, and P. B. Moore, J. Geophys. Res. 84(B3), 1113-15 (1979).

(29) L. C. Ming, M. H. Manghnani, T. Matsui, and J. C. Jamieson, Phys. Earth Planet. Inter. 23, 276-85 (1980).

(30) R. Hazen and L. W. Finger, J. Phys. Chem. Solids 42, 143-51 (1981).

(31) L. C. Ming and M. H. Manghnani, Geophys. Res. Lett. 5(6), 491-94 (1970).

(32) M. H. Rice, R. G. McQueen, and J. M. Walsh, Solid State Phys. 6, 1-63 (1958).

PHYSICS OF SOLIDS UNDER HIGH PRESSURE
J.S. Schilling, R.N. Shelton (editors)
© *North-Holland Publishing Company, 1981*

ELECTRONIC STRUCTURE AND PROPERTIES OF NONMETALS UNDER PRESSURE

Walter A. Harrison

Department of Applied Physics
Stanford University, Stanford, California 94305

Tight-binding theory, with universal parameters, provides elementary first-principles predictions of the electronic structure and of the dielectric and structural properties of ionic and covalent solids. It also gives, just as simply, predictions of the pressure dependence for all of these properties. The theory is outlined and the necessary parameters given. Then formulae are derived for the band gaps, dielectric susceptibility, and interatomic interactions in simple ionic compounds. The electronic structure and properties of covalent solids are formulated in terms of bond orbitals and formulae derived for the dielectric susceptibility and elastic rigidity. As an example, the dielectric susceptibility is predicted to decrease with pressure (linear with internuclear distance) in the elemental semiconductors such as Ge. That trend is predicted to decrease and finally change sign for their polar counterparts, such as GaAs, ZnSe and CuBr, and for ionic compounds. The variation of parameters with pressure in ionic compounds of transition metals is also discussed.

1. INTRODUCTION

Virtually all of the properties of solids are directly determined by their electronic structure. Similarly, the changes in the properties of solids under pressure are determined by the changes in the electronic structure under pressure. To guide in planning experiments and to assist in interpreting the experiment it can be very useful to have an understanding of the electronic structure, particularly if the understanding is simple enough to allow us to predict the various properties in terms of it.

In recent years just such a simple understanding has developed, not only for the simple metals, where pseudopotential perturbation theory provided the basis, but also in covalent and ionic solids, where tight-binding theory provides a more suitable basis. I attempted to provide a comprehensive description of the resulting theory in a recent text, Electronic Structure and the Properties of Solids.[1] The present discussion shall be restricted to ionic and covalent solids, with particular attention to the volume dependence of the properties of these systems. Fortunately the volume dependence enters the theory in an integral way. Fortunately also the theory has turned out to be so simple that almost all calculations, though ultimately from first principles, can be carried out without the aid of a computer and without specialized theoretical expertise. At least, what expertise is needed can easily be presented.

I should mention another theoretical development which is very important, but which is in very strong contrast to the theory I shall describe here. It appears to have become possible, with sufficiently extensive comput-

tation using pseudopotential and density functional techniques, to really predict the stable crystal structure and ground-state properties of crystalline solids, and of course therefore to predict the structural properties of these systems under pressure. Such studies of silicon and tin by M. L. Cohen and coworkers[2,3] are currently being extended to compounds. Even transition metal systems have become tractable using more sophisticated formulations than pseudopotentials.[4,5] These may in some sense be thought of as computer experiments. They only imitate the real system but they are less expensive than the real experiments. Like laboratory studies they do not directly provide physical insight into the systems studied, but as with laboratory experiments, insight can be derived from the results using more approximate theories such as I shall describe. Thus the computer experiments do not compete with the theory given here but they may hope to compete with the laboratory experiments which are the principal subject of this conference.

I would like as a first step to outline the tight-binding representation of the electronic structure and provide the parameters necessary to use it. The main concepts are very simple and are all that are needed to understand a variety of properties. I shall then classify non-metallic solids according to their electronic structure and position in the periodic table. We shall discuss first the electronic structure of simple ionic solids and see how their dielectric and bonding properties are easily understandable in terms of that electronic structure. I will then formulate the concept of the covalent bond orbital and see how a similar simple understanding of covalent solids can be based on that concept. Finally I will outline the extension to the theory to transition-metal compounds.

2.   TIGHT-BINDING THEORY

The basic assumption of tight-binding theory is that the electronic states of the solid can be written as linear combinations of atomic-like orbitals, $\psi_\alpha(\vec{r} - \vec{r}_i)$ centered on the positions $\vec{r}_i$ of the atoms. Such a linear combination may be written

$$\psi_k(\vec{r}) = \frac{1}{\sqrt{N}} \Sigma_\alpha u_\alpha \Sigma_i e^{i\vec{k}\cdot\vec{r}_i} \psi_\alpha(\vec{r} - \vec{r}_i) \quad (1)$$

This looks more complicated than it is. The fact that a wavevector $\vec{k}$ can be associated with any state in the solid follows from the translational symmetry of the solid and the coefficients of different equivalent orbitals are related to each other by the phase factor $e^{i\vec{k}\cdot(\vec{r}_i - \vec{r}_j)}$. A different constant coefficient u_α is required for each orbital type, and the square root of the number of equivalent orbitals in the crystal normalizes the state.

If there were only one orbit type in the crystal, its u would be one and the energy of the electronic state could be written down immediately,

$$E_k = \int \psi_k^*(\vec{r}) H \psi_k(\vec{r}) d^3r = \varepsilon_\alpha + \Sigma_j e^{i\vec{k}\cdot\vec{d}_j} H_{ij} \quad (2)$$

The first term, ε_α, comes from terms where the orbital from the right wavefunction is on the same atom as that on the left. The second contribution arises from terms where the orbitals on the two sides are centered on sites differing by $\vec{d}_j$; H_{ij} is called the matrix element of the Hamiltonian, H, between the two.

In real solids it is always necessary to include more than one orbital type and this requires the solving of a few simultaneous equations in the u_α rather than simply writing down the result. However, it turns out that for our purposes, the minimal basis set is adequate: only those valence orbitals which are occupied, or partially occupied, in the free atom. Furthermore, only matrix elements between nearest-neighbor atoms are required. This is a tremendous simplification since the only parameters which we will need for any calculation are the ε_α for each orbital - an s-state and a p-state energy for each atom type in simple compounds - and the matrix elements between each orbital type - these are written $V_{ss\sigma}$, $V_{sp\sigma}$, $V_{pp\sigma}$ and $V_{pp\pi}$ for simple compounds, where the σ indicates zero angular momentum around the internuclear axis and π represents an angular momemtum of $\hbar$. A few more are required for transition-metal compounds and we shall return to those later.

3.   THE TIGHT-BINDING PARAMETERS

Of course we can expect there to be systematic variations of these parameters over the periodic table and empirical rules could be developed to give values. It turns out in fact that nature made this very simple for us. By considering values obtained by fitting known band structures we find that to a reasonable approximation we may use free-atom term values for the ε_α and that each of the four interatomic matrix elements varies from system to system as the inverse square of the internuclear distance.[1]

The first point may not be so surprising since it has long been known that the charge density in most solids is rather close to a superposition of free-atom densities so that the Hamiltonian in the neighborhood of each orbital is not greatly modified from the free atom. The second point has now[6] been related to the fact that not only can the bands in covalent solids be represented as tight-binding bands but they also resemble free-electron bands, bands which would arise if the effect of the atomic potentials were very weak. Thus energy differences which arise from interatomic matrix elements in the one description can be equated to free-electron energy differences from the other, and the latter scale as the inverse square of the lattice distance.

These matrix elements do not change greatly (though the band structures do) when the atoms are rearranged from a covalent structure to an ionic crystal structure so we are led to a universal set of tight-binding parameters to be used for all simple non-metallic solids. (They are not adequate for the close-packed metallic structures, where pseudopotential theory becomes appropriate.) The interatomic matrix elements are given by

$$V_{ll'm} = \eta_{ll'm} \hbar^2/md^2 \quad (3)$$

where d is the internuclear distance. $\hbar^2/m$ is given by 7.62 eV-Å^2. We have recently refined the earlier analysis and proposed new universal coefficients[7]

$$\eta_{ss\sigma} = -1.32 \qquad \eta_{pp\sigma} = 2.22$$

$$\eta_{sp\sigma} = 1.42 \qquad \eta_{pp\pi} = -0.63 \quad (4)$$

We also choose to use atomic term values calculated in the Hartree-Fock approximation[8] as listed in Table 1.

We should note one very remarkable aspect of these parameters. Equation (3) predicts that the variation of the interatomic matrix element from one material to another (e.g., C to Si to Ge) follows the same formula as the variation in one material under pressure. This remarkable feature follows from the relation between tight-binding and free-electron descriptions and gives us direct relationships between pressure dependences and variations from one material to another. It is only approximately true and it

TABLE 1: Hartree–Fock term values after Fischer[8] in eV. The magnitude of ε_s values are given first for each element and of ε_p values next. ε_p values in the first two columns were obtained by extrapolation.

1	2	3	4	5	6	7	8	9	10	11
							He	Li		
							24.97	5.34		
	Be	B	C	N	O	F	Ne	Na		
	8.41	13.46	19.37	26.22	34.02	42.78	52.51	4.95		
	5.79	8.43	11.07	13.84	16.72	19.86	23.13	—		
	Mg	Al	Si	P	S	Cl	Ar	K	Ca	Sc
	6.88	10.70	14.79	19.22	24.01	29.19	34.75	4.01	5.32	5.72
	3.84	5.71	7.58	9.54	11.60	13.78	16.08	—	—	—
Cu	Zn	Ga	Ge	As	Se	Br	Kr	Rb	Sr	Y
7.72	7.96	11.55	15.15	18.91	22.86	27.00	31.37	3.75	4.85	5.34
2.37	4.02	5.67	7.33	8.98	10.68	12.43	14.26	—	—	—
Ag	Cd	In	Sn	Sb	Te	I	Xe	Cs	Ba	La
7.06	7.21	10.14	13.04	16.02	19.12	22.34	25.69	3.36	4.29	4.35
2.61	3.99	5.37	6.76	8.14	9.54	10.97	12.44	—	—	—
Au	Hg	Tl	Pb	Bi	Po	At	Rn			
6.98	7.10	9.82	12.48	15.19	17.96	20.82	23.78			
2.67	3.95	5.23	6.53	7.79	9.05	10.33	11.64			

can only be true near the equilibrium spacing. The tight-binding description must break down at sufficiently small spacing and the free-electron picture must break down at sufficiently large spacing. However, the region of overlapping validity includes most solids and most solids under attainable pressures.

Results obtained with these parameters are only of limited accuracy but it is extraordinarily valuable to have a set of parameters at the outset for the treatment of any system of interest. In addition to providing numbers it tells us a great deal immediately about the electronic structure and about its volume dependence.

4. THE CLASSIFICATION OF SOLID TYPES

Table 1 has been arranged in such a way as to clearly distinguish the two broad categories of nonmetallic solids. The nonmetallic elements are in columns 4 through 8. Column 4 contains the covalent elements C, Si, Ge, and Sn all in the diamond structure. The electronic structure consists of two-electron bonds. Each atom is surrounded by four nearest neighbors and contributes one electron to each bond. When one of the nonmetals in columns 5 through 7 forms a compound with an equal number of metal atoms from the columns 1 through 3 it is almost always in this tetrahedral structure, again with two-electron bonds. These bonds are now polar but it is still to be regarded as a covalent solid.

Column 8 contains the closed-shell inert gases. When one of the nonmetals in columns 5 through 7 forms a compound with an equal number of metal atoms from columns 9 through 11 to the right, it almost always forms in a more closely packed structure. With more than four neighbors it cannot be reasonably described as consisting of two-electron bonds. We should think of it as consisting of closed-shell ions, just as the elements of column 8, but now with a nuclear charge which differs from the charge of the electron cloud. There is coupling between the atoms and a corresponding covalent contribution to the bonding, but they are to be regarded as ionic solids and a different theory, based upon the same matrix elements but upon different approximations, is appropriate. We shall discuss the ionic solids, which are slightly simpler, first and in more detail.

5. THE PROPERTIES OF IONIC SOLIDS

It is helpful to consider an isoelectronic series such as Ar, KCl, CaS, and ScP (see Table 1). In argon the atomic 3s and 3p shells are filled and the energy difference up to the next shell (4s) is quite large, resulting in their inert chemical properties. We may conceptually construct KCl from argon by freezing the electronic states and transferring a proton from one Ar nucleus to that of a neighbor, making a chlorine nucleus of the first and a potassium nucleus of the second. This raises the energy levels on the first atom and lowers them on the second. We may then allow the electrons to relax slightly but the two resulting ions are quite inert. A similar transfer of a second proton makes CaS and lowers the energy still further on the positive ion and raises it on the negative but the system is qualitatively the same. The third proton ordinarily drops an empty level on the metal below a full level on the nonmetal and the compound is not stable. We will discuss only the monovalent case in detail but the theory applies equally to the divalent case.

5.1 The Electronic Structure

The brief discussion above suggests our starting view of the electronic structure. It is the traditional view given by Max Born[9] of closed-shell ions. The quantitative aspect, however, is quite different: we take the energies of the electronic states to be given by the <u>atomic term values</u>. In the traditional view it would be the energy in the free ion, corrected by a Madelung potential. These are very big corrections, each greater than ten electron volts but of opposite sign. It is their cancellation, because of the fact that at the observed spacing the charge density is nearly equal to that of superimposed atoms, as mentioned earlier, that allows the simpler view to make sense.

With this view the band gap in an alkali halide becomes the difference in term value between the alkali s-state energy and the halogen p-state energy, both of which may be read from Table 1. This gives for KCl a gap of 9.77 eV, to be compared with the observed 8.4 eV. A discrepancy of this magnitude is not atypical. The general magnitudes are roughly correct and, in addition, the principal trends from material to material are correctly given. This is not high accuracy but is impressive when recognized as a first-principles band calculation for the compound which did not even require a hand-held calculator.

This most elementary calculation is even good enough for discussing a property such as the cohesive energy. The energy gained in going from the free atom to the solid is just the energy 9.77 eV gained in transferring the electron between the two states. The experimental cohesive energy is 7.4 eV. However, this simplest view is not adequate for discussing other dielectric and elastic properties; these properties arise from the deviations from the free-ion electronic structure.

There is indeed coupling between the electronic states on neighboring ions, with matrix elements given by Eq. 3. To a first approximation the halogen s-state is deep enough to be neglected so the only coupling is between the alkali s-state and the halogen p-states, $V_{sp\sigma} = 1.42 \, \hbar^2/md^2$. It turns out that because of the high symmetry of the lattice, this coupling cancels out for electrons with $\vec{k} = 0$ (see Eq. 2) but the two levels repel each other at other wavenumbers. This means that the band gap (minimum separation) calculated above is not affected, but when averaged over all wavenumbers there is a raising of the conduction band and a lowering of one valence band (the other two valence bands remain at ε_p (halogen)

in this approximation) which may be calculated using perturbation theory as $6V_{sp\sigma}^2/(\varepsilon_p - \varepsilon_s)$, the six coming from the six neighbors in the rocksalt structure. This gives an average gap of

$$\overline{E}_g = \varepsilon_s - \varepsilon_p + 12V_{sp\sigma}^2/(\varepsilon_s - \varepsilon_p) \quad (5)$$

It is not difficult to plot up the bands themselves, $E(\vec{k})$, but the averages will be enough for describing the properties.

Note already a prediction of the theory concerning the effects of pressure: a change in volume does not change the atomic term values so that no change in the minimum gap with pressure is expected. On the other hand the average gap (because of the variation of $V_{sp\sigma}$ with d) has a change which we can estimate. The theory is crude and we may expect <u>some</u> change in minimum gap to occur, but it should be small compared to the change in $\overline{E}_g$ (Eq. 5) and small compared, say, to $\overline{E}_g$ times the fractional change in volume.

5.2 The Dielectric Properties

This coupling is necessary also for understanding the dielectric susceptibility since with our minimal basis set no empty electronic states are included for the halogen and all of the electrons are in halogen states; without the coupling between atoms there is no mechanism for the charge to redistribute in an electric field. Once we introduce the coupling, the calculation of the electronic contribution to the susceptibility χ is quite trivial. This is in fact the susceptibility which determines the refractive index $n = (1 + 4\pi\chi)^{1/2}$ and is therefore readily measurable. The calculation is in three steps.

First we calculate the charge transferred from a chlorine ion to a neighboring alkali due to the coupling, as illustrated in Fig. 1. The halogen p-state in the center is coupled by $V_{sp\sigma}$ to the s-state to the right; thus to first order the p-state becomes $|p> + (V_{sp\sigma}/(\varepsilon_p - \varepsilon_s))|s>$. The coefficient is squared and multiplied by two for spin to give a charge $-2eV_{sp\sigma}^2/(\varepsilon_p - \varepsilon_s)^2$ transferred to the right. An equal charge is transferred to the left. p-states of other orientations are not coupled to these two neighboring alkali atoms since their matrix elements vanish by symmetry.

Second we apply a field $\boldsymbol{\mathcal{E}}$ which lowers the energy of the s-state to the right, in comparison to the central p-state, by $e\boldsymbol{\mathcal{E}}d$. This changes the charge transfer, as obtained by taking the derivative of the original transfer with respect to the energy denominator, $\delta Q = 4V_{sp\sigma}^2e^2\boldsymbol{\mathcal{E}}d/(\varepsilon_p - \varepsilon_s)^3$. The corresponding dipole is δQd, and an equal contribution comes from the left.

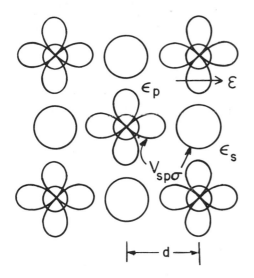

We should comment on the discrepancies with experiment. Within our minimal basis set and universal parameters, the most significant approximation has been treating the coupling as small and therefore, for example, dropping the second term in Eq. 5. Direct replacement of $\varepsilon_s - \varepsilon_p$ by Eq. 5 in Eq. 6 gives a cancelling contribution to the pressure dependence of the susceptibility considerably improving the agreement with experiment. However, it reduces the predicted susceptibilities in Table 2 and worsens the agreement there. The real point is that by making selected corrections at various stages we can greatly improve the agreement of some property with experiment but perhaps do not learn very much. If we are to take terms beyond the leading ones giving Eq. 6, we should probably take all of them, not just the one appearing in Eq. 5. Our prediction of $(\Omega/\chi)\, d\chi/d\Omega = -5/3$ has given the correct sign and general magnitude and will correctly explain the qualitative difference in the pressure dependence from that of semiconductors.

Table 2--Dielectric susceptibilities. The first entry is the nearest neighbor distance in Angstroms. The second is χ predicted from Eq. 6 and the third (in parentheses) is the experimental value.

	F	Cℓ	Br	I
Li	2.01	2.57	2.75	3.00
	0.07	0.10	0.12	0.16
	(0.07)	(0.14)	(0.18)	(0.22)
Na	2.32	2.82	2.99	3.24
	0.03	0.06	0.07	0.09
	(0.06)	(0.10)	(0.13)	(0.17)
K	2.67	3.15	3.30	3.53
	0.01	0.02	0.03	0.04
	(0.06)	(0.10)	(0.11)	(0.14)
Rb	2.82	3.29	3.45	3.67
	0.01	0.01	0.02	0.02
	(0.07)	(0.10)	(0.11)	(0.14)

The static susceptibility includes also contributions from the displacements of the ions by a static field. These are described by an effective ionic charge, called the transverse charge which also determines the coupling of vibrations with the infrared. This charge is derived in Ref. 1 using the kind of charge transfer calculation which we used for the susceptibility. It correctly gives a transverse charge slightly greater than one for the alkali halides. It is interesting that under pressure, though the ions become more nearly neutral, the transverse charge is predicted to increase. It would be surprising if experiments disagreed with that.

Figure 1: The valence p-states on the central halogen ion are coupled to the valence s-states on the neighboring alkali ions, transferring a total of $2V_{sp\sigma}^2/(\varepsilon_s - \varepsilon_p)^2$ electrons to each. An applied field modifies the energy difference $\varepsilon_s - \varepsilon_p$ by $e\mathcal{E}d$. This shifts the charges corresponding to an induced dipole moment of $8V_{sp\sigma}^2\, e^2\mathcal{E}d^2/(\varepsilon_s - \varepsilon_p)^3$ per halogen ion and the susceptibility given in Eq. 6.

We need finally to multiply that dipole by the halogen ion density, $1/(2d^3)$, and divide by the field to obtain the susceptibility,

$$\chi = \frac{4e^2 V_{sp\sigma}^2}{d(\varepsilon_p - \varepsilon_s)^3} \qquad (6)$$

As with the band gap, this gives susceptibilities of the right order of magnitude and reproduces the principal trends, as shown in Table 2, but the error is of the order of a factor of two. This is a first-principles calculation and to do a better one is a major theoretical effort and machine calculation. For understanding properties these elementary estimates are much more to the point. It also serves quite a different role than an empirical fit (such as sums of ion polarizibilities) which can give very accurate interpolations.

We may also discuss the pressure dependence of this susceptibility. Using this simplest formula we note that $V_{sp\sigma}$ varies as d^{-2} so the susceptibility has a variation of d^{-5}, or a volume dependence of $(\Omega/\chi)\, d\chi/d\Omega = -5/3$. Indeed this gives the correct sign, an increase in susceptibility with pressure, but too large a magnitude; the experimental values[10] range from -0.6 to -0.9.

5.3 The Elastic Properties

We may note that not all terms in the energy have been included in a sum over electronic energies. The band gap of Eq. 5 increases monatonically with decreasing volume and thus the sum of electronic energies decreases leading to instability. The principle term missing of course is the repulsion between closed shell atoms or ions which arises from the increased electronic kinetic energy as the electron distributions overlap. Nikulin[11] and Gordon and Kim[12] have given a direct way of computing this in terms of the free-atom electron distributions. More recently we have found it possible to calculate this overlap interaction analytically and shown that the total energy can to a good approximation be written simply as the sum of this overlap inter-action and the sum of electronic energies.[13] The overlap interaction takes the form,

$$V_0(d) = \eta_0 \frac{\hbar^2 \bar\mu^2}{2m} \bar\mu d e^{-5\bar\mu d/3} \qquad (7)$$

for two ions separated by d. For each ion μ is determined from the energy of the highest-energy occupied p-state from $\varepsilon_p = -\hbar^2\mu^2/2m$, and the average of the two used in Eq. 7. η_0 is a numerical constant depending only upon the row of the nonmetallic ion, 44 for the fluorine row, 86 for the chlorine row, 103 for the bromine row, and 146 for the iodine row. (These values are from Ref. 13. They should be modified to correct for the use of new matrix elements.) Use of this form and Eq. 5 (or a more accurate form) allows estimates of internuclear distance, the cohesion and the bulk modulus, and in fact the pressure depen-dence of the bulk modulus. The simple theory also gave the other elastic constants[13] and it would be easy to extend the calculations to the pressure dependence, but this has not been done. We may note in passing that empirically the bulk modulus is seen to vary as d^{-3} among the alkali halides.[14] This trend was only crudely reproduced in the theory and the variation for individual materials under pressure was not explored.

6.   THE PROPERTIES OF COVALENT SOLIDS

The closed-shell-ion picture does not make sense for the covalent solids. Every atom is identical in the diamond structure and there are only four electrons per atom. To obtain a meaningful starting description we must obtain bond states.

6.1 The Electronic Structure

We first transform the four atomic orbitals on each atom to sp^3 hybrids (e.g., $(|s> + \sqrt{3}\,|p>)/2$) directed at the four neighbors.

The energy associated with each hybrid is $\varepsilon_h = (\varepsilon_s + 3\varepsilon_p)/4$ and if the two hybrids directed toward each other have different energy, as in GaAs, we define a polar energy V_3 as half the energy difference. The matrix element between the two hybrids is a combina-tion of $V_{ss\sigma}$, $V_{sp\sigma}$, and $V_{pp\sigma}$ which is called the covalent energy V_2 and is given by $-3.22\,\hbar^2/md^2$ for our new parameters. If only these matrix elements V_2 are included in the calculation of the energies we can easily obtain levels

$$\varepsilon_{a,b} = \frac{\varepsilon_h^1 + \varepsilon_h^2}{2} \pm (V_2^2 + V_3^2)^{1/2} \qquad (8)$$

There are enough electrons to fill each of the lower states, the bond orbitals, and the upper, antibonding, states remain empty. This is our starting description, analogous to the closed-shell ions for ionic crystals. There are in fact matrix elements between neighboring bond orbitals due to a matrix element V_1 of the Hamiltonian between two hybrids on the same atom. That matrix element is called the metallic energy, $V_1 = (\varepsilon_p - \varepsilon_s)/4$. It and coupling between bonds through interatomic matrix elements broaden the bonding levels into bands, but if we make the bond orbital approximation of neglecting the matrix elements between bonding and antibonding orbitals we may directly calculate the dielectric and bonding properties in terms of the bond energies of Eq. 8. It is possible to systematically go beyond the bond orbital approximation using extended bond orbitals[1] and it would be interesting to do this for volume-dependent properties, but we shall stay with the simplest description here. Before going on to the dielectric and elastic properties however we may make a few comments about the energy bands, comments which do not depend upon the bond orbital approximation.

6.2 The Energy Bands

We consider first the series of covalent elements, diamond, silicon, germanium, and grey tin. The internuclear distance increases through this series so the interatomic matrix elements decrease, in fact by a factor of 0.3 from diamond to tin. If the atomic term values, and in particular $\varepsilon_p - \varepsilon_s$, were to scale by the same amount, then all parameters would scale together, and the bands would be of exactly the same form but with a reduced energy scale. We see from Table 1 that in fact $\varepsilon_p - \varepsilon_s$ changes by much less. The principal trend from the wide-gap covalent diamond to the semimetallic tin arises from the fact that the term values remain relatively stationary while the interatomic matrix elements are scaled down.

We may also draw implications about the trends in a single material under pressure. We have seen that the interatomic matrix elements vary as d^{-2}, both from material to material and for one material under pressure. We have also noted that $\varepsilon_p - \varepsilon_s$ does not vary greatly among the series C , Si , Ge, and Sn , nor, of course, does it change under pressure. This tells us that the change in the band-structure of silicon under pressure can be estimated by interpolating between the bands of silicon and those of diamond. Similarly any property, such as the refractive index, which can be directly calculated from the bands can also be interpolated between silicon and diamond. We not only know that the refractive index will decrease with pressure, but we also have an empirical estimate of by how much.

6.3 The Dielectric Properties

It is not difficult to make a direct prediction of the dielectric properties in terms of this tight-binding theory.[1] This is done by transforming the states from atomic orbitals to bond orbitals as outlined at the beginning of this section, without approximation, but then approximating the dielectric polarization as the sum of dipoles arising from individual bonds. The polarizability of the bonds may in fact be obtained directly from the change in the bond energy, Eq. 8, from the change in V_3 due to the application of a field as in the treatment of ionic compounds. This is then averaged over orientation and multiplied by the bond density to give

$$ \chi = \frac{\sqrt{3}e^2}{8d\ V_2} \frac{V_2^3}{(V_2^2 + V_3^2)^{3/2}} \qquad (9) $$

which may be compared with Eq. 6 for ionic compounds.

For the homopolar semiconductors, C, Si, Ge, and Sn, $V_3 = 0$ and the final factor is unity. This predicts a dielectric susceptibility proportional to d . (The dependence of the refractive index $n = (1 + 4\pi\chi)^{1/2}$ is a little more complicated.) Experimentally, the susceptibility increases more nearly as d^2 from diamond to tin; the principal error seems to arise from the bond-orbital approximation rather than errors in the parameters.[15] The pressure dependence is less certain experimentally but χ appears to decrease with pressure roughly as predicted.[16]

It is an interesting point that the decrease in the polarizability of the bonds under pressure is expected to, and does, outway the increased bond density and reduces the refractive index. This was not true for ionic solids (Eq. 6) and is not true for molecular solids nor filamentary solids such as selenium; for these, pressure brings the molecular units closer together without greatly changing the bond lengths and the susceptibility tends to vary as d^{-3}.

Three quite different dependences are predicted, and observed, for the three types of solids, d^{-5} , d , and d^{-3} .

This variation of properties of semiconductors with d and therefore with atomic number is described as an increase in <u>metallicity</u> with atomic number; we see that it arises directly from the decrease in interatomic matrix elements with increased bond length and thus that metallicity <u>decreases</u> <u>with</u> <u>pressure</u>.

A second important trend in semiconductors is illustrated by the increasing polarity in the isoelectronic series of germanium, gallium arsenide, zinc selenide, and copper bromide. This is made quantitative by defining a <u>polarity</u> as $\alpha_p = V_3/(V_2^2 + V_3^2)^{1/2}$, and the complementary quantity, <u>covalency</u>, as $\alpha_c = V_2/(V_2^2 + V_3^2)^{1/2}$. The covalency decreases through the series (1.00, 0.88, 0.68, 0.54) germanium to CuBr (as V_2 remains nearly constant but V_3 increases). We see from Eq. 9 that the susceptibility is given by the value for the homopolar semiconductor (since d down not vary appreciably through the series) times the cube of the covalency. This is quite accurately confirmed experimentally.[1,15]

Under pressure, the term values (and therefore V_3) remain fixed but the covalent energy increases. Thus <u>the covalency increases with pressure</u> contributing an increase to the susceptibility - opposite to the trend (proportionality to d or d^2) for the homopolar semiconductor. The net result of these two effects is easily evaluated giving[15]

$$ \frac{d}{\chi} \frac{\partial \chi}{\partial d} = 6\alpha_c^2 - 5 \qquad (10) $$

Note that it leads to the limit of unity obtained before for the homopolar semiconductors, $\alpha_c = 1$, and that it leads to - 5 in the polar limit, $\alpha_c = 0$, as we obtained for ionic crystals. Because of the cancellation between the two terms, the prediction is not accurate even as to sign for the intermediate cases; furthermore calculations of the pressure dependence using universal parameters but <u>no</u> bond orbital approximation differ considerably, and sometimes even in sign, from values deduced from Eq. 10. Even the experimental values vary greatly from one measurement to another.[15] This is a very sensitive property but for just that reason a systematic experimental study, combined with the simple theory, could be very informative.

The theory of effective charges in the semiconductors is also rather complete[1] and the predictions of the pressure dependence of them is direct. The transverse charge tends to be considerably larger in semiconductors than in ionic solids but a first look would suggest that they should be less sensitive to pressure.

6.4 The Elastic Properties

The elastic and other bonding properties of covalent solids are also determined by the electronic structure. The simplest such property is the elastic shear constant $(c_{11} - c_{12})$ giving the rigidity of the lattice under a shear $e_{yy} = -e_{xx}$. This particular shear misalines the two sp^3-hybrids making up a bond (in a manner to which the hybrids cannot adjust), thereby weakening the bond and increasing the elastic energy. The theory is quite straightforward and gives[1] (with the new parameters, Eq. 4)

$$c_{11} - c_{12} = 1.47 \, V_2/d^3 \qquad (11)$$

for homopolar semiconductors. This predicts (with experimental values in parentheses) 66.8 (95.1), 8.1 (10.2) and 6.7 x (8.0) x 10^{11} ergs/cm^3 for C , Si , and Ge. The predictions are some 20% too low but accurately reproduce the d^{-5} dependence observed among the series. We expect the same d^{-5} dependence to apply approximately to the variation of a single material under pressure.

The bond orbital approximation for the elastic constants of polar semiconductors modifies Eq. 11 by a factor of α_c , but corrections to the bond orbital approximation[17] give an additional factor of α_c^2 for a total variation as α_c^3 , of the same form as the variation of the dielectric susceptibility. (In contrast the cohesive energy varies only as α_c.[18]) This gives for the two compounds isoelectronic with germanium, GaAs and ZnSe, 4.6(6.5) and 2.1(3.2)x10^{11} ergs/cm^3. Again the values are too small but describe the trend well.

The pressure dependence of the elastic constant can be obtained immediately by differentiation,

$$\frac{d}{c_{11} - c_{12}} \frac{\partial(c_{11} - c_{12})}{\partial d} = -11 + 6\alpha_c^3 \qquad (12)$$

In contrast to the susceptibility, here the leading term dominates even for homopolar semiconductors and we may hope for reasonable accord with experiment although this has not yet been investigated.

The theory of the other elastic constants is not quite as direct. The bulk modulus, $(c_{11} + 2c_{12})/3$, describes the response to uniform compression and no misalinement of hybrids. The rigidity has a different physical origin, the excess electronic kinetic energy of the compressed system. If the semiconductor were like a uniform electron gas, as a metal is, we would predict a bulk modulus proportional to d^{-5}, the same as in Eq. 11, and to a first approximation this does describe covalent solids.[19] This trend is also understandable in terms of a

calculation of the direct overlap of the atomic charge distributions in the constituent atoms.[13] This uniform-gas view would also correctly suggest a very much weaker dependence upon polarity for the bulk modulus. The remaining independent elastic constant c_{44} involves both misalinement of hybrids and changes in inter-nuclear distance. The rigidity against changes in bond length is sufficiently large that the misalinement dominates and the behavior is similar to that for $(c_{11} - c_{12})/2$.

The general conclusion is that the variation of elastic constants from material to material among semiconductors is reasonably well understood. The corresponding predictions for the pressure dependence are also clear but have not been compared with experiment.

7.   TRANSITION METAL COMPOUNDS

All of our analysis has been based upon the variation of matrix elements among s- and p-states as d^{-2} , which derives from the free-electron-like nature of the semiconductor bands. The d-states in transition metals are in no sense free-electron-like and this variation is inapplicable. Nor is it possible to use free-atom term values for the electronic levels in the solids; the levels vary quite considerably from system to system. Nonetheless, a rather complete simple theory of these systems does exist.[20] We may indicate what that theory is but will not make any detailed applications.

The theory is based upon a requirement of consistency between transition-metal pseudo-potential theory[21] and Andersen's Muffin-Tin Orbital theory,[22] analogous to the consistency required between tight-binding theory and free-electron theory for non-transition solids. This leads to a variation of matrix elements among d-states as

$$V_{ddm} = \eta_{ddm} \, \hbar^2 r_d^3/md^5 \qquad (13)$$

This contains a variation of matrix elements as d^{-5} for a single material but the d-state radius r_d varies from element to element so that the variation of matrix elements from material to material is not given simply by a d-dependence. For coupling between d-states and s- or p-states, the matrix elements are given by

$$V_{\ell dm} = \eta_{\ell dm} \, \hbar^2 r_d^{3/2}/md^{7/2} \qquad (14)$$

Values for the coefficients η and for the r_d are given in Ref. 1.

Values for the d-state energy ε_d must be obtained self-consistently for each system; the procedure as rather simple and all necessary parameters have been given by Froyen.[20] The dominant effect is shifts in the energy such that the occupation of d-states does not deviate far from the free-atom occupation, as if the d-level were "floating" at the Fermi level.

A similar analysis of the actinides, in which f-bands arise, is under way which replaces the d^{-5} dependence of Eq. 13 by a d^{-7} dependence and corresponding changes in the other interatomic matrix elements.

Equations 13 and 14 give the variation of the matrix elements with internuclear distance for transition-metal systems. These have been tested for variations from material to material[24] but there has been almost no study of the variation in a single material under pressure. For example, Eq. 13 leads to the prediction that the band width in a pure transition metal under pressure should vary as d^{-5}. There are indications from earlier theoretical calculations[25] that this may be approximately correct but there appears not yet to have been experimental confirmation. Furthermore the variation of various dielectric and elastic properties of compounds, as discussed here for solids which do not contain transition metal atoms, has just begun to be explored. Preliminary studies of the cubic perovskites, such as strontium titanate, have been made and the theory appears to be meaningful.[24] This would seem to be an even more important avenue for experimental high-pressure study than the simple systems. The theoretical basis for the study of transition-metal systems seems established, and all of the necessary parameters are available. It remains for experiment to test the theory and perhaps to cause it to be refined and revised.

8.   ACKNOWLEDGEMENT

This work was supported by the National Science Foundation under Grant No. DMR77-22365.

REFERENCES:

[1]   Harrison, W. A., Electronic Structure and the Properties of Solids (Freeman, San Francisco, 1980).

[2]   Cohen, M. L. and Yin, M. T., Phys. Rev. Lett. 45, (1980) 1004.

[3]   Cohen, M. L. and Ihm, J., Phys. Rev. B23 (1980) 1576.

[4]   Moruzzi, V. L., Williams, A. R. and Janak, J. F., Phys. Rev. B15 (1977) 2854.

[5]   McMahan, A. K., Skriver, H. L. and Johansson, B., Phys. Rev. B23 (1981) 5016.

[6]   Froyen, S. and Harrison, W. A., Phys. Rev. B 20 (1979) 2420. Discussed also in Ref. 1.

[7]   Harrison, W. A., submitted to Phys. Rev. Rev. B.

[8]   Fischer, C. F., Atomic Data 4 (1972) 301.

[9]   Born, M., Ergebn. d. exakt. Naturw. 10 (1931) 387.

[10]   Bendow, B., Gianino, P. D., Tsay, Y. and Mitra, S. S., Appl. Opt. 13 (1974) 2382.

[11]   Nikulin, V. K., Zh. Tekhn. Fiz. XLI (1971) 41. [Sov. Phys.-Tech. Phys. 16 (1971) 28.]

[12]   Gordon, R. G. and Kim, Y. S., J. Chem. Phys. 56 (1972) 3122.

[13]   Harrison, W. A., Phys. Rev. B 23 (1981) 5230.

[14]   Ref. 1, pg. 311.

[15]   Ren, S.-Y. and Harrison, W. A., Phys. Rev. B 23 (1981) 762.

[16]   Ref. 1, pg. 115, 116. See also Ref. 15.

[17]   Ref. 1, pg. 188, 189.

[18]   Ref. 1, pg. 177.

[19]   Ref. 1, pg. 358.

[20]   Harrison, W. A. and Froyen, S., Phys. Rev. B 21 (1980) 3214; Froyen, S., Phys. Rev. B 22 (1980) 3119; and Ref. 1, Chapters 19 and 20.

[21]   Harrison, W. A., Phys. Rev. 181 (1969) 1036.

[22]   Andersen, O. K., Solid State Commun. 13 (1973) 133.

[23]   Harrison, W. A., unpublished.

[24]   Ref. 1, Chapter 19.

[25]   Pettifor, D. G., J. Phys. F7 (1977) 613.

PHYSICS OF SOLIDS UNDER HIGH PRESSURE
J.S. Schilling, R.N. Shelton (editors)
© North-Holland Publishing Company, 1981

X-RAY ABSORPTION SPECTROSCOPY AT HIGH PRESSURES

R. Ingalls, J. M. Tranquada, J. E. Whitmore and E. D. Crozier[*]

Department of Physics, FM-15, University of Washington
Seattle, Washington 98195, U.S.A.
[*]Department of Physics, Simon Fraser University
Burnaby, British Columbia V5A 1S6, Canada

We report x-ray absorption measurements on several binary compounds, CuBr, SmS, ZnSe, KBr and RbCl at high pressures using synchrotron radiation. Included are those features within 50 eV above an x-ray edge, the XANES (X-ray Absorption Near Edge Structure), as well as the region, 50 eV to 1000 eV above, namely the EXAFS (Extended X-ray Absorption Fine Structure). Both regions show a marked pressure dependence, especially when the sample undergoes a pressure induced phase transition. The XANES indicates the changing electronic structure, whereas the EXAFS clearly indicates the changing bond lengths and atom-pair mean-square displacements. Both features have proven to be useful for internal pressure calibration.

INTRODUCTION

Synchrotron radiation (SR) from the many modern, high energy storage rings throughout the world has revolutionized condensed matter science.(1,2) The availability of continuous, highly intense electromagnetic radiation, varying in energy from a few eV to perhaps 100 keV, has caused a renaissance in such areas as absorption, diffraction, reflectance, fluorescence, photoemission and Raman scattering.

In particular, SR has begun to find its way into high pressure research. X-ray diffraction research, particularly energy dispersive diffraction, has been undertaken by several groups.(3-5) Counting times have been reduced to minutes in obtaining lattice constants to accuracies of, say, 0.1%, which formerly took hours or days.

However, in contrast to making diffraction measurements merely more rapid than with conventional sources, SR has made possible the first x-ray absorption spectroscopy at high pressure.(6,7) Here we review some of our current results from a research program carried on at the Stanford Synchrotron Radiation Laboratory.

The fine structure appearing 0 to 50 eV above an x-ray absorption edge (Kossel structure) is currently called XANES (X-ray Absorption Near Edge Structure). The oscillatory fine structure (Kronig structure) appearing from 50 to perhaps 1000 eV is the EXAFS (Extended X-ray Absorption Fine Structure). The EXAFS function is well approximated by the expression:(8)

$$\chi(k) \sim \sum_j \left[\frac{N_j \exp(-2R_j/\Lambda)}{R_j{}^2} \right] \left[\frac{\exp(-2k^2\sigma_j{}^2)f_j(k)}{k} \right] \times$$
$$\sin(2kR_j + 2\delta_j) \qquad (1)$$

Here k is the photoelectron wave vector, R_j is the relative displacement from the absorbing atom to an atom in the jth shell of N_j neighboring atoms, Λ is an electron mean-free path, $\sigma_j{}^2$ is the mean-square fluctuation in R_j, $f_j(k)$ is the backscattering amplitude from an atom at R_j and $\delta_j(k)$ is a phase shift that depends upon both the central atom and its neighbor at R_j.

The XANES function is not so conveniently written, but promising theoretical progress is being made in characterizing it as well.(9) Both functions depend upon the matrix elements connecting the initial bound state with the photoelectron states. They thus depend upon the many-electron potential associated with a particular geometrical arrangement of atoms. The XANES, which includes contributions from excitonic and bound states, is more dependent on the actual angular arrangement within neighboring shells than the EXAFS. Clearly both functions aid in probing the environment of a given atomic species within a high pressure sample.

METHODOLOGY

Our method of pressure generation, as previously described (7), consists of hydraulically driven gasketed diamond or boron carbide anvils. Because of unwanted Bragg peaks from diamond, we routinely use an EXAFS calibrant, rather than the ruby fluorescence method, for determining the pressure.(10)

One of our most favorable pressure gauge calibrants has been the near neighbor bond length in RbCl measured using the EXAFS of the K-edge of the Rb. Figure 1 shows this absorption spectrum at 0 and 58 kbar. Since RbCl undergoes a B1 to B2 transition at approximately 5 kbar, the two spectra correspond to different crystallographic phases. The XANES region features clearly reflect this difference.

To determine pressure, the function $\chi(k)$ is extracted from the absorption curve after it is plotted as a function of k. The data is then Fourier transformed, giving the radial distribution function (Fig. 2). In principal, changes in the bond lengths with pressure could be obtained from such a plot. However, more accuracy is obtained by back-transforming the radial region of interest and plotting the difference of phase as a function of k (Fig. 3). The slope then provides the necessary ΔR, which can be obtained to perhaps an accuracy of 0.003 Å. From literature values of $\Delta V/V$ vs. P,(11) one can determine the pressure to an accuracy of perhaps one kilobar.

The above pressure measurement technique depends upon our ability to extract accurate ΔR values from the EXAFS. Indirectly this means the parameters $\delta_j(k)$ in Eq. 1 should be insensitive to pressure so that they cancel when the phase difference is plotted, as in Fig. 3. We have found this to be true for NaBr and Ge,(7) and are in the process of an in-depth check on RbCl. The latter is being done in several ways. The materials RbCl and NaBr are run simultaneously and their EXAFS compared at many pressures. RbCl has also been run with a diamond anvil so that its EXAFS may be checked against the ruby fluorescence standard. Analysis of such data is quite complicated because of the Bragg peaks from the diamond anvil (Fig. 4). As discussed previously changes in the XANES may also be used directly for a somewhat less precise pressure determination(7).

ALKALI HALIDES

We have performed measurements above and below the B1 – B2 transition in KBr and RbBr as well as in RbCl. Figures 1 and 2 are quite representative of the results, namely that there is a major change in the XANES region as well as changes in lattice spacings, and the disorder terms $\sigma_j{}^2$. The enhanced peak height in RbCl (Fig. 2) is mainly due to a decrease in $\sigma_1{}^2$ in the B2 phase from 5 to 56 kbar, although there is also a drop in this parameter at the B1 – B2 transition. (The increase in peak height in Fig. 2 also reflects a change in coordination number from 6 to 8.)

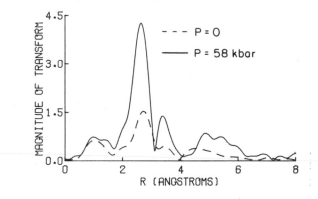

Figure 2 :   Fourier transforms of the Rb K-edge EXAFS in RbCl.

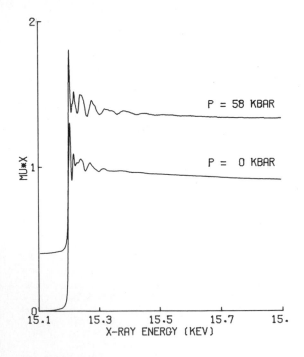

Figure 1: X-ray spectra of the Rb K-edge in RbCl.

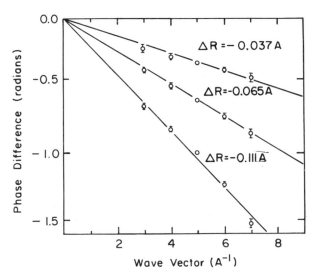

Figure 3 :   Phase difference of the back-transform of the R_1 shells of the Rb K-edge EXAFS in RbCl.

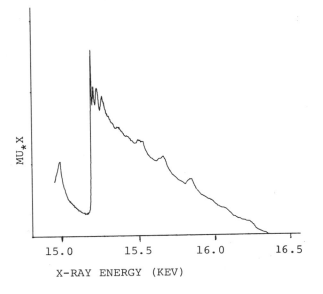

Figure 4 : X-ray spectrum of the Rb K-edge in
RbCl in a diamond anvil cell, 75 kbar

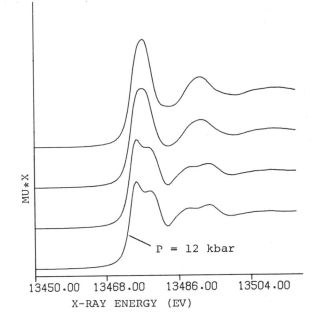

Figure 5 : XANES spectra of the Br K-edge in
KBr at several pressures around the
B1 - B2 transition, from 12 to 20 kbar

We are in the process of analyzing recent data
on the Br K-edge in KBr which undergoes the B1
to B2 transition at about 17 kbar. In Fig. 5
we show how the XANES of this edge changes as
one progresses through the phase transition. We
must await further theoretical analysis, perhaps
as in Ref. 9, before we can offer a detailed
interpretation of this behavior.

ZINC SELENIDE

We have measured the x-ray absorption spectra of
the Zn and Se K-edges in ZnSe. Changes in near
neighbor and next-nearest neighbor distances
were determined by a Fourier filtering of the
EXAFS data as discussed above. Using NaBr as a
pressure gauge, we find the distance changes
from both shells to be essentially consistent
with the literature data on P vs. $\Delta V/V$.(12)

Of perhaps more interest is our determination of
changes in the mean square relative displacements
of σ_1^2 and σ_2^2 as a function of pressure (Fig. 6).
These factors are different from the u^2 appear-
ing in the X-ray Debye-Waller factor, which is
the disorder in the position of the central atom
itself, rather than the disorder in various bond
lengths. Using an Einstein model to analyze the
change in σ_1^2 with volume, we find that the
Einstein mode Gruneisen parameter is 1.6 ± 0.3.
The fact that this value is considerably larger
than optical mode Gruneisen parameters measured
by Raman spectroscopy(13) ($\gamma_{LO} = 0.9 ± 0.1$ and
$\gamma_{TO} = 1.4 ± 0.1$) may be due to small anharmonic

effects. For the second shell, we see essentially no change in σ_2^2 with pressure, reflecting the influence of the TA phonons for which
the mode Gruneisen parameter is negative.

CUPROUS BROMIDE

The material CuBr has the zincblende structure,
CuBr III, at atmospheric pressure but transforms
to a tetragonal phase, CuBr V near 45 kbar and
to CuBr VI (B1 structure) near perhaps 60 kbar.
(14) In Fig. 7 we show the x-ray absorption
spectra of these three phases. Differences in
both the XANES and EXAFS are very apparent.

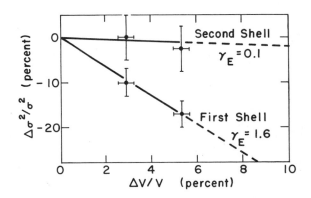

Figure 6 : Changes in the Zn-Se and Zn-Zn
mean-square relative displacements

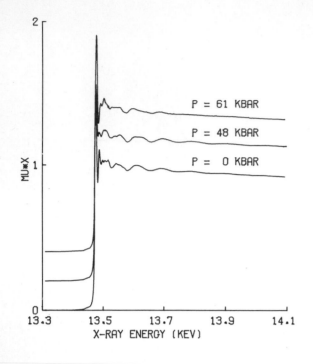

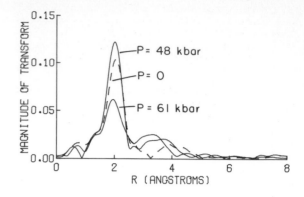

Figure 8 : Fourier transforms of the EXAFS data
            of Fig. 7, for the Br K-edge in CuBr

be much increased in the high pressure B1 phase
(CuBr VI) compared to the zincblende phase,
CuBr III. This is apparent in Fig. 8. The
factor σ_1^2 is determined mainly by the optical
phonon modes, and for the zincblende structure
its temperature dependence is well described by
an Einstein model. At room temperature σ_1^2 is
proportional to the inverse square of the
Einstein frequency. Neutron scattering

Figure 7 : X-ray spectra of the Br K-edge in
            CuBr for three different structural
            phases.

By the Fourier transform method we may compare
the radial distribution about the Br atoms in
each phase (Fig. 8). (The Cu K-edge data yield
similar results.) Although the tetragonal
structure has not been refined(15) our measurements indicate the near neighbor coordination
about both atomic species to be the same as in
the zincblende phase,'namely four. Although
the nearest neighbor distance shows a slight
reduction, the main change with pressure occurs
as a large decrease in second, and higher, shell
distances. These higher shells are difficult to
resolve, but they contain coordination number
information which may eventually be extracted.

The change with pressure in the XANES of the Br
edge in CuBr over a limited pressure range is
shown in Fig. 9. Such data helps to pinpoint
the various transition pressures. Our preliminary analysis has the transitions appearing at
somewhat lower pressures than listed in the
literature,(14,15) It should be emphasized that
theoretical advances are expected to eventually
enable an analysis of the XANES of CuBr V, for
example, facilitating a solution of its
structure.

We also consider that the nearest-neighbor mean
square relative displacement σ_1^2 appears to

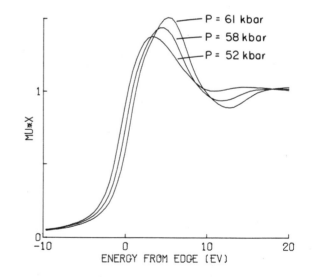

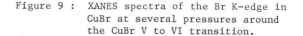

Figure 9 : XANES spectra of the Br K-edge in
            CuBr at several pressures around
            the CuBr V to VI transition.

measurements of the dispersion relations in AgBr
in the B1 structure(16) and AgI in the wurtzite
structure(17) indicate that the LO phonon band
is narrow and roughly at the same frequency in
NaCl-type and tetrahedral structures. However
the TO bands are much wider and extend much
lower in frequency in the B1 structure. If
similar considerations hold for CuBr, it would
not be surprising that the Einstein frequency
would be lower in the B1 structure, leading to
increased values of σ_1^2.

SAMARIUM CHALCOGENIDES

Earlier,(18) we reported measurements on L-edges
of Sm and the K-edge of Se in SmSe, which undergoes a structureless insulator-metal change(19)
in the range 20 to 40 kbar. The change ac-
companies an electronic change on the Sm ion
from the $4f^6$ state toward the configuration,
$4f^5 5d$. This work indicated a rapid decrease
of the first coordination shell R_1 about the Se
atom and also a rapid decrease of σ_1^2. The
latter indicated that there was a single average
nearest-neighbor bond length rather than separate static distances to Sm^{2+} and Sm^{3+} neighbors.

The XANES of the Sm L-edges in SmSe change
dramatically through the pressure-driven transition, reflecting directly the electronic
change. In Figure 10 we show our corresponding
data for the L_3-edge of Sm in SmS, which transforms to the mixed valence state at approximately 6.5 kbar. Although, from an electronic
structure viewpoint, the transition does not
appear to be sharp, further analysis of the
relevant bond lengths and disorder terms (σ_1^2)
is needed to rule out a mixed phase explanation.

A correct theoretical description of such systems
must yield such XANES profiles. X-rays corresponding to the Sm L-edges are rather low in
energy (~ 7 keV) and do not easily pass through
pressure cells. However the advent of wiggler
magnets at SSRL has given us sufficient flux
for reasonable EXAFS data at these energies.
Such data is being analyzed.

SUMMARY

Synchrotron radiation is a powerful new tool for
high pressure research. In particular it has
made possible the introduction of x-ray absorption spectroscopy into this field. Early
results with this technique show its power in
studying the geometrical, electronic and
vibrational structure of matter.

ACKNOWLEDGMENT

The authors are pleased to acknowledge the
following for their gracious assistance:
A.J. Seary, N. Alberding, and J.W. Allen. This
work was supported by the National Science
Foundation under Grants No. DMR 78-24995,
DMR 77-27489 (in cooperation with the U.S.
D.O.E.), and by the National Sciences and
Engineering Research Council of Canada.

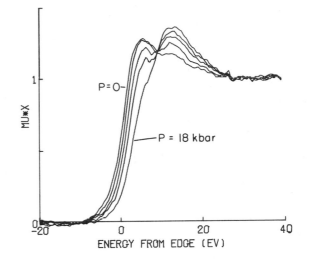

Figure 10 : XANES spectra of the Sm L_3-edge in
SmS at several pressures around its
6.5 kbar insulator-to-metal
transition

REFERENCES

[1] Sokolov, A.A. and Temov, I.M., Synchrotron
    Radiation (Pergamon, New York, 1980).
[2] Winick, H. and Doniach, S. (eds.),
    Synchrotron Radiation Research (Plenum,
    New York, 1980).
[3] Bourdillon, A.J., Glazer, A.M. Hidaka, M.
    and Bordas, J., J. Appl. Cryst. 11 (1978)
    684, and references therein.
[4] Buras, B., Staun-Olsen, J. and Gerward, L.,
    J. Appl. Cryst. 11 (1978) 137, and refer-
    ences therein.
[5] Skelton, E.F., private communication. See
    also the papers by Ruoff, A.L. and Spain,
    I.L., in this volume.
[6] Ingalls, R., Garcia, G.A. and Stern, E.A.,
    Phys. Rev. Lett. 40 (1978) 334.
[7] Ingalls, R., Crozier, E.D., Whitmore, J.E.,
    Seary, A.J. and Tranquada, J.M., J. Appl.
    Phys. 51 (1980) 3158.
[8] Stern, E.A., Sayers, D. and Lytle, F.,
    Phys. Rev. B11 (1975) 4836.
[9] Durham, P.J., Pendry, J.B. and Hodges, C.H.,
    Sol. St. Comm. 38 (1981) 159.
[10] Barnett, J.D., Block, S. and Piermarini,
    G.J., Rev. Sci. Instrum. 44 (1973) 1.
[11] Vaidya, S.N. and Kennedy, G.C., J. Phys.
    Chem. Solids 32 (1971) 951.
[12] Kennedy, G.C. and Keeler, R.N., in American
    Institute of Physics Handbook, 3rd ed.
    (McGraw Hill, New York, 1972).
[13] Weinstein, B.A. in Timmerhaus, K.O. and
    Barber, M.S., (eds.), High Pressure Science

and Technology (Plenum, New York, 1979).

[14] Pistorius, C.W.F.T., Prog. Solid State
     Chem. 11 (1976) 1.

[15] Meisalo, V. and Kalliomäki, M., High Temp.-
     High Press. 5 (1973) 663.

[16] Dorner, B., von der Osten, W. and Bührer, W.,
     J. Phys. C 9 (1976) 723.

[17] Bührer, W., Nicklow, R.M. and Bruesch, P.,
     Phys. Rev. B 17 (1978) 3362.

[18] Ingalls, R., Tranquada, J.M., Whitmore,
     J.E., Crozier, E.D. and Seary, A.J. in
     Teo, B.K. and Jay, D.C. (eds.), EXAFS
     Spectroscopy: Techniques and Applications
     (Plenum, New York, 1980).

[19] Jayaraman, A., Narayanamurti, V., Bucher,
     E. and Maines, R.G., Phys. Rev. Lett. 25
     (1970) 1430.

PHYSICS OF SOLIDS UNDER HIGH PRESSURE
J.S. Schilling, R.N. Shelton (editors)
© *North-Holland Publishing Company, 1981*

STRUCTURAL STUDIES AT HIGH PRESSURE AND TEMPERATURE USING SYNCHROTRON RADIATION

I. L. Spain,[*] S. B. Qadri,[*] C. S. Menoni,[*] A. W. Webb,[†] and E. F. Skelton[†]

[*]Department of Physics, Colorado State University
Fort Collins, Colorado 80523, U.S.A.

[†]Condensed Matter and Radiation Sciences Division, Naval Research Laboratory
Washington, DC 20375, U.S.A.

ABSTRACT

Ultra-rapid structural studies of materials at elevated pressures have been carried out using diamond anvil cells in white synchrotron radiation. A description of the technique using an energy dispersive detector is given. Examples of applications of the technique, including high temperature studies are discussed. Materials studied include Al_2O_3, Ge, Nb_3Si, chevrel phase superconductors, KCl and KI.

I. INTRODUCTION

The diamond anvil high pressure cell has revolutionized very high pressure measurements in the presure range of $\sim$0-1 Mbar (100 GPa). The cell lends itself to measurement of many physical properties, including structural studies using x-ray diffraction. However, when conventional x-ray sources are used, data collection times can be prohibitively long, particularly on lighter elements and compounds. Use of intense, high energy, synchrotron radiation from sources such as DESY in Hamburg, SPEAR at Stanford University, CHESS at Cornell University, provides the opportunity for very rapid data acquisition, which can be maximized using energy-dispersive detection techniques.

A brief review of x-ray diffraction techniques at high pressure using energy dispersive techniques has been given by one of us.[1] The first of such studies were carried out using synchrotron radiation by Buras et al[2] at DESY, Hamburg. These authors studied TeO_2 and FeS (fluorescence radiation from AgI was also studied). The present paper summarizes a series of experiments at the Stanford Synchrotron Laboratory (SSRL) to further assess the technique, and to report some novel results. A preliminary account of some aspects of this work can be found in Refs. 1 and 3.

II. SYNCHROTRON RADIATION CHARACTERISTICS

Experiments were conducted on the SSRL white radiation line. The polychromatic beam has a horizontal angular divergence of 1.0 mrad, with horizontal and vertical dimensions of 15 mm and 4 mm respectively at the radiation hutch. Highest intensity was found in a region $\sim$1 mm × 1 mm. It was found convenient to use the darkening of a film of NaCl on sticky tape when carrying out initial adjustments of the pressure cell in the beam.

The intensity of the white radiation peaks near a critical energy. When the storage ring is operated at 3 GeV, the critical energy of the spectral distribution is $\sim$5 keV ($\lambda \sim 2.5$ Å). With a typical ring current of 50 mA, the photon brilliance at this energy is $\sim$2.5 × 10^{14} photons sec^{-1} mrad^{-1}. White radiation levels are accordingly about 10^4 times higher than those obtained from a typical rotating anode source.

Apart from the advantages accruing from the high source brilliance and low angular divergence, another benefit arises from the well-defined energy dependence of the intensity. This allows quantitative analysis of diffracted radiation to be carried out readily. It is possible that the almost planar polarization of the beam could be used to more advantage in diffraction studies in the diamond anvil cell, but we have not done so.

Recent reviews of synchrotron radiation sources, their characteristics, and uses can be found in Refs. 4 and 5.

III. DETECTION TECHNIQUES

Diffracted radiation can be detected by any of the normal techniques used with conventional x-ray sources. Energy dispersive detectors[6] have the combined advantages of high detection efficiency and the ability to utilize photons with a wide range of energies. Although the efficiency of a Si(Li) detector falls off rapidly above $\sim$20 keV, and the synchrotron beam intensity above $\sim$5 keV, it is possible to detect diffraction peaks to at least 70 keV. Some improvement in the high energy region can be gained using a germanium detector, but "escape peaks" complicate the spectrum, making quantitative analysis more difficult. In the near future we hope to carry out measurements with both silicon and germanium detectors collecting data simultaneously.

Disadvantages of energy dispersive diffrac-
tometry include the pressure of fluorescence
peaks and the low energy resolution of the
detectors. Accuracy of the d-spacing will be
discussed later, and general comments on the
technique applied to the diamond cell can be
found in Ref. 7.

IV. EXPERIMENTAL DIFFRACTION ARRANGEMENT

A sketch of the experimental arrangement is
shown in Fig. 1. The diamond cell was positioned on an X-Y stage, and the Si(Li) detector
on an X-translation and ϕ-rotation stage. A
collimator was not used in our work to date,
so that angular adjustment of the diamond cell
was not needed. A fine, adjustable collimator
is to be used in subsequent work, so that
diffraction lines from the gasket can be eliminated from the spectrum.

The detector was located approximately 520 mm
from the scattering center and viewed the
scattered radiation through a 130 μm wide
aperture; this corresponds to a window 0.25
mrad wide. At a typical detection angle of
14° in 2θ, this represents an angular resolution of 0.1%. Since this is well below the
energy resolution throughout the entire
measured energy range (measured as 149 ± 3 eV
at 5.894 keV), it is predominantly the energy
resolution which dominates the measurable
resolution in terms of interatomic spacings.

Data were stored in a 1024 channel analyzer
(MCA), calibrated against 18 characteristic x-
ray peaks. Data could be transferred to
magnetic tape for computer analysis. We are
anticipating on-line analysis of data during
runs in the near future.

A typical spectrum is illustrated in Fig. 2,
consisting of fluorescence and diffraction
peaks from the gasket, sample and pressure
calibrant. The fluorescence peaks occur at
fixed photon energy, and diffraction peaks at
energies governed by the Bragg equation

$$Ed_{hk\ell}\sin\theta = hc/2 \ .$$

where E is the photon energy, $d_{hk\ell}$ the spacing
between planes of Miller indices (hkℓ), θ is
the Bragg angle, h is Planck's constant and c
the speed of light. Accordingly, the two
types of peaks can be distinguished by
recording data at two different Bragg angles.
At the start of a run, the Bragg angle can be
adjusted until diffraction peaks lie well away
from fluorescence peaks. It is important to
choose gasket, sample and calibrant so that
this can be done effectively. Typical angles
used were 2θ ∿ 6-8°. It was difficult to
collect data at diffraction angles much above
20° in 2θ due to the reduction in intensity
due to polarization.

Experiments using intense synchrotron radiation
have to be carried out in a radiation hutch
with safety locks. For this reason the position and rotation stages were remotely controlled via stepping motors. Also the diamond
anvil cell was designed so that it could be
operated remotely.

V. THE DIAMOND ANVIL HIGH PRESSURE CELL

A cross-section of the anvil region of the
pressure cell is given in Fig. 2. The upper
diamond was seated on a "hemispherical" rocker,
with center coincident with the center of the
anvil face. Angular adjustment could be made
by four set screws. The lower diamond could
be translated by four set screws acting on the
cylindrical base. All alignment adjustments
could be made externally.

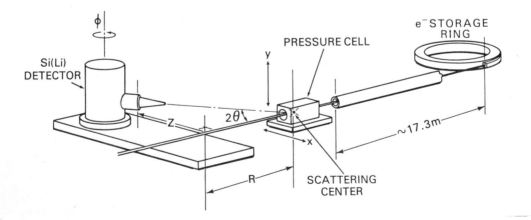

Fig. 1. Sketch of the experimental setup for diffraction.

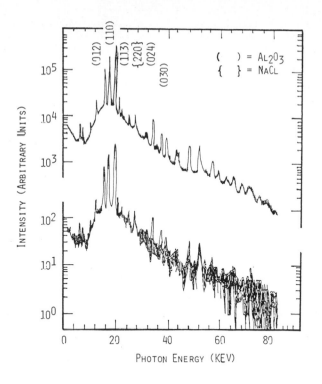

Fig. 2. Typical spectrum of Al_2O_3 and NaCl pressure calibrant at P = 2.5 GPa upper curve 1000 secs, lower curve 10 secs.

The sample calibrant and pressure fluid were located in a hole in the gasket of diameter ∿0.15 mm. The Inconel 718 gasket was of initial thickness 0.45 mm, indented to ∿0.15 mm before drilling, and was heat-treated to ∿43 R_c. This material is a high temperature alloy of high toughness. The synchrotron beam was roughly collimated before striking the sample and gasket with a lead plug. It was determined that the incident radiation did not change the temperature of the sample by more than about 1°C.

The furnace was designed to reach 600°C and was similar to a design reported by Sung.[8] Its body was constructed from a machinable ceramic (Maycor-Corning Glass Company). Heater windings were of 0.075 mm nichrome wire, of resistance ∿8Ω. Temperature was measured with a chromel-alumel thermocouple, spot-welded to the gasket. It was calculated that the temperature difference between the sample and thermocouple was less than 5°C. It was found that a temperature of 600°C could be reached with heater power of ∿60 watts. Similar results have been obtained by Sung.[8]

The body, piston, rocker, and translator (Fig. 2) were all constructed of Inconel 718

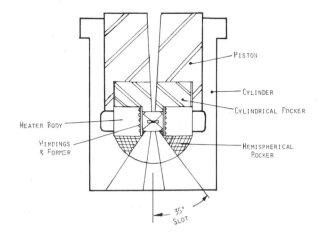

Fig. 3. Cross-section of the diamond anvil cell (high pressure region).

hardened to 43 R_c. At 600°C, the yield stress of this material is ~170,000 psi (1.2 GPa) so that diamonds with a table diameter above 4 mm were needed for operation to 500 kbar (50 GPa).

The diamond anvil cell was built to be hydraulically or pneumatically operated. A ram [Miller Fluid Power, 38 mm diameter piston, 3,000 psi (~20 MPa) maximum pressure] was used to operate a lever arm with 4:1 force ratio. It was found more convenient to operate with gas from a storage bottle than hydraulically.

V.1 Force-Pressure Relationship in the Diamond Cell

The force applied to the diamond anvils could be assessed easily from the pressure applied to the ram. The "efficiency" of the device could then be assessed, where:

$$\text{Efficiency} = \frac{\text{Load applied to anvil tips}}{\text{Load applied to piston}} .$$

For the purposes of calculation it was initially assumed that the stress remained constant over the anvil tips. The efficiency was then measured to be 49%. Bundy[9] used a criterion for anvil failure based on considerations of shear strength. From this analysis it may be concluded that a diamond anvil should not be capable of reaching a pressure above ~300 kbar for this efficiency value. Since gasketted diamond-anvil cells are almost routinely used to 500 kbar it may be concluded that the stress is not uniform over the diamond anvil, so that the stress gradient in the regions outside the anvils is reduced, allowing higher pressures to be reached.

VI. APPLICATION OF THE TECHNIQUE TO Al_2O_3 – CALIBRATION OF TECHNIQUE

Several runs were made with a variety of materials in order to characterize the technique. An example of a spectrum obtained with Al_2O_3 and NaCl calibrant has been given in Fig. 2. During the scans of 10,100 and 1000 secs duration, the beam energy was 3.5 GeV (x-ray critical energy 7.5 keV), and the beam current decayed from 44 to 41 mA.

The spectra were computer analyzed for peak energies and background-corrected, integrated areas. These data for the five most prominent peaks are given in Table 1, together with the same information for the most prominent peak of NaCl (220). In all cases the variance in the measured peak intensities are within the energy resolution of the detection system. It may be concluded that the d-spacing may be resolved to an accuracy of better than 0.4% after only 1 sec and better than 0.1% after 100 secs. The accuracy of the integrated areas increases with counting time.

Based on analysis of the shift in NaCl peaks with pressure using the equation of state of Decker the variance of the pressure is ±0.05 GPa up to about 10 GPa. Above this pressure, line broadening due to nonhydrostatic stresses dominates the accuracy attainable.

A measurement of 10 secs is therefore sufficient for routine compressibility measurements. Larger exposure times are required for more detailed structural analyses.

It should also be noted that at the longer counting times, peaks can be resolved out to 70 keV, where the detector efficiency is

		Scan Duration		10 Sec.			100 Sec.			1000 Sec.	
Peak	I.D.	E(keV)	d(Å)	Area*	E(keV)	d(Å)	Area*	E(keV)	d(Å)	Area*
1	Al_2O_3 - (012)	12.686	3.375	0.68	12.673	3.427	0.73	12.680	3.425	0.80
2	- (110)	18.518	2.345	0.20	18.548	2.342	0.29	18.540	2.343	0.29
3	- (113)	21.157	2.053	0.94	21.137	2.055	0.94	21.133	2.055	0.92
4	- (024)	25.297	1.717	0.28	25.321	1.715	0.29	25.314	1.716	0.30
5	- (030)	32.020	1.356	0.10	32.033	1.356	0.08	32.036	1.356	0.08
c	NaCl - (220)	22.420	1.927	0.23	22.425	1.937	0.31	11.297	1.929	0.15

Scan conditions: $2\theta = 16.41°$; Pressure = 2.5 GPa; Beam = 3.42 GeV; 44 - 41 mA.

*Area = normalized area after background correction.

Table 1. Comparison of peak energies, and integrated, background-corrected intensities.

reduced to less than 15% and the incident photon intensity reduced by a factor of 100 from its value near the critical energy. At the other extreme, it was found difficult to detect diffraction peaks below about 15 keV, probably due to absorption by the diamond anvils.

VII. HIGH PRESSURE BEHAVIOR OF Ge

Early work on the high pressure behavior of Germanium[11] showed that the resistivity exhibited a broad peak at ∿5.0 GPa then dropped about six orders of magnitude at ∿10.5-11.0 GPa (corrected pressure[12]). Jamiesen[13] showed that the cubic form of Ge (phase I) transformed into the tetragonal form (phase II) at the semiconductor-metal transition. It was later shown[14,15] that a third phase (III) was formed on release of pressure from above 12 GPa. This phase (III) was relatively dense compared to the cubic (I), was semiconducting, and tetragonal. No structural studies have been made to our knowledge as a function of pressure, although the present authors[16] have carried out such measurements on many III-V compounds using conventional x-ray sources and photographic detection techniques.

Two runs have been made on germanium - the first with NaCl as calibrant. NaCl and Ge (220) and (311) lines occur at roughly the same energies and could not be resolved. However, the Ge (111) and NaCl (200) peaks, respectively, could be resolved reliably, since the Ge (200) and NaCl (111) peaks are relatively weak. Because of these difficulties, a second experiment was also carried out with Au as a pressure calibrant.

In both cases, the bulk modulus extrapolated to zero pressure (B_o) was within experimental error of the tabulated value from elastic moduli (7.6×10^{10} GPa). However, new lines appeared above about 5 GPa, and increased in strength as the pressure increased. This new phase was labelled Ge IV and possessed a different structure from Ge III.[14,15] Upon holding at any pressure between the onset (5.0-5.5 GPa) and 8.5 GPa, the strength of the lines did not change within a one-hour time period. Upon release of pressure from 8.5 GPa to room pressure, new lines appeared which could be indexed as Ge (II) (β-Sn structure) while lines due to phase IV persisted, and those due to phase I reappeared.

These results have also been confirmed using a conventional x-ray source and photographic detection of diffracted radiation. It is clear that a major re-examination of high pressure behavior of these "simple" materials is called for. The use of synchrotron radiation will be invaluable in assessing the effect on the observed transitions of such factors as the hydrostaticity of the pressure medium, speed of pressure increase, etc. It is stressed that one of the above runs occupied ∿50 minutes. Conventional techniques would have taken at least one month.

VIII. HIGH PRESSURE POLYMORPHS OF Nb₃Si

Both theoretical and empirical predictions have been made that Nb_3Si, if formed in the A-15 structure, might have an extremely high superconducting transition temperature.[17-20] Arguments have also been made that such a form of Nb_3Si might be favored at very high pressures.[21] Limited success was achieved in two earlier shock-wave experiments,[22,23] while two recent experiments have been more successful.[24,25] Stoichiometric Nb_3Si with the A-15 structure obtained from these measurements was found to have a disappointingly low value of $T_c \sim 18.5$ K.

Many polymorphs of Nb_3Si have been obtained by sputtering. Waterstraat[24] has summarized data obtained on deposited films and zone-melted material as shown in Table 2.

Table 2. Summary of superconductivity transition temperatures of Nb_3Si in various polymorphs.

Phase	Ti_3P	bcc	Amorphous	Hexagonal	A-15
T_c(K)	0.3	2.8	3.0	5.5	>15 18.5

A sample of Nb_3Si was taken to 25 GPa and 250°C in the diamond cell while structural information was being collected. Under these conditions a new set of diffraction lines appeared, which persisted when temperature and pressure were lowered. The rather broad diffraction lines could not be indexed on the basis of any of the above structures. The small sample was then placed in a pair of a.c. susceptibility coils capable of detecting a superconducting transition in a sample with less than 1% of the sample volume. No transition could be detected between 1.6 and 35 K.

Further work is in progress to analyze the structure using Debye-Scherrer techniques.

IX. CHEVREL PHASE SUPERCONDUCTORS

The Chevrel phases consist of a group of ternary molybdenum chalcogenides which have the same rhombohedral structure, and which may have as the second metal, elements (M) from the transition, post-transition or rare-earth series.[27-29] Many are superconducting, some with superconducting transition temperatures, T_c above 10K, and some have the highest dT_c/dP known.[28-31]

In seeking the fundamental parameters which relate the superconducting properties to the structure both the volume of the metal atom in $M_xMo_6S_8$ and the cluster valence electron concentration has been found to be significant.[29] Complicating the picture is the fact that both S defects and excess Mo can occur, which makes a careful characterization of the materials of primary importance.

Yvon has concluded[31] that the delocalization of M from the central position is significant. This delocalization increases with decreasing size of M, and also causes variation in the rhombohedral cell angle α. T_c has been studied as a function of pressure[30] as has the bulk compressibility.[32] To determine the effect of pressure on α, however, x-ray diffraction studies were needed, which were carried out in the system described above.

Chevrel phase materials which had been properly characterized were provided by K. Yvon. Thus far data have been collected and analyzed for $PbMo_6S_8$, $Sn_{1.2}Mo_6S_8$ and $Cu_{2.4}Mo_6S_8$. The results for the copper compound are shown in Fig. 4. The hexagonal cell parameters are shown, as well as the derived rhombohedral angle α, and the hexagonal cell volume. The heavy curve gives the bulk volume compressibility as determined previously.[32] The decrease of 0.9° noted in α contrasts with the increase of 0.2° found upon cooling from ambient to ∼190 K.[31] The variation of α for $PbMo_6S_8$ is 0.2° upon pressurization to 3.0 GPa, while for the tin ternary α increased 0.8°. Upon cooling to 5–10 K both decreased by 0.2°.

These results can be combined with the earlier T_c data[30] to obtain the change in T_c as a function of the change in α (with pressure as the common parameter). Figure 5 gives the result, and shows an extended region of linear relationship, which implies that decreasing α for the tin ternary would lead to an increase in T_c. Yvon notes that $GdMo_6S_8$ has a smaller α value than $SnMo_6S_8$, and Fisher found that the initial addition of Gd increased T_c in the tin compound. This would confirm α as being a significant parameter affecting the superconducting properties. It also confirms Yvon's idea[29] that T_c is favored by minimizing the thermal vibration delocalization of M both for large M (Pb and Sn) and for small M (Cu).

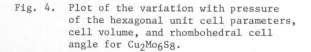

Fig. 4. Plot of the variation with pressure of the hexagonal unit cell parameters, cell volume, and rhombohedral cell angle for $Cu_2Mo_6S_8$.

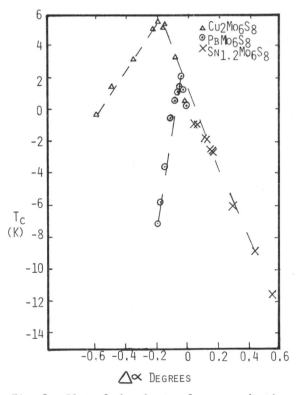

Fig. 5. Plot of the change of superconducting transition temperature correlated against change in the rhombohedral cell angle.

Further work needs to be carried out to extend these measurements to other systems and to determine whether the abrupt change in T_c ($\Delta\alpha$) is caused by a (subtle) phase transition.

X. PHASE TRANSITION IN KCl AND KI

A structural transition occurs in both KCl and KI from the normal rocksalt phase (B1) to a CsCl (B2) phase. As reported by Nomura et al[34] the transition occurs at 2.10 GPa for KCl at room temperature, and KI at 1.88 GPa. Work on these transitions was initiated to ascertain the capabilities of the above system for studying kinetic factors. The pressure dependence of the transition rate has recently been studied for KCl by Akimoto, Hamaya and Yagi.[35] Our work is essentially an expansion of their results to study the temperature dependence of the hysteresis loop in both KCl and KI, and to develop kinetic models. Space limitations prevent a detailed account.

Since all the phases of interest here are cubic, i.e., the NaCl calibrant and each phase of the potassium halides, only one diffraction peak from each material is required to determine the pressure and volume of each phase. Therefore, in order to optimize the data collection efficiency by minimizing the deadtime of the MCA, the MCA energy window was adjusted to include only the three diffraction peaks of interest: NaCl-(200), K(Cl or I) B1-phase (200) and K(Cl or I) B2-phase (110). At a 2θ setting of 12.00°, this corresponds to an energy range from 18 keV to 23 keV for KCl and at 2θ = 9.78°, from 20.7 keV to 27 keV for KI.

Similar reactions are observed with increasing pressure for both KCl and KI and, as an example, the energy-intensity profiles are plotted as a function of increasing pressure in Fig. 6. These spectra were recorded as the pressure was increased from 0.83 GPa to 3.17 GPa. Below P_t only the B1 phase is seen; at a pressure of about 18 GPa, the KI enters the two phase region and both the B1-(200) and the B2-(110) peaks are recorded simultaneously. As the pressure is increased above this, the transition is completed and only the B2 phase is seen.

It should be noted that each of the 10 spectra shown in Fig. 6 were recorded for only 100 sec, thus the total time required to obtain these data was less than 17 minutes. As noted above, with conventional radiation sources and photographic techniques, we would expect that from one to three days would be needed to acquire this amount of information. So in this case, an experiment that might have required from 10 to 30 days was, in fact, performed in less than 17 minutes.

The results of the compressibility measurements are plotted for KCl and KI in Figs. 7 and 8, respectively. The curves running

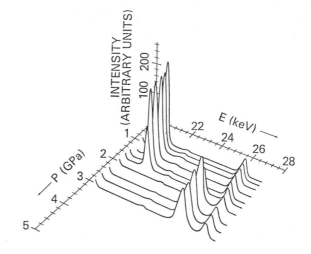

Fig. 6. Three dimensional plot of x-ray intensity versus energy for different pressures.

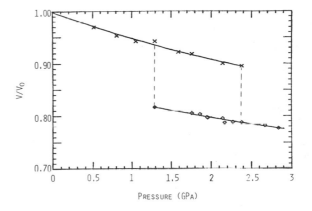

Fig. 7. Fractional change in the unit cell volume of KCl as a function of pressure to 3 GPa.

through the data represent second-order least squares fits. These compression data are in agreement with those previously reported by Nomura et al.[34] Our hysteresis loops are somewhat wider than those measured by Nomura, i.e., 1.07 GPa for KCl and 0.88 GPa for KI measured in this work, as compared with 0.87 GPa and 0.35 GPa, respectively, determined by Nomura. However, as stated above, we believe that the width of the hysteresis loop, since it relates to a nonequilibrium phenomenon, is dependent on the specifics of the pressure system as well as the temperature. The width of the loops reduces with increase in tempera-

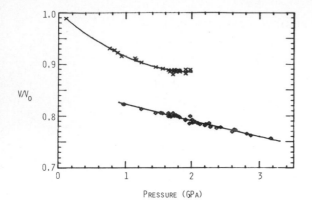

Fig. 8. Fractional change in the unit cell volume of KI as a function of pressure to 3.2 GPa.

ture, and the data are being analysed to suggest mechanisms by which the transformation occurs.

ACKNOWLEDGEMENTS

Thanks are due to the Stanford Synchrotron Laboratory for providing beam time, and for the excellent help given by their support staff. Thanks are also due to Mr. Roy Carpenter of the Naval Research Laboratory and Miss Carol Riggs of the Department of Physics, CSU for technical assistance. This work was supported by a grant from the Office of Naval Research, and CSU personnel also have been supported by a grant from NASA Lewis Research Center.

REFERENCES

[1] E. F. Skelton, accepted for publication in High Pressure Research in Geophysics.
[2] B. Buras, J. Staun Olsen, L. Gerward, G. Will, and E. Hinze, J. Appl. Cryst. 10, 431 (1977).
[3] E. F. Skelton, J. Kirkland, and S. B. Qadri, accepted for publication in J. Appl. Cryst.
[4] C. Kunz, "Synchrotron Radiation Technique and Application," p. 442 in "Topics in Current Physics," Vol. 10, Springer Verlag, N.Y. (1979).
[5] H. Winick and S. Doniach, "Synchrotron Radiation Research," Plenum Press, N.Y., (1979).
[6] A bibliography of energy dispersive x-ray diffraction can be found in E. Laine and I. Lahteenmäki, J. Mat. Sci. 15, 269 (1980).
[7] E. F. Skelton, I. L. Spain, S. C. Yu, and C. Y. Liu, Rev. Sci. Instr. 48, 879 (1977).

[8] C. M. Sung, Rev. Sci. Instr. 47, 1343 (1976).
[9] F. P. Bundy, Rev. Sci. Instr. 48, 591 (1977).
[10] D. L. Decker, J. Appl. Phys. 42, 3239 (1971).
[11] S. Minomura and H. G. Drickamer, J. Phys. Chem. Solids 23, 451 (1962).
[12] H. G. Drickamer, Rev. Sci. Instr. Notes 42, 1667 (1971).
[13] J. C. Jamieson, Science 139, 764 (1963).
[14] F. P. Bundy and J. S. Kasper, Science 139, 340 (1963).
[15] J. S. Kasper and S. M. Richards, Acta Cryst. 17, 752 (1964).
[16] S. C. Yu, I. L. Spain and E. F. Skelton, Sol. St. Comm. 25, 49 (1978).
[17] L. Gold, Phys. Stat. Solidi 4, 261 (1964).
[18] B. T. Matthias, Physics Today 24, 23 (1971).
[19] D. Dew Hughes, Cryogenics 15, 435 (1973).
[20] D. Dew Hughes and V. K. Rivlin, Nature (London) 250, 435 (1975).
[21] E. F. Skelton, D. U. Gubser, J. O. Willis, R. A. Hein, S. C. Yu, I. L. Spain, R. M. Waterstraat, and A. R. Sweedler, Phys. Rev. B20, 4538 (1979).
[22] V. M. Pan, V. P. Alekseevskii, A. G. Popov, Y. I. Beletskii, L. M. Yupko, and V. V. Yarosh, Zh. ERP, Pis. Red. 21, 494 (1975) (JETP Lett., 21, 228 (1975)).
[23] D. Dew Hughes and V. D. Linse, Brookhaven National Lab. Rep. #24546 (1978).
[24] R. M. Waterstraat, Ph.D. Thesis, Univ. of Geneva, #1871 (1978).
[25] B. Ollinger and L. R. Newkirk, Sol. St. Comm. 37, 613 (1981).
[26] S. Ohshima, N. Sone, T. Wakiyama, T. Goto, and Y. Syono, Sol. St. Comm. 38, 923 (1981).
[27] R. Chevrel, M. Sergent, and J. Prigent, J. Sol. State Chem. 3, 515 (1971).
[28] O. Fisher, Appl. Phys. 16, 1-28 (1978).
[29] K. Yvon, in "Current Topics in Materials Science," Vol. 3, E. Kaldis, ed., (North-Holland Publ. Co., 1979), pp. 53-129.
[30] R. N. Shelton in "Proc. 2nd Rochester Conf. on Superconductivity in d- and f-band Metals," D. H. Douglass, ed., (Plenum Press, N.Y., 1976), pp. 137 ff.
[31] K. Yvon, Sol. St. Commun. 25, 327-31 (1978).
[32] A. W. Webb and R. N. Shelton, J. Phys. F: Met. Phys. 8, 261-9 (1978).
[33] O. Fisher, Colloques Inter CNRS, No. 242 79-85 (1975), in B. W. Roberts, NBS Tech. Note 983, 32 (1978).
[34] M. Nomura, Y. Yamomoto, N. Nakagiri, Y. Shirai, and H. Fujiwara, in "High Pressure Science and Technology," p. 757, ed. Timmerhaus and Barber, Plenum Press, N.Y. (1979).
[35] S. Akimoto, N. Hamaya, and T. Yagi, "High Pressure Science and Technology," p. 194, ed. B. Vodar and Ph. Marteau, Pergamon Press, N.Y.

PHYSICS OF SOLIDS UNDER HIGH PRESSURE
J.S. Schilling, R.N. Shelton (editors)
© *North-Holland Publishing Company, 1981*

ENERGY DISPERSIVE X-RAY DIFFRACTION AT HIGH PRESSURE IN CHESS

Arthur L. Ruoff and Millard A. Baublitz, Jr.*

Department of Materials Science and Engineering
Cornell University, Ithaca, NY 14853, USA

Energy dispersive x-ray techniques were used with a diamond anvil cell in the Cornell
High Energy Synchrotron Source (CHESS). It was shown that quantitative relative intensity measurement could be made when the pressure was hydrostatic and the crystals
were relatively defect free. The crystal structures of the high pressure polymorphs
of Ge, GaAs, GaP, and AlSb were studied. Ge exhibits the β-tetragonal structure as
found by Jamieson; however, the transition pressure is 80 ± 5 kbars. GaAs exhibits
an orthorhombic structure above 172 ± 7 kbars, GaP the β-Sn structure above 215 ± 8
kbars, and AlSb an orthorhombic structure above 77 ± 5 kbars.

1. INTRODUCTION

In energy dispersive x-ray diffraction, a heterochromatic x-ray source is used. The x-rays
diffracted from the sample are detected by a
solid state detector with excellent energy resolution; the detector is held at one fixed
angle, 2θ. Because of the ability of this detector to resolve the different energies of the
diffracted x-rays, the entire diffraction pattern of the polycrystalline sample is obtained
at a single angle. The diffraction lines appear at specific energies, E, which are related
to the corresponding interplanar spacings, d,
by the Bragg equation

$$Ed = \frac{hc}{2\sin\theta} = \frac{6.1993\ keV\ \text{Å}}{\sin\theta} . \qquad (1)$$

Detailed discussions of energy dispersive x-ray
diffraction have been published.(1-3) In addition, at least four different research groups
have used this technique during the past five
years to study specimens at high pressure.(4-7)
In our first studies,(6) the Bremsstrahlung
from commercial x-ray tubes was utilized as the
heterochromatic (white) x-ray sources while in
the next experiments a 12 kw, 60 kV, 200 mA
Rigaku rotating anode x-ray generator was utilized. While these experiments provided some
useful information on high pressure germanium,
the results were not nearly as good as those
recently obtained with synchrotron radiation.
Conventional x-ray sources were also used by
Skelton et al.(5) and by Piermarini (7) for
high pressure energy dispersive x-ray diffraction experiments. All of these efforts have
been hindered, at least somewhat, by the following disadvantages of the radiation from conventional sources: (a) the Bremsstrahlung has
a relatively low intensity, (b) the range of
useful intensities is limited, (c) the polarization of the x-rays is both complicated and
difficult to determine, (d) the divergence of
the primary beam is usually rather large, and

(e) undesirable characteristic lines are present. Buras et al. (4) have utilized synchrotron radiation from Deutsches Elektronen -
Synchrotron (DESY) for high pressure studies of
TeO_2 and have reported a pressure-induced phase
transition in TeO_2 from a tetragonal to an orthorhombic phase. However, no detailed structural information is given for the high pressure phase.

In this paper, results of high pressure energy
dispersive experiments conducted at the Cornell
High Energy Synchrotron source (CHESS) are described. CHESS is ideally suited for this type
of experiment because of the exceptionally high
intensity of x-rays between 8 keV and 70 keV.
As an example, when the Cornell Energy Storage
Ring (CESR) is filled with a current of 10 mA
of 5.5 GeV electrons, the synchrotron radiation
beam at CHESS at 50 keV in a 10% bandwidth has
5×10^{12} photons sec^{-1} per horizontal milli-
radian. A typical spectra of synchrotron radiation for CHESS is shown in Figure 1.(8) In
the future it is expected that CHESS will operate at still higher energies and will produce
intensities at 50 keV of seventy or more times
that shown. Each fill of the ring lasts about
two to six hours before the beam is seriously
degraded.

In the present paper we will first briefly describe the experimental apparatus and procedure. Next, the method of obtaining integrated intensities is discussed. Finally, we describe the results we have obtained for the high
pressure phases of germanium, gallium arsenide,
gallium phosphide, and aluminum antimonide.
This will include results for the crystal
structures and for the volume change at the
transition and for the transition pressure of
each of these materials.

2. EXPERIMENTAL

A schematic of the diamond anvil cell used in

---

* Now at Polaroid Corporation, 750 Main Street, 1A, Cambridge, MA 02139 USA

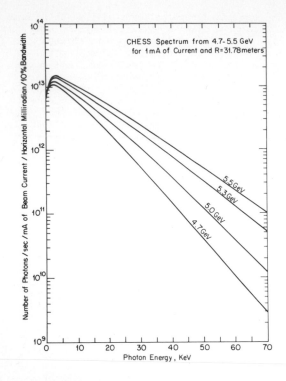

Figure 1 : Spectrum of CHESS

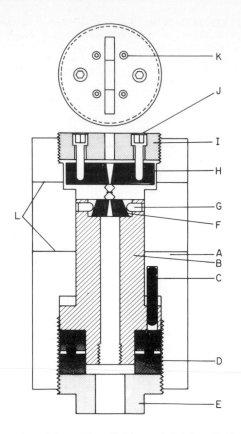

this work is shown in Figure 2. Its operation is described in the legend of Figure 2. Detailed drawings for each of the components are presented elsewhere.(9) The device shown is a modification of one used by Takemura et al.(10) which in turn is an improvement on the earlier design of Bassett et al.(11) This pressure cell has been used to 275 kbars with no problems. This is the highest pressure attempted with it since in the experiments described herein we saw no need to go to higher pressure. The pressures were generated between 1/3 carat diamonds with face culets of 0.6 mm. From our design calculations, failure of the tilting screws would occur at substantially higher pressures. Thus, to achieve very much higher pressures, either the design would have to be modified or diamonds with smaller tips would have to be used. We found that the alignment process for this diamond cell was many times easier than that of the cell constructed by A. Van Valkenburg which was identical to that used by Mao and Bell.(12) The details of mounting the diamonds and the alignment procedures are given elsewhere.(9)

The polycrystalline sample consists of a very fine powder which is mixed with water, methanol: ethanol (4:1 by volume), or hexane and placed in the hole in a metal gasket in order to obtain more nearly hydrostatic pressure on the powder sample.(13) Nimonic 80A, a nickel superalloy, (14) seems preferable to all other materials which we have tested for use as high pressure gasket materials.(15) The starting gasket

Figure 2 : Schematic of Diamond Cell - Bottom - Cross-Sectional View. Top - Top View of Support Nut. A. Container block 5.0 cm x 5.0 cm x 9.0 cm. B. Piston. C. Guide pin. This prevents rotation of the piston as the loading nut, E, is tightened. D. Thrust ball bearing. E. Loading nut. F. Stage for diamond. Stage can be moved in two directions orthogonal to piston axis by screws G. Stage is a frustum of a regular pyramid with a square base. Hole in stage is conical. G. Screws for translation of stage F. There are four such screws. H. Tiltable support plate for diamond. This plate is slotted for the emerging x-rays. Slit has rectangular bases of different dimensions. I. Support nut for tiltable plate. J. Lifting screws. These screws hold plate H in position but do not provide axial support for plate H. The lifting screws pass through the support nut, I, but are threaded into the tiltable support plate, H. K. Tilting screws. These screws are threaded into the support nut, I, and exert an axial thrust on plate H. Manipulation of these tilting screws allows rapid alignment of the diamond faces. L. Window. This is a slotted groove 2.5 cm long and 6.5 mm wide which allows access to the translation screws G. General - The block contains a number of screw holes on its external surfaces which are not shown which can be used for mounting collimators, etc.

material is 0.15 mm thick with gasket holes about 0.19 mm in diameter produced by electrical discharge using a tungsten electrode with a diameter of 0.15 mm. In some experiments we tried high strength metallic glasses (amorphous) as the gasket material with the hope that their amorphous nature would require less strict collimation of the beam; these materials have thus far proven to be disappointing as gasket materials.

The pressure measurement technique of Barnett, Block, and Piermarini (16) utilizing the ruby optical fluorescence shift was used. A tiny chip of ruby was present in the pressurized cell. This has almost no effect upon the x-ray diffraction pattern. The focused beam of an argon ion laser was used to excite the ruby fluorescence. With our arrangement, as used at present, it is necessary to remove the diamond anvil cell from its yoke in the synchrotron radiation beam line and to place it on an optical bench for each pressure measurement. Our design is such that when the diamond anvil cell is returned to the yoke it is precisely located in the same position it had before removal. Experiments at two pressures can be performed for each time the diamond anvil cell is mounted in the yoke.

Figure 3 is a schematic of the experimental arrangement used for energy dispersive x-ray diffraction studies of high pressure samples.

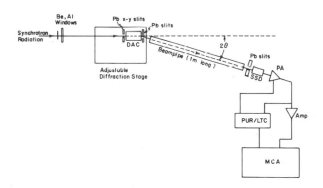

Figure 3 : Schematic of energy dispersive x-ray diffraction apparatus.

Photons of energy less than about 10 keV are absorbed by the beryllium and the aluminum windows. This greatly reduces the heating of the sample. The remainder of the primary beam then passes through the remotely-controlled collimator slits into the diamond anvil cell. The diamond anvil cell is securely mounted in a yoke on the remotely-controlled adjustable diffraction stage. This stage is firmly attached to a remotely-controlled variable height mechanical table. The collimator is fastened directly on the diamond anvil cell. There is also a remote-

ly-controlled vertical exit slit which is bolted onto the diamond anvil cell at the downstream end. The diffracted x-rays pass through this exit slit at an angle, 2θ, which is often chosen to be 15°. The angle, 2θ, must be less than 31° because of geometrical constraints of the diamond anvil cell. A beam pipe extends between the pressure cell and the solid state detector. When filled with helium, this pipe reduces background radiation owing to scattering of x-rays by air.

Either a Si(Li) solid state detector with an active area of 50 mm^2 or an intrinsic Ge detector with an active area of 25 mm^2 was used. The signal from the detector is amplified by a preamplifier and a Princeton Gamma Tech 340 amplifier. This signal is then processed by the Tracor Northern 1710 multichannel analyzer. The PGT 343 pileup rejector/live time corrector prevents two separate signals which occur nearly simultaneously from being summed and counted as a single photon. The intrinsic Ge detector has a higher efficiency for x-ray energies above 35 keV and hence is more likely to detect higher Miller index reflections (smaller d-spacings). However, the intrinsic Ge detector has the disadvantage of having much more intense escape peaks than the Si(Li) detectors.

The vertical dimension of the beam in CHESS is about one millimeter and the angle of divergence is 19 seconds of arc. Since the diameter of the sample (and of the gasket opening) is about 0.2 mm, the diamond cell must be capable of being aligned by remote control after having been approximately positioned. This is accomplished with the remotely-controlled diffraction stage shown in Figure 4; the operation of this stage is described in the legend. The diamond cell is held in a vise-like yoke. The pressure can be increased without removing the cell. When the cell is removed, a dowel pin makes accurate relocation possible. The pressure vessel may be remotely positioned as follows: (a) vertical translation in steps of 0.04 mm, (b) rotation about a horizontal axis in steps of 6 seconds of arc, and (c) rotation about a vertical axis in steps of 2 seconds of arc. The positioning process is carried out until either the x-ray fluorescence and/or the diffraction signal from the sample is maximised on the multichannel analyzer.

In the experiments described in this paper, the primary beam was collimated by using remotely-controlled x-y lead slits which are bolted onto the front side of the pressure cell. Each pair of slits can be opened to one centimeter and closed down to less than 50 μm. Each of the four lead tablets can be moved independently by motors which drive a 2-56 screw at 2½ rpm; this allows incremental motion (without change of direction) of 4 μm. These slits are positioned until all extraneous peaks due to the diamond anvil cell materials or the gasket materials vanish. These slits are mounted about 75 mm

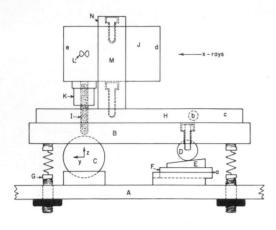

Figure 4 : Schematic of Side View (Opposite Operator) of Pressure Vessel Positioner. A. Horizontal table which is moved vertically (z direction) by motor drive. Drive mechanism, etc., not shown. B. Plate supported by three balls (two C balls about which rotation of plate B about the x-axis occurs and one D ball which is rigidly attached to plate B). C. Two balls (one directly behind the one shown) with their centers 14 cm apart. The origin of the coordinate system is at the center of the ball nearest the reader. The x-direction is into the paper. D. Ball whose x-coordinate is half the distance between the two C balls. E. Wedge (6°). F. Sliding micropositioner. A motor driven micrometer (not shown) applies a force at "a" which advances the wedge E in the y-direction thus rotating plate B (and everything above it) about the x-axis. Motion at "a" of 0.04 mm in the y-direction causes an angular rotation of 6 seconds of arc about x. G. Spring arrangements (there are four) which pull plate B downward toward plate A. H. Plate which rests on plate B and can rotate about shaft I. This rotation is caused by application of force by a ball at "b" (from behind the plate). The ball at "b" is pushed in the negative x-direction (toward the reader) by a wedge mechanism (by a micropositioner such as F but rotated 90° about the y-axis); this is not shown in the drawing. A spring, not shown, is attached at "c" and pulls on plate H in the positive x-direction (away from the reader). The angular sensitivity about the z-axis is 2 seconds of arc. I. Shaft which is press fitted into plate B. The axis of this shaft passes through the center of ball C. (The ball nearest the reader.) J. Pressure vessel block (square cross section) inner details not shown. K. Aligner. This is attached by screws to J. The aligner hole has a close fit with shaft I. When the pressure vessel is readied for use, the diamond anvils, L, are centered about the center of the aligner hole by temporarily placing a close fitted rod with a pointed tip (centered) in the aligner hole. M. Pressure vessel yoke holder. Attached to H with two screws. N. Clamp. When removing the pressure vessel the two screws which attach N to M are removed. The pressure vessel and attached aligner can then be lifted from the yoke; then the pressure can be changed, the ruby shift can be measured, and the vessel (J & K) can then be returned to the original position. d and e designate respectively the location where the slit holders are attached. For simplicity these are not shown.

away from the sample, but, because the beam divergence of CHESS is only 19 seconds of arc, the collimated beam at the sample is still well collimated so there is relatively little scattering of the x-rays from the gasket. The vertical slit at the exit side excludes from the detector 80-90% of the Compton scattering from the diamonds. The two lead tablets of this exit slit are remotely controlled in the same way as the x-y entrance slits.

Typical data count rates with the Ge detector for our samples were 1000-7000 counts sec^{-1} over the 10-70 keV range for a CESR current of 5-10 mA of 5.5 GeV electrons. Our detection system optimally processes about 5000 counts sec^{-1} with 20% dead time. (If the intensity of high energy radiation at CHESS is increased substantially as planned, a more sophisticated electronics detector system will be required to obtain optimum efficiency.) With our present experimental arrangement the most intense lines of a high pressure sample in a diamond anvil cell are observed in one or two seconds.

3.   ERROR ANALYSIS

If the errors are small, the full width of an energy peak at half maximum E_{FWHM} can be described by

$$E_{FWHM} = [(\Delta E_{SSD})^2 + E^2 cot^2\theta (\Delta\theta)^2 + (\frac{E}{d})^2 (\Delta d_p)^2]^{\frac{1}{2}}. \quad (2)$$

Here ΔE_{SSD} is the intrinsic energy resolution of the solid state detector. For the Si(Li) detector

$$(\Delta E_{SSD})^2 = (0.145 \text{ keV})^2 (\frac{E}{5.9 \text{ keV}}) . \quad (3)$$

The second term in Equation (2) obtained by differentiating the Bragg equation is the geometrical factor due to the finite collimation of the incident and diffracted x-ray beams. The third term in Equation (2) is the broadening of the diffraction line due to nonhydrostatic pressure. As an example we consider a germanium sample at 80 kbars; there is a peak at 21.08 keV at θ = 6.823°; also $\Delta\theta$ = 1.7 milliradians. The ruby fluorescence peaks were quite sharp so the pressure appeared to be hydrostatic and thus we assumed Δd_p = 0. From Equations (2) and (3) we obtain E_{FWHM} = 0.405 keV in excellent agreement with the observed result. We note that at 200 kbars, the pressure broadening term becomes the larger term (with either the methanol-ethanol

solution or with water). Eventually it is likely that helium will be used as a hydrostatic
pressure fluid to several hundred kilobars. The
centroid of the diffraction line for which E_{FWHM}
is 0.2 to 0.4 keV can usually be used to determine a lattice parameter within ± 0.002 Å.

4. INTEGRATED INTENSITIES

The integrated intensities for the diffraction
peaks were calculated from the expression obtained by Baublitz et al. from the kinematic
approximation of diffraction.[15]

$$P(\theta,E)=K'|F|^2 p''I_o(E)\frac{A(\theta,E)\eta(E)(e^{-\frac{\mu\tau}{\cos 2\theta}} - e^{-\mu\tau})}{E^2 \mu (1 - \frac{1}{\cos 2\theta})} \quad (4)$$

This simplified expression can be used only if
relative diffracted intensities are measured at
a specific angle with a fixed detector aperture
such as used in the present conditions. Here
K' is a constant (which we need not know when
computing relative intensities), F is the structure factor, p'' is the multiplicity factor,
$I_o(E)$ is the beam intensity and can be obtained
from Figure 1, $A(\theta,E)$ accounts for the absorption by the air and the diamonds in the cell,
$\eta(E)$ is the detector efficiency, μ is the absorption coefficient of the sample, and τ is
the total sample thickness. We note that Equation (4) does account for polarization; the
polarization of radiation at CHESS must be
virtually independent of the x-ray energy so
the polarization factor is essentially a constant for a fixed diffraction angle. When the
actual I/I_o are calculated, corrections are also
made for the Debye-Waller temperature effect.[15]

5. RESULTS

5.1 Germanium

A phase transition in germanium has been observed by other research groups using electrical
resistivity (100-105 kbars, corrected),[17] optical measurements (109 kbars),[18] and x-ray
diffraction (100-105 kbars, corrected).[19]
Jamieson, using tungsten carbide anvils, showed
that the structure was the tetragonal β-Sn
structure.[19] It seemed reasonable for us to
begin our work with a known material. At the
time we carried out this work our collimation
and alignment procedure had not been perfected.
Consequently we will not present quantitative
I/I_o results for this material.

A semiconductor grade crystal of germanium was
ground into a fine powder, mixed with water, and
placed in a gasket made from Allied Chemical
Corporation metallic glass 2866A. At 80 ± 5 kbar
and ambient temperature (28°C) the diamond cubic
material had been compressed to 0.915 ± 0.010 of
its initial volume when the crystal structure
began to transform to the high pressure β-Sn
type tetragonal structure. The fractional volume
decrease associated with the transition was

0.194 ± 0.009. Figure 5 shows the results at
P = 1 bar and Figure 6 shows the results at P =
86 ± 2 kbars obtained with the Si(Li) detector.

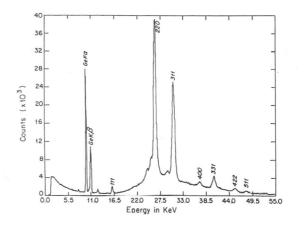

Figure 5 : Diffraction pattern from Ge at
P = 1 bar and T = 28 ± 1°C.

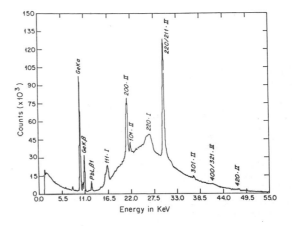

Figure 6 : Diffraction pattern from Ge at
P = 86 ± 2 kbar and T = 28 ± 1°C.

At 86 kilobars, the tetragonal structure is the
dominant phase. Table 1 lists the diffraction
data from the high pressure tetragonal phase of
Ge at 86 ± 2 kbars and 28 ± 1°C. The observed
values of the diffraction lines are compared
with the calculated values assuming a = 4.950Å
and c = 2.728Å. The c/a ratio is in agreement
with that of Jamieson. The transformation
pressure is, however, substantially less than
that observed in other static experiments and
in shock experiments. In shock experiments the
corrected mean pressure at the transition has
been computed to be 114-122 kbars.[20-21] On
the basis of the shock data of Gust and Royce,
[22] Ruoff [23] has calculated a shock transi-

TABLE 1

Ge Diffraction Pattern at P = 86 ± 2 kb and
θ = 6.830° Tetragonal β-Sn Structure with a =
4.950A, c = 2.728A. CESR Conditions: 4.98 GeV,
4 mA. Si(Li) Detector

hkℓ	d(Å) (observed)	d(Å) (calculated)	I/I$_0$ [**] (observed)
200	2.475	2.475	S
101	2.389	2.389	M
220*	1.750	1.750	VS
211*	1.718	1.719	M
301	1.411	1.412	W
112	1.261	1.271	W
400 }	1.230	1.238	W
321 }		1.226	W
420	1.102	1.107	W

*The peaks 220, 211 overlap substantially. The
data for these peaks result from a mathematical deconvolution of the peaks.

**S - Strong, VS - Very Strong, M - Medium,
W - Weak

tion pressure of 96 kbars. We note that the
static experiments only give the pressure where
the transition proceeds at a sufficient rate to
be observed. The equilibrium transformation
may be substantially below this. The mechanism
of this transformation is not known. However,
the β-Sn to α-Sn transition is known to be a
diffusion-controlled nucleation and growth process.(24) That reaction proceeds at an optimal
rate at about -35°C (the transition temperature
is 13°C). Even at -35°C, crystal growth only
proceeds at about 1 mm per hour. This temperature corresponds to about 0.56 of the melting
temperature of α-tin estimated by Van Vechten.
(25) At the transition pressure of germanium
the melting point is about 960°K;(25) our experiments are carried out at only 0.31 of this
temperature and so, assuming that the transformation in germanium in the static experiments
is diffusion controlled, substantial overpres-
surization would have to occur to drive the
reaction. Thus the equilibrium transformation
pressure must be lower than the pressure of 80
± 5 kbars at which we observed the transition.

5.2 Quantitative I/I$_0$ Measurements Below the Transformation Pressure for GaP

Table 2 shows results for GaP at P = 53.5 ± 1.5
kbars and 28 ± 1°C obtained with water (ice) as
the pressure transmitting medium.

A function R, which is a measure of the deviation between the experimental and the theoretical relative integrated intensities, is defined in Equation (5). The summations in Equation (5) extend over all the observed diffraction lines.

$$R = \sum_h \left| \frac{I_{obs}(h)}{\sum_{h'} I_{obs}(h')} - \frac{I_{calc}(h)}{\sum_{h''} I_{calc}(h'')} \right| . \qquad (5)$$

TABLE 2

GaP Diffraction Pattern at P = 53.5 ± 1.5 kb
and θ = 8.854° Cubic Zincblende Structure with
a = 5.353Å. CESR Conditions: 5.236 GeV, 8 mA

hkℓ	E$_{obs}$ (keV)	E$_{calc}$ (keV)	I/I$_0$ (observed)	I/I$_0$ (calculated)
111	13.06	13.03	4.1	2.5
200	15.05	15.05	6.9	2.9
220	21.28	21.28	100	100
311	24.92	24.95	99	85
222	26.06	26.06	11.8	7.8
400	30.08	30.09	14	15.9
331	32.82	32.79	31	23.1
422	36.82	36.86	17	17.9
511, 333	39.09	39.09	7.8	4.5
440	42.52	42.56	1.9	2.9
531	44.56	44.51	4.7	4.6
620	47.61	47.58	2.3	2.1

The results in Table 2 have an R value of 13%.
For GaAs at 151 ± 6 kbars with 4:1 methanol-
ethanol as the pressure medium the R value was
26%.(9)

Four possible reasons for these differences
are: (a) there is a preferred orientation in
the powder specimen due to the relatively small
number of crystals; (b) the powder achieves a
preferred orientation due to the application of
nonhydrostatic pressures; (c) there are uncertainties in the sample absorption and in the
initial spectral distribution of the synchrotron radiation; (d) the presence of nonhydrostatic pressures may have introduced defects
into the crystal structure which would selectively diminish certain diffraction lines. The
fact that R increases as the pressure on the
sample was increased above 130 kbars in each
case suggests that (b) and (d) may be important.

5.3 The GaAs Transition

The pressure induced phase transition has been
studied by electrical resistivity,(17) by a
recent EXAFS experiment,(26) and optically.(27)
Attempts have been made to determine the high
pressure phase of GaAs with conventional x-ray
powder diffraction,(19,28) but these attempts
were less than wholly successful. Yu et al.
tentatively suggested, on the basis of preliminary findings, the possibility of an orthorhombic structure.(28)

A high purity single crystal of GaAs was utilized for this work. It was obtained from Dr.
Colin Wood of the National Research and Resource
Facility for Submicron Structures at Cornell.
Each of the three experimental samples was prepared by grinding a piece of the GaAs crystal
into a fine powder. This was mixed with the
pressure fluid and placed within the hole of a
Nimonic 80A gasket. In two experiments water
was used as the pressure medium and in the
third experiment methanol:ethanol (4:1 by vol-

ume) was used. In these experiments at CHESS, GaAs began to transform from the cubic zinc-blende type structure to an orthorhombic crystal structure at 172 ± 7 kbars and 28 ± 1°C. This transition pressure is in good agreement with that reported by Piermarini at 180 ± 8 kbars (27) and Yu et al. at 170 ± 5 kbars.(28)

The cubic phase had been compressed to 0.874 ± 0.006 of its initial volume when the transition began. The fractional volume decrease from the cubic to the orthorhombic phase was 0.173 ± 0.008. As the transformation was approached the pressure was increased in small increments. This procedure also was followed above the transformation. There was no sign of an intermediate phase at about 190 kbar between the low pressure cubic phase and the high pressure orthorhombic metallic phase as previous electrical resistivity measurements had indicated there might be.(17,29) Indeed, no other phases were present to 250 kbars.

Figures 7, 8, and 9 show the diffraction pat-

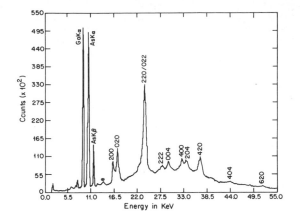

Figure 9 : Diffraction pattern of GaAs at P = 209 ± 7 kbar and T = 28 ± 1°C.

terns for the following respective pressures: P = 0, 178 and 209 kbars. We consider now the analysis of the data at 209 kbars. This is shown in Table 3. The diffraction lines are indexed on the basis of an orthorhombic cell with a = 4.946Å, b = 4.628Å, and c = 5.493Å.

TABLE 3

GaAs Diffraction Pattern at P = 209 ± 7 kb and θ = 8.842° Orthorhombic Structure with a = 4.946Å, b = 4.628Å, c = 5.493Å. CESR Conditions: 5.384 GeV, 9 mA

hkl	d_{obs} (Å)	d_{calc} (Å)	I/I_{220} (observed)	I/I_{220} (calculated)
200	2.471	2.473	19	19
020	2.314	2.314	29	30
022*	1.772	1.770	5	106
220	1.690	1.690	100	100
222	1.443	1.439	3	130
004	1.373	1.373	9	27
400	1.238	1.237	14	16
204	1.200	1.201	10	27
040	not seen	1.157	≤ 0.1	11
402	1.128	1.128	0.7	19
420	1.092	1.091	21	15
240	1.047	1.048	0.4	11
422	1.013	1.014	1.1	18
006)	0.9156	(0.9155)	3	(2.0
404)		(0.9189)		(4.1
424)	0.8478	(0.8540)	0.6	(1.3
440)		(0.8448)		(1.9
600	0.8244	0.8243	0.5	0.8
620	0.7781	0.7765	2	0.9
604	0.7064	0.7068	0.3	0.3

* The 022 line overlapped with the 220 line, but a mathematical deconvolution technique was used to obtain this data for 022.

It should be noted that, in attempts to index the high pressure phase, GaAs II, tetragonal and hexagonal crystal structures were also

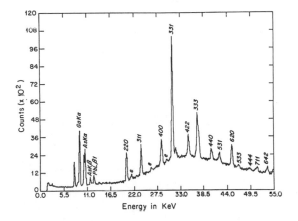

Figure 7 : Diffraction pattern of GaAs at P = 1 bar and T = 28 ± 1°C.

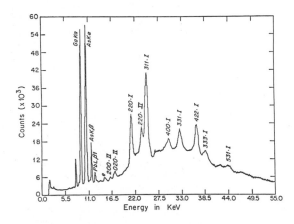

Figure 8 : Diffraction pattern of GaAs at 178 ± 4 kbar and T = 28 ± 1°C.

systematically considered, but all of these structures were very unsatisfactory and were rejected. Of the 17 observed peaks 11 corresponded extremely closely to specific planes, two were overlapped but could be separated by a mathematical deconvolution technique and two other high index pairs could not be separated. The standard deviation of the interplanar spacings was only 0.0005Å.

The theoretical intensities shown in Table 3 were calculated on the basis of a distorted sodium chloride cell which had been compressed along the b-axis and stretched along the a-axis. This structure belongs to space group Fmmm. It is important to note that the atomic scattering factors of Ga and As are very similar and so the atoms will be effectively indistinguishable with energy dispersive diffraction. Thus we could not say whether the metallic phase is ordered as assumed or a random solid solution. The existence of random solid solutions in metals is common.(24) If the structure is ordered, then each Ga atom has approximately six-fold coordination with pairs of As atoms situated at distances of 2.473Å, 2.314Å, and 2.746Å from the Ga atom. These values yield a mean Ga-As interatomic distance of 2.511Å at 209 ± 7 kbars, which is in excellent agreement with the EXAFS data of Shimomura et al. (26)

It should be noted that there are some glaring discrepancies in the I/I_o values in Table 3. The immediate reaction is to question whether the proposed structure is correct. It might be suggested that the bond angles in GaAs II differ from the 90° angles in the face centered orthorhombic structure given here, but careful inspection of the diffraction lines from high pressure GaAs does not seem to permit this option. If any of the bond angles differed significantly from 90° in the orthorhombic unit cell, then there would be additional diffraction lines with all odd or mixed indices; these are demonstrably absent in the experimental patterns. A more likely possibility is that GaAs II has the face centered orthorhombic structure but that there are defects present in the crystal structure which selectively diminish certain diffraction lines. The results in Table 3 indicate substantial agreement between the experimental data and the theoretical intensities for the diffraction lines 200, 400, 600, 020, 220, 420, and 620, but the 040 line is missing or at least very weak and all of the lines with $\ell \neq 0$ are weaker than expected. Baublitz (9) has proposed that stacking faults on the (001) planes could lead to this behavior. It should be noted that we do not know the mechanism of this transformation. At the transformation pressure, the melting temperature is about 800°K (25) so the transformation was studied at 0.38 of this melting temperature. In the case of tin (where this fraction is 0.56), the transformation [which is a nucleation and growth process (24)] leads to heavily deformed and highly defective crystals,(24) presumably because of the

large volume change. A thorough treatment of the GaAs results will be given elsewhere.

5.4   The GaP Transition

The phase transition in GaP was first determined by electrical resistivity measurements.(30-32) The transition pressure based on the ruby scale is about 220 kbars.(28,33) There have been at least two unsuccessful attempts to determine the crystal structure of this pressure phase. (28,34) In the present experiments GaP began to transform from the cubic phase to a tetragonal phase at 215 ± 8 kbars on the ruby scale in essential agreement with similar studies.(28,33) When the transition began the cubic phase had been compressed to 0.864 ± 0.010 of its initial volume. The fractional decrease in volume at the transition was 0.175 ± 0.010.

A single high purity crystal of GaP was obtained from Texas Instruments. Two polycrystalline specimens were prepared by grinding pieces of the GaP crystal into a fine powder. One of the GaP powder specimens was mixed with water and the other was mixed with methanol:ethanol (4:1 by volume). The diffraction data obtained with the intrinsic Ge detector from the two specimens were nearly identical. Both samples were pressurized with Nimonic 80A gaskets. A pattern of the fully transformed GaP II is shown in Figure 10. Table 4 lists the associated

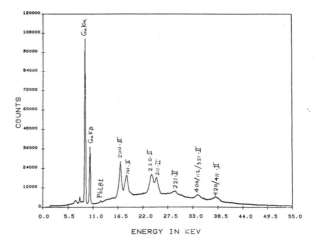

Figure 10 : Diffraction pattern of GaP at P = 235 ± 10 kbar and T = 28 ± 1°C.

numerical data and the corresponding lines assuming a tetragonal unit cell with a = 4.720Å and c = 2.468Å. Because the ratio a/c is almost two, several of the diffraction lines lie close together and are difficult to resolve. The agreement of the 12 observed peak spacings with the calculated spacings is good.

The theoretical relative intensities are computed on the basis of a tetragonal cell with Ga and P atoms randomly distributed on the sites

TABLE 4

GaP Diffraction Pattern at P = 240 $\pm$ 10 kb and
θ = 8.891° Tetragonal Structure with a =
4.720Å, c = 2.468Å. CESR Conditions: 5.260 GeV,
8 to 11 mA

$hk\ell$	d_{obs}(Å)	d_{calc}(Å)	I/I_0 (observed)	I/I_0 (calculated)
200	2.359	2.360	100	97
101	2.187	2.187	84	100
220	1.674	1.669	86	54
211	1.603	1.604	70	93
221	1.384	1.382	12	0
301	1.319	1.327	1	18
400⎫		⎧1.180⎫		⎧ 8.4
112⎬	1.181	⎨1.157⎬	34	⎨15
321⎭		⎩1.156⎭		⎩15
420	1.056	1.055⎫	34	⎧ 7.4
411	1.038	1.038⎭		⎩ 6.5
421⎫	0.9623	⎧0.9704⎫	2	⎧ 0
312⎭		⎩0.9511⎭		⎩ 6.3
501,	0.8827	0.8817	1	2.4
431				
512⎫		⎧0.7405⎫	0.7	⎧0.6
611⎭	0.7406	⎩0.7402⎭		⎩0.3

(000; 0 ½ ¼; ½ 0 ¾; ½ ½ ½); this is the β-Sn
structure with the symmetry of the space group
I4₁/amd. Again we note that it is quite common
for electrically conducting phases to have disordered structures.(24) The absence of 110, 310,
002, and 202 peaks seems to rule out the ordered
structure of Ga at (000; ½ ½ ½) and As at
(0 ½ ¼; ½ 0 ¼). Of the 17 Miller indices listed
in Table 4, the data are reasonably consistent
with all but three of the lines. These are the
very low intensity of the 301 peak which should
have medium intensity, the existence of the 221
peak which should be absent, and the probable
existence of the 421 peak which should be absent. Baublitz (9) has noted that this discrepancy might be due to twinning on the (301)
planes. It should be noted that such twinning
does occur in the metallic phase of tin.(35)
Such twinning would decrease the (301) intensities and would introduce peaks not otherwise
present. We note once again that the large volume change in this transformation produces substantial defects in the case of tin. (24) We
also note that for our experiments the transformation occurs at 0.38 of the melting temperature
of 800°K (at the transition pressure),(25) a
fraction substantially lower than that at which
the α→β tin or β→α tin transformation proceeds.
A more detailed analysis of the GaP results will
be presented elsewhere.

5.5 The AlSb Transition

AlSb exhibits a phase transition which has been
studied electrically (17) and by x-ray diffraction.(19,28) Jameison found that high pressure
AlSb was tetragonal,(19) while Yu et al. found
that it exhibited the sodium chloride type
structure. (28)

In the present case, two specimens of 99.999%
initial purity were studied. They were prepared in a dry, argon atmosphere. One specimen
was mixed with hexane, the other with methanol:
ethanol (4:1 volume). Both specimens began to
transform to an orthorhombic phase at 77 $\pm$ 5
kbars and 28 $\pm$ 1°C. Yu et al. found a transition pressure of 80 $\pm$ 3 kbars.(28) The fractional decrease in volume at the transition was
0.196 $\pm$ 0.015. There was no evidence in the
data for any other high pressure phase. A
thorough analysis of these results will be presented elsewhere.

6. SUMMARY

1. Table 5 compares the fractional volume
changes and transition pressures found in this
work with the theoretical predictions of Van
Vechten.(25)

TABLE 5

Fractional Volume Changes & Transition Pressures

	$\dfrac{\Delta V_{\alpha\beta}}{V_\alpha(P_t)}$ (exp.)	$\dfrac{\Delta V_{\alpha\beta}}{V_\alpha(P_t)}$ (theory) Ref.25	P_{trans}. (kb) (exp.)	P_{trans}. (kb) (theory) Ref.25
Ge	-0.194$\pm$.009	-0.209	80$\pm$5	92
GaAs	-0.173$\pm$.008	-0.192	172$\pm$7	153
GaP	-0.175$\pm$.01	-0.188	215$\pm$8	216
AlSb	-0.196$\pm$.015	-0.185	77$\pm$5	122

2. We have demonstrated that quantitative
relative intensities can be obtained with the
energy dispersive technique at high pressures
in the diamond anvil cell.

3. The upper bound for the transformation
for Ge is 80 $\pm$ 5 kbars on the ruby scale, lower
than previously obtained by other methods.

4. The transformed III-V compounds and group
IVB elements appear to be heavily deformed with
important defects which strongly affect intensity ratios of particular peaks.

5. GaAs II orthorhombic. It may have a
face-centered structure. From the present
results alone we cannot show whether the structure is ordered or disordered.

6. GaP II is tetragonal, probably β-Sn with
a disordered structure (with equal probability
of Ga or P on any atom site).

7. AlSb II is orthorhombic. The structure
is probably disordered (with equal probability
of Al or Sb on any atom site).

8. Because all of these transformations are
observed at only one-third the melting temperature, there is no assurance that the phase

being produced is the stable phase at that pressure and temperature. Moreover, even if the high pressure phases are stable phases, there is no assurance that the transition pressure observed is the equilibrium pressure (because of kinetic reasons).

9. Energy dispersive x-ray diffraction in CHESS is a powerful tool for research at high pressure.

Acknowledgments. The authors wish to thank the National Aeronautics and Space Administration and the National Science Foundation (through the Cornell Materials Science Center) for support of this work. They also wish to acknowledge the National Science Foundation for the support of CHESS and B. W. Batterman, D. H. Bilderback and the entire staff of CHESS for their help. We also wish to thank W. A. Bassett for loaning the Ge detector to us and Volker Arnold for his considerable technical assistance.

References

(1) C. J. Sparks, Jr., D. A. Gedcke, Advances in X-ray Analysis 15, 240 (1972).

(2) M. Mantler, W. Parrish, Advances in X-ray Analysis 20, 171 (1977).

(3) A.J.C. Wilson, J. Appl. Cryst. 6, 230 (1973).

(4) B. Buras, J. Staun Olsen, L. Gerward, G. Will, E. Hinze, J. Appl. Cryst. 10, 431 (1977).

(5) E. F. Skelton, C. Liu, I. L. Spain, High Temp.-High Press. 9, 19 (1977).

(6) M. Baublitz, Jr., A. L. Ruoff (unpublished).

(7) G. J. Piermarini, F. A. Mauer, S. Block, A. Jayaraman, T. H. Geballe, G. W. Hull, Jr., Solid State Comm. 32, 275 (1979).

(8) D. H. Bilderback, T. Motyka (unpublished).

(9) M. Baublitz, Jr., Ph.D. thesis, Cornell University, Ithaca, NY (in preparation).

(10) K. Takemura, O. Shimomura, K. Tsuji, S. Minomura, High Temp.-High Press. 11, 311 (1979).

(11) W. A. Bassett, T. Takahashi, P. W. Stook, Rev. Sci. Inst. 38, 37 (1967).

(12) H. K. Mao, P. M. Bell, Carnegie Inst. Washington Yearbook 74, 402 (1975).

(13) G. J. Piermarini, S. Block, J. D. Barnett, J. App. Phys. 44, 5377 (1973).

(14) Nimonic 80A is made by Henry Wiggin and Co., Ltd., Hereford, England.

(15) M. A. Baublitz, Jr., Volker Arnold, A. L. Ruoff, Cornell Materials Science Center Report #4479, Ithaca, NY, June 1981. (Accepted for publication in Review of Scientific Instruments.)

(16) J. D. Barnett, S. Block, G. J. Piermarini, Rev. Sci. Inst. 44, 1 (1973).

(17) S. Minomura, H. G. Drickamer, J. Phys. Chem. Solids 23, 451 (1962).

(18) B. Welber, Rev. Sci. Instrum. 47, 183 (1976).

(19) J. C. Jamieson, Science 139, 762, 845 (1963).

(20) R. A. Graham, O. E. Jones, J. R. Holland, J. Phys. Chem. Solids 27, 1519 (1966).

(21) M. N. Pavlovskii, Soviet Phys.-Solid State 9, 2514 (1968).

(22) W. H. Gust and E. B. Royce, J. Appl. Phys. 43, 4437 (1972).

(23) A. L. Ruoff in High Pressure Science and Technology, K. D. Timmerhaus, M. S. Barber (eds.), Plenum, New York, 1979, vol. 1, p. 525.

(24) C. S. Barrett, T. B. Massalski, Structure of Metals, McGraw-Hill, New York (1966).

(25) J. A. Van Vechten, Phys. Rev. B, 7, 1479 (1973).

(26) O. Shimomura, T. Kawamura, T. Fukamachi, S. Hosoya, S. Hunter, A. Bienenstock, in High Pressure Science and Technology, B. Vodar, P. Marteau (eds.), Pergamon Press, Oxford, 1980, vol. 1, p. 534.

(27) G. J. Piermarini, private communication.

(28) S. C. Yu, I. L. Spain, E. F. Skelton, Solid State Comm. 25, 49 (1978).

(29) L. Merrill, J. Phys. Chem. Ref. Data 6, 1205 (1977).

(30) A. Onodera, N. Kawai, K. Ishizaki, I. L. Spain, Solid State Commun. 14, 803 (1974).

(31) F. P. Bundy, Rev. Sci. Inst. 46, 1318 (1975).

(32) C. G. Homan, D. P. Kendall, T. E. Davidson, J. Frankel, Solid State Commun. 17, 831 (1975).

(33) G. J. Piermarini, S. Block, Rev. Sci. Inst. 46, 973 (1975).

(34) G. J. Piermarini, private communication.

(35) A. Kelly, G. W. Groves, Crystallography and Crystal Defects, Addison-Wesley, Reading, 1970, p. 304.

PHYSICS OF SOLIDS UNDER HIGH PRESSURE
J.S. Schilling, R.N. Shelton (editors)
© *North-Holland Publishing Company, 1981*

RECENT ADVANCES IN THE STUDY OF STRUCTURAL PHASE TRANSITIONS AT HIGH PRESSURE[*]

G. A. Samara

Sandia National Laboratories[†]
Albuquerque, New Mexico 87185 USA

The role of pressure in the study and understanding of some structural phase transitions in solids is discussed. Results on a number of crystals exhibiting displacive,
order-disorder and cooperative Jahn-Teller transitions are presented. The occurrence
of each type of these transitions is determined by a balance between competing interactions, and pressure alters this balance and allows tests of our understanding of the
physical phenomena involved. Some emphasis is on displacive transitions especially
with regard to incommensurate structures and the behavior near the so-called displacive
limit. This limit is the special case where the transition temperature $T_c = 0$ K. As
T_c decreases and approaches 0 K quantum effects can be expected to come into play.
Pressure turns out to be an excellent variable to study the crossover from classical
to quantum behavior expected on approaching the displacive limit.

1. INTRODUCTION

This paper deals with the role of hydrostatic
pressure in the study and understanding of some
structural phase transitions in insulators. The
occurrence of many such transitions is determined
by a delicate balance between competing lattice
interactions. It is a main theme of this paper
that pressure is one of the best, and often the
only, variable which can significantly alter
crystal potentials and thereby this balance between forces. This in turn allows tests of our
understanding of the phenomena involved.

Several aspects of this subject have been reviewed elsewhere.[1] In this paper we restrict the
presentation to recent and new results. In the
space available it is not possible to give anything like a detailed account of these results.
Thus we restrict ourselves to the following
topics: (i) Displacive soft mode transitions
(Sec. 2), (ii) coupled electron-phonon (or cooperative Jahn-Teller) transitions (Sec. 3),
(iii) coupled proton-phonon transitions (Sec. 4)
and (iv) quantum effects associated with displacive transitions (Sec. 5).

All of the pressure measurements reported in this
paper were made in an apparatus using helium as
the pressure transmitting medium. The reader is
referred to the papers by the author referenced
both here and in Reference 1 for details of the
apparatus and techniques.

2. DISPLACIVE SOFT MODE TRANSITIONS

This class of structural phase transitions is
one of the best understood. It includes a large
number of ferroelectric, antiferroelectric and
other antiferrodistortive transitions which are

*This work was supported by the U.S. Department
of Energy under contract DE-AC04-76-DP00789.
†A U.S. Department of Energy facility.

driven by soft phonon modes.[1,2] By a soft mode
we simply mean a normal mode of vibration whose
frequency decreases (and thus the lattice literally softens with respect to this mode) and approaches zero as the transition point is
approached.

Inherent in the soft mode picture of phase transitions is the premise that the crystal is unstable in the harmonic approximation with respect
to the soft mode. Specifically, the square of
the harmonic frequency, ω_0^2, is presumed to be
sufficiently negative (i.e., ω_0 is imaginary)
that this mode cannot be stabilized by zero-point
anharmonicities alone. Thermal fluctuations
then renormalize ω_0 and make it real at finite
temperature, thereby stabilizing the lattice.
The transitions temperature is the temperature
where the renormalization is complete. Even
though thermal fluctuations (anharmonicities)
provide the stabilization at high temperatures,
it should be strongly emphasized that the interesting and unusual features of the soft mode are
not so much a consequence of the large anharmonicities as they are a result of the extremely
small (or imaginary) harmonic frequencies.[1,2]
For a given crystal, this harmonic frequency is
determined by a very delicate balance between
competing short-range and Coulomb forces. The
key to the understanding of the pressure effects
on displacive transitions rests on the fact that
the short-range forces are much more strongly
dependent on interionic distance than are the
Coulomb forces.[1,2]

In general, the soft mode can be either an optic
or an acoustic phonon, of either long or short
wavelength and either commensurate or incommensurate with the lattice. Earlier pressure
studies of soft mode transitions have dealt
largely with commensurate ferroelectric and antiferroelectric (or more generally antiferrodistortive) transitions.[1,2] In these cases the soft

mode is a long wavelength ($q \approx o$) transverse
optic phonon for ferroelectrics and a short wavelength optic phonon for the other transitions.
The different pressure effects observed on these
two types of transitions were explained in terms
of a reversal in the roles of the competing
short-range and Coulomb forces in the lattice
dynamics of the two cases.[1,2] This conclusion,
based on pressure results, is supported by some
detailed lattice dynamical calculations.

More recently there has been growing interest in
incommensurate displacive transitions and several
have been studied under pressure. The best understood of these is potassium selenate (K_2SeO_4), and
we shall now present some of our recent results[3]
on this crystal.

K_2SeO_4 exhibits two structural phase transitions
which have attracted a great deal of recent attention.[4,5] At room temperature the crystal is
paraelectric and orthorhombic (space group Pnam).
The structure can be visualized as consisting of
two distinct types of planes separated by a half
lattice vector along the c-axis with K^+ ions and
$SeO_4^=$ tetrahedra embedded in each plane. At 129K
the crystal undergoes a second-order displacive
transition which is driven by a soft phonon propagating along the [100], or Σ, direction with
wave vector $q = \frac{1}{3}[(1 - \delta)a*]$, where $a*$ is the
first reciprocal lattice vector and δ is a small
deviation.[4] Thus, this new phase is incommensurate with the lattice, and K_2SeO_4 has become
the most studied material which undergoes an
incommensurate structural phase transition.

On further cooling the deviation δ decreases and
vanishes discontinuously at 93K (= T_c) where the
crystal locks into a commensurate superstructure
whose a axis is triple the size of the room
temperature phase. The low temperature phase
(space group Pna2$_1$) is an improper ferroelectric.[4]

Recent lattice dynamical calculations have shown
explicitly how the phonon instability results
from the cancellation between Coulomb and short-
range forces, and that the soft mode frequency
is especially sensitive to some of the short-
range force constants.[5] Another finding of this
work is that there is nothing special about the
wave vector at which the instability occurs in
relation to a reciprocal lattice vector, and it
is thus not surprising that the crystal makes a
transition to an incommensurate phase. This is
an important result for our purposes because it
shows that there are no essential differences in
the balance of forces (and thereby the pressure
effects) between commensurate and incommensurate
transitions.

The very delicate balance of competing forces
for the soft mode branch in K_2SeO_4 suggests that
the soft mode frequency, and thereby the transition temperatures, should be strongly pressure
dependent. This expectation is confirmed
experimentally.[3]

Figure 1 shows the temperature dependence of the
dielectric constant ε_c measured at different
pressures. At atmospheric pressure ε_c exhibits
a relatively large λ - type anomaly at the incom-
mensurate-to-ferroelectric transition (T_c = 93K)
and a more subtle, but well-defined, anomaly
(see inset) at the paraelectric-to-incommensurate
transition (T_i = 129K). Both transition temperatures decrease rapidly with pressure. It is also
seen that the ferroelectric transition vanishes
at a pressure somewhat above 0.72 GPa. The 0.72
GPa isobar, shows a rounded, but well-defined,
transition at $\sim$ 9K. The 0.75 GPa isobar, on the
other hand, shows no sign of any transition to
the low temperature ferroelectric state, i.e.,
the ferroelectric state has vanished.

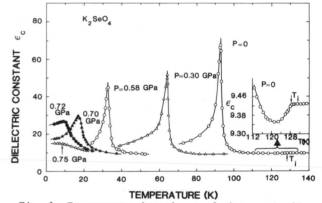

Fig. 1   Temperature dependence of the c-axis dielectric constant of K_2SeO_4 at different
pressures. Data were taken on the cooling cycle. The inset shows the response
at the incommensurate transition.

Figure 2 shows the low temperature-pressure
phase diagram for K_2SeO_4. The paraelectric-to-
incommensurate transition is second order and
has no thermal hysteresis; however, the incom-
mensurate-to-ferroelectric transition is first
order and is accompanied by a thermal hysteresis
of $\sim$ 2K between cooling and warming cycles at
zero pressure. This hysteresis increases to
$\sim$ 6K at 0.7 GPa (see shaded region).

The initial pressure derivatives of T_i and T_c
(on cooling) are $dT_{i,c}/dP$ = -65.5 and -96.0
K/GPa, respectively. We note that the $T_i(P)$ and
$T_c(P)$ shifts are non-linear, and that the ferroelectric phase vanishes at $\sim$ 0.73 GPa. It
appears that T_c may go to 0 K with infinite
slope, i.e. $dT_c/dP \rightarrow -\infty$ as $T_c \rightarrow$ 0 K. This
feature will be discussed later. The measure-
ments did not extend to a sufficiently high
pressure to completely suppress the incommensurate phase, but we expect $T_i(P)$ to be qualitatively similar to $T_c(P)$. This is depicted by
the dashed line in Fig. 2 which suggests that
T_i should go to 0 K at $\sim$ 1.13 GPa. (see Eq. 3,
Sec. 5).

In terms of the soft mode picture, the decrease
T_i (and T_c) with pressure results from an increase (hardening) of the soft mode frequency[1,2]

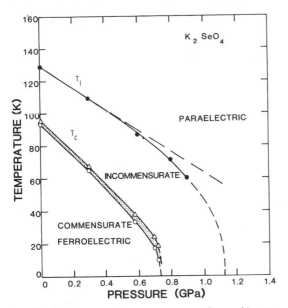

Fig. 2 Low temperature-pressure phase diagram
for K_2SeO_4.

As is generally well know,[1,2] and has been specifically demonstrated for K_2SeO_4,[4] vanishing of
this frequency at the transition results from
the cancellation between Coulomb and short-range
forces in the lattice dynamics of the crystal.
The lattice dynamical calculations on K_2SeO_4
(harmonic model) show that the soft Σ_2 branch is
stabilized by the Coulomb forces and destabilized
by the short-range forces, and that this desta-
bilization is lowered by decreasing r (or increasing pressure).[5] The latter result implies
a lowering of the transition temperature with
pressure,[1,2] as is observed.

The negative sign of the pressure derivative
dT_i/dP in K_2SeO_4 deserves some comment. First,
results on two other crystals should be mentioned.
Rb_2ZnCl_4 is isomorphous with K_2SeO_4 and its incommensurate transition occurs at T_i = 303K.
T_i increases (non-linearly) with pressure.[6]
$BaMnF_4$ exhibits a well-known incommensurate transition at T_i = 247K. Its T_i first increases
with pressure (dT_i/dP = 33K/GPa), reaches a
maximum at $\sim$ 1 GPa and then decreases at a rate
faster than the increase below 1 GPa.[7] We had
earlier noted[1,2] the generalization that for all
known (at that time) displacive transitions
associated with short wavelength optic modes,
the transition temperature increases with pressure, and that this result can be explained on
the basis that, for these transitions, the short-
range interaction is negative (i.e. attractive)
whereas the Coulomb interaction is positive (i.e.
repulsive). The opposite is true in the case of
soft mode transitions associated with long wavelength optic phonons where T_c decreases with
pressure. The above results on Rb_2ZnCl_4 and
$BaMnF_4$ (below 1 GPa) agree with this generalization, but the results on K_2SeO_4 and $BaMnF_4$

(above 1 GPa) appear to disagree. Nevertheless
it can be argued that the generalization still
holds. The key to this is the observation that
with increasingly higher pressure and as the ions
get closer and closer, the short-range interaction cannot be expected to continue to be
attractive. Ultimately it should become repulsive. If this were the case, then a reversal in
the sign of dT_c/dP might be expected.[2,7] The
sign reversal in $BaMnF_4$ may thus be a manifestation of this phenomenon. In this context, the
case of K_2SeO_4 (as especially contrasted with
Rb_2ZnCl_4) is particularly interesting because
the pressure results imply that for this crystal
the reversal has already occurred at atmospheric
pressure. As possible support for this argument
we note that the lattice dynamical calculations
on K_2SeO_4 show that for the soft Σ_2 phonon branch
the dominant short-range force constant
$[\phi''_{oo}(7)]$ becomes softer or less negative by
decreasing the interionic separation (i.e. increasing pressure).[5] This is opposite to what
is generally expected[1,2] and it explains the
decrease of T_i with pressure.

While the above argument is plausible, it should
be cautioned that in any given crystal the
specific balance between competing forces depends,
of course, on the details of the crystal structure. In the case of complex structures or
coupled mode interactions general trends may not
be obvious.

3. COUPLED ELECTRON/PHONON OR CO-OPERATIVE JAHN-TELLER TRANSITIONS

In systems with degenerate or nearly degenerate
low-lying electronic states there can be coupling
of these levels to the lattice vibrations producing excitations of mixed electron/phonon (or
so-called vibron) character and, in certain circumstances, a spontaneous distortion. The cooperative Jahn-Teller (CJT) transition is a phase
transition which is driven by this interaction
between localized orbital electronic states and
the phonons. It involves the simultaneous splitting of the electronic states and a symmetry-
lowering distortion of the lattice.[8,9] The transition is accompanied by a softening of the
vibron, but the actual soft mode is usually an
elastic excitation, or an acoustic mode. There
can also be additional coupling to optic phonons.
The phase transition can result in parallel
alignment of all the distortions (i.e. ferrodis-
tortive) or in a more complicated arrangement
(e.g. antiferrodistortive as in $DyVO_4$).

Much of the early experimental evidence for such
effects occurred in cubic compounds of transition
metal ions (e.g. AB_2O_4 spinels), but more recently
the emphasis of work in this areas has shifted
to rare-earth compounds, specifically rare-earth
vanadates and arsenates (e.g. $DyVO_4$ and $DyAsO_4$).
This shift was due to the availability and relative simplicity of the latter compounds. The
electronic energy level sprectrum of rare-earth
ions (with their deep-lying 4f electrons) is
simpler than that of transition metal ions, and,

additionally, the electron phonon coupling is much weaker in rare-earth compounds than in transition metal compounds. Both factors combine to make the theory simpler and the results more amenable to interpretation in the case of the rare-earth compounds.

Formally, the CJT transition problem is quite analogous to the well known coupled proton/phonon (or KH_2PO_4-type) transition. Namely, the total Hamiltonian consists of an electronic part, a lattice part (taken to be harmonic) and a coupling term (usually taken to be bilinear in the electron and phonon coordinates).[8,9] The electronic system is described by an Ising model $(-\frac{1}{2} \Sigma J_{ij} \sigma_i^z \sigma_j^z)$ where the two levels of the crystal field doublet for the ion are represented by the two states $\sigma^z = \pm \frac{1}{2}$ of a psuedo spin operator. In cases where there is electronic energy splitting in the high temperature phase and in systems, such as in $DyVO_4$, which have two Kramers doublets (but where the system can still be considered to be effectively a two-level system), the electronic part of the Hamiltonian will have an additional term $- \Sigma \Gamma \sigma^x$, i.e. a transverse field. Γ provides a measure of the splitting in the high temperature phase.

When the total Hamiltonian is solved in the molecular field approximation, the transition temperature, T_d, is found to be

$$4\Gamma/\tilde{J} = \tanh (\Gamma/kT_d) , \qquad (1)$$

where $\tilde{J}$ is the phonon-mediated dipolar interaction between electronic states. The physical picture represented by Eq. (1) is as follows: T_d is determined by the competition between two fields, (i) a dipolar field represented by $\tilde{J}$ which tends to order the spins and induce the distortion, and (ii) a transverse field represented by Γ which is proportional to the splitting and stabilizes the high temperature phase. At atmospheric pressure the dipolar field dominates, so that $(4\Gamma/\tilde{J}) < 1$, and there is a finite T_d. For $(4\Gamma/\tilde{J}) \geqslant 1$ there is no transition. Pressure can be expected to change the balance between these two fields and thereby influence T_d.

We have investigated the effects of pressure on the CJT phase transitions in three crystals: dysprosium vanadate ($DyVO_4$), dysprosium arsenate ($DyAsO_4$) and uranium dioxide (UO_2). $DyVO_4$ and $DyAsO_4$ are isomorphous, their crystal structure being tetragonal (D_{4h}) above T_d and orthorhombic (D_{2h}) below T_d. The Dy^{3+} ion is a Kramers ion whose ground state consists of two doublets split above T_d, by ~ 9 cm^{-1} in $DyVO_4$ and ~ 5 cm^{-1} in $DyAsO_4$. The splittings increase below T_d. The properties of the two crystals, including pressure effects, are qualitatively similar except that the transition in $DyVO_4$ ($T_d = 14.6K$) is second order whereas that in $DyAsO_4$ ($T_d = 11.1K$) is first order.

Figure 3 shows the temperature dependence of the c-axis static dielectric constant of $DyVO_4$ near

the CJT transition at two pressures. There is a well-defined dielectric constant anomaly at T_d. The inset shows the shape of the anomaly over a broader temperature range. It is seen that the transition shifts to lower temperatures with increasing pressure, and Fig. 4 shows that this shift is nonlinear as is expected on the basis of Eq. (1).

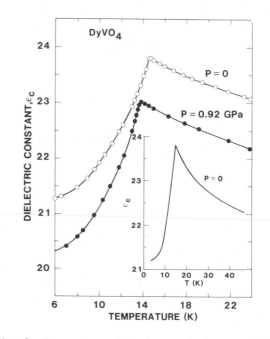

Fig. 3   Temperature dependence of the c-axis dielectric constant of $DyVO_4$ at two pressures near the transition. The inset shows the response at $P = 0$ over a broader temperature range.

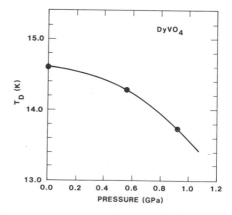

Fig. 4   Pressure dependence of the cooperative Jahn-Teller transition in $DyVO_4$.

In terms of the model discussed above, the decrease in T_d results from an increase in the ratio $(4\Gamma/\bar{J})$ which in turn results from either an increase in the electronic energy splitting, and thereby Γ, in the high temperature phase or from a decrease in $\bar{J}$, or both. As in the analogous case of KDP, both Γ and $\bar{J}$ are expected to be pressure dependent, but we suspect that the pressure dependence of Γ may be the more dominant effect. The initial pressure derivative of T_d is relatively small, but this is not unexpected because the 4f electrons of the Dy^{3+} ion are deep lying (or shielded by outer electrons), and it takes a large volume change to perturb them. The magnitude of dT_d/dP increases with pressure, and ultimately Eq. (1) predicts that $dT_d/dP \rightarrow -\infty$ as $T_d \rightarrow 0$, or $(4\Gamma/\bar{J}) \rightarrow 1$. It will be desirable to extend the measurements to the much higher pressures needed to test these predictions. However, the experiment will be difficult as it is necessary to maintain hydrostatic pressure.

In the case of the $DyAsO_4$, T_d decreases with pressure also, but the magnitude of the effect is considerably smaller than in $DyVO_4$. This might be expected on the basis of the much lower initial doublet splitting in $DyAsO_4$. This contrast between $DyVO_4$ and $DyAsO_4$ also provides indirect support to the presumption that it is the change in Γ with pressure that is the more dominant factor in determining $T_d(P)$.

UO_2 has the cubic fluorite structure (Fm3m) at high temperature. At $T_d \simeq 30K$ it undergoes a simultaneous CJT distortion and antiferromagnetic ordering resulting from the triplet ground state of the U^{4+} ion.[10] Unfortunately, the strong exchange interactions which lead to the antiferromagnetism in this crystal complicate the picture and make the interpretation of the CJT transition difficult. Neutron scattering results[11] have shown that the CJT distortion involves a zone-boundary phonon mode, and, thus, the strongest electron-phonon interaction must occur at the zone boundary: however, there must also be significant coupling to the zone center acoustic modes since the transition is accompanied by an anomaly in the elastic constant C_{44}.

We find that between 0 and 1 GPa T_d increases linearly with pressure with a slope $dT_d/dP = 0.59K/GPa$. Because of the mixed and complicated nature of the interactions leading to the transition it is difficult to interpret the sign of this pressure shift. The CJT distortion energy and the exchange field, both of which can lift the collective ground state degeneracy of the U^{4+} ion, are nearly equal in UO_2 at atmospheric pressure.[10] Pressure can, of course, favor one of these energies over the other. In this regard it is worth noting that pressure usually causes an increase in the Néel temperature (which is the same as T_d for UO_2) of antiferromagnets. This results from an increase in the exchange interaction J. Thus, the pressure results may imply the dominance of the exchange forces in determining the transition in UO_2.

4. COUPLED PROTON/PHONON TRANSITIONS

This class constitutes groups of hydrogen-bonded crystals in which the proton plays an essential role in determining the phase transitions and related physical properties. This role is most evidenced by the unusually large effects on both the static and dynamic properties observed on deuteration. The protons undergo order-disorder transitions in double-well potentials along the H-bonds, and this motion of the protons is coupled to the rest of the lattice in such a way that the resulting low temperature phases are either ferroelectric, or antiferroelectric.[1]

Pressure studies have been very important to the understanding of the nature of the soft modes and phase transitions in these crystals. The early pressure work was primarily on KH_2PO_4 (KDP) and some of its isomorphs and has been reviewed elsewhere.[1] More recently interesting new results have been reported on CsH_2PO_4[12] and $H_2C_4O_4$.[13] In this Section we summarize these results briefly.

Unlike other members of the KDP family, CsH_2PO_4 has a monolithic rather than a tetragonal structure. The crystal has strong one-dimensional character and the low temperature phase is characterized by chain-like ordering of the protons in the short $O-H\cdots O$ bonds along the polar b-axis. The ordering is identical in all chains and this results in a ferroelectric state, i.e. there is parallel ordering in adjacent chains. At normal conditions the transition temperature is 153K and upon deuteration T_c increases to $\sim 267K$, thus again emphasizing the role of the protons in the transition.

Pressure decreases T_c of CsH_2PO_4, and above 0.33 GPa, the ordering of the chains changes from parallel to antiparallel resulting in an antiferroelectric phase.[12] A qualitatively similar result obtains for the deuterated crystal CsD_2PO_4. The change in the character of the transition can be explained, in the context of an Ising spin model, by a pressure-induced change in the sign of the relatively weak interchain interaction parameter.[14] This has been attributed to the 1-D nature of the ordering and the weak 3-D ferroelectric correlations among the chains.[13,14]

Square acid, $H_2C_4O_4$, is a relatively new hydrogen-bonded crystal. It consists of layers of strongly H-bonded planar C_4O_4 groups, each linked to four neighboring C_4O_4 groups in the plane by linear $O-H\cdots O$ bonds. The crystal undergoes a continuous structural phase transition at 375K.[13] The transition is especially interesting because it is one of the simplest and very few known 2-D structural transitions. There is a strong deuteration effect on T_c ($T_{c,D}$ = 528K), again emphasizing the important role of the protons in the transition. The transition does involve an order-disorder transition of the protons in the H-bonds. Above T_c

the protons move in a symmetrical double-well potential whereas below T_c the protons order preferentially in one of the wells. The ordering induces a lattice distortion. The molecule possesses an electric dipole moment, and, in the low T phase the molecules are arranged in such a way that the structure consists of planar ferroelectric layers, with the layers antiferroelectrically stacked along the unique monoclinic b-axis, so there is no net polarization.

The planar natue of $H_2C_4O_4$ and $D_2C_4O_4$ molecules and their crystal structure suggests that the main effects of pressure should be an increase in the interlayer interaction and a shortening of the H bond length, R.[13] The decrease in R should lead to a large change in the potential seen by the proton as it moves along the bond. Specifically, the shorter the bond, the lower is the energy barrier and the closer is the separation of the minima. Both effects should lead to a lower T_c. Experimentally it is found that T_c decreases with pressure with $dT_c/dP = -105$ and $-102 K/GPa$ for $H_2C_4O_4$ and $D_2C_4O_4$, respectively. Additionally, there is a pressure-induced transition in $D_2C_4O_4$ at ~ 0.8 GPa. A similar transition can be expected in $H_2C_4O_4$ at a pressure above the highest pressure (~ 0.7 GPa) of the available experiments. Although the nature of this transition is not known, we suspect that it may reflect a pressure-induced change in the inter-layer coupling.[13]

By analogy to KDP, it is tempting to attribute the large isotope effect in $H_2C_4O_4$ to proton tunneling, the much larger tunneling probability of the proton relative to the deuteron being responsible for the lower T_c for $H_2C_4O_4$. However, we have shown[13] that on the basis of the pressure results and simple bond length considerations, it is possible to provide a satisfactory account for the H isotope effect on T_c. There is no need to invoke tunneling. In this regard $H_2C_4O_4$ is markedly different from KDP where such arguments do not hold, and where tunneling has to be invoked to account for the H isotope effect on T_c and on dT_c/dP.[1] It should be also noted that $H_2C_4O_4$ and KDP differ in other respects which are relevant for the present argument. Specifically (i) $H_2C_4O_4$ has a considerably longer $O-H\cdots O$ bond length R (2.56 vs. 2.48A) and (ii) $H_2C_4O_4$ has a much higher transition temperature T_c (375 vs. 122K). The longer R leads to deeper potential minima along the $O-H\cdots$bond, which in turn lead to higher T_c. The rather high T_c of $H_2C_4O_4$ (compared with other H-bonded crystals) suggests that tunneling should be less (and thermal excitations should be more) important in this crystal than in KDP.

5.   QUANTUM EFFECTS

Although the study of displacive structural phase transitions has been an active subject for many years, in almost all cases studied the transition temperatures (T_c) have been sufficiently high that there are no experimentally discernible quantum-mechanical effects. Consequently, most of the theoretical treatments of such transitions have been based on classical mechanics. As T_c decreases and approaches 0 K, however, quantum effects can be expected to come into play and influence the experimental response. A particularly important case in this regard is the so-called quantum-mechanical displacive limit where quantum effects become extremely important. This limit is defined as the special case where the transition occurs at $T_c = 0$ K. There has been considerable recent effort devoted to the study of the displacive limit especially with regards to the critical behavior.[15-18] It is found that quantum fluctuations cause the critical exponents at the displacive limit to be quite different from those at high T_c.

In the classical regime the soft mode is stabilized by thermal fluctuations.[1,2] These fluctuations decrease with decreasing T and ultimately this stabilization vanishes at some T_c. However, when T enters into the regime of quantum lattice motions, decreasing T does not appreciably decrease the fluctuations. Quantum fluctuations cause T_c to decrease below its classical value, and in some cases they can completely suppress the transition.[15-18] One consequence of this supression in ferroelectric crystals is the formation of a quantum paraelectric state.[19] This state is characterized by a high static susceptibility (or low soft mode frequency) that is constant at low temperature over an extended T range.

One of the important results of the recent quantum theoretical treatments[15-18] is that ε should depend on T as

$$\varepsilon - \varepsilon_\infty = B \ (T - T_c)^{-\gamma} \ , \qquad (2)$$

where ε is the high T limiting value of ε and B is constant. In the classical limit Eq. (2) reduces to the Curie-Weiss law $\gamma = 1$; however, at the quantum displacive limit ($T_c = 0$) the recent theories lead to the specific prediction $\gamma = 2$ for a T regime near 0 K. The quantum displacive limit is thus a special critical point whose occurrence implies that with decreasing T there will be a crossover regime from $\gamma = 1$ to $\gamma = 2$.

Another prediction of the recent theories is that close to the displacive limit T_c is given by

$$T_c \ (x) \ \propto \ (x-x_c)^{1/\phi} \qquad (3)$$

with the critical exponent, $\phi = 2$. In the classical regime $\phi = 1$. In Eq. (3) x is a general interaction parameter, and x_c is its value at $T_c = 0$. For our purposes, and under certain assumptions, x can be replaced by pressure, P.[20]

One method of experimentally studying the onset of quantum effects and the behavior at the displace limit is to lower T_c by chemical substitution, and recent such work[18,19] has led to interesting results and confirmation of some of the predictions of the recent theories. However, changing the chemistry introduces randomness on a local scale which changes some of the interaction parameters of the system and causes each mixed crystal sample to have a distribution of T_c's. Thus the behavior is not as "clean" as desirable.

We believe that a much "cleaner" way, perhaps the cleanest way, of studying the onset of quantum effects and the behavior at the displacive limit, and thereby testing the theoretical predictions is by the application of hydrostatic pressure. Indeed, there are many phase transitions whose T_c's decrease with pressure, and we have shown that in several crystals T_c decreases with pressure, and we have shown that in several crystals T_c decreases to 0 K at readily accessible pressures.[1-3] These crystals include KH_2PO_4 and $NH_4H_2PO_4$, RbH_2PO_4, $SbSI$, and $K_{.72}Na_{.28}TaO_3$, and K_2SeO_4. Results on these crystals afford an excellent means for studying the onset of quantum effects and the behavior at the displacive limit.[20] Here we shall present some of the results on K_2SeO_4 and $K_{.72}Na_{.28}TaO_3$ to illustrate this point.

Figure 5 shows logarithmic $(\varepsilon - \varepsilon_\infty)^{-1}$ vs $(T - T_c)$ plots for K_2SeO_4 at various pressures. For this crystal the region of interest for this purpose is that bounded by the two transitions (see Figs 1 and 2) and is thus limited in temperature extent. It is seen that at atmospheric pressure (T_c = 93K) the response is classical with γ = 1.0. For the 0.72 and 0.75 GPa isobars, on the other hand, where T_c is very close to the quantum displacive limit T_c = 0 K, there is a distinct γ = 2.0 regime, as is predicted by the theory. Although, the γ = 2.0 regime is predicted theoretically strictly for the quantum displacive limit, it is also expected and has been observed[18] for cases where T_c is near 0 K. For the 0.58 GPa isobar where T_c = 33K, the response is clearly in the crossover regime, and here we find γ = 1.2, which is compatible with the value expected for the mode-coupling model for this regime.[19] The flattening of $(\varepsilon - \varepsilon_\infty)^{-1}$ at very low temperatures is commonly observed in ferroelectrics and has to do with the very strong quantum mechanical stabilization of the paraelectric phase. Unfortunately, this flattening reduces the accessible range of critical behavior. Ultimately, the quantum stabilization is complete (i.e. no transition), resulting in a quantum paraelectric state as appears to be observed at 0.75 GPa.

Figure 6 shows the pressure dependence of T_c of a crystal of $K_{.72}Na_{.28}TaO_3$. The transition vanishes at pressures > 0.66 GPa, and the quantum displacive limit is then T_c = 0 K at $P_c \cong$ 0.66 GPa. Also shown is $T_c^2(P)$ which is seen to be linear for T_c's up to $\sim$ 30K. Both the $T_c(P)$ and $T_c^2(P)$ responses in Fig. 6 are in agreement with the theoretical prediction (Eq. 3). Logarithmic $(\varepsilon - \varepsilon_\infty)^{-1}$ vs. $(T - T_c)$ isobars for this crystal extend over a much broader temperature range than was possible for K_2SeO_4, and each isobar

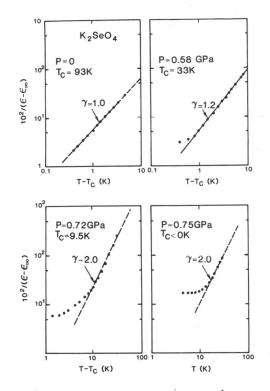

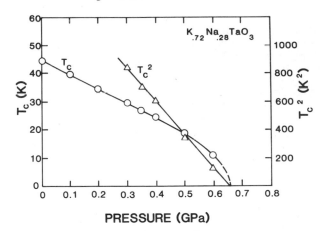

Fig. 5 Logarithmic plots of $(\varepsilon - \varepsilon_\infty)^{-1}$ vs $(T - T_c)$ for K_2SeO_4 at various pressures showing the transition from the classical γ = 1.0 regime to the quantum γ = 2.0 regime for $T_c \approx$ 0 K.

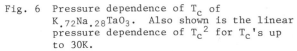

Fig. 6 Pressure dependence of T_c of $K_{.72}Na_{.28}TaO_3$. Also shown is the linear pressure dependence of T_c^2 for T_c's up to 30K.

shows different γ regimes.[20] At sufficiently high temperatures all isobars exhibit classical Curie-Weiss (γ = 1) behavior, but at low temperatures deviations are noted. At pressures very close to the quantum displacive limit ($P_c \approx 0.66$GPa) the predicted $\gamma = 2.0$ regime is observed. On either side of this limit, a crossover regime in γ is observed.

Acknowledgment: It is a pleasure to acknowledge the expert technical assistance of B. E. Hammons. Some of the work on K_2SeO_4 was in collaboration with N. E. Massa and F. G. Ullman and some of the work on $DyVO_4$ and $DyAsO_4$ was in collaboration with D. R. Taylor and S. H. Smith. More detailed accounts of the work on these materials will be published elsewhere. The UO_2 crystals were kindly provided by H. J. Anderson.

References

1. Samara, G. A. and Peercy, P. S. in Advances in Solid State Physics, Edited by D. Turnbull, et al (Academic Press, New York), Vol. 36 (in press).

2. Samara, G. A. in High-Pressure and Low-Temperature Physics, Edited by C. W. Chu and J. A. Woollam) Plenum Publishing Corp. 1978) p. 255.

3. Samara, G. A., Massa, N. E. and Ullman, F.G., Ferroelectrics (to be published).

4. Iizumi, M., Axe, J. D., Shirane, G. and Shimaoka, K., Phys. Rev. B15, 4392 (1977)

5. Haque, M. S. and Hardy, J. R., Phys. Rev. B21, 245 (1980).

6. Aleksandrova, I. P., Ferroelectrics 24, 135 (1980).

7. Samara, G. A. and Richards, P. M., Phys. Rev. B14, 5073 (1976).

8. Elliott, R. J., Harley, R. T., Hayes, W., and Smith, S. R. Rp., Proc. R. Soc. Lond. A. 328, 217 (1972) and references therein.

9. Gehring, G. A., and Gehring, K. A., Rep. Prog. Phys. 38, 1 (1975).

10. Allen, S. J., Phys. Rev. 166, 530 (1968); 167, 492 (1968).

11. Faber, J. Jr., and Lander, G. H., Phys. Rev. B14, 1151 (1976).

12. Yasuda, N., Fujimoto, S., Okamoto, M., Shimizu, H., Yoshino, K. and Inuishi, Y., Phys. Rev. B20, 2755 (1979); Gesi, K., and Ozawa, K., Jap. J. Appl. Phys. 17, 435(1978).

13. Samara, G. A. and Semmingsen, D. J. Chem. Phys. 71, 1401 (1979).

14. Blinc, R. and SaBarreto, F. C., J. Chem. Phys. 72, 6031 (1980).

15. Operman, R. and Thomas, H., Z. Phys. B22, 387 (1975).

16. Schneider, T., Beck, H., and Stoll, E., Phys. Rev. B13, 1123 (1976).

17. Morf, R., Schneider, T. and Stoll, E., Phys. Rev. B16, 462 (1977).

18. Hochli, U. T. and Boatner, L. A., Phys. Rev. B20, 266 (1979) and references therein.

19. Müller, K. A. and Burkard, H., Phys. Rev. B19, 3593 (1979).

20. Samara, G. A., Proc. of the 8th AIRAPT Conf., Uppsala, 1981 (to be published).

PHYSICS OF SOLIDS UNDER HIGH PRESSURE
J.S. Schilling, R.N. Shelton (editors)
© *North-Holland Publishing Company, 1981*

NEW WAYS OF LOOKING FOR PHASE TRANSITIONS AT MULTI-MEGABAR DYNAMIC PRESSURES*

J. W. Shaner

Los Alamos National Laboratory
Los Alamos, New Mexico 87545

Shock wave techniques have provided much of our information about very high pressure phase transitions. However, because of the specific thermodynamic region accessible to simple shock wave experiments, phase transitions with very small density changes are difficult to detect. Those phase changes with a density decrease at constant pressure, such as normal melting, are virtually impossible to detect in a shock experiment.

Three new techniques have been applied to thermodynamic measurements behind a shock wave, with the result that several previously undetected phase boundaries have been measured. First, optical pyrometry has been used to measure the temperature of shocked SiO_2. We find a temperature anomaly indicating a phase change, probably melting, between 0.6 and 1.1 Mbar. Secondly, magnetic probes have been used to measure the velocity of sound in shock compressed metals. The loss of a longitudinal elastic component of the sound wave is indicative of melting. Finally, a novel optical analyzer technique, exploiting the sensitivity of thermal radiation to small temperature changes, has also been used to measure the sound velocity in shock compressed metals. With the extra sensitivity provided by this derivative measurement of an equation of state surface, we have been able to detect both solid-solid and melting transitions of iron above 2 Mbar.

1. INTRODUCTION

The use of shock waves in the investigation of material properties at very high pressures has already had a long history. Subject to the validity of a few simplifying assumptions, a shock wave produces a reasonably well defined thermodynamic state of both elevated pressure and elevated temperature, in samples which may be several cubic centimeters or larger in size. The temperature rise is an important aspect of the irreversible compression in a shock wave. It may be as much as several thousand degrees for a 100 GPa shock. Therefore we explore a very different region of the equation of state (EOS) surface with high pressure shocks than is accessible with static techniques. In fact, if one is looking for phase changes, the last first order transition we expect as temperature and pressure rise is melting. Much of this paper will be devoted to measurements of high pressure melting phenomena.

We show in Fig. 1 the magnitude of the uncertainty in the high pressure behavior of the melting line for a simple material, lead. The thermodynamically correct prescription for the melting line is the locus of points where the solid and liquid Gibbs free energies are equal [1]. However, evaluation of the free

energies with sufficient accuracy for a reliable phase line is notoriously difficult. Various empirical models, like the Kraut-Kennedy [2] or Simon [3] equations, or semi-empirical models, like the Lindemann Law [4], have also been used. We stress that all of these formulations have been constrained to match static data in the low pressure region indicated in Fig. 1.

In the next sections we describe some of the assumptions and problems in doing shock wave experiments and several new techniques which get around many of the difficulties.

2. BACKGROUND

With the assumption of a steady propagating shock wave, the conservation of mass, momentum, and energy lead respectively to the three Hugoniot equations relating the state of the shock compressed material to that of the uncompressed material:

$$V = V_0 \ (1 - U_p/U_s) \qquad (1)$$

$$\sigma_\ell - \sigma_{\ell 0} = \rho_0 U_s U_p \qquad (2)$$

$$E - E_0 = \frac{1}{2} \ (\sigma_\ell + \sigma_{\ell 0}) \ (V_0 - V) \qquad (3)$$

* This work was performed under the auspices of the United States Department of Energy.

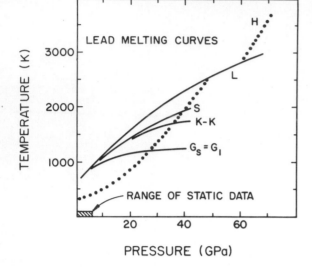

Fig. 1. High pressure melting lines for lead. Models are equal free energy – $G_s=G_\ell$; Simon–S; Kraut Kennedy–KK; and Lindmann. The curve labeled H is a calculated Hugoniot.

In equations 1 through 3, U_s is the shock velocity, U_p is the material velocity behind the shock, V is the specific volume, σ_ℓ is the stress component along the shock propagation direction, and E is the internal energy. The measurement of U_s and U_p along with the initial conditions of the material are sufficient to describe the thermodynamic state behind the shock front. The locus of such states starting from a common initial condition is referred to as the Hugoniot curve, and it lies on, or close to the equilibrium equation of state surface.

The assumption that a strong shock propagates as a steady wave is generally consistent with measurement. Weak shock waves, where the longitudinal elastic precursors are not overdriven, generally are not steady, although portions of even these wave fronts may achieve a steady state [5]. The phenomena described in this paper all relate to strong shock conditions, where the steady wave assumption is good.

Another assumption is that thermodynamic equilibrium is established behind the shock front. Although the time intervals are very short, experimental evidence indicates that equilibrium is very quickly established, with perhaps two notable exceptions. Some phase transitions, such as that from graphite to diamond, are sluggish. This transition is seen in shock waves, but at substantially higher pressures and temperatures than is required to

convert graphite to diamond statically. Also, in samples recovered after shock loading, we find evidence for a non-equilibrium concentration of defects. Most shock recovery experiments involve complex unloading and reloading processes, so any deduction concerning the state of materials behind the first shock should be taken with caution.

The stress in equation (2) refers to a longitudinal component and not the pressure. The fact that non-zero stress deviators may exist behind a shock wave in a solid was recognized from the very beginning of such experiments [6]. The problems of non-hydrostatic stress can be understood with the help of Fig. 2. For a planar shock wave we are concerned only with one-dimensional strains and principle stress components along and transverse to the shock propagation direction – σ_ℓ and σ_t, respectively. Writing

$$\sigma_\ell - \sigma_t = 2\tau \tag{4}$$

where τ is the maximum resolved shear stress, and

$$P = \frac{1}{3}(\sigma_\ell + 2\,\sigma_t), \tag{5}$$

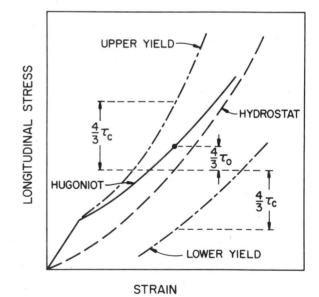

Fig. 2. Compression processes for materials with shear strength.

we have that

$$\sigma_\ell = P + \frac{4}{3}\tau, \qquad (6)$$

where P is the pressure. The hydrostat represents the compression curve for hydrostatic stress, or pressure. The upper and lower yield surfaces deviate from the hydrostat by 4/3 τ_c, where τ_c is the shear strength for the solid. This parameter increases with compression and work hardening, but it decreases with temperature and disappears upon partial melting [7]. The Hugoniot curve deviates from the hydrostat by 4/3 τ_o where τ_o is the magnitude of the shear stress behind the shock. This shear stress increases in an initial elastic region to τ_c at a stress called the Hugoniot elastic limit (HEL). Above the HEL τ_o equals τ_c in the ideal elastic-plastic model. The magnitude of τ_o is a significant source of uncertainty in comparing static and shock data [8].

Although we have accumulated data over the past two decades which indicate the expected increase in shear strength with compression, at least until high temperature effects dominate [9, 10], we are only now beginning to get quantitative data concerning the magnitude of τ_o. Recent experiments by Asay have shown that in the case of aluminum shocked to over 21 GPa the state of the material appears to be within 0.2 GPa of hydrostatic conditions [11]. We can interpret the low value of the shear stress in a shocked metal in terms of the adiabatic shear model developed by Grady [12]. In this model, the shearing occurs locally in bands, where the internal energy increases unstably. As a result, the material may have little macroscopic shear strength. Although the available experimental evidence supports these ideas, the data necessary for a complete quantitative model of the pressure dependence of stress deviators are still incomplete. With more complete data we should be able to make consistent solid strength corrections to all of our shock wave data.

One further limitation of shock waves as a means of studying phase changes at high pressure is a result of the particular track the Hugoniot curve makes in thermodynamic space. To show this we sketch in Fig. 3 a compression curve with a phase change to higher density, but without the heating associated with the shock. Using equations (1)–(3) one may show that the shock velocity is given by

$$U_s = V_o\left[\frac{P-P_o}{V_o-V}\right]^{\frac{1}{2}} = V_o j^{\frac{1}{2}} \qquad (7)$$

where j is minus the slope of the line connecting initial and final states, the Rayleigh line. The normal situation is one in which the compressibility decreases with compression, so U_s increases monotonically with pressure, or U_p. For the phase change illustrated in Fig. 3a, there is a pressure range between points b and c for which a single shock wave is unstable. A first wave brings material to point b, while a second, slower wave, further compresses material to the final pressure. The map of this process in the U_s–U_p plane is shown in Fig. 3b. The offset in U_p is simply given by

$$\frac{\Delta U_p}{U_p} = \frac{V_b-V_c}{V_o - \frac{V_c+V_b}{2}} \qquad (8)$$

Since offsets of 2% can be measured in shock wave data, phase changes involving volume decreases of 1–2% can be detected.

For the case of melting, however, the volume normally increases. The effects of increasing entropy and increasing pressure compete such that the phase change is difficult, if not impossible to detect in pressure-volume or pressure-energy representations. Urlin and Ivanov have shown that the change in slope of the U_s–U_p curve in passing through the two-phase region can be of either sign depending on the slope of the phase line [13]. In either case the changes are subtle for normal melting. As an indication of the subtlty, we show in Fig. 4 the Hugoniot curve for lead in the region where melting should be taking place. The experiments give no indication of a melting transition predicted by any of the curves in Fig. 1.

The new techniques described in the next sections overcome some of the shortcomings of shock waves as a means of studying materials at high pressures. In particular, we describe three new techniques which allow one to detect new phase transitions and to measure new details of the thermal and mechanical behavior of shock compressed materials.

3. OPTICAL PYROMETRY

From equation 3 we see that the thermal parameter measured in a shock wave experiment is internal energy. In his early work, Kormer

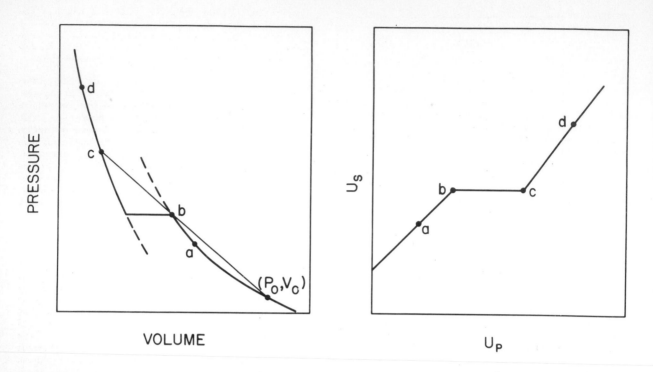

Fig. 3.   a) Compression curve for a material with a density increasing phase change.   b)Map   of   the compression curve of a into the U_s-U_p plane.

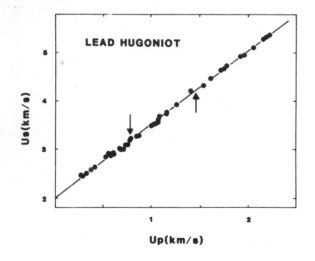

Fig. 4.   Shockwave   data   for   lead.   The region where we expect melting to occur is between the arrows.

showed   that   one   might   be   able   to   measure   the temperature   as   well   in   a   shock   compressed transparent   material   by   high   speed   optical pyrometry [14].   In addition   to   providing   heat capacity data, certain phase transitions showed up   much   more   clearly   in   the

pressure-temperature representation than in the pressure-energy plane.

This work formed the conceptual basis   for   two sets   of   pyrometric   measurements   on   shocked SiO_2, one set by McQueen at Los Alamos [15] and another   by   Lyzenga   at   Cal   Tech   [16].   When a strong   shock   propagates   into   a   transparent sample,   the   thermal   radiation   from   the   hot compressed   material   can   pass   through   the uncompressed   material   and   into   a   fast   optical pyrometer.   In   these   experiments,   McQueen   used a   rotating   mirror   streaking   camera   with   a detonation   wave   in   nitromethane   as   a calibration   standard.   Lyzenga   used   silicon photodiodes   and   color   filters   calibrated against   a   tungsten   ribbon   lamp.   The   data   from both   sets   of   measurements   are   shown   in   Fig. 5. In   order   to   make   the   comparison,   average emissivities   from   Lyzenga's   measurements   were used.   The   agreement,   within   5%   in   absolute temperature,   is   excellent   for   such   measurements above   5000   K   and   with   time   resolution   of   less than 10 ns.

Several   surprises   appeared   in   this   work. Firstly,   even   at   shock   temperatures   below 5000 K,   the   rise   time   of   the   optical   signal   was less   than   10   ns.   Therefore   we   may   say   that   a layer   of   shocked   material   less   than   100 µm thick   is   opaque.   We   expect   that   a   wide   gap insulator   like   stishovite   should   still   be

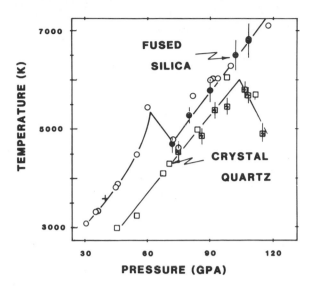

Fig. 5. Analyzed pyrometric data for shock compressed SiO_2. Solid symbols are from Lyzenga [16]. Open symbols are from McQueen [15]. The cross is from Kondo, et. al. [17].

optically thin under these conditions, unless there are many defect electronic states in the gap. If that is true, we worry about whether the radiation is grey body and whether the apparent temperatures represent equilibrium. Measurements by Lyzenga [18] indicate a good fit to a grey body spectrum, so at least the radiation appears to be thermal. For normal solid densities coulomb scattering should equilibrate the electron and ion temperatures in picoseconds or less, so we might have expected the radiation to be thermal.

The next surprise was the region for both fused silica and crystal quartz where increasing shock pressure (and internal energy) decreases the apparent temperature. This indication of a phase change cannot be detected in conventional shock measurements on SiO_2. The anomalous behavior can be explained in two ways. The Hugoniot may follow an equilibrium phase line with negative dT/dP. The shock temperatures are such that we suspect the new phase change to be melting, in which case the silica melt is denser than the solid. On the other hand, the discrepancy in location of the phase boundary for the fused silica and the quartz suggests complex high temperature phase behavior or non-equilibrium effects. The latter explanation has been proposed by Lyzenga [16]. Both transitions are then supposed to be melting from a superheated solid. Although one rarely can superheat a solid, and certainly not by 1000 K under normal conditions, in these experiments we are looking less than 100 μm

into the shocked material. At this depth, the material is only a few nanoseconds behind the onset of the shock. These experiments may in fact be the first observation of a limit on the time scale necessary for a melting transition.

In order to resolve the uncertainty in the interpretation of the SiO_2 results we must pursue both more carefully controlled shock waves with faster pyrometry, and similar experiments around better known phase transitions. Examples of the latter would be the melting transitions in the alkali halides. On the other hand, optical pyrometry has already proven to be a viable technique, as well as a means to observe high pressure phase behavior which is not accessible by any other means.

4. THE OPTICAL ANALYZER

We have seen that the radiation from shocked transparent materials may be thermal and that the optical depth may be very small. These ideas lead to a scheme for measuring release wave velocities in shock compressed media. The information contained in this data includes longitudinal and bulk sound velocities. The optical analyzer scheme, developed by McQueen [19], is based on the use of a short shock illustrated in Fig. 6. If a thin flyer plate impacts a target, shocks propagate forward in the target and backward in the flyer. The interaction of the shock with the flyer free surface results in a release wave coming forward again with the velocity of sound in the compressed medium. Typically this rarefaction travels faster than the shock, which in turn is supersonic with respect to the uncompressed medium. When the rarefaction catches the shock

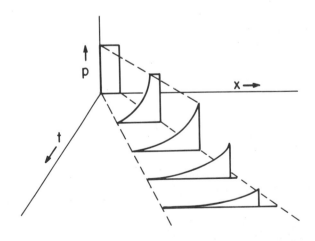

Fig. 6. Propagation of a shock of short duration. The shock front moves with constant velocity until the release catches up.

front, the peak pressure starts to diminish. At this point two measurable physical phenomena occur. First the shock velocity decreases with the peak pressure. In the limit of zero pressure increase, the disturbance travels with the longitudinal elastic velocity in the uncompressed material. Secondly, the temperature drops with the peak pressure.

Since the thermal radiation varies with a high power of the temperature ($> T^4$ on the blue side of the black body peak) and temperature varies between the first and second power of the stress, small changes in stress make large changes in light intensity. Also, if the source is optically thick, we will see the decrease in thermal radiation within a few nanoseconds of the catch-up time.

For use with metal targets, the technique consists of propagating a short shock through varying thicknesses of target backed with a transparent radiator. For a thin target, the short shock travels a longer time in the radiator with a constant peak pressure and radiation intensity before catch-up. A thicker target will result in a shorter time of constant radiation intensity. Actual thermal radiation signals recorded by photomultipliers

are shown in Fig. 7 for two different target thicknesses and three analyzer materials.

If we plot the time interval of constant thermal radiation against target thickness, we get a straight line. We have shown this to be rigorously true from similarity arguments [19]. By extrapolating down to zero time we get the target thickness required for the release to catch up at the target analyzer interface. By this means we avoid all corrections for wave interaction at the interface. The data from Fig. 7 is plotted in Fig. 8 to show that the measurements do not depend on the analyzer material. From this target thickness and the known flyer thickness we can calculate the release wave velocity.

The kind of release wave structure we are looking for has been described previously by Asay at pressures below 10 GPA. In both porous aluminum [20] and bismuth [21], he found that the leading edge of the release wave moving into solid material propagated with the longitudinal sound velocity, $c_\ell^2 = (B + 4\mu/3)/\rho$, where B is the bulk modulus and μ is the shear modulus. In each case when the shock was intense enough to give as little as 20% melt fraction, the leading edge of the release wave had slowed to the calculated bulk sound velocity, $c_B^2 = B/\rho$ Both of these

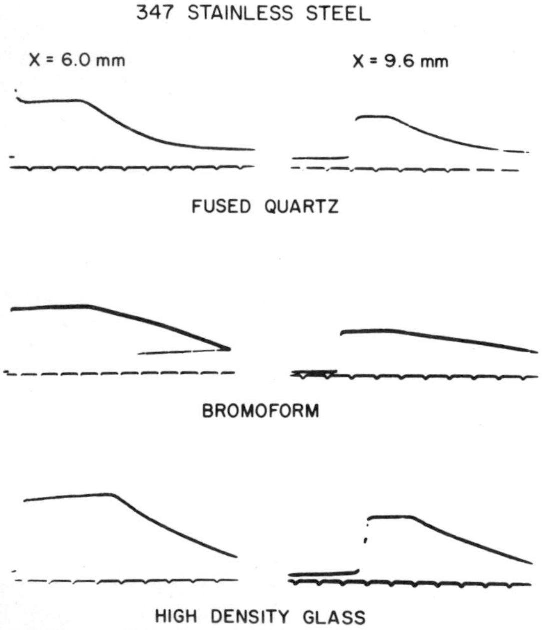

347 STAINLESS STEEL

X = 6.0 mm   X = 9.6 mm

FUSED QUARTZ

BROMOFORM

HIGH DENSITY GLASS

Fig. 7. Thermal radiation signals, recorded by photomultipliers, from three different optical analyzers and two target thicknesses of stainless steel. The time marks are at 100 ns intervals.

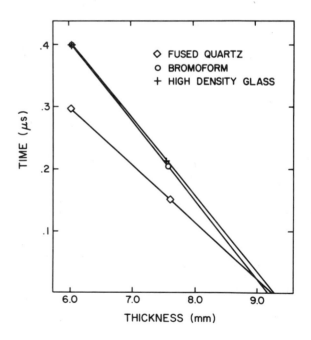

Fig. 8. Duration of constant thermal radiation intensity as a function of target thickness for the experiments shown in Fig. 7.

experiments were done at sufficiently low pressures that there was little uncertainty about the location of the melting transition. Since the optical analyzer has none of the high pressure limitations of the interface velocity interferometer used by Asay, we could extend this kind of measurement to pressures in excess of 100 GPa.

The first application of the optical analyzer has been to the melting region for iron by Brown and McQueen [22]. They measured the velocity of the beginning of the release wave in iron shocked to pressures as high as 260 GPa. Surprisingly, in experiments using thin iron flyer plates driven by both explosives and our two-stage light gas gun, they obtained consistent results showing two breaks in the release velocity as a function of shock pressure. These data are shown in Fig. 9. Since we expect no sharp changes in the thermophysical properties of fluid iron, we have interpreted the upper break as the melting of iron and the lower break as a solid-solid phase change. This interpretation is made more plausible by Fig. 10, where the iron Hugoniot is plotted in the known and extrapolated phase diagram for iron. The Hugoniot may cross the

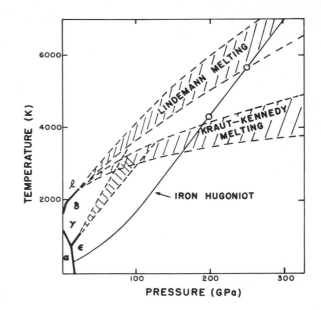

Fig. 10.
Phase diagram for iron. The known region is indicated by solid lines in the lower left corner. We indicate the ranges for both Lindemann and Kraut-Kennedy melting models.

ε (hcp) to γ (fcc) phase boundary before melting occurs. The upper break would then represent the γ to liquid solidus. We also show in Fig. 10, if our interpretation is correct, that melting of iron even at 260 GPa and 5500 K agrees reasonably well with a best calculation using the Lindemann melting criterion. Use of the Kraut-Kennedy criterion gives a melting curve at much lower temperatures.

These experiments show the optical analyzer technique to be a very effective means to detect high pressure phase transitions. Neither of those found in iron can be detected in the conventional shock wave measurements. The particular sensitivity of the thermal radiation makes small changes in driving stress easy to detect, and the extrapolation procedure eliminates the need for corrections to account for stress wave interactions with the interfaces. However, the extreme non-linearity of the thermal radiation vs stress makes it very difficult to extract anything but the velocity of the head of the rarefaction wave. More quantitative results for the high pressure mechanical properties are described in the next section.

5. AXIALLY SYMMETRIC MAGNETIC PROBE

The axially symmetric magnetic (ASM) probe is another new technique we are exploiting in

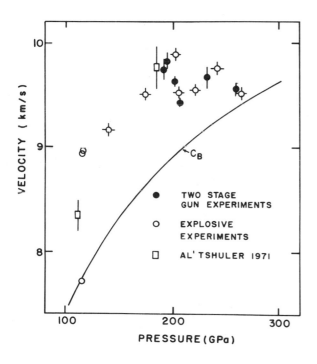

Fig. 9. Velocity of the head of the release wave as a function of initial shock pressure for iron. The highest pressure data points are on the calculated bulk wave velocity.

order to overcome the pressure limitations on velocity interferometry. This technique was originally developed by Fritz, et al. [23, 24]. The ASM measurement consists of a static magnetic field set up by a cylindrical magnet surrounded by a coaxial loop antenna. When a metal sample is shocked and starts to move, the magnetic field distribution is deformed in such a way that voltage is induced in the pick-up loop. Thus, we measure U_p of the metal behind the shock wave.

A typical assembly for an explosive driven experiment is shown in Fig. 11. As in the interferometer measurements a non-conducting window is placed over the sample free surface to prevent uncontrolled spray coming off and confusing the measurement. The window, if matched in impedance to the metal, also minimizes confusion due to wave interaction at the interface. However, for the ASM probe there is no requirement that the window stay transparent or have a calibrated refractive index as a function of stress. The only requirement is that the resistivity stay above 10^4-10^5 $\mu\Omega$-cm, so no flux will be trapped during the several hundred nanoseconds of the experiment. On the other hand, calibration of the system is more difficult than the interferometer, particularly if corrections for flux diffusion in the metal or trapping in the window are necessary.

We show in Fig. 12 a typical wave profile for a 2024 aluminum target backed by a teflon window, shocked to 55 GPa [25]. The time t_3 denotes the arrival of the shock at the aluminum-teflon interface. Similarly, t_8 and t_{10} denote the arrival times of the longitudinal elastic and bulk parts of the release wave at this same interface. The time t_d denotes demagnetization of the magnet when the shock wave impacts it. From the U_p measured before t_8 and the U_s

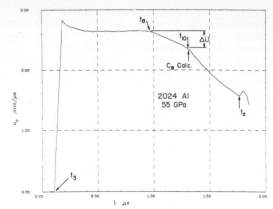

CEM 22 B1   6.368 2024 AL/9.009 2024 AL//10.980 TEFLON//25.497

Fig. 12.
A typical analyzed material velocity record for aluminum shocked to 55 GPa, backed by a teflon window.

measured by the time interval t_3-t_d we can obtain the Hugoniot state in the teflon, and by standard shock impedence matching techniques we can tell the Hugoniot state of the aluminum. We can also measure the Hugoniot state in the aluminum independently, so the initial shock state is well characterized.

We show in Fig. 13 some preliminary ASM probe data on the change of longitudinal elastic wave velocity in aluminum as a function of Hugoniot stress [25]. The other curves shown are the

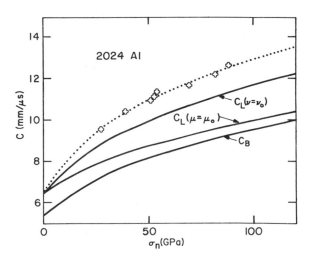

Fig. 13. Longitudinal elastic wave velocity in aluminum as a function of pressure. Other curves are the bulk wave velocity (C_B), and the longitudinal elastic wave velocity calculated using constant shear modulus and constant Poisson ratio.

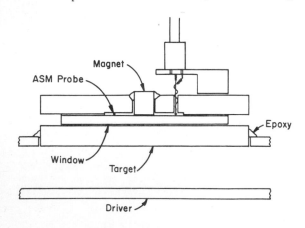
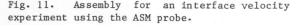

Fig. 11. Assembly for an interface velocity experiment using the ASM probe.

calculated bulk wave velocity and longitudinal velocities derived from the bulk assuming a constant shear modulus ($\mu=\mu_0$) or a constant Poisson ratio (μ/B=const.). Evidently the data show that the shear modulus is increasing faster with compression than the bulk modulus for aluminum. This point is consistent with ultrasonic data at low compressions [26], as well as shock wave data as analyzed by Romain, et. al. [27], using the formulation of Vaschenko and Zubarev relating the volume dependences of Grüneisen's parameter and Poisson's ratio [28]. Similar results for copper show a slight decrease of μ/B with compression.

From the change in U_p between t_3 and t_{10}, ΔU_p, we can calculate the change in shear stress accompanying this quasi-elastic part of the release wave. This calculation is virtually the same as that reported by Asay for use with velocity interferometers [11]. If we know the shear stress in the Hugoniot state, then the shear stress change upon release is a measure of the critical flow stress, or yield strength, of the compressed material. Since we now believe that the Hugoniot stress includes a shear component which may be dependent on compression and temperature, the change in stress associated with the quasi-elastic release is $4/3 (\tau_0 + \tau_c)$. In Fig. 14 we show preliminary data on this parameter as a function of shock pressure for copper (OFHC) and aluminum (2024). In each case, $(\tau_0 + \tau_c)$ is decreasing at high pressure. However, since τ_0 and τ_c may have different dependences on Hugoniot pressure, it is risky at this point to extrapolate to zero shear stress, where melting has occurred. On the other hand further measurements of the profiles of both reloading and release waves should allow us to unravel

these complex phenomena and have an independent measurement of melting by a strong shock.

6. SUMMARY

We have shown three new techniques developed to increase our knowledge about the high pressure-high temperature behavior of materials behind strong shock waves. These have proven to give new information about the thermal and mechanical behavior of compressed materials. The pyrometry and elastic wave measurements also provide very sensitive indicators of both solid-solid and melting phase changes which cannot be detected by conventional shock techniques. With these new means, we should be able to answer many more important questions about the properties of materials at pressures greater than a megabar.

7. ACKNOWLEDGMENTS

The author wishes to thank R. G. McQueen, J. N. Fritz, J. M. Brown, and C. E. Morris for many conversations and for sharing as yet unpublished ideas and data.

8. REFERENCES

1. "Shock Wave Techniques for the Examination of Phase Transitions," W. J. Carter in *Phase Transformations - 1973*, L. E. Cross ed. (Pergamon Press, NY 1973).
2. E. A. Kraut and G. C. Kennedy, Phys. Rev. 151, 668 (1966).
3. S. E. Babb, Rev. Med. Phys. 35, 400 (1963).
4. For a modern derivation, see M. Ross, Phys. Rev. 189, 233 (1969).
5. J. N. Johnson and L. M. Barker, J. Appl. Phys. 40, 4321 (1969).
6. D. C. Pack, W. M. Evans, and H. J. James, Proc. Phys. Soc. 60, 1 (1948).
7. D. J. Steinbert, S. G. Cochran and M. W. Guinan, J. Appl. Phys. 51, 1498 (1980).
8. H. K. Mao, P. M. Bell, J. W. Shaner, and D. J. Steinberg, J. Appl. Phys. 49, 3276 (1978).
9. C. E. Morris and J. N. Fritz, J. Appl. Phys. 51, 1244 (1980).
10. L. V. Al'tshuler, M. I. Brazhink and C. S. Telegin, J. Appl. Mech. Tech. Phys. 12, 921 (1971).
11. J. R. Asay and L. C. Chhabildas in "Shock Waves and High-Strain-Rate Phenomena in Metals," M. A. Meyers and L. E. Murr, eds., Plenum, NY 1981.
12. D. E. Grady, J. Geophys. Res. 85, 914 (1980).
13. V. D. Urlin and A. A. Iranov, Sov. Phys.-Doklady, 8, 380 (1963).
14. S. B. Kormer, M. V. Sinitsyn, C. A. Kirillov, and V. D. Urlin, Soviet Physics JETP 21, 689 (1965).

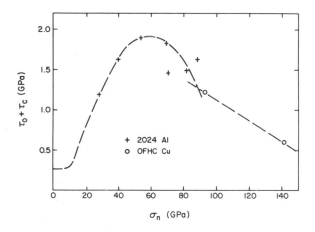

Fig. 14. Change in shear stress in the quasi-elastic part on a release wave following a strong shock.

15. R. G. McQueen,   J. N. Fritz,   and
    J. W. Hopson, EOS <u>60</u>, 952 (1979).

16. G. A. Lyzenga and T. J. Ahrens, Geophys.
    Res. Lett. <u>7</u>, 141 (1980).

17. H. Sugiura, K. Kondo,   and   A. Sawaoka,
    Rev. Sci. Inst. <u>51</u> 750 (1980).

18. G. A. Lyzenga and T. J. Ahrens, Rev. Sci.
    Inst. <u>50</u>, 1821 (1979).

19. R. G. McQueen,   J. W. Hopson,   and
    J. N. Fritz   to   be   published   in
    Rev. Sci. Inst.

20. J. R. Asay,   J. Appl. Phys. <u>46</u>,   4797
    (1975).

21. J. R. Asay,   J. Appl. Phys. <u>48</u>,   2832
    (1977).

22. J. M. Brown and R. G. McQueen, Geophys.
    Res. Lett. <u>7</u>, 533 (1980).

23. J. N. Fritz,   R. S. Caird,   and
    R. G. McQueen, "The Use of a Magnetic
    Field   to   Measure   Particle   Velocity,"
    LA-4545 MS (1970).

24. J. N. Fritz   and   J. A. Morgan,
    Rev. Sci. Inst. <u>44</u>, 245 (1973).

25. C. E. Morris,   J. N. Fritz   and
    B. L. Holian, to be published.

26. "Single Crystal Elastic Constants   and
    Calculated   Aggregate   Properties: A
    Handbook," G. Simmons and H. Wang eds.,
    MIT Press, Cambridge, MA (1971).

27. J. P. Romain,   A. Migault   and
    J. Jacquesson, J. Phys. Chem. Sol <u>37</u>, 1159
    (1976).

28. V. Vaschenko   and   V. N. Zubarev,
    Sov. Phys. Solid State <u>5</u>, 653 (1963).

PHYSICS OF SOLIDS UNDER HIGH PRESSURE
J.S. Schilling, R.N. Shelton (editors)
© *North-Holland Publishing Company, 1981*

STUDIES ON PHASE TRANSITIONS IN LIQUID CRYSTALS UNDER HIGH PRESSURE

R.Shashidhar*

Department of Chemistry, Institute of Physical Chemistry,
Ruhr-University of Bochum, 4630 Bochum 1, FRG

This paper describes some recent results obtained using the diamond anvil cell on the phase transitions occurring in liquid crystals under the influence of pressure. They are: (i) Identification of the pressure induced mesophases by observing the optical 'textures', (ii) pressure dependence of the smectic A layer spacing of 8OCB in relation to its re-entrant behaviour, (iii) variation of the maximum pressure of occurrence of the re-entrant nematic phase with concentration for 6OCB/8OCB mixtures and (iv) the search for a 'Lifshitz point' in a single component liquid crystalline system.

1. INTRODUCTION

Thousands of organic compounds when heated do not show a single transition from solid to liquid, but instead, they melt into the 'liquid crystalline phase' which on further heating transforms into the isotropic liquid phase. Liquid crystals possess mechanical as well as symmetry properties which are intermediate between those of a liquid and those of a crystal [1]. Depending on the molecular arrangement, liquid crystals or mesophases can be broadly classified into three types: nematic, cholesteric and smectic. The molecules in the nematic phase are spontaneously oriented about a mean direction called the director (Figure 1a). The molecular centres are however free to diffuse throughout the system so that translational invariance is not destroyed. The cholesteric liquid crystal is also basically of the nematic type except that it is composed of optically active molecules as a consequence of which the structure has a screw axis superimposed normal to the mean molecular axis (Figure 1b). In the smectic phase the molecules are arranged in layers, various types of arrangements being possible within a layer. For instance, in smectic A the molecules are upright, but the centres of molecules are irregularly spaced within each layer (Figure 1c). In smectic B the molecules are not only normal but their centres are arranged in a hexagonal close packed manner within a layer. In the smectic C phase, which is a tilted form of smectic A, the molecules are tilted with respect to the layer (Figure 1c). Several other smectic

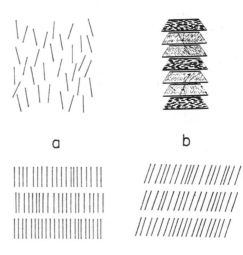

Figure 1. Schematic representation of molecular order in the (a) nematic, (b) cholesteric and (c) smectic A and (d) smectic C phases.

modifications are also known [2 - 5], but their structures are not yet fully understood.

Often a substance exhibits not one but several liquid crystalline modifications. Normally the sequence of transitions on heating are:

*Humboldt Fellow, on leave of absence from the Raman Research Institute, Bangalore - 560 080, India

solid → smectic B → smectic C → smectic A
                                              ↓
                    isotropic ← nematic

(We shall see later that there are some
remarkable exceptions to this sequence).
The heats associated with these transitions vary very widely. Typically, the
heats are about 25-30 kJoules (or kJ)/
mole for the solid - smectic B transition,
3-5 kJ/mole for the smectic B - smectic C
transition, ∿ 0 kJ/mole for the smectic
C - smectic A transition, 0.1 - 0.8 kJ/
mole for the smectic A - nematic transition and 1-3 kJ/mole for the nematic-
isotropic transition. It is therefore
obvious that techniques like differential thermal analysis (DTA) or volumetric
are not sufficient to study such a complex variety of transitions under pressure. This is exactly where the versatility of the diamond anvil cell comes in
useful.

The use of the diamond anvil cell in the
study of solids under pressure is well
known [6]. In these studies the solid
under investigation is immersed in a
pressure transmitting fluid which in turn
is compressed by the diamond anvils. It
is therefore not difficult to envisage
that the diamond anvil cell can also be
used to study liquid crystals at high
pressures, the liquid crystal serving
not only as the substance under investigation, but also as the pressure transmitting medium itself. This has been
done of late with a considerable degree
of success. This paper describes some of
these recent results.

2. EXPERIMENTAL

A schematic diagram of the cell used is
shown in Figure 2a while the anvils, used
in the gasketted configuration, are shown
in Figure 2b. The constructional details
of this cell have already been described
elsewhere [7] and will not be repeated
here. The pressure was estimated by an
indirect but nevertheless accurate
method, ie, the 'in-situ' transition temperature of the sample in the cell was
determined by optical transmission technique and with a previous knowledge of
any phase boundary (as obtained by DTA
[8]) for the same substance, the corresponding pressures could be read off.
(For details regarding the pressure calibration procedure see reference 7).

3. RESULTS

3.1 Identification of the Mesophases at
    High Pressure by Observation of Tex-
    tures

The energy required to deform a liquid

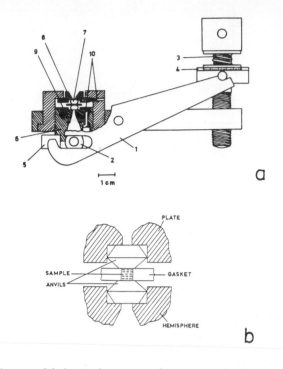

Figure 2(a). Schematic diagram of the
diamond anvil cell: 1) lever arm,
2) pressure plate bearing, 3) spring,
4) graduated screw head, 5) pressure
plate, 6) extended piston, 7) diamond
anvils, 8) translating plate, 9) hem-
isphere, 10) adjusting screws. (For
enlarged view of the anvils, see Fig.2b)

Figure 2(b). Sample encapsulated in
aluminium gasket between diamond anvils

crystal is so small that even the
slightest perturbation, caused either
by a dust particle, or by a surface inhomogeneity, can distort the structure
quite profoundly. Thus when a liquid
crystal sandwiched between the surfaces
of the diamond anvils is examined under
a polarizing microscope, we see quite
often rather complex optical patterns -
called 'textures'. Identification of
the various mesophases under pressure
can therefore be done by textural studies. Such a study, though essentially
qualitative in nature, is nevertheless
important, especially when dealing with
a complicated phase diagram. We shall
illustrate this by taking an example.

Figure 3 shows the P-T diagram of a
polymesomorphic substance, viz, 4-n-
hexyloxyphenyl - 4'-n-decyloxybenzoate
(6OPDOB for short) [9]. This compound
exhibits 4 smectic phases - E,B,C and A
in addition to the nematic phase. Also,

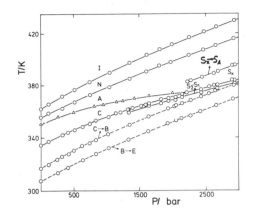

Figure 3. P-T diagram of 6OPDOB. The dashed lines indicate monotropic transitions

there are two pressure induced phases marked S_1 and S_X in the P-T diagram (see Figure 3). The $S_2 - S_1$, $S_1 - S_X$ and $S_X - S_A$ phase boundaries have been obtained by DTA. The DTA experiments indicate that S_1 is a pressure-induced solid phase while S_X could be a pressure-induced liquid crystalline phase. In order to identify this S_X phase textural photographs were taken using the diamond anvil cell by keeping the pressure at about 2.8 kbar and decreasing the temperature from the isotropic phase. As we come to the nematic phase, a typical 'schlieren' texture is seen (figure 4). As the temperature is

Figure 5. Ellipses and the focal conic texture of the smectic A (S_A) phase of 6OPDOB, pressure = 2.8 kbar, temperature = 130°C, X 125. From Herrmann, ref. 7.

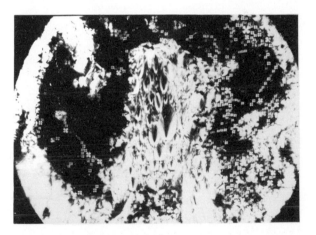

Figure 6. Same texture as in figure 5, but of the S_X phase of 6OPDOB, pressure = 2.8 kbar, temperature = 115°C X 125, crossed polarizers. From Herrmann, ref. 7.

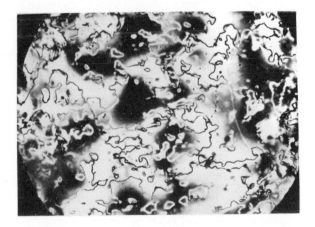

Figure 4. The schlieren texture of the nematic phase of 6OPDOB, pressure = 2.8 kbar, temperature = 151°C, X 125, crossed polarizers. From Herrmann, ref.7.

reduced further, S_A phase shows the ellipses of the focal conic texture (Figure 5). As the sample is cooled further, no change in the texture is seen even when the $S_X - S_A$ phase boundary is crossed. The focal conic texture with ellipses continues unchanged (Figure 6) throughout the S_X phase till the $S_X - S_B$ transition occurs after which the mosaic texture of smectic B is seen (Figure 7). Thus the evidence concerning the $S_A - S_X$ transition, viz, a first order transition with no textural

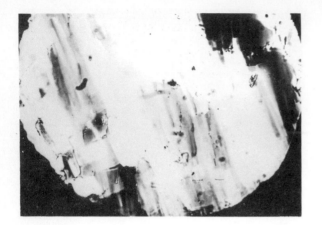

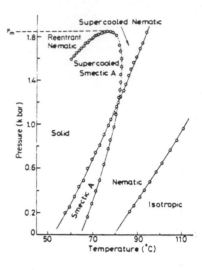

Figure 7. Mosaic texture (with some para-morphic focal conics) of the smectic B phase of 6OPDOB, pressure = 2.8 kbar, temperature = 107°C, X 125, crossed polarizers. From Herrmann, ref.7.

Figure 8. P-T diagram of 4'-n-octyloxy-4-cyanobiphenyl showing the re-entrant nematic phase in the pressure range of 1.6 - 1.8 kbar

change, seems to strongly indicate that this may be nothing other than the smectic A - smectic A (A-A) transition observed first by Sigaud et al. [10] in mixtures and later by Hardouin et al. [11] in a pure compound, both these observations being at atmospheric pressure. This is however the first observation of an A-A transition at high pressure.

3.2 Studies_on_'Re-entrant'_Nematic Substances

While studying the P-T diagram of a strongly polar liquid crystal, viz, 4'-n-octyloxy-4-cyanobiphenyl (8 OCB), Cladis et al. [12] made the interesting discovery that in the pressure range of 1.6 to 1.8 kbar, the sequence of transitions on cooling is:

isotropic → nematic → smectic A →

nematic → solid.

The lower temperature nematic phase has been designated as the 're-entrant' nematic phase in analogy with similar phenomena observed in super conductors [13] and ^{3}He [14]. The P-T diagram of 80CB is shown in Figure 8. The main features of this diagram are: (i) the smectic A - nematic phase boundary has an elliptic shape, (ii) the re-entrant nematic and the smectic A (beyond 1 kbar) are super cooled phases and (iii) there is a maximum pressure (P_M) at which the smectic A and hence the re-entrant nematic phases exist.

3.2.1 High pressure X-ray Studies

The compound 8 OCB has a strongly polar cyano group attached to one end of the molecule which gives rise to strong antiparallel correlations [15] between neighbouring molecules, which in turn results in a bilayer structure. The antiparallel arrangement has been recently confirmed experimentally by Lead-better et al. [16]. Assuming that the polar-polar and non polar-non polar interactions must be maintained to ensure the stability of the bilayer smectic A phase under pressure, Cladis et al.[11] proposed that there might be an expansion of the layer spacing with increasing pressure. High pressure x-ray studies were undertaken on 8 OCB to verify this prediction.

The X-ray diffraction maxima were recorded photographically. As the aim of the experiment was to measure small changes in the layer spacing, the specimen-to-film- distance was kept quite large (about 290 mm) and Ni-filtered CuK_α radiation was used despite the fact that there was relatively large attenuation of the intensity of the X-ray beam on passage through diamonds. Each exposure was around 100

hours. The temperature was maintained
constant to within ±0.25°C throughout
the exposure. After the end of each exposure the 'in-situ' transition temperatures of the sample in the cell were
redetermined and were found to be exactly the same as those measured before
the commencement of the exposure showing thereby that the pressure was maintained constant throughout the exposure.
The photograph at every pressure was
taken 3°C below the normal nematic-
smectic A transition point to minimise
the effect ot temperature on layer
spacing. The specimen-to-film distance
was calibrated using the 100 reflection
of p-decanoic acid [17]. The internal
consistency of the layer spacing evaluated from the different diameters of any
one diffraction ring was found to be
better than ±0.1 Å. It is therefore
estimated that the relative accuracy of
the spacing at different pressures is
±0.1 Å. The layer spacing was brought to
an absolute scale using the spacing of
decanoic acid. The value of d at atmospheric pressure was 32.0 ± 0.1 Å which is
in exact agreement with the most recent
determination for this compound.

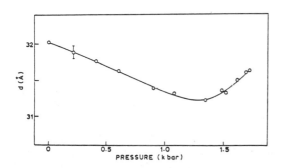

Figure 9. The variation of the smectic A
layer spacing (d) with pressure. The
error bar indicates the precision of the
measurements

Figure 9 shows the plot of the layer spacing (d) of 80CB with pressure [18]. It is
seen that the spacing decreases more or
less linearly at the rate of about 0.6 Å/
kbar till the pressure attains 1.4 kbar
beyond which it increases. The question
may be asked as to what is the contribution of the pure temperature variation
of the layer spacing to the curve shown
in Figure 9. To ensure that the contribution due to this effect is negligible,
the layer spacing of 8 OCB was measured
at many temperatures covering the entire
smectic A phase. There was no detectable
change in 'd' with temperature. Thus we
can conclusively say that the re-entrant

behaviour of 8 OCB which occurs at
higher pressures is associated with an
expansion of the smectic A layer spacing
with increasing pressure. It should
however be mentioned that the origin
of this effect is still to be understood in terms of molecular interactions.

3.2.2 Variation of P_M with concentration for re-entrant nematic mixtures

We saw in the beginning of § 3.2 that
for 8 OCB there exists a maximum pressure (P_M) of occurrence of the smectic
A and the re-entrant nematic phases. It
is interesting to study the variation
of P_M with concentration in binary mixtures. We have made by optical microscopy a pressure study of the mixtures
of 4'-n-hexyloxy-4-cyanobiphenyl
(6 OCB) and 8 OCB [19] with particular
reference to the smectic A - nematic
phase boundary. These results are
shown in Figure 10 [20]. With increasing
molar concentration of 6 OCB, P_M
decreases until for a concentration of
about 23.7% the phase boundary is almost flat. These results show that the
addition of 6 OCB molecules drastically
affects the polarity of the system (and
hence P_M) until the system is no longer
polar enough to support the formation of
the smectic A phase. The plot of P_M versus
concentration is shown in Figure 11. It is
seen that initially P_M decreases rather
rapidly as the concentration of 6 OCB
is increased, but the rate of decrease
becomes smaller at higher concentrations.
An extrapolation of this curve shows that
P_M should be zero for a concentration of
about 30%.

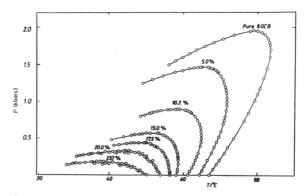

Figure 10. Pressure-temperature diagram
for 8 OCB and mixtures of 6 OCB and
8 OCB. The concentrations are given in
molar percent of 6 OCB

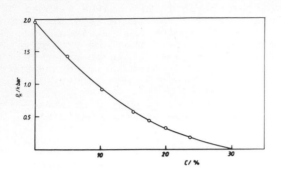

Figure 11: Variation of P_M, the maximum pressure of occurrence of the smectic A phase, with concentration (molar percent) of 6 OCB for the 6 OCB - 8 OCB mixtures

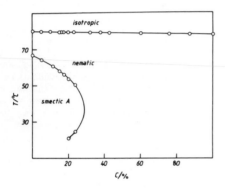

Figure 12: Temperature-concentration diagram for 6 OCB - 8 OCB mixtures. The transition temperatures are those determined at 1 bar. Concentration is given as molar percent of 6 OCB

This is in fact exactly what is observed experimentally also as seen in the temperature-concentration diagram (Figure 12) evaluated at atmospheric pressure, ie, there is no smectic A phase and hence no re-entrant nematic phase for concentrations beyond 30%.

3.3 The Search for a 'Lifshitz Point' in a Single Component System

As stated in § 1 the nematic, smectic A and smectic C phases have structural differences and transitions are possible between all these phases. The nematic-isotropic (N-I) transition has a reasonably large heat associated with it. There are different opinions on the nature of the smectic A - nematic (A-N)

transition. Theoretical considerations indicate that the director fluctuations should make the transition first order [21,22] and that if these fluctuations are ignored, the A-N transition can be second order with helium exponents [23-26]. As regards the experiments, volumetric measurements [27] indicate a first order transition while light scattering and A.C calorimetric studies [28,29] show that the transition is more second order like. The smectic C - smectic A (C-A) transition is generally accepted to be a second order transition with helium exponents [26,30-32]. The smectic C-nematic (C-N) transition, though much less studied, is always first order. A Landau-Ginzburg mean field theory has been developed by Chen and Lubensky [33] for a case where in all these three transitions occur. This theory yields the result that the C-A, A-N and C-N transitions can meet at a point. This polycritical point, which is the point of intersection of two second order (C-A and A-N) and one first order (C-N) phase boundaries, is referred to as the 'Lifshitz point' [34]. Such a Lifshitz point has indeed been observed experimentally in the composition-temperature diagrams of certain liquid crystalline mixtures [35,36]. Pressure experiments were undertaken on some pure compounds with a view to locating a Lifshitz point in a single component system.

The compound studied was N- 4-n-pentyloxybenzylidene -4'-n-hexylaniline (50.6) which exhibits at atmospheric pressure 5 smectic phases in addition to the nematic phase. We shall however be concerned here only with B,C,A and the nematic phase. In all, six different sets of measurements were carried out, one set with the DTA cell and the rest with the diamond anvil cell using the optical transmission technique. The B-C phase boundary determined with the DTA cell was used to bring the diamond anvil data to an absolute scale. The internal consistency of the different sets of measurements was checked by plotting all the data (on an absolute scale) for the A-N phase boundary and the agreement was found to be excellent.

The complete P-T diagram of 50.6 is shown in two parts for convenience of representation of the data; the measurements up to 4 kbar are shown in Fig.13, while those for pressures between 4kbar and 8 kbar are shown on an enlarged temperature scale in Figure 14 [37].

The temperature range of the smectic A phase diminishes with increasing pressure, becoming as small as about 0.7ºC

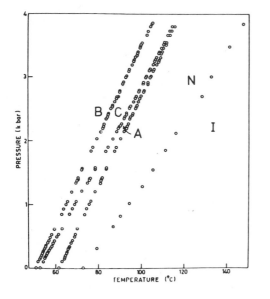

Figure 13. P-T diagram of 50.6 up to
4 kbar

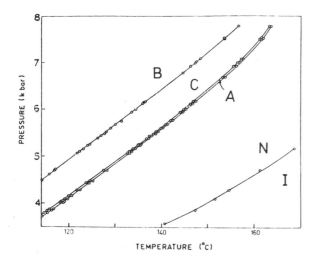

Figure 14. P-T diagram of 50.6 from
4 - 8 kbar

at 4 kbar (Fig. 13). At higher pressures
the range of the A phase decreases still
further, but the rate of decrease is
very slow; at 8 kbar the range is about
0.4°C (Fig.14). Had the resolution of
our experimental set up been less than
what it is at present, we might have
concluded that the C-A and N-A phase

boundaries intersect at about 4 kbar.
However, the precision of our data
shows that this is not in fact the case
and we can only infer that a Lifshitz
point will probably occur at a pressure
higher than 8 kbar.

It must be mentioned that since the
Lifshitz point is the intersection of
two second order and one first order
phase boundaries, it is expected that
the A-N transition line should have a
tricritical point before it intersects
the C-A line. To see if this is the
case, special care was taken to see
from the DTA experiment if the A-N
transition becomes second order. It
was observed that the first order
character of this transition diminishes
with increasing pressure up to about
4 kbar, but beyond this pressure the
change is hardly detectable and the
transition remains first order, though
weakly so, right up to 8 kbar. This is
perhaps partly the reason why the
Lifshitz point is not observed in the
range of pressure investigated.

The award of a fellowship by Alexander
von Humboldt-Stiftung is gratefully
acknowledged. Thanks are also due to
Frau Ingelore Mildt for the preparation of the camera-ready copy.

4. REFERENCES

[1] see eg, Chandrasekhar, S., Liquid
    Crystals (Cambridge University
    Press, Cambridge 1977).

[2] Sackmann, H. and Demus, D., Mol.
    Cryst. Liquid Cryst. 21(1973)239.

[3] Goodby, J.W. and Gray, G.W., Mol.
    Cryst. Liquid Cryst. Letters
    49(1979)165.

[4] Demus, D., Goodby, J.W., Gray,
    G.W., and Sackmann, H., Mol.
    Cryst. Liquid Cryst. Letters
    56(1980)311.

[5] Gane, P.A.C., Leadbetter, A.J. and
    Wrighton, P.G., Proc. 8th Inter-
    national Liquid Cryst. Conference,
    Kyoto, 1980.

[6] see eg, Block, S. and Piermarini,
    G.J., Physics Today (September
    1976)44

[7] Shashidhar, R. and Rao, K.V., in
    Chandrasekhar, S. (ed.), Liq.Cryst.
    - Proc. Int. Liq. Cryst. Conf.,
    Bangalore (Heyden, London, 1980)
    115; Herrmann, J. (unpublished)

[8] Reshamwala, A.S. and Shashidhar,
    R., J.Phys. E - Sci. Instrum.
    10(1977)180.

[9] Shashidhar, R. and Kleinhans, H.D., Mol. Cryst. Liq. Cryst. Letters 64(1981)217.

[10] Sigaud, G., Hardouin, F., Achard, M.F. and Gasparoux, H., J. de Physique 40(1979) C_3- 356.

[11] Hardouin, F., Levulut, A.M. and Sigaud, G., J. de Physique 42(1981)71.

[12] Cladis, P.E., Bogardus, R.K., Daniels, W.B. and Taylor, G.N., Phys. Rev. Letters 39 (1977)720.

[13] see eg, Fertig, W.A., Johnston, D.C., De Long, L.E., McCallum, R.W., Maple, M.B. and Mathias, B.T., Phys. Rev. Letters 38(1977)987.

[14] Lounasmaa, O.V., Experimental principles and methods below 1^{OK} (Academic Press, New York, 1974).

[15] Madhusudana, N.V. and Chandrasekhar, S. (ed.), Liquid Crystals - Proc. Int. Liq. Cryst. Conf., Bangalore, 1973, p. 57.

[16] Leadbetter, A.J., Frost, J.C., Gaughan, J.P., Gray, G.W. and Mosley, A., J. de Physique 40(1979) 375.

[17] Guillon, D. and Cladis, P.E. (private communication).

[18] Chandrasekhar, S., Shashidhar, R. and Rao, K.V., in Proc. Third Liq. Cryst.Conf. Socialistic countries, Budapest, 1979.

[19] Guillon, D., Cladis, P.E. and Stamatoff, J., Phys. Rev. Letters 41(1978)1598.

[20] Shashidhar, R., Kleinhans, H.D. and Schneider, G.M., Mol. Cryst. Liq. Cryst. Letters (submitted).

[21] Halperin, B.I. and Lubensky, T.C., Solid State Commun. 14 (1974) 997.

[22] Halperin, B.I., Lubensky, T.C. and Ma, S.K., Phys. Rev. Letters 32(1974)292.

[23] McMillan, W.L., Phys. Rev. A4 (1971)1238.

[24] Kobayashi, K.K., Mol. Cryst. Liq. Cryst., 13(1971)137

[25] de Gennes, P.G., Solid State Commun. 10(1972)753.

[26] de Gennes, P.G., Mol. Cryst. Liq. Cryst. 21 (1973) 49.

[27] Torza, S. and Cladis, P.E., Phys. Rev. Letters 32(1974)1406.

[28] Chu, K.C. and McMillan, W.L., Phys. Rev. A13 (1975) 1059.

[29] Garland, C.W., Kasting, G.B. and Lushington, K.J., Phys. Rev. Letters 43(1979)1420

[30] McMillan, W.L. Phys. Rev. A8 (1973)1921.

[31] Wulf, A., Phys. Rev. A11(1975)365.

[32] Priest, R.G., J. de Physique 36(1975)437.

[33] Chen, J. and Lubensky, T.C., Phys. Rev. A14(1976)1202.

[34] Hornreich, R.M., Luban, M. and Shtrikman, S., Phys. Rev. Letters 35(1975)1678.

[35] Johnson, D.L., Allender, D., de Hoff, R., Maze, C., Oppenheim, E. and Reynolds, R., Phys. Rev. B16(1977)470.

[36] Sigaud, G., Hardouin, F., and Archard, M.F., Solid State Commun. 23 1977)35.

[37] Shashidhar, R., Kalkura, A.N. and Chandrasekhar, S., Mol. Cryst. Liq. Cryst. Letters 64(1980)101.

PHYSICS OF SOLIDS UNDER HIGH PRESSURE
J.S. Schilling, R.N. Shelton (editors)
© North-Holland Publishing Company, 1981

INTER- AND INTRA-MOLECULAR DISORDER AND VOLUME CONTRIBUTIONS
TO THE ENTROPY OF MELTING

Antun Rubčić and Jasna Baturić-Rubčić

Faculty of Science, University of Zagreb, Yugoslavia

Melting processes of different substances are considered by means of correlation between entropy of melting and relative volume change under high pressure. Four groups of substances are recognized: with spherical, linear, generally-shaped rigid and non-rigid molecules. The homologous series of linear chain hydrocarbons is chosen as representative of the last group.
Disorder and volume entropy contributions including both, inter- and intra-molecular disorder activated at melting, are resolved. An explanation of these entropy contributions by means of molecular degrees of freedom is discussed.

1.   INTRODUCTION

A certain uniformity of melting processes for different substances has recently been considered [1]. Three groups have been pointed out: with spherical, linear and generally-shaped molecules. The correlation between relative volume change $\Delta V/V_s$ and associated melting entropy S_m for the first two groups, at normal and increased pressure, has been described by the simple expression

$$S_m = a + b\,\Delta V/V_s = S_d + S_v \qquad (1)$$

From the slope of such a straight line defined by eq. (1), the volume dependent contribution to the melting entropy S_v has been established. The coefficient b equal to $(7 \pm 2)R$ and $(10.5 \pm 1.5)R$ have been obtained for spherical and linear molecules, respectively [1]; R being the gas constant. A preliminary analysis indicated that for the group with generally-shaped molecules, simple correlation of eq. (1) for different $\Delta V/V_s$ values could be obtained only for particular substance under increased pressure [1] due to a greater variation in coefficient b.

Values of pure disorder contributions S_{do} (the remaining of melting entropy in hypothetical condition $\Delta V/V_s = 0$) have been established to be close to 1.4, 5.0 and 5.5 e.u., respectively for each of the considered groups. These values have been correlated with molecular degrees of freedom [1]. Positional and orientational components have also been distinguished:

$$S_{do} = S_{dt} + S_{dr} = R\ln(t+1) + R\ln(r+1) \qquad (2)$$

with t and r as numbers of translational and rotational degrees of freedom, respectively, which may be activated at melting. Eq. (2) follows from basic entropy relation $S_d = R\ln(N_1/N_s)$, where N_1/N_s represents the numbers of states per molecule in dynamically disordered liquid compared to those in crystal structure of solid at melting temperature T_m [1-3]. A generally-shaped rigid molecule may have the same positional and orientational state in the liquid as in the

solid, but also three new orientational states due to its more or less hindered rotations about one, two or three axes of the molecule (r=3). At the same time, molecule may execute equally likely one-, two-, or three-dimensional translations (t=3). Thus it leads to (3+1)(3+1)=16 possible combinations, i.e. to 16 more distinguishable positional and/or orientational microstates per molecule in the liquid than in the solid. An associated entropy is $S_{do}=R\ln 16=5.49$ e.u. [1,2]. Linear molecules, for example, can activate only two new orientational states, resulting in $S_{do}=R\ln 12$ in accordance with experimentally deduced value [1]. A special situation appears for the substances with atoms or with almost spherically-shaped molecules, which may activate no significant additional rotations at all (r=0). Therefore, translations towards any possible position are mutually undistinguishable. Due to such a degeneracy, translation may be considered as quasi-one-dimensional (t=1), resulting in $S_{do}=R\ln 2$ [1,2], again in accordance with experimental value for this group [1,4,5].

Necessary data for compounds with generally-shaped molecules are not abundant [6]. Thus disorder term $R\ln 16$ could be indicated for relatively small number of substances by means of ΔV-S_m diagrams [2]. Moreover, in many liquids with arbitrarily shaped molecules intra-molecular changes may also be realized. Then additional disorder components of melting entropy should be taken into account. Also, an appropriate change of volume is necessary to accomodate the each new type of molecular motions, resulting again in an additional increase of entropy. Generally, new disorder and associated volume contributions to melting entropy should be involved, i.e.

$$S_m = \sum_{i=0}^{m} S_{di} + \sum_{i=0}^{m} S_{vi} \qquad (3)$$

which would complicate an analysis.

However, a homolog series of such compounds might be very suitable to check the assumption that basic (positional and orientational) disorder term S_{do} is influenced only by general shape of molecules. Such a series might also confirm the particular volume contribution to the melting entropy is associated to each type of inter- or intra-molecular motions. Linear chain hydrocarbons C_nH_{2n+2} represent such an appropriate series. The conformational changes, due to the onset of hindered internal rotations around C-C bonds of a molecule, depend on the length of the chain, i.e. on the number n of its carbon atoms. It results in a certain regularity of their overall melting behaviour (3,6).

The purpose of this work is to resolve the individual terms of eq.(3) by means of data under high pressure. These entropy terms will then be compared with theoretically expected values and with those experimentally deduced for the other substances at normal pressure, searching thus for more general insight into the melting process of non-simple substances.

2. ANALYSIS OF DATA UNDER HIGH PRESSURE

2.1 Substances with rigid molecules

Correlations of $\Delta V/V_s$ and S_m values under high pressure to about 10 kbars for some representatives of substances with rigid molecules are shown in Figure 1. For the first group, Ar (4,5), Na (4), N_2 and CH_4 (7) are included. For the second group, with linear-like molecules, data for CS_2 and CO_2, and for the third group, data for $CHCl_3$ and $CHBr_3$ are given.

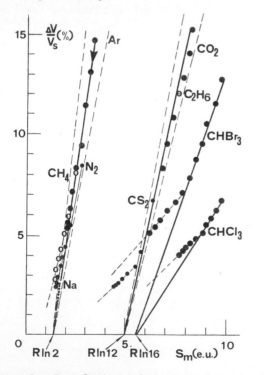

Figure 1 : Correlations of the relative volume change $\Delta V/V_s$ with melting entropy S_m under high pressure for substances with rigid molecules

The melting entropy is obtained by means of the Clausius-Clapeyron equation from T_m, P_m and ΔV data at melting. The curves with available ΔV-S_m data (9) have the similar properties as those, where relative volume change $\Delta V/V_s$ is used.

Extrapolation of straight lines, or initial parts of the curves at the beginning of pressure increase, for considered three groups, intersect the entropy axis at the values close to Rln2, Rln12 and Rln16, respectively. Data under increased pressure remain inside expected values of S_{do} and b, for the first two groups (dashed lines in Figure 1), which have been estimated previously, based only on melting data at normal pressure for series of compounds (1,8). Compounds of the third group give relatively higher values for b equal to about 21R and 39R for bromoform and chloroform, respectively.

From ΔV-S_m diagrams, coefficient b cannot be determined. However, the same values of S_{do} follow. Thus for benzene $S_{do} \approx Rln12$ (2) indicates a possibility of molecular rotations around the axis normal to the molecular plane already in the solid state. It means, that benzene would behave like a linear molecule at melting.

The recent melting $\Delta V/V_s$-S_m data for methane CH_4 up to 10.5 kbars (7), as well as ΔV-S_m data for CCl_4 (6), as representatives of plastic crystals, fall in diagram of the first group. It is known that both compounds have the orientational solid-to-solid (s-s) transition (3,6), resulting in quasi-spherical shape of their molecules. Therefore, only positional disorder can be expected at melting, with S_{do} Rln2, as it is indeed obtained by extrapolation of initial parts of their curves. Associated slope of the curve for CH_4 b≈8.4R, also falls inside expected values of (7 ± 2)R.

At very high pressures, many of considered curves including those of CH_4, CS_2 etc., actually deviate towards rather smaller (for CCl_4 even to zero) values (6) of residual entropies. It indicates, that some kinds of molecular motions may not be possible at melting (6), or that these motions could be activated already in the solid (3), both resulting in decrease of S_{do} values. The same deviations appear also for compounds of the third group, as shown in Figure 1. Moreover, for some other of them, "normal" behaviour of ΔV-S_m curves (6) with extrapolations towards Rln16 have been appeared in medium range of pressure, with deviations to smaller entropy values at both, small and very high pressures (as for example, for C_6H_5Br or $C_6H_5NH_2$) (6,9). Thus the last disorder entropy term is based on relatively small number of substances, including ΔV-S_m diagrams in addition to only two ones with more general $\Delta V/V_s$-S_m data. Therefore, there is a need for a more rigorous experimental confirmation of this value by means of additional measurements under high pressure, including substances with general rigid and non-rigid molecules.

2.2 Substances with non-rigid molecules

For some compounds, like benzophenone or p-tolu-idine, initial slopes of ΔV-S_m curves (9) give S_d around 10 e.u., which is rather greater than $R\ln 16$. This means that molecules could not be rigid any more and that some additional motions should be confined within molecules themselves. The analogous increase of entropy appears also for the hydrocarbons, especially for longer ones. However, necessary melting data are scanty, not always very precise and sometimes even contoversial (6,9-11).

General $\Delta V/V_s$-S_m correlation for particular compound remains as in eq.(1), although S_m may involve quite a complex balance of contributory terms in eq.(3). Melting data in a certain region of pressure would be desirable for coefficient b to be determined and total volume contribution S_v resolved from disorder one S_d. Unfortunately, vital informations at enhanced pressure are seldom available as yet (6).

No melting data under high pressure have been found for shorter hydrocarbons, except the mentioned ones for the first member methane. The melting point of the second member-ethane C_2H_6 at normal pressure, given in Figure 1, falls in the group of substances with linear-like molecules. The data under increased pressure are, of course, desirable to confirm, whether its S_{do} is indeed $R\ln 12$. The third member-propane C_3H_8 has even in normal pressure unknown volume change (6) but it may be supposed that it belongs to the group with the generally-shaped rigid molecules.

No data under high pressure are available also for butane-C_4H_{10}, the first hydrocarbon in which rotations around the central C-C bond may result in three different conformations of a molecule: so called trans T (with negligible rotations of the C-C bond) and gauche G^+ and G^- (with bond rotations of about $\pm 120°$) (12,13). The deepest minimum of potential energy U corresponds to extended zig-zag form of T conformation, which is the stable form for all hydrocarbon in solid state at lower temperatures. In liquid state the minimum of G form is by ΔU of about (0.5-0.8) kcal mole^{-1} higher than T one, which have been experimentally evaluated by means of Raman and IR spectroscopy (14-16). Rotations about the end C-C bond remain a molecule in an unchanged conformation. Thus generally, only (n-3) bonds are effective. However, for higher members (with n greater than 4), possible conformations become rather complex, due to different contrary effects (the original papers should be consulted for details) (13,17) and their number increases to be greater than $3^{(n-3)}$. Thus melting entropy will rise, more or less, with n, as its conformational part S_c will do (3,6,10).

Available melting data under adequately increased pressure for longer hydrocarbons were, fortunately, found for few ones from dodecane C_{12} ($C_{12}H_{26}$) to tetracosane C_{24} (11). $\Delta V/V_s$-S_m pairs of values up to ten kbars are shown in Figure 2. Melting entropies are calculated from T_m, P_m and ΔV directly measured data, as in Sec. 2.1. Associated $\Delta V/V_s$ data are deduced by means of liquid phase specific volumes V_1' (cm^3/gr) and compressibilities of C_{12}, C_{15} and C_{18} in temperature-pressure regions well away from their freezing point, by extrapolation to T_m (11).

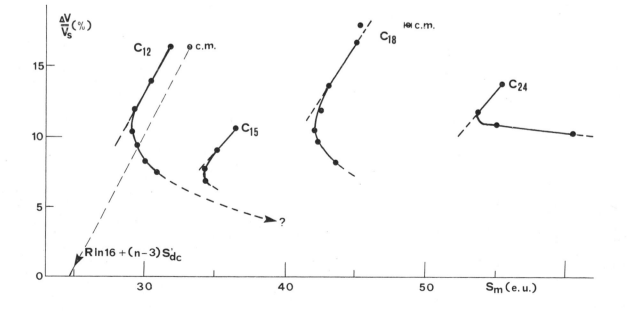

Figure 2 : Dependence of the relative volume change $\Delta V/V_s$ on the melting entropy S_m for some hydrocarbons under high pressures

$V_1'(n)$ values for a given melting pressure differ mutually only about 1%, with the same order of an uncertainty (11), enabling thus estimations of V_1' also for neighbouring hydrocarbons.

Only C_{12} and C_{18} have no s-s transitions. Their curves at relatively low pressures can be approximated by straight lines of definite slopes, resulting in coefficient b of about 29R and 30R, respectively. Other hydrocarbons in Figure 2 have s-s transition from orthorombic (C_{15}) or monoclinic (C_{24}) crystal structure to hexagonal one before melting at lower pressures, instead of direct melting from triclinic structure (for C_{12} and C_{18}) (10). However, these transitions disappear at higher pressures and certain similarity of all curves in Figure 2 becomes then evident. Thus the short straight parts of these curves under high pressures could contain an information about coefficients b. Their slopes correspond to values close to 33R and 43R for C_{15} and C_{24}, respectively. Thus b evidently rise with a complexity of molecules (see Sec. 2.1) and consequently S_v values do, but deduced quantities are not enough complete for more detail conclusions.

The way in which conformational disorder entropy S_{dc}' per additional CH_2 group rises, is also not possible to obtain exactly from displacements of such few curves. Moreover, heats of melting obtained by calorimetric methods at normal pressure differ by an amount of 2.8 up to 7.6% (11) from those calculated by means of Clausius-Clapeyron equation, as presented in Figure 2 (with points marked by c.m.). By assuming that S_{dc}' is approximately constant for n 1, it follows

$$S_m(n) = S_{do} + (n-3)S_{dc}' + S_v \qquad (4)$$

and one may estimate an approximate value of S_{dc}' from entropy difference ($\Delta S_m - \Delta S_v$). Basic disorder term S_{do} between about 4 and 6 e.u. then results, using C_{12} and C_{18} data. This is an acceptable value expected for that term. However, in the lack of adequate data under high pressure it must be supported by relevant melting parameters at normal pressure.

At very high pressures, deviations of the curves in Figure 2 have just opposite tendency of those in Figure 1. It may be expected, because temperature of melting is then correspondingly higher and the probability of molecular G conformations due to $RT_m \gtrsim \Delta U$ is also higher (18). This results in greater conformational part of melting entropy. It is interesting to know that minimum value of S_m appears to be for all hydrocarbons investigated under high pressure at approximately the same temperature T_m close to 100 °C (11). This leads to RT_m of about 740 cal/mole which is a very good approximation to ΔU values estimated by spectroscopic methods (14-16).

3.   MELTING PARAMETERS AT NORMAL PRESSURE

Volume contribution to the melting entropy may be calculated according to relation (1,2,19)

$$S_v = \int_{V_s}^{V_e} (\partial S/\partial V)_{T_m} dV \approx (\partial S/\partial V)_{T_m} \Delta V = \gamma_m \Delta V \qquad (5)$$

The gradient of a P-T isochore is the thermal pressure coefficient $\gamma = (\partial P/\partial T)_V = (\partial S/\partial V)_T$. A set of γ values have been determined for liquid even hydrocarbons with n equal to 6, 8, 16, 22 and 36, but not at freezing temperature T_m. By means of Orwol's and Flory's least-square (l-s) equations (20) for lines drawn through data points, extra- and inter-polated values of γ_m are obtained. They are in reasonable agreement with data of other authors (2o,21), for n equal to 14, 20 and 32, and with another set of γ_m data for odd hydrocarbons (11 n 19). The latter data (19) have been obtained also by an extrapolation technique from measured P-T isochores above T_m and elevated pressures. A regularity of γ_m behaviour is well pronounced in its dependence on melting temperature as shown in Figure 3.

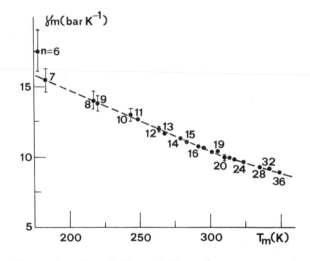

Figure 3 : Correlation of thermal pressure coefficient γ_m with melting temperature of liquid linear chain hydrocarbons

For lower n, the temperature extrapolations in γ_m evaluation are too high (about 100 °C), introducing greater uncertainty which must be taken into account. Otherwise, shorter hydrocarbons should be excluded from consideration.

Volume changes at melting at normal pressure p_o are not available for all of the considered hydrocarbons (up to n=32). However, certain regularities of T_m (10), V_1' (6,11), ΔV, $\Delta V/V_s$ and the total relative volume change (including s-s transitions) in dependence of n may be used. According to that, inter- and extra-polations from known data (which will be discussed elsewhere) result in reasonable estimations of necessary volume change values ΔV.

Approximate volume contributions S_v to melting entropies are then calculated by means of eq.(5).

Within experimental errors, the linear correlation

$$S_v = S_{vo} + (n-3)S'_{vc} \qquad (6)$$

is obtained, both for the hydrocarbons without and with s-s transitions; S_{vo} being about 3.8 and 3.5 e.u. with S'_{vc} 0.55 and 0.28 e.u., respectively. Finally, the associated coefficients b are calculated according to eq.(1). Taking into account estimated uncertainties of data, again a nearly linear dependence

$$b = b_o + (n-3)b' \qquad (7)$$

follows, as shown in Figure 4.

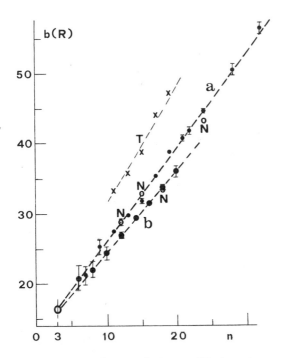

Figure 4 : Dependence of the coefficient b on the number of C atoms n of the hydrocarbons with (a) and without s-s transition (b)

The l-s method gives, for members without and with s-s transitions, 16.25 and 16.40R as b_o with about 1.17 and 1.36R as b', respectively. The values of b evaluated from melting data under high pressure (points marked with N in Figure 4) are in good agreement with those calculated by means of eqs.(5) and (1) bearing in mind the respective approximations.

Volume contributions S_v have been systematically evaluated previously only for restricted group of five odd hydrocarbons, with s-s transition, from C_{11} to C_{19} (19). However, an integral form of eq.(5) has been used by means of an unreal hypothetical melting, which developed through an isochore melting (including only conformations) and

after that through an isothermal expansion of liquid up to its normal V_1 value at T_m (19). Thus $\gamma(V)$ values at $V_s < V < V_1$ have been associated to $P > P_o$, and also extrapolated to T_m they remain higher than γ_m at P_o. Therefore, relatively higher values of S_v have been calculated. Discrepancies in b, with overestimations of about 20%, are evident in Figure 4 (points marked with T). They suggest that one will do probably smaller error using approximate relation $\gamma_m \Delta V$ in accordance with $b\Delta V/V_s$ for the S_v evaluation, than by the use of more exact integral form of eq.(5), but with unreasonable values of $\gamma(V)$ originated far from freezing. It has been experimentally verified by Raman spectroscopy (22), that population of G conformations in respect to T ones rises with pressure, altering probably values of γ also.

4. DISORDER CONTRIBUTION TO MELTING ENTROPY OF LINEAR CHAIN HYDROCARBONS

Once the volume contribution S_v to melting entropy is determined, the remaining difference $S_m - S_v$ should be the pure disorder contribution S_d. However, this term will now be related to intra-molecular conformational changes in addition to translations and rotations of generally-shaped molecules themselves.

Data for temperatures and heats of melting at normal pressure for different members of investigated series belong to many authors (23,24,25). The most complete and the "best" values have been systematited by Broadhurst (10), achieving overall consistency with total accumulation of published data. Evaluated melting entropies can be considered quite accurate (inside 1 to 2%) for n<20. Using these S_m and S_v values, evaluated in Sec. 3, a linear dependence of S_d on n arises, which leads to correlation

$$S_d = S_{do} + (n-3)S'_{dc} \qquad (8)$$

analogous to associated volume correlation of eq.(6).

Three associated straight lines appear as shown in Figure 5, for even hydrocarbons (up to n=20) without s-s transition (a), and for odd (b) and even members (c) both with s-s transition. The slope of the $S_d(n)$ straight line corresponds to approximately constant disorder contribution S'_{dc} per additional C-C bond for the each of three groups. The steepest slope is obtained for the group (a), indicating the greatest conformational disorder contribution to the melting entropy. Values of γ_m and ΔV are directly evaluated for C_{16}, and b values from both sources (under increased-Figure 2 and under normal pressure-eq.(5)) agree well for C_{18} (see Figure 4). Therefore, S_d for these two compounds may be expected to be the most accurate. Difference $S_d = S_d(18) - S_d(16)$ leads to S'_{dc} 2.1 e.u., which introduced in eq.(8), gives S_{do} 5.7 e.u. with $\pm$ 0.1 e.u. and $\pm$ 0.6 e.u. rms and maximum error, respectively. The l-s method for data of eq.(8) from C_{10}

to C_{18}, presented in Figure 5, gives $S_{do}=5.99$ e.u. and $S'_{dc} \approx 2.07$ e.u.

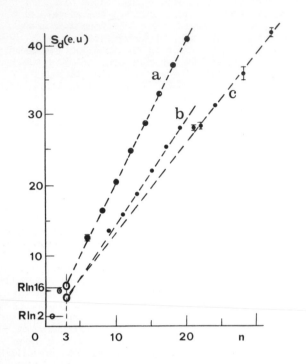

Figure 5 : Dependence of disorder entropy contribution S_d on the number of C atoms n of linear chain hydrocarbons: for even members without s-s transition (a), and for odd (b) and even (c) ones with s-s transition

With both pairs of values, the melting entropies are fitted within an error less than 1% by means of equation

$$S_m(n) = S_{do} + (n-3)S'_{dc} + S_{vo} + (n-3)S'_{vc} \qquad (9)$$

which is analogous to eq.(3) and follows from eqs.(6) and (8). Thus assumed relation of eq.(4) is confirmed inside experimental errors. Volume contributions to the melting entropy appear to be associated to particular disorder entropy contributions, as expected.

By the same procedure of the l-s method, rather smaller specific conformational entropies S'_{dc} of about 1.5 and 1.3 e.u. with smaller resultant basic disorder components S_{do} of about 3.7 and 4.6 e.u. follow for members of (b) and (c) groups, respectively. The values of these entropy terms may be considered only as approximate ones, because of too small number of even uncertain experimentally evaluated points. Nevertheless, they undoubtfully suggest, that smaller disorder would be activated at the melting of the linear chain hydrocarbons with s-s transition than without it.

Conformational entropies have been previously experimentally deduced only for the odd hydrocarbons,

for which the volume contributions S_v has been also evaluated (see Sec.3). However, the value $S_o R$ of the so called "communal entropy", connected with the wondering of each molecule across the whole volume of the melt phase, has been assumed and used as basic disorder entropy term (19). This term is a much controversial one and discussion of it for the disordered liquid state has a long history (26-28). Although this value should be rejected, as an universal constant for all kinds of molecules (27), it has been still applied in the lack of the better one (19,29). Too small basic disorder term S_o S_{do} and too high volume contribution S_v (see Se.3) at the same time (19), have been mutually compensated. This results in a surprising coincidence between the conformational disorder entropies obtained by our approach and previous ones (19) for odd hydrocarbons from C_{11} to C_{19}.

5. CONCLUSION AND DISCUSSION

The basic disorder entropies S_{do} for substances with spherical, linear and generally-shaped rigid molecules, explained by means of molecular degrees of freedom (1,2) and equal to Rln2, Rln12 and Rln16, respectively, are rather well verified with melting data under high pressures (Figure 1). The last entropy term is also confirmed, within estimated deviation, for non-rigid molecules of linear chain hydrocarbons. These results are deduced by means of melting data both at normal (Figure 5) and high pressure (Figures 1 and 2). This presents an additional support to our assumption that basic disorder entropy is determined only by the general effective shape of molecules. Consequently, the associated dimensionality of translational and rotational motions defines the number of new distinguishable states in the liquid after melting (1,2). Thus non-rigidity of molecules have no influence on the value of S_{do}.

Above statement is valid, according to Figure 2, for melting at relatively low pressures (up to about 3 kbars for C_{12} as an example). Although molecules, in this region of pressure, are distorted with many various molecular conformations, they may be represented by means of an average "rigid-like" molecule. By further increase of pressure, the sense of this average rigid molecule becomes less and less valuable. It may be justified by deviations of the curves for hydrocarbons in Figure 2. Moreover, at very high pressures and correspondingly high T_m, the melting entropies tend probably towards nRln16 values. It means that rotations of C-C bonds are more and more favoured. In the limit of infinite pressure, it seems that CH_2 and/or CH_3 groups could be considered like independent monomers, each contributing Rln16 per one mole of the monomer. More comprehensive and detail measurements under very high pressures are, of course, needful, to support such an assumption.

Extrapolated values $S_{do} < Rln16$ for odd and even hydrocarbons (b) and (c) in Figure 5 may be explained by their known s-s transitions to hexagonal crystal structure before melting (10).

Molecules in this structure have high degree of rotational freedom around their chain axes (30), while for triclinic structure (of even members up to C_{20}) these rotations are greatly restricted before the melting. Analogously, smaller values of S'_{dc} suggest that conformational disorder should also be partly consumed already in premelting s-s transition, which has been theoretically possible (31), but not experimentally confirmed as yet.

It should be also emphasize, that generally, positional and orientational disorder are mutually correlated. Thus, associated basic disorder entropy S_{do} is not the same for all compounds, and an eventual use of "communal entropy" $S_o \approx R$, instead of S_{do}, should rise to uncorrectness in evaluation of other entropy contributions (19, 29).

Pure disorder conformational entropy S_{dc} is independently determined only by the flexibility of molecules themselves, resulting in its infinite growth with increasing chain length of hydrocarbons, according to eq. (8). The importance of associated volume entropy contributions S_{vc} has been emphasized before (6), expecting S_{vc} to be even dominant. This is obviously not true, according to the quantitatively resolved terms of the final eq. (9). It follows that S'_{dc} is about four times greater than S'_{vc} at normal pressure. This ratio will still be more pronounced under high pressure (Figure 2).

Theoretical evaluation of S_{dc} has been based on more or less simplified models (18,19,25), using the assumed effective energy difference ΔU between the trans T and gauche G forms. However, various modifications of such calculations lead to smaller values of S_{dc}, than those experimentally evaluated. Detail discussion of these problems would exceed the scope of this work. Here it may be emphasized that our simple calculations in very good agreement with experimental value of S'_{dc}, indicate that more than two effective gauche conformations should exist, which is theoretically possible (13,31). It follows also that some conformations should be forbidden due to the intramolecular excluded volume effect (18). Contrary to that, the solvent effect which favourizes gauche conformations in liquid compared to those of non-interacting chains (17) are also taken into account (which will be published elsewhere). Thus the traditional model of only three states for the each C-C bond, extended from butane to other members, should be an oversimplification of the rotational isomeric states of linear chain hydrocarbons (13,31). An exact calculation of S_{dc}, as well as S_{vc}, is an enormously difficult problem still to be solved. A more comprehensive data under high pressure would surely be a great help for a better theoretical approach to melting process generally.

ACKNOWLEDGMENT
We wish to thank Dr.John Cooper for enlightening discussions and support throughout the work.

REFERENCES

(1) A.Rubčić and J.Baturić-Rubčić, Phys.Lett., 72A(1979) 27.
(2) A.Rubčić and J.Baturić-Rubčić, Fizika,12, Suppl.1(1980) 253; J.Baturić-Rubčić and A. Rubčić, Fizika 12,Suppl.1(1980) 259.
(3) D.Fox,M.M.Labes and A.Weissberger, Physics and Chemistry of the Organic Solid State, (Wiley,1963).
(4) S.M.Stishov, Usp.Fiz.Nauk, 96(1968) 467; 114 (1974) 3; S.M.Stishov, J.N.Makarenko, V.A.Ivanov and A.M.Nikolaenko, Phys.Lett.,45A(1973) 18; 47A(1974) 75.
(5) R.K.Crawford and W.B.Daniels, Phys.Rev.Lett., 21(1968) 367; V.M.Cheng, W.B.Daniels and R.K.Crawford, Phys.Lett.,43A(1973) 109.
(6) A.R.Ubbelohde, The Molten State of Matter, (Wiley,1978).
(7) V.M.Cheng, W.B.Daniels, R.K.Crawford, Phys. Rev.,B11(1975) 3972; R.K.Crawford, W.B.Daniels and V.M.Cheng, Phys.Rev.,A12(1975) 1690; R.L.Mills, D.H.Liebenberg and J.C.Bronson, J.ChemPhys.,63(1975) 4026.
(8) M.Lasocka, Phys.Lett.,51A(1975) 137.
(9) Landolt-Börnstein, Mechanisch-termische Zustangrössen (Springer, 1961 and 1971); Spravochnik Himika (Himia,Leningrad,Moscow, 1964).
(10) M.G.Broadhurst, J.Res.Natl.Bur.Stand.,(U.S.) 66A(1962) 241.
(11) R.R.Nelson, W.Webb and J.A.Dixon, J.Chem.Phys. 33(1960) 1756.
(12) K.S.Pitzer, J.Chem.Phys., 8(1940) 711
(13) R.A.Scott and P.A.Scheraga, J.Chem.Phys.,44 (1966) 3054.
(14) Q.J.Szasz, N.B.Sheppard and D.M.Rank, J.Chem. Phys.,16(1948) 704.
(15) R.G.Snyder, J.Chem.Phys.,47(1967) 1316.
(16) J.R.Scherer and R.G.Snyder, J.Chem.Phys.,72 (1980) 5798.
(17) D.Chandler and L.R.Pratt, J.Chem.Phys.,65 (1976) 2925; L.R.Pratt, C.S.Hsu and D.Chandler, J.Chem. Phys.,68(1978) 4202; C.S.Hsu, L.R.Pratt and D.Chandler, J.Chem. Phys.,68(1978) 4213; J.P.Ryckaert and A.Bellemans, Chem.Phys.Lett. 30(1975) 123.
(18) R.P.Smith, J.Polym.Sci.,A-2(1966) 869.
(19) A.Turturro and U.Bianchi, J.Chem.Phys.,62 (1975) 1668; 65(1976) 697.
(20) R.A.Orwoll and P.J.Flory, J.Am.Chem.Soc.,89 1967) 6814.
(21) D.Sims, Polymer, 6(1965) 220.
(22) P.E.Schoen, R.G.Priest, J.P.Sheridan and J.M.Schnur, J.Chem.Phys.,71(1979) 317.
(23) H.L.Finke, M.E.Gross, G.Waddington and H.M. Huffman, J.Am.Chem.Soc.,76(1954) 333.
(24) A.A.Schaerer, C.J.Busso, A.E.Smith and L.B. Skinner, J.Am.Chem.Soc.,77(1955) 2017.
(25) R.H.Aranov, L.Witten and D.H.Andrews, J.Chem. Phys.,62(1958) 812.
(26) H.Eyring et al., Statistical Mechanics and

Dynamics (Wiley,1965).

(27) W.G.Hoover and F.H.Ree, J.Chem.Phys.,49
     (1968) 3069.

(28) E.J.Jensen, W.D.Kristensen and R.M.J.Cotte-
     rill, J.Non-Crystalline Solids,24(1976) 737.

(29) H.W.Starkweather and R.H.Boyd, J.Phys.Chem.
     64(1960) 410.

(30) J.D.Hoffman, J.Chem.Phys.,20(1952) 541.

(31) L.D'Ilario and E.Giglio, Acta Cryst.,B30
     (1974) 372.

PHYSICS OF SOLIDS UNDER HIGH PRESSURE
J.S. Schilling, R.N. Shelton (editors)
© *North-Holland Publishing Company, 1981*

OPTICAL PROPERTIES OF CESIUM AND IODINE UNDER PRESSURE

K. Syassen, K. Takemura, H. Tups, and A. Otto

Physikalisches Institut III, Universität Düsseldorf
D-4000 Düsseldorf 1, Federal Republic of Germany

The optical reflectivity of cesium and iodine has been measured from 0.5 to 3.5 eV at pressures up to 300 kbar. Pressure induced changes of the reflectivity of cesium deliver direct spectroscopic evidence for the electronic 6s - 5d transition in this material. The molecular metallic phase of iodine has a very low carrier density. The monatomic metallic phase of iodine exhibits free electron-like behaviour and a strong 5p - 5d interband absorption at around 1.3 eV.

I. INTRODUCTION

Solid cesium and iodine exhibit drastic changes of their electronic properties under pressure. Cesium undergoes a well known electronic transition below 100 kbar, which manifests itself in low bulk modulus, phase transitions, and resistance anomalies /1-3/. Iodine is as yet the only diatomic molecular crystal with an insulator metal transition at pressures readily available in the laboratory /4/. Cesium and iodine have in common that their ground state resp. excited electronic state properties at high pressure are strongly affected by 5d bands.

Recent band structure calculations for cesium suggest that both the isostructural transition and the unusual softness of the low pressure isotherm are related to the evolution of valence electrons from 6s to 5d states /5/.

An experimental verification of the theoretical model can be obtained from spectroscopy of the electronic structure. The purpose of the present work on Cs is to report results on the pressure dependence of the optical reflectivity and to discuss the relationship between optical interband absorption and the electronic s-d transition.

Iodine is normally an insulator with a band gap of 1.3 eV. The experimental evidence from high pressure resistance /4/ and structural /3, 6/ studies implies that iodine first undergoes an insulator to metal transition in the molecular phase and then dissociates to a monatomic metallic phase. In the present work on iodine we report recent results on the optical properties of the two metallic phases and draw some conclusions about their electronic properties. Crystal structure aspects of both iodine and cesium under pressure are discussed in the foregoing paper in this volume /3/.

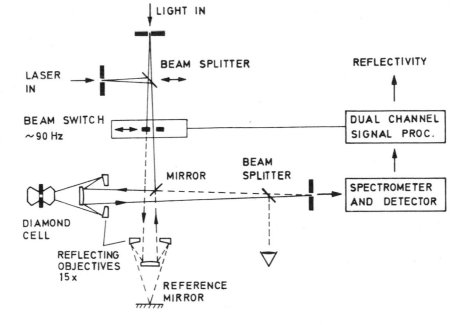

Fig. 1: Microoptical double beam system for reflectivity measurements with a diamond anvil cell.

II. EXPERIMENT

Reflection spectra of cesium and iodine were measured between 0.5 and 3.5 eV using a diamond anvil cell and the ruby pressure scale. In each case the gasket hole was completely filled with the sample material. Molten cesium of 99.95% purity was pushed from a syringe into the gasket hole under argon atmosphere. Iodine crystals were oriented with the bc plane normal to the incident light (parallel to the diamond faces). The reflectivity was measured at the diamond-sample interface. A schematic drawing of the micro-optical double beam system is shown in Fig. 1. A chopper switches the light between sample beam and reference beam. The reflectivity (ratio of the two reflected intensities) was corrected for absorption losses in the diamond window and for reflection losses at the diamond air interface. It should be clearly stated that all reflection spectra shown in this paper give the reflectivity against a diamond window with index of refraction n_{dia} = 2.4.

III. CESIUM

III.1 Results

The optical reflectivity of cesium was measured between 3 kbar and 136 kbar. Fig. 2 shows the reflection spectra for the bcc phase (Cs I). The spectrum at 3 kbar is almost identical with the 1 bar spectrum calculated from the optical constants given by N. V. Smith /7/. This agreement proves that we do not have to be concerned about sample purity and surface contamination. When pressure is increased to 9 kbar, the reflection spectrum changes considerably. A sharp decrease in reflectivity develops at 1.2 eV. The structure becomes even more pronounced at 19 kbar where the reflectivity drops to 35% at 1.2 eV followed by a broad peak.

The bcc-fcc phase transition at 23 kbar is easily detected by visual observation. Fig. 3 shows reflection spectra for the fcc-phase (Cs II). The reflectivity minimum moves from 1.2 eV in bcc to 1.6 - 1.9 eV in fcc. While the phase

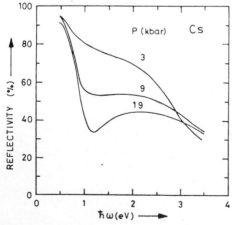

Fig. 2:
Reflectivity of cesium in the bcc phase (Cs I).

transition from Cs II to Cs III could not be detected visually, the occurrence of the tetragonal phase Cs IV can be clearly seen due to the optical anisotropy of Cs IV. Crystallites of different orientation have a slightly different colour. The most important change on passing from Cs II to Cs IV is an abrupt drop of the reflectivity to less than 60% at the low energy side of the spectral range. In contrast to Cs IV, the appearence of Cs V is that of an optically isotropic material. The reflection spectrum at 136 kbar is structureless with an average reflectivity of 50%.

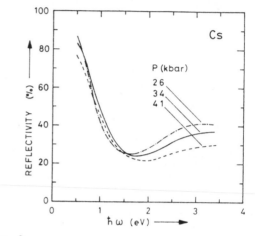

Fig. 3:

Reflectivity of cesium in the fcc phase (Cs II).

III.2 Discussion

The optical properties of cesium at 1 bar (3 kbar) are not described adequately by a simple Drude model /7/. The Drude like optical conductivity is given by

$$\sigma(\omega) = \varepsilon_o \cdot \omega_p^2 \cdot \Gamma / \omega^2 \qquad (1)$$

where Γ = $1/\tau$ and ω_p = $N\,e^2/\varepsilon_o\,m$ · ω_p is the plasma frequency, τ is the optical scattering time, and N is the carrier density. The deviation of the optical conductivity of cesium from the Drude behaviour is seen in Fig. 4 which shows the result of a Kramers-Kronig-analysis of the optical reflectivity at 3 kbar and 19 kbar. The peak in $\sigma(\omega)$ at roughly 1 eV (3 kbar) can be explained by interband absorption. Fig. 5a and b show the relevant part of the bcc bandstructure calculated by Ham /8/. An interband transition threshold occurs between occupied s-like states near N_1 and excited p-states near N_1 (Wilson-Butcher model) and d-states near N_2. At 9 kbar the interband absorption has increased (see Fig. 2) indicating that d-hybridization has pushed the state N_1 below the Fermi level. Now optical transitions become possible between flat bands at the Brillouin zone edge. The further increase of the absorption at 19 kbar is explained by a larger joint density of states. The bands near N_1 and N_2 become parallel over a large part of the Brillouin zone. The assignment of the pres-

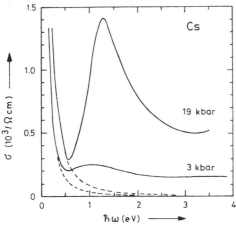

Fig. 4

Optical conductivity of cesium at 3 kbar and 19 kbar. The dashed lines correspond to the Drude like conductivity. Drude parameters are derived from DC conductivity and volume compression.

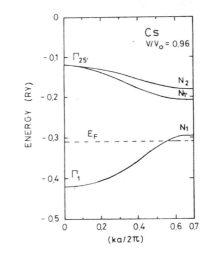

Fig. 5 a

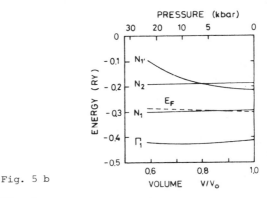

Fig. 5 b

Fig. 5:

Bandstructure of bcc cesium /8/.
a) Energy bands along ΓN at $V/V_0 = 0.96$.
b) Volume dependence of selected critical points.

sure enhanced 1.2 eV absorption to $N_1 - N_2$ transitions becomes obvious from Fig. 5b, where only the energy difference between states N_1 and N_2 is almost independent of volume change.

In fcc cesium the corresponding interband transition between occupied s-d hybridized states near X_1 and the excited pure d states near X_3 also is almost independent of volume change /5/. The depth of the reflectivity minimum at 1.6 - 1.9 eV does not depend on pressure. This indicates that the density-of-states-effect on the optical transitions has saturated.

In Cs IV the decrease of the optical reflectivity at 0.5 eV may be explained by a drop of the interband transition threshold to below 0.5 eV. The spontaneous occurrence of such a low threshold can be understood within the framework of the current theoretical model for the volume collaps at the II-III phase transition: if the state X_3 drops discontiually below the Fermi level, interband transitions are no longer possible at the X point. Transitions between the two lowest bands move toward the center of the Brillouin zone, where the excitation threshold would be smaller than 1.5 eV /9/.

In summary, the pressure dependence of the optical reflectivity gives a direct spectroscopic evidence of the continuous and discontinuous aspects of the 6s - 5d transition in cesium.

IV. IODINE

IV.1 Experimental results

Reflection spectra of iodine were measured between 42 kbar and 298 kbar off the bc plane. At all pressures the ruby fluorscence lines stayed remarkably sharp. Iodine seems to have a very low shear strength even at the highest pressure. Some reflection spectra are depicted in Fig. 6.

Below 210 kbar the low level reflectivity rises continuously with increasing photon energy and also with increasing pressure. Between 210 and 240 kbar the reflectivity at the low energy side rises steeply. The metallic like reflectivity above 230 kbar shows a pronounced minimum at 1.2 eV.

IV.2 Discussion

The first occurrence of the low energy structure in the reflection spectrum at 217 kbar coincides with the structural phase transition from molecular to monatomic iodine at around 210 kbar /3, 6/. The low energy structure is due to a Drude like behaviour, whereby the carrier density increases with pressure. At 217 and 225 kbar the plasma edge (which depends on the square root of the carrier density) lies below 1 eV. At higher pressures the Drude behaviour is masked by strong absorption processes. However, the increasing reflectivity between 1.5 and 3 eV indicates that $\hbar\omega_p$ has quickly moved

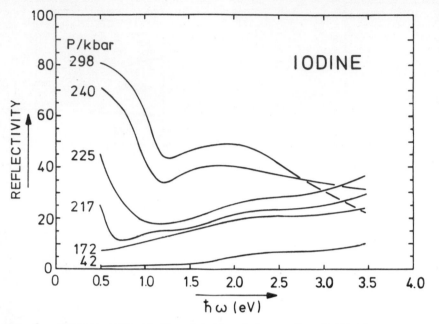

Fig. 6: Optical reflectivity of iodine between 42 and 298 kbar.

up into the near UV, e.g. the carrier density increases considerably. The optical spectra suggest that most of the transition to the monatomic high carrier density metal occurs between 217 and 240 kbar.

Atomic iodine has a $5p^5$ electronic configuration. The expected carrier density in the monatomic phase is one per atom, the effective mass of the p band carriers may be as high as $2 m_e$, and the lowest estimate for $\hbar\omega_p$ is then 5 eV. If we assume that at 298 kbar the reflectivity at 0.5 eV is determined only by intraband scattering we obtain an optical scattering time $\tau = 2\times10^{-15}$ s. This low value of τ could be due to defects induced at the structural phase transition. A rough estimate of the minimum electrical conductivity results in a value of $\sigma_{dc} = 1.5\times10^4/\Omega cm$. If the highest reflectivity at 298 kbar is not only limited by intraband scattering but also by interband absorption, the DC conductivity would be higher.

A Kramers-Kronig-analysis of the 298 kbar spectrum, based on the above Drude parameters, is shown in Fig. 7. The 1.3 eV interband absorption can be interpreted in at least two different ways: transitions between 5p subbands or transitions to the bottom of the excited 5d band. The band structure for a hypothetical fcc lattice reported by McMahan et al. /10/ shows 5p subband splittings near E_F in the order of 1 eV and direct 5p-5d transition energies of roughly 5 eV. Of course, we cannot expect these numbers to apply to body centered orthorhombic iodine, but the large calculated 5p-5d splitting makes it somewhat difficult to assign the 1.3 eV absorption feature to 5p-5d transitions. However, one strong argument in favour of the 5p-5d assignment

is the very high intensity of the absorption band. For 5p-5d transitions a high intensity can be easily explained by the large joint density of states of flat energy bands near the Brillouin zone egde. We therefore tentatively assign the 1.3 eV transition to 5p-5d excitations. A more definite interpretation could be given, when band structure calculation for the appropriate crystal structure become available. We like to point out, that our assignment is not in controversy with the interpretation of the Hugoniot data by McMahan et al. /10/.

In a simple picture the metallic conductivity in the molecular phase occurs, when occupied π^* states begin to overlap in energy with unoccupied σ^* states. At small overlap the number of carriers is small and the Drude like plasma edge lies at low energy outside of the present spectral range. The number of carriers in the two metallic phases differs by at least two orders of magnitude. The remaining point to discuss then is: Why does the electrical DC conductivity only change by a factor of about 3 /4, 11/ between 170 and 240 kbar? Two out of several possible explanations are: (i) At small overlap of the π^* and σ^* states in the molecular phase the phase-space volume for scattering is small and therefore the scattering time is long compared with the monatomic phase. (ii) The electrical conductivity of the monatomic phase is anisotropic /4/. After the phase

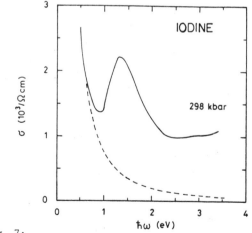

Fig. 7:

Optical conductivity of iodine at 298 kbar. The Drude like conductivity corresponds to $\hbar\omega_p$ = 5 eV, $\tau = 2\times10^{-15}$ s.

transition with 5% volume decrease the crystallites in the sample are expected to be more randomly oriented. Electrical conductivity measurements will then result in resistance value which tend toward the resistance of the less conductive crystal direction. In addition the defect structure of the new phase possibly limits the electrical conductivity.

Acknowledgement

K. T. would like to thank the Alexander von Humboldt Stiftung for the postdoctoral fellowship during the course of which part of this work was done.

REFERENCES

/1/ Hall, H.T., Merill, L., and Barnett, J.D., Science 146, 1297 (1964)

/2/ McWhan, D.B., Bloch, D., and Parisot, G., J. Phys. F 4, L69 (1974)

/3/ Takemura, K., Minomura, S., and Shimomura, O., in: Schilling, J.S., and Shelton, R.N. (eds), Proc. Int. Symp. Physics of Solids under High Pressure (North Holland, Amsterdam 1981)

/4/ Balchan, A.S., and Drickamer, H.G., J. Chem. Phys. 34, 1948 (1961); Riggleman, B.M. and Drickamer, H.G., J. Chem. Phys. 37, 446 (1962); 38, 2721 (1963)

/5/ Glötzel, D. and McMahan, A.K., Phys. Rev. B20, 3210 (1979)

/6/ Takemura, K., Minomura, S., Shimomura, O., and Fujii, Y., Phys. Rev. Lett. 45, 1881 (1980)

/7/ Smith, N.V., Phys. Rev. B2, 2840 (1970)

/8/ Ham, F.S., Phys. Rev. 128, 82 (1962); 128, 2524 (1962)

/9/ Louie, S.G., and Cohen, M. L., Phys. Rev. B10, 3237 (1974)

/10/ McMahan, A.K., Hord, B.L., and Ross, M., Phys. Rev. B15, 726 (1977)

/11/ Sakai, N., Tsuji, K., Takemura, K., Kajiwara, T., and Minomura, S., private communication

PHYSICS OF SOLIDS UNDER HIGH PRESSURE
J.S. Schilling, R.N. Shelton (editors)
© North-Holland Publishing Company, 1981

STRUCTURE OF CESIUM AND IODINE UNDER PRESSURE

Ken-ichi Takemura[+] and Shigeru Minomura

Institute for Solid State Physics, The University of Tokyo, Tokyo 106, Japan

Osamu Shimomura

National Institute for Researches in Inorganic Materials, Ibaraki 305, Japan

[+]Present address: Physikalisches Institut III, Universität Düsseldorf, F.R.G.

The crystal structure of the high-pressure phase of cesium (IV) has been determined from powder x-ray diffraction with a diamond cell and a position-sensitive detector. The structure is a tetragonal lattice with four atoms in a unit cell. The crystallographic data at 80 kbar are a = 3.349 ± 0.006, c = 12.487 ± 0.030 (Å), space group D_{4h}^{19}-I4$_1$/amd. The atomic radius of cesium decreases drastically at the III-IV phase transition, suggesting a discontinuous s-d transition. The crystal structures of molecular and monatomic iodine are also briefly discussed.

I. CESIUM

I.1 Introduction

Cesium is the most compressible material of all elements. In fact, the relative volume of cesium at 100 kbar is about 30% of that at normal pressure. The softness of this material arises in part from the well known s-d electronic transition. The crystal structure of cesium changes from bcc (phase I) to fcc (phase II) at 23.3 kbar and to fcc (phase III) at 42.5 kbar. It transforms further to phase IV at 43 kbar and to phase V at about 110 kbar. The isomorphic transition from phase II to III is also discussed with relation to the s-d transition. According to the recent energy band calculation, the s-d transition takes place in two steps [1]. The lowest d-bands are located near the X-point in the Brillouin zone. The X_1-subband has the same symmetry as that of the s-band at Γ-point and hybridizes with the s-band. As the pressure is increased, the X_1-subband moves down relative to the s-band at the Γ-point. The d-character of the conduction electrons increases continuously with pressure, that is, the spatially spread s-electrons are transferred to more localized d-state. This results in the large compressibility of cesium at low pressure.

On the other hand, the X_3-subband, which lies higher than the X_1-subband, has a different symmetry as that of the s-band and no hybridization is possible with the s-band. Therefore, the s-d transition to the X_3-band occurs abruptly when the bottom of the X_3-band touches the Fermi level. This corresponds to the II-III (fcc-fcc) transition or the II-IV transition at low temperature. The phase III exists in a rather small region of pressure-temperature diagram. Glötzel and McMahan have explained that the lattice vibration plays an essential role in the II-III isostructural transition at room temperature [1].

In order to investigate the nature of s-d electronic transition in cesium, it is important to know the crystal structure of phase IV which might consist of more d-like cesium atoms. In fact, what kind of structure occurs after the close packed fcc lattice? In previous experiments by Hall et al. (x-ray diffraction [2]), McWhan et al. (neutron diffraction [3]) and Inoue (x-ray diffraction with SSD [4]), it was not possible to determine the crystal structure of Cs IV. The present high-pressure powder x-ray diffraction experiments were made with a diamond cell [5] and a position-sensitive detector (PSD) [6].

I.2 Experimental

Cesium with purity of 99.95% was melted in silicone oil. The gasket was set on the diamond anvil and covered with silicone oil. The melted cesium was drawn into a syringe and pushed out on the gasket hole. The cesium droplet was quickly squeezed into the gasket hole. The colour of cesium prepared in this way was golden and there were no diffraction lines of cesium oxide in the x-ray photograph. The pressure was determined by ruby R_1 line.

X-ray photographs of cesium at low pressure were spotty and it was very difficult to get a powder pattern of good quality. At higher pressure, however, the size of the cesium crystallites was reduced effectively according to the large deformation of the gasket and sample. X-ray photographs of the phase IV (at about 80 kbar) showed continuous Debye-rings, from which it was possible to estimate the relative intensities. The measurements with the PSD were made after the photographs had proved the polycrystalline nature of the sample.

I.3 Result and discussion

The x-ray patterns of cesium (IV) are shown in Fig. 1. The pattern (c) was taken on the process of up-pressure. The pattern (a) was taken on the process of down-pressure from the phase V. The both patterns seem different. Two

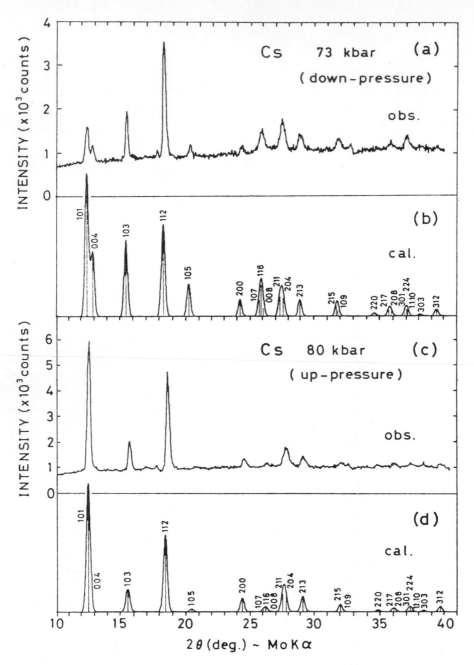

Fig. 1   Observed and calculated x-ray diffraction patterns of cesium IV.

peaks at $2\theta = 13.0°$ and $20.5°$ are missing in pattern (c) and peak intensities in both patterns do not coincide. Note that the Debye-rings in x-ray photographs on the up-pressure process were more homogeneous and one could obtain the reproducible peak intensities. On the down-pressure process, however, the Debye-rings were inhomogeneous and the relative intensities of reflection peaks were not reproducible.

Considering the above situation, six dominant peaks in pattern (c), which are also ob-

served in pattern (a), were used for indexing. The indexing procedure is based on the method developed by Aoki et al. /7/. Among the cubic, tetragonal, hexagonal and orthorhombic systems, only a tetragonal lattice with axial ratio of $c/a = 3.729$ and $z = 4$ gives a reasonable set of indices which satisfy the extinction rule: $h + k + l = 2n$, $2h + l = 2n + 1$ or $4n$. The space group is uniquely determined to be $D_{4h}^{19}-I4_1/amd$. In this space group, four atoms

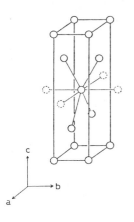

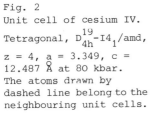

Fig. 2
Unit cell of cesium IV.
Tetragonal, D_{4h}^{19}-I4$_1$/amd,
z = 4, a = 3.349, c =
12.487 Å at 80 kbar.
The atoms drawn by
dashed line belong to the
neighbouring unit cells.

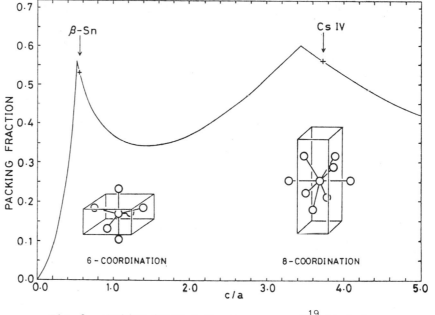

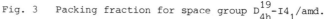

Fig. 3   Packing fraction for space group D_{4h}^{19}-I4$_1$/amd.

occupy the special positions; (O,
O,O), (1/2,1/2,1/2), (O,1/2,1/4),
(1/2,O,3/4). The calculated intensity profile for this structure is shown in Fig. 1 (b). The
peak positions agree well with
those in the pattern (a) and (c),
moreover the two peaks at 2θ =
13.0° and 20.5° can be indexed
as OO4 and 1O5, respectively.
The problem is now to explain
the pattern (c). The missing or
weak peaks in the pattern (c)
contain the OO1 reflections or
reflections from the planes nearly parallel to (OO1). This
suggests the possibility of preferred orientation. If we assume
that the c-axis lies parallel to
the incident x-ray beam, which is
also the direction of uniaxial
stress in the present diamond
cell, the OO1 reflections become
very weak. The calculated
pattern in Fig. 1 (d) includes
the effect of preferred orientation. Here the c-axis is assumed
to distribute around the incident
x-ray beam according to a Gaussian function.
The agreement with the pattern (c) is excellent.
Thus both patterns on up- and down-pressure cycle
can be explained by this crystal structure. The
unit cell of Cs IV is shown in Fig. 2. One cesium
atom is coordinated with eight neighbouring
atoms. The space group D_{4h}^{19}-I4$_1$/amd is the same
as that of the β-Sn type structure. The compa-

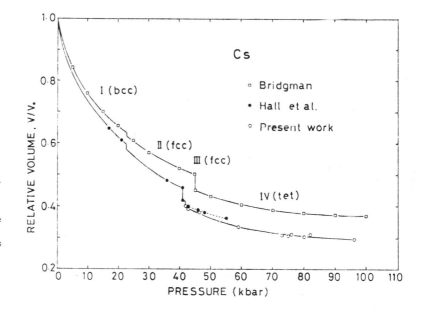

Fig. 4   Volume compression of cesium.

rison of Cs IV structure with the β-Sn type
structure is shown in Fig. 3. The curve shows
the packing fraction for this space group as a
function of axial ratio, c/a. The β-Sn type
structure is located near the cusp at c/a =
0.516, where the atom takes an ideal 6-coordina-
tion. The Cs IV structure is located near the
cusp at c/a = 3.464, where the ideal 8-coordina-
tion is realized. That is to say the Cs IV

structure is a typical 8-coordinated
structure. The axial ratio of c/a stays
constant with pressure in the phase IV.

The pressure-volume diagram calculated from the lattice constants is
shown in Fig. 4 together with the data
by other workers. The data by Bridgman
/8/ have been obtained from direct volume measurement in a piston-cylinder
apparatus. The data by Hall et al. /2/
are based on the x-ray diffraction.
They assigned one line of the phase IV
(possibly 103 reflection of tetragonal
lattice) to a fcc lattice (possibly as
200 reflection) and calculated the
volume. In Fig. 4, their data for the
phase IV are corrected to the tetragonal lattice assuming the same c/a ratio. The present result continues well
with the data by Hall et al. A dramatic
change is seen in the atomic radius of
cesium which is defined as a half of
nearest neighbour distance (Fig. 5).
It decreases as much as 10% at the III-
IV transition. If the II-III-IV (at
room temperature)or the II-IV (at low temperature) transition is induced by the s-d transition concerning the X_3-band, the abrupt collapse
of atom could be well explained.

Cesium transforms further to the phase V
after a wide region of coexistence with the
phase IV (100-120 kbar). The diffraction pattern
of the phase V is more complicated than that of
IV. The structure analysis of Cs V is now in
preparation.

II. IODINE

II.1 Introduction

The most interesting characteristics of iodine under pressure are the continuous metallization and the molecular dissociation. Both characteristics are thought to be rather common na-
tures of molecular crystals including solid hydrogen. Previous experiments by the present
authors revealed a structural change of iodine
under pressure /9/ /10/ /11/.
During the course of metallization, iodine stays
in a molecular phase, while it dissociates to a
monatomic phase just after the metallization is
completed /11/. In this chapter, the structural
change of iodine is described briefly so as to
give a background to the optical study, which
follows this paper. A diamond cell /5/ and a
PSD /6/ were used in the x-ray diffraction
experiments.

II.2 Molecular iodine

The lattice constants and the atomic parameters were calculated from the positions and
intensities of diffraction peaks by least-squares
method. The obtained structure of metallic molecular phase of iodine is shown in Fig. 6 (b), as
a projection onto the b-c plane (molecular plane).
If we compare it with the structure at normal
pressure [(a)], the following results are derived:
(1) intramolecular distance is unchanged,

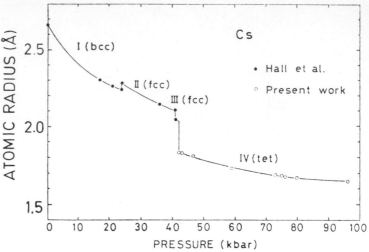

Fig. 5   Atomic radius of cesium

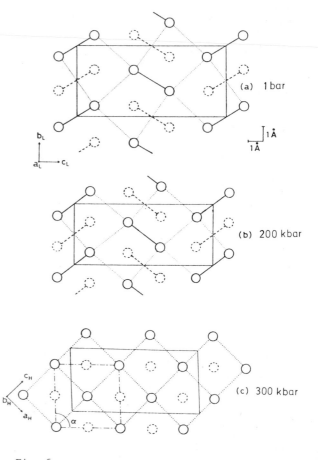

Fig. 6
Projection of atomic positions of iodine onto
the molecular plane.

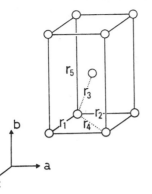

Fig. 7
Unit cell of monatomic iodine. Orthorhombic,
D_{2h}^{25}-Immm, z = 2, a = 3.031, b = 5.252, c = 2.904
Å at 300 kbar.

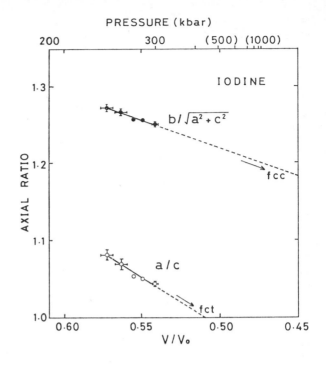

Fig. 8 Variation of axial ratio with pressure.

(2) the angel between the molecular axis
    and the c-axis increases with pressure.
Consequently, the intra- and inter-molecular
distances become comparable which causes the
increase of charge transfer between adjacent
molecules. The metallic conduction can be explained by this charge-transfer model. The intermolecular distance along the a-axis (perpendicular to the molecular plane) is about 20% longer
than that in the molecular plane at 200 kbar.
The conductivity along the a-axis is supposed to
be much lower than that in the plane.

II.3 Monatomic iodine

    Iodine undergoes a structural phase transi-
tion at about 210 kbar /10/. The diffraction pattern of the high-pressure phase has been well
interpreted by a body-centered orthorhombic
lattice which is formed by iodine atoms /11/.
That is, iodine molecule dissociates to atoms at
210 kbar. The structure of metallic monatomic
phase of iodine is shown in Fig. 6 (c). The
dissociation is understood as the culmination
of the structural change in molecular phase.
The unit cell of body-centered orthorhombic lattice is shown in Fig. 7. The difference between
the c- and a-axis is about 9% at 210 kbar, while
it is as small as 4% at 300 kbar. The axial ratio
a/c and $b/\sqrt{a^2+c^2}$ are plotted as a function of
relative volume or pressure in Fig. 8. If the
a/c equals to 1, the structure is a body-centered
tetragonal (= face-centered tetragonal). A
simple extrapolation of the plot predicts the
occurrence of such a structure at 450 kbar. If
the $b/\sqrt{a^2+c^2}$ further equals to 1, the structure
is a face-centered cubic. The plot suggests
the occurrence of this simple lattice at an
extremely high pressure.

Acknowledgement

    The work on iodine was performed with Y.
Fujii of Brookhaven National Laboratory.

References

/1/ Glötzel,D. and McMahan, A.K., Phys. Rev.
    B20 (1979) 3210-3216
/2/ Hall, H.T., Merrill, L. and Barnett, J.D.,
    Science 146 (1964) 1297-1299
/3/ McWhan, D.B., Bloch, D. and Parisot, G.,
    Rev. Sci. Instrum. 45 (1974) 643-646
/4/ Inoue, K., (unpublished)
/5/ Takemura, K., Shimomura, O., Tsuji, K. and
    Minomura, S., High Temp.-High Press. 11
    (1979) 311-316
/6/ Fujii, Y., Shimomura, O., Takemura, K.,
    Hoshino, S. and Minomura, S., J. Appl.
    Cryst. 13 (1980) 284-289
/7/ Aoki, K., Shimomura, O. and Minomura, S.,
    J. Phys. Soc. Japan 48 (1980) 551-556
/8/ Bridgman, P.W., Collected Experimental
    Papers, (Harvard University Press, Cam-
    bridge, 1964) 6, 61
/9/ Shimomura, O., Takemura, K., Fujii, Y.,
    Minomura, S., Mori, M., Noda, Y. and
    Yamada, Y., Phys. Rev. B18 (1978) 715-719
/10/ Takemura, K., Fujii, Y., Minomura, S. and
    Shimomura, O., Solid State Commun. 30
    (1979) 137-139
/11/ Takemura, K., Minomura, S., Shimomura, O.
    and Fujii, Y., Phys. Rev. Lett. 45 (1980)
    1881-1884

PHYSICS OF SOLIDS UNDER HIGH PRESSURE
J.S. Schilling, R.N. Shelton (editors)
© *North-Holland Publishing Company, 1981*

INTERATOMIC POTENTIALS FOR SOLID ARGON AND NEON AT HIGH PRESSURES

Guantian Zou, Ho Kwang Mao, Larry W. Finger,
Peter M. Bell, and Robert M. Hazen

Geophysical Laboratory
Carnegie Institution of Washington
Washington, D.C. 20008

New, single-crystal volume data of argon and neon were obtained by x-ray diffraction for the first time in situ at high pressure, 293 K.[1] The data in the pressure range 11.5-82 kbar for argon and 47.5-144 kbar for neon were corrected to 0 K for comparison with theoretical models. The corrected (0 K) data coincided closely with data previously obtained at low pressure.[2] An equation of state derived from an interatomic potential function was fit to the data with good agreement.

INTRODUCTION

Recently a new series of high pressure experiments on the solid, crystalline forms of the rare gases, argon and neon, were done at 293 K by x-ray diffraction.[1,3] Single crystals of argon and of neon were confined in separate experiments in diamond-window high-pressure cells,[4] that were positioned on an automated, 4-circle x-ray diffraction goniometer. X-ray difffraction data were obtained for argon and neon in the pressure ranges 11.5-82 kbar and 47.5-144 kbar respectively, at 293 K.

Rare gas solids are relatively compressible, and their equations of state, as deduced by precise specific volume versus pressure measurements at high pressure are useful for comparison with fundamental theoretical calculations of their properties.[5]

Unlike most experiments on rare gas solids in the past, the present study was conducted at high temperature (constant, 293 ± 1 K) rather than at cryogenic temperatures. Thus the data can be used to provide a high temperature point of reference for comparison (after temperature correction) with low temperature data that had been obtained under a limited range of pressure.[2] An advantage in the present experiments was the ease with the pressure could be varied from point to point at high precision. In each experimental sample a few small crystals of ruby were included for reference to the calibrated special shift of the R_1 fluorescent line with pressure.[6] The pressure measurement was precise to $\pm$ 0.5 kbar. The results are in the form of unit cell parameters and volumes at each pressure measured.[3] The techniques employed in these experiments are described in detail elsewhere.[4]

DATA REDUCTION TO ZERO K

A useful comparison of the present high-temperature data with previous cryogenic data could be made by correcting the 293 K values to 0 K and calculating the static pressure. The method was to subtract the thermal pressure and evaluate the zero-point vibration pressure as follows. The equation of state of a rare gas solid is assumed to be represented by

$$P(V,T) = P_0(V) + P_t(V,T)$$
$$P_0(V) = P_s(V) + P_z(V) \tag{1}$$

where P is pressure, V, volume; T, absolute temperature; the subscripts s, z, and t refer to components of the pressure at V and T originating from the static lattice, 0 K vibrations, and thermal energy, respectively. The zero subscript is referenced to 0 K. The thermal pressure (P_t) is given by

$$P_t(V,T) = (\gamma/V)3RTD(\Theta/T) \tag{2}$$

where γ is the Gruneissen parameter; R, the gas constant; $\Theta(V)$, the Debye temperature; D, the Debye function of the form

$$D(\Theta/T) = 3(T/\Theta)^3 \int_0^{\Theta/T} \frac{u^3 du}{e^u - 1} \tag{3}$$

In the Debye approximation, the 0 K vibration pressure (P_z) can also be expressed by

$$P_z(V) = \frac{9\gamma}{8V} R\Theta(V) \tag{4}$$

These relations can be utilized to obtain the static pressure correction by defining the Gruneissen parameter as independent of volume ($\gamma = (V/V_0)\gamma_1 + \gamma_\infty$, as suggested by Holt and Ross,[7] where

$$\gamma = \gamma_\infty = 1/2 \text{ as } V \rightarrow 0).$$

Then the Debye temperature is given by

$$\Theta(V) = \Theta_0(V_0/V)^{\frac{1}{2}} e^{\gamma_1(1 - V/V_0)} \tag{5}$$

Equations 1-5 were applied to the present va-

lues[3] to yield curves for the thermal and
static lattice pressures as functions of the
specific volume. The curves for argon and
neon given in Fig. 1, are correction functions
for reducing the 293 K values to 0 K.

The data are of sufficient precision to justi-
fy the application of potential functions at 0
K. Numerous theoretical techniques have been
derived to treat the intermolecular potentials
for argon and neon [5] and thus there is sufficient basis to develop a potential for these
data. The fit of a potential function at the
low pressures can be compared with previous
calculations.

Syassen and Holzapfel[8] in a low temperature
study of xenon described the atomic interaction of a solidified rare gas in the
zeroth-order approximation as an additive
pairwise central force model.

For solid xenon, however, contributions to the
potential function from many-body interactions
are relatively large (Syassen and Holzapfel[8]
reported a difference in compression of 30%
between observed data and values calculated
from a pair-potential at 100 kbar, 85 K).

The potential energy function of a usable
model is a combination of assumed short-range
repulsive forces and long-range Van der Waals
type interactions calculated from quantum mechanics and from empirical experimental data.
The objective in the present calculation was
to fit the data to a theoretical potential
function of the attractive forces and to a
parameterized model of the repulsive forces
partly constrained by the boundary conditions
of the potential energy minima for all atoms
in a crystal of argon or of neon. The present
approach can be summarized as follows. The
potential energy (U) per mole (N atoms) is

$$U(\vec{r}_1, \vec{r}_2, \cdots \vec{r}_N) = \sum_{i<j=1}^{N} U_2(\vec{r}_i, \vec{r}_j)$$
$$+ \sum_{i<j<k=1}^{N} U_3(\vec{r}_i, \vec{r}_j, \vec{r}_k) \tag{6}$$

where U_2 and U_3 are the pair and triplet potentials, and the $\vec{r}$ are atomic potential
vectors (interatomic distances).

The atomic potentials ($\emptyset$) are given by

$$\emptyset/\epsilon = [A_1 + A_2(1-x)]e^{\alpha(1-x)} + (A_3 + A_4(1-x))e^{\frac{\alpha(1-x)}{d}}$$
$$- \sum_{k=6,8,9,10} (-1)^k C_k/x^k \tag{7}$$

where $X = R/R_m$, $C_k = C_k^*/\epsilon R_m$
and where A_1, A_2, A_3, A_4, α, and d are constants, the C_k^* are constant dispersion coefficients[5] and R_m is the ratio of interatomic
distance to the interatomic distance at the
potential minimum, whose value is $-\epsilon$.
In this formulation, the negative summation
term on the right-hand side of equation 7 is
the large distance potential derived from

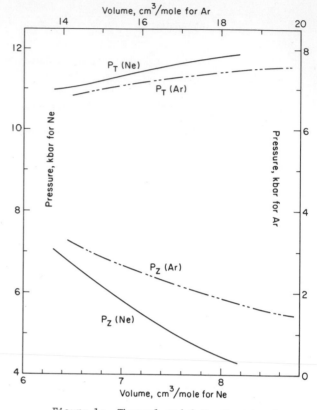

Figure 1: Thermal and 0 K vibrational
pressure correction curves

quantum mechanical perturbation theory.[5] The
positive terms are derived from the
self-consistent-field (SCF) Hartree-Fock repulsion for empirical calculation of the adjustable parameters. Two exponential terms
(instead of one) provided a better fit to the
data over the large pressure range.

The total potential energy can thus be expressed.

$$U/\epsilon = \frac{N}{2}\left\{ \sum_n [A_1 + A_2(1-nX)]e^{\alpha(1-nX)} + \sum_n [A_3 + A_4(1-nX)]e^{\frac{\alpha(1-nX)}{d}} \right.$$
$$\left. - \sum_k (-1)^k N_k C_k/X^k \right\} \tag{8}$$

where $n = 1, \sqrt{2}, \sqrt{3}, 2$, and $k = 6, 8, 9, 10$
The equation of state as a thermodynamic function, related to the potential, is given by

$$P = -\frac{\partial U}{\partial V} = -\frac{1}{3NR^2}\frac{\partial U}{\partial R} = -\frac{1}{3NR_m^3 X^2}\frac{\partial U}{\partial X} \tag{9}$$

where P is the pressure, and V is the volume.

Non-linear least squares procedures were employed in a computational program to fit the
functions to the data. The adjustable parameters are D, R_m, A_1, A_2, A_3, A_4, two of which
are eliminated by the boundary conditions of
equation 7, where $R = R_m$. These boundary conditions are

$$\emptyset/\epsilon = A_1 + A_3 - \sum_{k=6,8,9,10}(-1)^k C_k = -1 \quad \text{and}$$
$$\partial\emptyset/\partial X = 0 \tag{10}$$

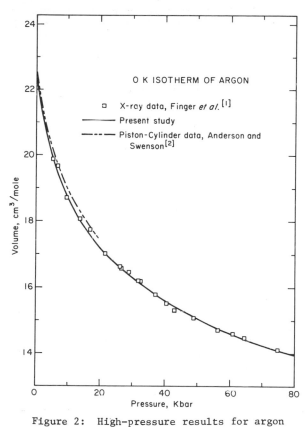

Figure 2: High-pressure results for argon

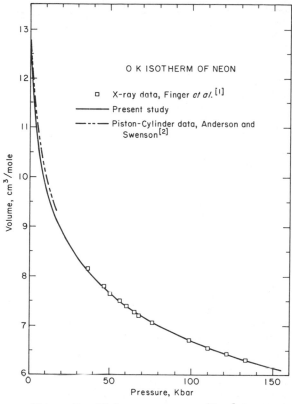

Figure 3: High-pressure results for neon

Table I.   Interatomic Potential Parameters
for Neon and Argon

Parameter	Neon	Ref.	Argon	Ref.
Θ_0, K	75.1	(9)	93.3	(10)
γ_1	2.05	(9)	2.20	(11)
$C_6 * 10^{-12}$ erg Å^6	6.589	(12)	64.04	(12)
$C_8 * 10^{-11}$ erg Å^8	1.980	(12)	31.54	(12)
$C_9 * 10^{-12}$ erg Å^9	4.634	(12)	189.1	(12)
$C_{10} * 10^{-11}$ erg Å^{10}	7.245	(12)	219.9	(12)
N_6	14.454	(12)	14.454	(12)
N_8	12.802	(12)	12.802	(12)
N_9	15.259	(13)	15.259	(13)
N_{10}	12.311	(12)	12.311	(12)
ϵ/k, K^{-1}	42.75	(5)	141.3	(5)
α	14		14	
d	2.2		5.0	
R_m, Å	3.25		3.95	
A_1	1.406		0.7362	
A_2	-1.302		0.0629	
A_3	-1.115		-0.2873	
A_4	-11.291		-9.5656	

The data points and the fitted curves for argon and neon are shown in Figures 2 and 3. Constants for the curves are listed in Table 1.

The rare gas solids are highly compressible and thus can be used to test theory and to provide a better basis for understanding the compressibility behavior of solids in general.

The results for argon and neon are within the uncertainty limits of the data of Anderson and Swenson.[2] The curves fit the recent data of Finger et al.[1] with good agreement over the extended pressure range.

REFERENCES:

(1) Finger, L. W., Hazen, R. M., Zou, G., Mao, H. K., and Bell, P. M., Structure and compression of crystalline argon and neon at high pressure and room temperature, in press, J. Appl. Phys. Lett. 39 (1981).

(2) Anderson, M. S., and Swenson, C. A., Experimental equation of state for the rare gas solids, J. Phys. Chem. Solids. 36 (1975) 145-162.

(3) Zou, G., Finger, L. W., Hazen, R. M., Bell, P. M., and Mao, H. K., Isothermal equations of state for neon and argon, Carnegie Inst. Wash. Year Book 80, in press (1981).

(4) Mao, H. K. and Bell, P. M., Design and operation of a diamond-window, high-pressure cell for the study of single-crystal samples loaded cryogenically, Carnegie Inst. Wash. Year Book 79 (1980) 409-411.

(5) Barker, J. A., Interatomic potentials for inert gases from experimental data, in Klein, M. L. and Venables, J. A., eds., Rare Gas Solids, Vol. 1 (Academic Press, New York, 1976).

(6) Mao, H. K., Bell, P. M., Shaner, J. W., and Steinberg, D. J., Specific volume measurements of Cu, Mo, Pd, and Ag and calibration of the ruby R_1 fluorescence pressure gauge, J. Appl. Phys. 49 (1978) 3276-3283.

(7) Holt, A. C., and Ross, M., Calculation of the Gruineisen parameter for some models of the solid, Phys. Rev. B1 (1970) 2700-2705.

(8) Syassen, K. and Holzapfel, W. B., High-pressure equation of state for solid xenon, Phys. Rev. B18 (1978) 5826-5834.

(9) Fugate, R. Q., and C. A. Swenson, Specific heats of solid natural neon at five molar volumes and of the separated neon isotopes at $\underline{P}$ = 0, J. Low Temp. Phys. 10 (1973) 317-342.

(10) Finegold, L., and N. E. Phillips, Low-temperature heat capacities of solid argon and krypton, Phys. Rev. 177 (1969) 1383-1391.

(11) Tilford, C. R., and C. A. Swenson, Thermal expansions of solid argon, krypton, and xenon above 1 K, Phys. Rev. B5, (1972) 719-732.

(12) Bell, R. J. and I. J. Zucker, Long-range forces in Klein, M. L. and Venables, J. A., eds., Rare Gas Solids, Vol. 1 (Academic Press, New York, 1976).

(13) Doran, M. B., The multipole long range (Van der Waals) interaction coefficients for neon, J. Phys. B5, (1972) L151-L152.
Doran, M. D., The multipole long range (Van der Waals) interaction coefficients for neon, argon, krypton and xenon, J. Phys. B7 (1974) 588-569.
Doran, M. B. and Zucker, I. J. Higher order multipole three-body Van der Waals interactions and stability of rare gas solids, J. Phys. C4 (1971) 307-312.

PHYSICS OF SOLIDS UNDER HIGH PRESSURE
J.S. Schilling, R.N. Shelton (editors)
© North-Holland Publishing Company, 1981

INVERSION OF THE Γ AND X CONDUCTION BANDS IN InP UNDER HIGH PRESSURE

Toshihiko Kobayashi, Takamasa Tei, Kazunori Aoki,
Keiichi Yamamoto and Kenji Abe

Department of Electrical and Electronic Engineering
Faculty of Engineering, Kobe University
Rokkodai, Nada, Kobe 657, Japan

Photoluminescence measurements on InP samples have been made under hydrostatic pressure up to the phase transition ($\sim$100 kbar). The direct E_O gap shifts sublinearly with pressure. At pressures above 70 kbar the emission intensity decreases exponentially. This can be explained in terms of electron transfer from the Γ_{1c} minimum to the descending X_{1c} minima. The critical pressure for the inversion of the two conduction bands is found to be $\sim$80 kbar. Results show for the first time that the X_{1c} minima move *sublinearly* towards the Γ_{1c} minimum with pressure and the Γ–X energy separation is about 0.8 eV at normal pressure.

1. INTRODUCTION

The experimental investigations of the influence of hydrostatic pressure on the optical properties of semiconductors have attracted special interest lately. Hydrostatic pressure can change the energies of the conduction bands with respect to the valence band without changing the symmetry of crystal structure and can provide an independent means of studying band structure, especially the position of the higher-lying conduction-band minima. The general trends of the dependence of the energy gaps of diamond- and zinc-blende-type semiconductors on hydrostatic pressure have been known in previous measurements with conventional large-volume pressure cells at relatively low pressure (typically up to $\sim$10 kbar).

In GaAs, from transmission measurements at room temperature with a diamond-anvil cell, Welber et al.[1] have shown a sublinear dependence of the E_O direct gap energy on pressure up to the phase transition ($\sim$180 kbar). Yu and Welber [2] have confirmed this sublinear variation of E_O with pressure by means of photoluminescence and resonant Raman scattering also in a diamond-anvil cell for pressures up to 72 kbar. At normal pressure the lowest energy gap in GaAs is direct, so photoexcited carriers recombine radiatively to produce the observed luminescence. When the pressure is increased above 40 kbar, due to the difference of the pressure coefficients of the Γ_{1c} and X_{1c} minima, an inversion of the two conduction-band minima takes place. The lowest energy gap becomes indirect and the band structure resembles that of GaP. Therefore the recombination of photoexcited carriers becomes almost nonradiative and the emission intensity decreases. At 45 kbar the intensity is about three orders of magnitude smaller than its initial value. By fitting the pressure dependence of the emission intensity with a theoretical expression, the critical pressure P_O for the inversion of the Γ_{1c} and X_{1c} minima was found to be $\sim$33 kbar for n-type GaAs with carrier concentration $<10^{16}$ cm^{-3} and the linear pressure coefficient of the X_{1c}

minima (-2.7×10^{-3} eV/kbar) was obtained for the first time. Similar effects on the emission intensity in heavily doped p- and n-type GaAs at 120 K were also reported by Olego et al.[3].

In InP, the absorption and photoluminescence measurements at room temperature have been recently performed by Müller et al.[4] under hydrostatic pressure up to the phase transition ($\sim$100 kbar). The E_O gap also exhibits a sublinear dependence on pressure as in the case of GaAs at room temperature [1,2]. The emission intensity is almost constant to 95 kbar and at 101 kbar it is only about 0.25 times the intensity at lower pressures. Beyond this pressure the E_O emission completely disappeared. By fitting this decrease in the pressure range 95–101 kbar with a theoretical expression, the critical pressure P_O was estimated to be 104 ± 1 kbar, this indicating that an inversion of the Γ_{1c} and X_{1c} minima may take place at pressure close to the phase transition. This is sufficiently different from the behavior found for GaAs [2,3]. Further the linear pressure coefficient for the X_{1c} minima was found to be $-(3 \pm 1) \times 10^{-3}$ eV/kbar. Recently, Kobayashi [5] and Tei et al.[6] have pointed out that the emission intensity is almost constant to 70 kbar and beyond this pressure it suddenly decreases exponentially. At around 100 kbar the intensity is three or four orders of magnitude lower than the values at lower pressures. The observed variation of emission intensity with pressure is in distinct contrast to that found by Müller et al.[4].

According to the results of electrical transport measurements by Pitt et al.[7,8,9], it is expected to change the lowest band gap of InP from direct to indirect at pressures above 70 kbar. Pitt and Vyas [9] have assumed that the Γ_{1c} minimum moves away linearly from the valence band at a rate of 9.5×10^{-3} eV/kbar over the whole pressure range (0–80 kbar) of their measurements, which is higher than the experimental value of 8.4×10^{-3} eV/kbar obtained from 100 kbar measurement [4].

To estimate the value of P_o, reliable information concerning the pressure dependence of the Γ_{1c}, L_{1c} and X_{1c} minima as well as their energy separations at normal pressure is required. At this stage, however, the data have not been fully refined to give the exact critical pressure P_o or the pressure coefficients of higher-lying minima. For a more rigorous discussion about the conduction band structure, considerable experimental and theoretical studies in InP are still required. In this paper we present more detailed experimental results on the luminescence across the E_o direct energy gap of InP with several electron concentrations ($10^{15} - 10^{19}$ cm^{-3}) for pressures up to about 105 kbar, with the intention to observe distinct changes in the emission intensity when the Γ_{1c} minimum is shifted above the X_{1c} minima. By using empirical pseudopotential form factors, band-structure calculations are also carried out to place some constraints on the pressure coefficients of the indirect X_{1c} minima and the Γ-X energy separation at normal pressure.

2. EFFECT OF HYDROSTATIC PRESSURE ON THE LUMINESCENCE INTENSITY

The application of hydrostatic pressure does not affect the symmetry of the crystal. The pressure causes simply a decrease in the crystal volume and hence a decrease in lattice constant. Therefore only band edges may be moved with respect to one another.

The pressure dependence of the emission intensity can be generally analized by the luminescent efficiency for a two-valley semiconductor system wherein the total electrons occupy the direct and indirect conduction-band minima simultaneously. The efficiency expression can be derived in terms of radiative and non-radiative lifetimes and also intervalley scattering times in each valley of the conduction band. A schematic

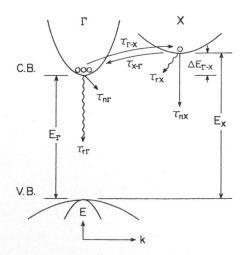

Figure 1 : Schematic diagram showing the band structure of a two-valley semiconductor

diagram of the band structure of the two-valley semiconductor and the lifetimes relevant to the two valleys (Γ_{1c} and X_{1c} minima) are shown in Figure 1. The lifetimes τ_r and τ_n are the radiative and non-radiative recombination times, respectively. The subscripts Γ and X in the lifetimes denote the minimum from which such recombinations are taking place. Similarly, $\tau_{\Gamma-X}$ and $\tau_{X-\Gamma}$ represent the intervalley scattering times from the Γ_{1c} to X_{1c} minima and vice versa, respectively.

The transfer rates of carriers for the Γ_{1c} and X_{1c} minima can be written as [10]

$$I_o + (n_X/\tau_{X-\Gamma}) = n_\Gamma(1/\tau_{r\Gamma} + 1/\tau_{n\Gamma} + 1/\tau_{\Gamma-X}), \quad (1)$$

and

$$n_\Gamma/\tau_{\Gamma-X} = n_X(1/\tau_{X-\Gamma} + 1/\tau_{nX} + 1/\tau_{rX}), \quad (2)$$

where I_o is the net rate of generation of carriers in the case of photoluminescence, n_Γ and n_X the carrier concentrations in the Γ_{1c} and X_{1c} minima, respectively. The generation rate I_o was put into Eq.(1) for the Γ_{1c} state transfer rates. In view of momentum conservation, this is strictly valid for photoluminescence excitation of direct-gap semiconductors up to the pressure at which the Γ_{1c} and X_{1c} minima become degenerate ($E_\Gamma \simeq E_X$). For higher pressures where the inversion of the Γ_{1c} and X_{1c} minima occurs ($E_\Gamma > E_X$), it implicitly assumes that the recombination times τ_r and τ_n are longer than the intervalley scattering times so that thermal equilibrium between the Γ_{1c} and X_{1c} minima can be reached and thus Eq.(1) still gives the correct results. The rate of photon generation I in the semiconductor under pressure is expressed by

$$I = n_\Gamma/\tau_{r\Gamma} + n_X/\tau_{rX} . \quad (3)$$

From Eqs.(1)-(3) one can easily obtain the luminescent efficiency

$$\eta = I/I_o$$
$$= (n_\Gamma + n_X \frac{\tau_\Gamma}{\tau_X} \frac{n_X}{n_\Gamma})(1 + \frac{\tau_\Gamma}{\tau_X} \frac{n_X}{n_\Gamma})^{-1}, \quad (4)$$

where

$$1/\tau_\Gamma = 1/\tau_{r\Gamma} + 1/\tau_{n\Gamma} , \quad (5a)$$

$$1/\tau_X = 1/\tau_{nX} + 1/\tau_{rX} , \quad (5b)$$

and η_Γ, η_X are the efficiencies at the Γ_{1c} and X_{1c} minima, respectively, defined by

$$\eta_\Gamma = \tau_\Gamma/\tau_{r\Gamma} , \quad (6a)$$

$$\eta_X = \tau_X/\tau_{rX} . \quad (6b)$$

If $\eta_\Gamma \simeq 1$ (mostly radiative) and $\eta_X \simeq 0$ (non-radiative) as in the case of GaAs or InP, then the luminescent efficiency (4) can be simplified as

$$\eta = 1/(1 + \frac{\tau_{r\Gamma}}{\tau_{nX}} \frac{n_X}{n_\Gamma}) \; . \qquad (7)$$

It is seen that under conditions of thermal equilibrium and in the absence of other carrier transfer, the ratio n_X/n_Γ in Eq.(7) can be represented by the usual Boltzmann distribution

$$n_X/n_\Gamma = N_X/N_\Gamma \, (m_X/m_\Gamma)^{3/2} \exp(-\Delta E_{\Gamma-X}/kT) \; , \qquad (8)$$

where N_Γ (=1) and N_X account for the degeneracy of the Γ_{1c} and X_{1c} minima, respectively, m_Γ is the density of states mass of Γ_{1c} minimum and m_X that of X_{1c} minima, and $\Delta E_{\Gamma-X} = E_X - E_\Gamma$ is the subband energy gap between the two minima.

In InP the pressure dependence of E_Γ (i.e., the E_o direct gap energy) at room temperature exhibits a sublinearity similar to those observed for GaAs [1,2], as will be shown in subsequent section, and can be written as

$$E_\Gamma = E_\Gamma(0) + \alpha_\Gamma P + \beta_\Gamma P^2 \; , \qquad (9)$$

where α_Γ and β_Γ the linear and quadratic pressure coefficients of E_Γ, respectively, and P the pressure. As to E_X, recent theoretical calculations [11] suggest the sublinear pressure dependence. If we write

$$E_X = E_X(0) + \alpha_X P + \beta_X P^2 \; , \qquad (10)$$

where α_X and β_X the linear and quadratic pressure coefficients of E_X, respectively, then the luminescent efficiency (7) is given by

$$\eta(P)$$
$$= \left\{ 1 + A \exp\left[\frac{(\alpha_\Gamma - \alpha_X)(P-P_o) + (\beta_\Gamma - \beta_X)(P^2 - P_o^2)}{kT} \right] \right\}^{-1}$$
$$\qquad (11)$$

where

$$A = N_X/N_\Gamma \, (m_X/m_\Gamma)^{3/2} \frac{\tau_{r\Gamma}}{\tau_{nX}} \; , \qquad (12)$$

P_o the critical pressure for the inversion of the Γ_{1c} and X_{1c} minima. Fitting the experimental pressure dependence of emission intensity with Eq.(11), we can determine the pressure P_o, the pressure coefficients α_X and β_X, and parameter A. If the Γ_{1c} and X_{1c} minima are assumed to shift linearly with pressure to a first approximation, Eq.(11) becomes the same as that used by Yu and Welber [2].

3. EXPERIMENT

The photoluminescence measurements under high pressure were performed in the back scattering configuration using the diamond-anvil pressure cell described by Kobayashi [5]. The InP samples, cut from single crystals and polished down to a thickness of about 30 μm, were broken into sufficiently small pieces. A fragment suitable to fit into the hole (∼100 μm diameter) of the pressure cell gasket was selected under the microscope and inserted into the hole. In the present experiment, a 4:1 mixture of methanol and ethanol was used as a pressure-transmitting medium, which ensures a truly hydrostatic pressure up to about 100 kbar. The pressure inside the diamond-anvil cell was determined by the ruby fluorescence method.

The photoluminescence was excited by the 488 nm line of a Coherent CR-4 Ar$^+$ laser and analyzed by a detection system consisting of a Spex 0.85 m double spectrometer, an RCA 31034 or 7102 photomultiplier and a photon-counting electronics. All measurements were made at room temperature.

4. RESULTS AND DISCUSSION

4.1 Photoluminescence Spectra

Figure 2 shows typical photoluminescence spectra

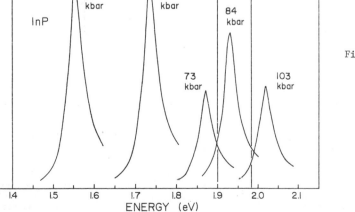

Figure 2 : Typical photoluminescence spectra of n-InP (1.5×10^{16} cm^{-3}) for various pressures. With increasing pressure the lines shift to higher energies. The intensity shows a rapid decrease for pressures above 70 kbar.

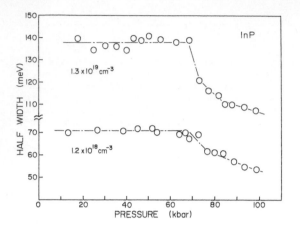

Figure 3 : Pressure dependence of the half
width of the emission line for the heavily
doped n-InP samples (1.2×10^{18} and 1.3×10^{19} cm^{-3}). The half width exhibits a
steplike dependence on pressure.

of n-type InP with 1.5×10^{16} cm^{-3} at various
pressures. At room temperature the luminescence
spectrum is dominated by a single peak due to
band-to-band recombination at the lowest E_O energy gap. With increasing pressure the peak shifts
to higher energies, this corresponding to the
rise in the Γ_{1c} conduction-band minimum relative
to the top of valence band: the E_O direct gap
increases. At pressures around 70 kbar the emission intensity for each sample suddenly decreases.
This decrease at higher pressures is attributed
to the proximity in energy or the inversion of
the Γ_{1c} and X_{1c} minima and to the electron transfer from the Γ_{1c} minimum to the descending X_{1c}
minima where radiative recombination must be indirect and thus very weak (i.e., $\eta_X \simeq 0$).

For the heavily doped samples (1.2×10^{18} and 1.3×10^{19} cm^{-3}) the shape of the emission line
changes drastically for pressures above which
the material is indirect. Figure 3 shows the
half width of the emission line as a function of
pressure. When the pressure is increased above
70 kbar the lines become narrower and the half
width exhibits a steplike dependence on pressure.
This effect can be related to the carrier transfer from the Γ_{1c} minimum to the descending X_{1c}
minima. The position of the Fermi level with
respect to the bottom of the Γ_{1c} conduction band
rapidly decreases. The width of the emission
lines at higher pressures may be interpreted as
the natural line width in undoped n-type InP.

The emission disappears for pressures beyond
103.5 ± 0.5 kbar. This is found to be reversible
with pressure, and is related to the metallic
phase transition. In fact, under the microscope
with white light it is also observed that the
crystals turn opaque beyond this pressure, followed by a complete disappearance of luminescence
in InP.

4.2 Pressure Coefficients of the E_O Direct Gap

Figure 4 shows typical pressure dependence of
the emission peak across the E_O direct gap of n-
InP. Its pressure dependence is the same as that
of E_O as reported for GaAs [2,3] and the maximum
of the intensity of the spectra was taken as the
direct gap E_O. There is no doubt that the emission peak energy is shifted sublinearly to higher energies with increasing pressure.

The experimental data have been fitted with Eq.
(9) and the resulting coefficients α_Γ and β_Γ are
presented in Table I. No significant carrier-
concentration dependence was observed for samples
used in the present experiments. Clearly the
results are in close agreement with the previous
work by Müller *et al.*[4], which are also listed
in Table I for comparison.

The sublinear behavior observed at room temperature can be partly interpreted in terms of the
nonlinearlity of the change in lattice constant
with hydrostatic pressure [4]. When the energy
shift caused by a hydrostatic compressive pressure is plotted as a function of relative lattice
compression ($-\Delta a/a_o$), the observed direct gap
energies exhibit a linear variation within the
experimental error (see Figure 5),

$$E_\Gamma = E_\Gamma(0) + 3a(-\Delta a/a_o) , \qquad (13)$$

where a the deformation potential of the Γ_{1c} minimum, also listed in Table I. In our experiments
the average value for the deformation potential
a was foumd to be -6.47 eV. This is again in
close agreement with the experimental value of
(-6.35 ± 0.05) eV obtained by Müller *et al.*[4].

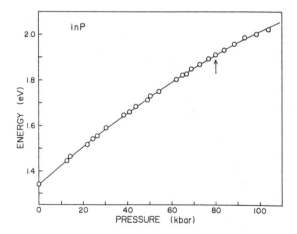

Figure 4 : Typical pressure dependence of the
emission peak across the E_O direct gap for
n-InP (1.5×10^{16} cm^{-3}). The solid line is
a fit performed with Eq.(9). The critical
pressure for the inversion of the two conduction bands is indicated by an arrow.

Table I : Pressure coefficients α_Γ and β_Γ, and deformation potential *a* for several electron concentrations.

n (cm^{-3})	α_Γ (eV/kbar)	β_Γ (eV/kbar^2)	a (eV)
6.7×10^{15}	8.50×10^{-3}	-1.81×10^{-5}	-6.27
1.5×10^{16}	8.75	-2.24	-6.28
5.8×10^{17}	8.48	-1.78	-6.45
1.2×10^{18}	8.63	-1.92	-6.49
1.3×10^{19}	8.71	-2.47	-6.87
$N_D \sim 5 \times 10^{15}$ *	8.4 ± 0.2	$-(1.8 \pm 0.3)$	$-(6.35 \pm 0.05)$

*Müller *et al.*[4]

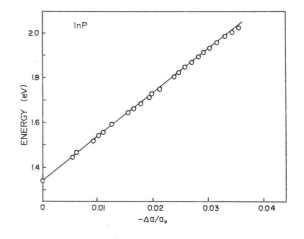

Figure 5 : Typical E_0 direct gap of *n*-InP (1.5×10^{16} cm^{-3}) as a function of lattice constant. The solid line represents a least-squares fit.

4.3 Pressure Dependence of the Emission Intensity

Figure 6 shows typical luminescence intensity as a function of pressure for *n*-InP. The emission intensity is almost constant within the experimental error at lower pressures. Beyond 70 kbar, however, all samples show an exponential decrease in the intensity, in good agreement with the preliminary measurements by Kobayashi [5] and Tei *et al.*[6]. At around 100 kbar the emission intensity is three or four orders of magnitude smaller than the values at lower pressures. Clearly the results are in marked contrast to the recent work of Müller *et al.*[4] for high purity InP ($N_D \approx 5 \times 10^{15}$ cm^{-3}), but are somewhat similar to the behavior observed by Yu and Welber [2] for *n*-type GaAs ($< 10^{16}$ cm^{-3}) at room temperature or by Olego *et al.*[3] for a heavily doped *n*-type GaAs (7×10^{18} cm^{-3}) at 120 K up to about 50 kbar, which is well below the pressure for the metallic phase transition (~ 180 kbar) in GaAs.

We have fitted the pressure dependence of emission intensity with Eq.(11). The resulting fit is represented by the solid line in Figure 6. From the fitting procedure the parameters A, α_X, β_X and P_0 are determined and listed in Table II. For comparison the corresponding values, except β_X, reported by Müller *et al.*[4] are also tabulated. In spite of the apparent sublinear pressure dependence of the direct E_0 gap or the Γ_{1c} conduction minimum, the effect of quadratic pressure coefficient β_Γ on the emission intensity was not included in their calculations, in a manner similar to that reported by Yu and Welber [2] for GaAs. In addition they implicitly assumed the linear pressure dependence of the X_{1c} minima.

From the data in Table II, no significant carrier concentration dependence of linear and quadratic pressure coefficients α_X and β_X is observed for samples used here. We find the average values of α_X and β_X at room temperature: $\alpha_X = -(2.7 \pm 0.5) \times 10^{-3}$ eV/kbar and $\beta_X = -(0.6 \pm 0.5) \times 10^{-5}$ eV/kbar^2. To our knowledge this is the first direct measurement of the *sublinear* pressure dependence

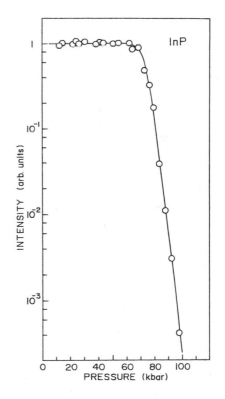

Figure 6 : Typical pressure dependence of the emission intensity for *n*-InP (1.5×10^{16} cm^{-3}). The solid line represents a fit performed with Eq.(11).

Table II : Parameters obtained with Eq.(11) by fitting the pressure dependence of
the emission intensity.

n (cm^{-3})	A	P_O(kbar)	α_X(eV/kbar)	β_X(eV/kbar^2)	$\Delta E_{\Gamma-X}$(eV)
6.7×10^{15}	8	80 ± 1	$-(2.50 \pm 0.5) \times 10^{-3}$	$-(0.61 \pm 0.5) \times 10^{-5}$	0.80 ± 0.07
1.5×10^{16}	8	81 ± 1	$-(2.75 \pm 0.5)$	$-(0.64 \pm 0.5)$	0.83 ± 0.07
5.8×10^{17}	8	80 ± 1	$-(2.52 \pm 0.5)$	$-(0.68 \pm 0.5)$	0.81 ± 0.07
1.2×10^{18}	8	81 ± 1	$-(2.87 \pm 0.5)$	$-(0.56 \pm 0.5)$	0.84 ± 0.07
1.3×10^{19}	8	78 ± 1	$-(2.80 \pm 0.5)$	$-(0.60 \pm 0.5)$	0.78 ± 0.07
$N_D \sim 5 \times 10^{15}$ *	8	104 ± 1	$-(3 \pm 1)$	———	1.0 ± 0.2

*Müller et al.[4]

of the indirect X_{1c} conduction-band minima at
room temperature. One can see from Eqs.(9) and
(10) that the sub-band energy gap between the Γ_{1c}
and X_{1c} minima ($\Delta E_{\Gamma-X}$) also shows a sublinear
pressure dependence. Our results for P_o, the
pressure at which the inversion of the Γ_{1c} and
X_{1c} minima occurs, are about 80 kbar well below
the pressure for the metallic phase transition
(~ 104 kbar), but considerably lower than the
value of 104 ± 1 kbar estimated by Müller et al.
[4] for high purity InP at the same temperature.
These experimental results are consistent with
our pressure-dependent band-structure calculations for InP using empirical pseudopotential
form factors [11].

4.4 Conduction Band Structure of InP

The higher band minima L_{1c} and X_{1c} at normal
pressure have been the subject of some discussion and at present considerable uncertainty
still exists. Values in the literature for the
sub-band energy gaps $\Delta E_{\Gamma-L}$ and $\Delta E_{\Gamma-X}$ center round
$0.5 - 0.6$ eV and $0.7 - 1.0$ eV, respectively. The
sub-band energy gap $\Delta E_{\Gamma-X}$ at normal pressure can
now be determined by using the experimental value
of the critical pressure P_o and extrapolating the
sublinear pressure dependence of the X_{1c} minima.
The values of $\Delta E_{\Gamma-X}$ so obtained at normal pressure are about 0.8 eV in all samples measured in
the present experiments (see Table II), in reasonable agreement with that estimated by Pitt and
Vyas [9]. This result is also supported by the
experimental and theoretical work on high-field
transport properties in InP under high pressure
by Kobayashi et al.[12,13].

In Figure 7 we plot typical variation of the conduction band structure in InP with pressure.
The energy was measured from the valence band
maximum at k = 0. The corresponding results of
Pitt and Vyas [9] derived from electrical transport measurements under pressure are also shown
for comparison. As can be seen in Figure 7, the
critical pressure P_o for the inversion of the Γ_{1c}
and X_{1c} minima in our experiments is about 80
kbar which is slightly larger than the value of

~ 70 kbar estimated by Pitt and Vyas [9]. This
discrepancy has mainly resulted from difference
in pressure dependence of the Γ_{1c} conduction
minimum.

In order to explain the above data, we have performed band structure calculations for InP under
pressure, by using empirical pseudopotential
form factors [11]. In the calculations 89 plane
waves are included. The form factors under hydrostatic pressure are needed at different wave
vectors. Using the values determined by Cohen
and Bergstresser [14] as a guide, we chose a
wide range of values appropriate for describing
the observed dependence of the E_o direct energy
gap with pressure. Figure 8 shows the calculat-

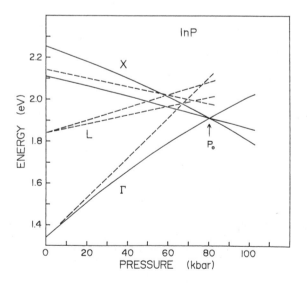

Figure 7 : Change of the conduction bands Γ_{1c}
and X_{1c} with pressure. The dashed lines
are the results obtained by Pitt and Vyas
[9].

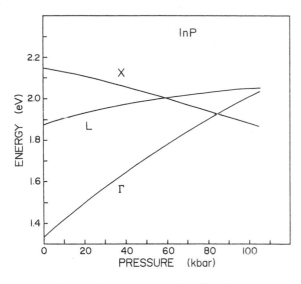

Figure 8 : Calculated results for the change
of the conduction bands Γ_{1c}, L_{1c} and X_{1c}
with pressure.

ed results for the pressure dependence of the Γ_{1c},
L_{1c} and X_{1c} minima. At normal pressure the calculated sub-band energy gap $\Delta E_{\Gamma-X}$ is about 0.82 eV
and the pressure P_o at which the inversion of the
two bands takes place is about 85 kbar. The overall shape of each curve is in good agreement with
experimental results. We find the theoretical
pressure coefficients for the Γ_{1c} minimum : $\alpha_\Gamma =$
8.5×10^{-3} eV/kbar and $\beta_\Gamma = -1.8 \times 10^{-5}$ eV/kbar^2.
The corresponding values for the X_{1c} minima are
$\alpha_X = -2.4 \times 10^{-3}$ eV/kbar and $\beta_X = -0.20 \times 10^{-5}$ eV/
kbar^2, respectively, in reasonable agreement with
those obtained experimentally.

As to the L_{1c} conduction-band minima, detailed
information concerning the pressure dependence
of them and the sub-band energy gap $\Delta E_{\Gamma-L}$ at normal pressure cannot be obtained by the results
from present experiments alone. Taking $\Delta E_{\Gamma-L}$ of
0.5 eV at normal pressure, the pressure coeffi-
cient of $\alpha_L = 2 \times 10^{-3}$ eV/kbar for the L_{1c} minima
[9] and using the pressure coefficients α_Γ and
β_Γ for the Γ_{1c} minimum determined here, an inversion of the Γ_{1c} and L_{1c} minima may be expected
to occur at around 110 kbar, which is somewhat
higher than the pressure for the metallic phase
transition in InP. If this inversion takes place
in the pressure range 70-100 kbar, additional
decrease in the emission intensity should occur.
However, this was not observed in our experiments. Therefore, the pressure for the inver-
sion of the Γ_{1c} and L_{1c} minima proposed by Pitt
and Vyas [9] may have to be revised significantly
upwards, if other parameters used here are essentially correct.

[1] Welber, B., Cardona, M., Kim, C.K. and
Rodriguez, S., Dependence of the direct
energy gap of GaAs on hydrostatic pressure,
Phys. Rev. B 12 (1975) 5729-5738.
[2] Yu, P.Y. and Welber, B., High pressure photoluminescence and resonant Raman study of
GaAs, Solid St. Commun. 25 (1978) 209-211.
[3] Olego, D., Cardona, M. and Müller, H., Photoluminescence in heavily doped GaAs II. Hydrostatic pressure dependence, Phys. Rev. B 22
(1980) 894-903.
[4] Müller, H., Trommer, R., Cardona, M. and
Vogl, P., Pressure dependence of the direct
absorption edge of InP, Phys. Rev. B 21
(1980) 4879-4883.
[5] Kobayashi, T., Diamond-anvil pressure cell
and photoluminescence measurements on InP,
Memoirs of the Faculty of Engineering, Kobe
Univ., No.26 (March 1980).
[6] Tei, T., Shimazu, M., Aoki, K., Kobayashi, T.
and Yamamoto, K., Inversion of the conduction bands Γ-L-X of InP under hydrostatic
pressure, TGSSD80-129, Technical Group on
Semiconductors and Semiconductor Devices
(Feb. 20, 1981) Inst. Electronics Comm.
Engrs. Japan.
[7] Pitt, G.D., The conduction band structure of
InP from a high pressure experiment, Solid
St. Commun. 8 (1970) 1119-1123.
[8] Pitt, G.D., Lees, J., Hoult, R.A. and
Stradling, R.A., Magnetophonon effect in
GaAs and InP to high pressure, J. Phys. C:
Solid St. Phys. 6 (1973) 3282-3294.
[9] Pitt, G.D. and Vyas, M.K.R., Pressure effects on the threshold for high-field instabilities in InP, J. Phys. C: Solid St. Phys.
8 (1975) 138-146.
[10] Hakki, B.W., Theory of luminescent efficiency of ternary semiconductors, J. Appl. Phys.
42 (1971) 4981-4995.
[11] Kobayashi, T., Tei, T., Aoki, K. and
Yamamoto, K., in preparation.
[12] Kobayashi, T., Kimura, T., Yamamoto, K. and
Abe, K., Effect of hydrostatic pressure on
transferred electron instabilities in n-InP,
in: Wilson, B.L.H. (eds.), Physics of Semiconductors (The Institute of Physics,
Bristol and London, 1979).
[13] Kobayashi, T., Mori, S., Hirata, Y., Aoki,
K. and Yamamoto, K., Transferred electron
effects in InP under high pressure, J. Phys.
C: Solid St. Phys. 13 (1980) L29-35.
[14] Cohen, M.L. and Bergstresser, T.K., Band
structure and pseudopotential form factors
for fourteen semiconductors of the diamond
and zinc-blende structures, Phys. Rev. 141
(1966) 789-796.

PHYSICS OF SOLIDS UNDER HIGH PRESSURE
J.S. Schilling, R.N. Shelton (editors)
© *North-Holland Publishing Company, 1981*

ABSORPTION SPECTRA AND PRESSURE DEPENDENCE OF PSEUDODIRECT GAPS IN ZnSiP$_2$ CRYSTAL

Tsuguru Shirakawa and Junkichi Nakai

Department of Electronic Engineering,
Faculty of Engineering, Osaka University
Osaka, Japan

Absorption coefficients near pseudodirect gaps in ZnSiP$_2$ crystals were measured using the polarized light at 77K. The spectra obtained were analyzed with the absorption theory by the Wannier exciton. The results indicate that the spectrum near pseudo-direct gaps is similar to that of the direct gap in III-V semiconductors.
The values of three pseudodirect gaps between the lowest Γ_6^c conduction band and three valence bands are 2.145, 2.257 and 2.299 eV at 77K, respectively.
The hydrostatic-pressure dependence of pseudodirect gaps also was investigated at 300K under the pressures up to 56 kilobar. The pressure coefficients of three pseudo-direct gaps are $-2.0_5 \times 10^{-6}$, $-1.9_2 \times 10^{-6}$ and $-1.8_8 \times 10^{-6}$ eV/bar, respectively. These values are very close to that of the $\Gamma_{15} - X_1$ indirect gap in III-V semiconductors.

1. INTRODUCTION

Ternary II-IV-V$_2$ semiconductors with the chalco-pyrite(CP) structure, which are structural and electronic analogues of binary III-V compounds with the zincblende(ZB) structure, have attracte tracted the attention of many researchers because of the extensive range of properties more than those of elemental (Ge, Si) and binary semiconductors[1-3]. ZnSiP$_2$ crystal is a electronic analogue of GaP or AlP and is a semiconductor of which the band gap is about 2.05 eV at room temperature. This crystal has attracted special interest in the applications to opto-electronic devices at the visible range and nonlinear optic devices at the infrared region.

The electronic band structure in ZnSiP$_2$ crystal is unique because of the appearance of "pseudo-direct gaps" at Γ point in the Brillouin zone which originate in the transformation from the indirect band in III-V binary analogue into the direct one in compounds with CP structure by doubling of unit cell. The pseudodirect gap can not be observed in III-V semiconductors. The studies on the band structure nea the pseudodirect gaps have been performed with the measurements of absorption coefficients[4-8], photoconductivity[9] and photovoltage[10]. However, the theoretical analysis for the shape of the absorption spectra are not found in previous papers except Ref.7, although it is shown that three inflections observed in the spectra are caused by optical transitions be-tween pseudodirect gaps on the basis of the quasi-cubic model.

The purpose of this work is to make the band structure in ZnSiP$_2$ clear by comparing the shape of absorption spectra with the theoretical one at low temperature and by investigating the hydrostatic-pressure dependence of pseudo-direct gaps at room temperature.
We have performed the measurements of absorption

spectra using the polarized light at 77K and analyzed the spectra with the excitonic absorption theory. The results show that the absorptioncoefficients at pseudodirect gaps are considerably smaller than those at the direct gap in III-V semiconductors, but the shapes of spectra is similar to that near the direct gap in III-V semiconductors.
It has been known that the pressure coefficient of the band gap energy in elemental annd binary semiconductors depends on the position in k-space of the conduction band extremum, but is independent of the semiconductor type[11].
We have measured the pressure coefficients of three pseudodirect gaps under the pressures up to 56 kilobar using a diamond anvil cell.
The values obtained are very close to that of the indirect gap in III-V semiconductors.
We also discuss the pressure dependence of the splitting of the valence band.

2. ELECTRONIC BAND STRUCTURE

ZnSiP$_2$ crystal has a tetragonal system shown in Fig.1, in which two cations (Zn and Si) are substituted for Ga atom in GaP crystal with the ZB structure. It is expected that the band structure of ZnSiP$_2$ is similar to that of III-V semiconductors. The Brillouin zone in the crystal with the CP structure is one-fourth of that in the ZB structure because the lattice constant in z-direction is twice that in x-direction (doubling of unit cell). Therefore, the approximate band structure of compounds with the CP structure can be obtained by embedding the energy levels in the zincblende Brillouin zone into a quater zone.
As shown in Fig.2, the lowest conduction band Γ_6^c in ZnSiP$_2$ is derived from X$_1$ band in semi-conductors with the ZB structure, and the most upper valence band is from the Γ_{15} valence band in the ZB-structure which splits into three bands (Γ_6^v, Γ_7^v, Γ_6^v) due to the crystal field and the spin-orbit interaction. "pseudodirect gaps"

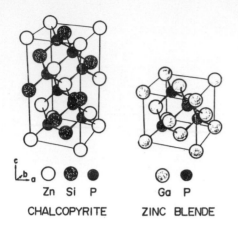

Fig. 1. Chalcopyrite and zincblende crystal structures. Zn and Si atoms in the chalcopyrite crystal are substituted for Ga atom in GaP crystal.

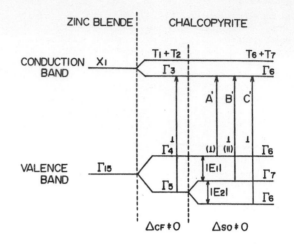

Fig.2. Band structure near the pesudodirect gaps in a chalcopyrite crystal and selection rule for the transition.

originate in the transformation from the indirect band of III-V binary analogue (GaP/AlP) into the direct one in semiconductors with the CP structure.

According to Hopfield's quasi-cubic model[12], the energies of two Γ_6^V levels relative to the Γ_6^V level in valence bands are given by

$$E_{1,2} = +\frac{1}{2}(\Delta_{so} + \Delta_{cf}) \pm \frac{1}{2}[(\Delta_{so} + \Delta_{cf})^2 - \frac{8}{3}\Delta_{so}\Delta_{cf}]^{1/2}, \qquad (1)$$

where Δ_{so} is the spin-orbit splitting in the cubic field and Δ_{cf} is the crystal field splitting. Δ_{cf} is described as a function of the compression rate c/a along z-axis by the following equation,

$$\Delta_{cf} = \frac{3}{2}b(2 - \frac{c}{a}) , \qquad (2)$$

where b is the deformation potential and c and a are the lattice constants in z- and x-directions, respectively. Δ_{so} is presented by

$$\Delta_{so} = \frac{1 - f_i}{2}\frac{\Delta(II) + \Delta(IV)}{2} + \frac{1 + f_i}{2}\Delta(V) , \qquad (3)$$

where f_i is the ionicity of the chemical bond, and $\Delta(N)$ is the spin-orbit splitting energy of a free N atom[13].

3. EXPERIMENTAL

ZnSiP$_2$ crystals used in this work were grown by slowly cooling a Zn solution containing 3 mole % ZnSiP$_2$ at the rate of 7 °C/hr from 1200 °C, and the crystals obtained have a small needle like shape (10 x 2 x 0.5 mm^3) with $(\bar{1}\bar{1}2)$

and $(\bar{1}01)$ faces, growing along the [111] direction.

The samples were prepared by mechanically lapping and polishing to about 0.05∼0.1 mm thick paralell to the $(\bar{1}01)$ face, and then etching with the mixture of HF : HNO$_3$ = 1 : 1 and HF for 5 minures and 15 minutes, respectively. The experimental apparatus used for the measurement of the absorption coefficient is described in Ref.8.

The absorption coefficient α was calculated by using the following equation,

$$\alpha = \frac{1}{d}\ln[\frac{1}{2(I/I_0)}(\frac{1 - R}{R})^2 \\ \times[-1 + (1 + \frac{4R^2(I/I_0)^2}{(1 - R)^4})^{1/2}]], \qquad (4)$$

where d is the sample thickness, R is the reflectivity and I_0 and I are the incident light intensity and the transmitted one, respectively. The reflectivity R is 0.28∼0.32 at the spectral range measured in tis work[14] and so we used $R = 0.30$ for the calculation.

4. RESULTS AND DISCUSSION

4.1 Absorption Spectra at Low Temperature

In Fig.3, we show the absorption spectrum obtained at 77K using the light polarized with its E vector perpendicular to the c-axis of the crystal ($E \perp c$). Three peaks are found at A'$_{n=1}$, B'$_{n=1}$ and C'$_{n=1}$ in the spectrum. Similar structure has been observed in the spectra of absorption coefficients[4-8], photoconductivity [9] and photovoltage[10] at the temperatures from 77K to 300K. These peaks are attributed to the optical absorption by the free excitons formed from the Γ_6^C conduction band and three valence bands (named A', B' and C' excitons), because the crystal field splitting Δ_{cf} and the spin- orbit splitting Δ_{so}, calculated from three peak energies using eq.(1), are in good agreement with those calculated theoretically[13, 15].

In order to confirm that the A'$_{n=1}$ peak in Fig.3 originates in the absorption by the exciton formed from the Γ_6^c band and the most upper valence band, we have investigated the dependence of the A'$_{n=1}$ peak height on the polarization of the light. The measurements were performed on the ($\bar{1}$01) face. The relation between the directions of the light beam and the c-axis of the crystal is shown in Fig.4. γ is the angle between the light beam direction and the c-axis which is 62.66° in the present case, and θ is the angle between the E vector of the light and the a-axis of the crystal. A plot of the A'$_{n=1}$ peak height as a function of θ is shown in Fig.5, where θ = 90° corresponds with the $E \perp c$ geometry, but θ = 180° does not correspond with the $E /\!/ c$ (see Fig.4). In the configuration shown in Fig.4, the absorption coefficient is described by the following equation,

$$\alpha = \alpha_{\perp}(1 - \sin^2\gamma\cos^2\theta) + \alpha_{/\!/}\sin^2\gamma\cos^2\theta \quad , \quad (5)$$

where $\alpha_{\perp}$ and $\alpha_{/\!/}$ are the absorption coefficients for the light polarized with $E \perp c$ and $E /\!/ c$, respectively. According to the selection rule for the optical transition, the transition between the Γ_6^c band and the most upper valence band is forbidden for the light polarized with $E /\!/ c$, and so $\alpha_{/\!/}$ = 0. Therefore the theoretical curve is decribed by $\alpha = \alpha_{\perp}(1 - \sin^2\gamma\cos^2\theta)$. The calculated curve of $\alpha/\alpha_{\perp}$ also is shown in Fig.5. The experimental result is in good agreement with the calculated curve and satisfies the selection rule for the optical transition.

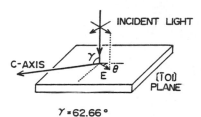

Fig.4. Relations between the directions of the light beam, the E vector and the c-axis of the crystal.

The spectrum for the light polarized with $E /\!/ c$ also is shown in Fig.3. This spectrum was obtained by eliminating the component for $E \perp c$ from the spectrum at θ = 180°, on the basis of eq.(5). Since the transitions between the pseudodirect gaps are forbidden for the light polarized with $E /\!/ c$, the spectrum for $E /\!/ c$ originates in the transition into the other conduction band from the valence band. The energy of the direct gap between the Γ_7^c band and the valence band in ZnSiP$_2$ ias been known to be 2.96 eV at room temperature from the electroreflectance spectrum[16]. So the spectrum for $E /\!/ c$ can be attributed to the transition from the valence band to the (Γ_6^c + Γ_7^c) conduction band (indirect band). The (Γ_6^c + Γ_7^c) band also is derived from the X$_1$ band in semiconductors with the ZB structure and its energy level is comparable with that of the Γ_6^c band. If the spectrum for $E /\!/ c$ does not depend on the polarization of the light for the reason of the transition at the indirect gap, the absorption spectrum due to only the transition at pseudodirect gaps can be obtained by eliminating the component of the spectrum for $E /\!/ c$ from the spectrum for $E \perp c$. The result is shown in Fig.6.

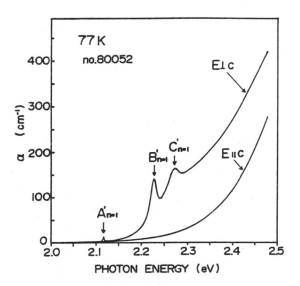

Fig.3. Absorption spectra near pseudodirect gaps at 77K.

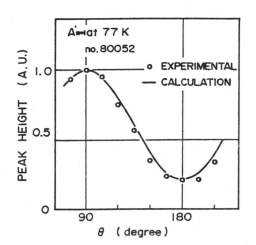

Fig.5. A plot of the peak height due to the A'$_{n=1}$ exciton as a function of θ.

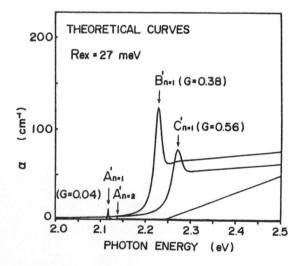

Fig.6. Spectrum due to only the transitions at three pseudodirect gaps.

We have attempted to fit a theoretical curve to that shown in Fig.6, on the basis of the Elliot's theory[17].
According to the Elliot's theory, the total absorption coefficient due to a Wannier exciton plus the direct band-band transition is described by the following equation,

$$\alpha(h\nu) = \frac{(2\mu)^{3/2} e^2 f_{cv}}{Nc\hbar m_0} [\sum_n 4\pi R_{ex}^{*\,3/2} \frac{\delta(h\nu - E_n)}{n^3} + \frac{2\pi R_{ex}^{*\,1/2} u(h\nu - E_g)}{1 - \exp[-2\pi(\frac{R_{ex}^*}{h\nu - E_g})^{1/2}]}] , \tag{6}$$

THEORETICAL CURVES

$R_{ex} = 27$ meV

Fig.7. Theoretical curves of the absorption coefficient due to the transitions between each pseudodirect gap.

Table 1. Parameters used for the curve-fitting.

	A'	B'	C'
Band Gap E_g (eV)	2.145	2.257	2.299
G	0.04	0.38	0.56
Ratio of Peak Height	0.012 :	1 :	0.85
(Values from eq.(8))	0.027 :	1 :	0.98

$$E_n = E_g - \frac{R_{ex}^*}{n} , \tag{7}$$

where $h\nu$ is the photon energy, μ the reduced mass, f_{cv} the oscillator strength, N the refractive index, c the light velocity, m_0 the free electron mass, R_{ex}^* the effective Rydberg, $\delta(h\nu - E_n)$ the delta function, n the integer, E_g the band gap energy and $u(h\nu - E_g)$ the unit step function.
The realistic curve which can be compared with the experimental curve can be obtained using the curve-fitting procedure proposed by Sell and Lawaety[18] in which they took the broadening effect into theoretical curves by convoluting $\alpha(h\nu)$ with a Lorentzian function $\Gamma\pi^{-1}((h\nu)^2 + \Gamma^2)^{-1}$, where Γ is the half width at half-maximum.
The curve-fitting was tried for various values of the broadening parameter $G = \Gamma/R_{ex}^*$ and the height of A'$_{n=1}$, B'$_{n=1}$ and C'$_{n=1}$ peaks, using $R_{ex}^* = 27$ meV and $\mu = 0.24 m_0$. The values of R_{ex}^* and the reduced mass μ have been obtained from the absorption peak energies by the n = 1 and n = 2 quantum states of A' exciton in Ref.8.
The calculated curve, which was obtained using the values of G and the band gap energies tabulated in Table 1, is shown in Fig.6. This theoretical curve consists of three curves and a straight line shown in Fig.7. The origin of the straight line is not clear, although it seems that it is caused by the polalization dependence of the absorption at the indirect gap.

The ratio of the oscillator strength at A' and C' gaps to that at B' gap are presented by

$$\frac{2}{2 + (2 - \frac{3E_{1,2}}{\Delta_{so}})} , \tag{8}$$

where E_1 and E_2 are the energy differences between the Γ_7^v and two Γ_6^v valence bands shown in Fig.2. The calculated values of ratio, which were obtained by substituting the experimental values of E_1, E_2 and Δ_{so} into eq.(8), are in good agreement with the values determined from curves in Fig.7 tabulated in Table 1.
It is shown from the result of the curve-fitting that the absorption spectrum near the pseudodirect gaps is similar to that near the direct gap in semiconductors with the ZB struc-

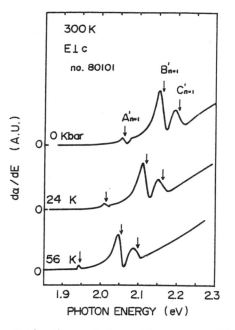

Fig.8. Derivatives of absorption curves with respect to the photon energy under the pressure.

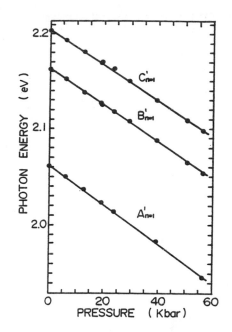

Fig.9. Energy shifts of absorption peaks by three excitons under the pressure.

ture, which is desribed by eq.(6), although the absorption coefficients in ZnSiP$_2$ are very small (< 200 cm^{-1}) in contrst to that of III-V semiconductors ($\sim 10^4$ cm^{-1}).

4.2 Absorption Spectrum under Hydrostatic Pressure

We have measured the absorption coefficients under the hydrostatic pressures up to about 56 kilobar at 300K using a diamond anvil cell. Absorption peaks due to the excitons can be observed even at 300K because of the large exciton binding energy (27 meV), although the peaks considerably broad. To obtained the precise values of peak energies, the derivative of the absorption curve with respect to the photon energy (dα/d($h\nu$)) was obtained with the numerical calculation using the mini-computer. Three spectra of dα/d($h\nu$) at 0, 24.0+0.3 and 56.0+0.3 kilobar are shown in Fig.8. The energy shifts of the absorption peaks with the pressure are shown in Fig.9. The energies of three pseudodirect gaps decrease linearly with the pressure within the limits of this experiments. Assuming that the exciton binding energy does not depend on the pressure, we obtain $-2.0_5 \times 10^{-6}$, $-1.9_2 \times 10^{-6}$ and $-1.8_8 \times 10^{-6}$ eV/bar as the pressure coefficients dE_g/dP of A', B' and C' pseudodirect gaps, respectively. These values are in good agreement with those reported in Ref.6 in which the measurements were performed under the pressure of only 4 kilobar. As mensioned in Sec.1, The pressure coefficients of the band gap in semiconductors with the ZB structure depends on the symmetry of the electron wave function in the conduction band

extremum[11]. For III-V semiconductors, the typical pressure coefficients of ($\Gamma_{15}^{v} - \Gamma_{1}^{c}$), ($\Gamma_{15}^{v} - L_{1}^{c}$) and ($\Gamma_{15}^{v} - X_{1}^{c}$) gaps are (+8.5 ~11.7) x 10^{-6}, ~+5.0 x 10^{-6} and (-1.0~-1.5) x 10^{-6} eV/bar, respectively. The coefficients of three pseudodirect gaps in ZnSiP$_2$ obtained in this work are very close to that of ($\Gamma_{15}^{v} - X_{1}^{c}$) indirect gap in III-V semiconductors. This result indicates. that the Γ_{6}^{c} conduction band in ZnSiP$_2$, which is derived from the X$_1$ band in compounds with the ZB structure, show the similar characteristic to the X$_1$ band in the ZB structure, under the hydrostatic pressure.

The pressure dependence of the crystal field splitting Δ_{cf} is shown in Fig.10. The pressure coefficeints of Δ_{cf} is about 0.14 x 10^{-6} eV/bar. Assuming that the deformation potential b does not change under the pressure used in the

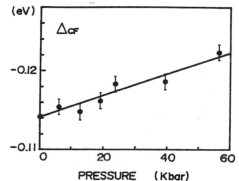

Fig.10. Dependence of the crystal field splitting Δ_{cf} on the pressure.

present work, we can know the pressure dependence of c/a. The deformation potential b under zero pressure has been determined to be -1.16 eV by using $\Delta_{cf} = -0.114$ eV and $c/a = 1.934$ which was determined by the measurements of X-ray diffraction.

Under thre pressure of $56.0+0.3$ kilobar, c/a becomes to be 1.92_9 with $\Delta_{cf} = -0.122$ eV in Fig.9 and $b = -1.16$ eV. This result shows that the compression rate c/a in the z-direction increases linearly with pressure.

The spin-orbit splitting Δ_{so} change scarcely under the pressure. The pressure coefficient of Δ_{so} is about 0.04×10^{-6} eV/bar.

It is necessary for the deeper understanding of the band structure to measure the absorption spectrum under higher pressures, in addition to measurements of the compression rate c/a under the pressure by X-ray diffraction.

This work was performed with support by the Special Budget of the Ministry of Education, Science and Culture of Japan.

REFERENCES

[1] Borshchevskii, A.S., Goryunova, N.A., Kesamaly, F.P. and Nasledov, D.N., Semiconducting $A^2B^4C_2^5$ compounds, Phys. Stat. Sol. 21 (1967) 9-55.

[2] Prochukham, V.D. and Rud', Yu.V., Potential practical applications of $II - IV-V_2$ semiconductors, Sov. Phys. Semicon. 12 (1977) 121-135.

[3] Shay, J.L. and Wernick, J.H., Ternary chalcopyrite semiconductors: Growth, electric properties and applications (Pergamon Press, 1975).

[4] Bendorus, R., Prochukham, V.D. and Sileika, A., The lowest conduction band minima of $A^2B^4C_2^5$-type semiconductors, Phys. Stat. Sol. (b) 53 (1972) 745-752.

[5] Babonas, G., Ambrazevicius, G., Grigoreva, V.S., Neviera, V. and Sileika, A., Wavelength-modulated absorption spectra of pseudodirect gap $A^2B^4C_2^5$ compounds, Phys. Stat. Sol. (b) 62 (1974) 327-334.

[6] Babonas, G., Bendorius, R. and Sileika, A., Pressure dependence of the forbidden energy gap in $ZnSiAs_2$ and $ZnSiP_2$, Phys. Stat. Sol. (b) 62 (1974) K13-K14.

[7] Humphreys, R.G. and Pamplin, B.R., Optical absorption spectrum in $ZnSiP_2$, Jrnl. Phys., C3 (1975) 156-162.

[8] Shirakawa, T., Okamura, K., Nishida, H. and Nakai, J., Exciton absorption spectrum in $ZnSiP_2$ crystal, Japan Jrnl. Appl. Phys. 19 (1980) L618-L620.

[9] Kuhnel, G., Siegel, W. and Ziegler, E., Photoconductivity of $ZnSiP_2$ near the absorption edge, Phys. Stat. Sol. (a) 30 (1975) K25-K27.

[10] Shirakawa, T., Okamura, K and Nakai, J., Photovoltaic effect of $ZnSiP_2$ crystals from Zn melt, Phys. Lett. 73A (1979) 442-444.

[11] Zallen, R. and Paul, W., Effect of pressure on interband reflectivity spectra of Germanium and related semiconducotrs, Phys. Rev. 155 (1967) 703-711.

[12] Hopfield, J.J., Fine structure in the optical absorption edge of anisotropic crystal, Jrnl. Phys. Chem. 15 (1960) 97-107.

[13] Hubner, K. and Unger, K., Spin-orbit splitting in $II- IV-V_2$ compounds, Phys. Stat. Sol. (b) 50 (1972) K105-K107.

[14] Ambrszeviciud, G., Babonas, G. and Sileika, A., Reflectance spectra of $CdSiP_2$ and $ZnSiP_2$ Phys. Stat. Sol. (b) 95 (1979) 643-647.

[15] Willey, J.D., Valence-band deformation potentials for the III-V compounds, Sol. Stat. Commun. 8 (1970) 1865-1868.

[16] Shay, J.L., Tell, B., Buehler, E. and Wernick, J.H., Band structure of $ZnGeP_2$ and $ZnSiP_2$-ternary compounds with pseudo-direct energy gaps, Phys. Rev. Lett. 30 (1973) 983-986.

[17] Elliot, R.J., Intensity of optical absorption by excitons, Phys. Rev. 108 (1957) 1384-1389.

[18] Sell, D.D. and Lawaetz, P., New analysis of direct exciton transition: Application to GaP, Phys. Rev. Lett. 26 (1971) 311-314.

PHYSICS OF SOLIDS UNDER HIGH PRESSURE
J.S. Schilling, R.N. Shelton (editors)
© *North-Holland Publishing Company, 1981*

ORDERED GROUND STATES OF METALLIC HYDROGEN AND DEUTERIUM

N. W. Ashcroft

Laboratory of Atomic and Solid State Physics
Cornell University, Ithaca, N.Y. 14853

Metallic hydrogen and metallic deuterium are both predicted to be liquids in their ground states over a wide range of densities. They may be described as quantum liquid metals, but since the proton and deuteron fluids obey Fermi-Dirac and Bose-Einstein statistics, respectively, their physical properties can be markedly different. In both there is the likelihood of superconductivity through the usual coupling via the density fluctuations of the ions. But in addition the deuteronic fluid admits of the possibility of Bose condensation and may develop superfluid order. Finally, the protons may themselves pair at extremely low temperatures.

1. INTRODUCTION

Under conditions of modest compression and temperature hydrogen and deuterium form diatomically ordered insulating states. This familiar situation changes radically at very high densities where the dense phases of these systems undergo an insulator-metal transition to conducting states. Though this is expected on very general grounds, what cannot yet be foretold with any degree of certainty is the nature of the resulting high density phases, where the many possibilities that might be expected differ in energy by exceedingly small amounts. Perhaps the most interesting aspect of both metallic hydrogen and metallic deuterium is that they seem to fall securely within the class of quantum solids and liquids [1]. Combined with their specifically metallic character, this places them in the somewhat rare category of quantum metallic solids (or as we shall see later, liquids). This in itself is noteworthy: however, at low temperatures there can be long-range ordering in both the electronic and ionic systems, and these too are imbued with unusual characteristics.

The purpose of this article is to discuss the physical attributes of some of the more physically distinct ordered states of metallic hydrogen and metallic deuterium at T = 0 and nearby. In many respects the experimental fabrication of these metals still represents one of the major goals of high pressure physics. As noted by Ginzburg [2], the production of and understanding of new materials and substances is a fundamental challenge. In this context the metallization of hydrogen has been and continues to be as difficult as it is fundamental. The conditions required for its attainment involve serious problems associated with the ultimate strengths of materials. Thus the spirit of this report is very much directed to 'what might be' and in cataloging the remarkable properties below one is nevertheless very conscious of the fact that the ease of theoretical conjecture bears here, apparently, an inverse relationship to difficulty of experimental reality. With this cautionary remark in mind, we shall see that the inferences that can be made from relatively well accepted principles lead to some interesting predictions. These depend in a striking way on the quantum statistics of the systems under consideration and for this reason, among others no doubt, both common isotopes of hydrogen are worthy of detailed experimental examination in the high pressure context.

2. THE FUNDAMENTAL HAMILTONIAN

We begin by considering a neutral assembly of electrons (e) and protons or deuterons (i = p or d), under conditions of density that span, on the one hand, the normal condensed insulating phases of identifiable molecules and on the other high density phases exhibiting metallic properties. This range of densities can be conveniently expressed in terms of the familiar linear measure r_s, defined by $r_s = (3/4\pi\overline{\rho}_e a_0^3)^{1/3}$, where $\overline{\rho}_e$ is the single-particle density for the electrons averaged over the system. From the known density of condensed hydrogen and deuterium at 1 atmosphere, we find $r_s = 3.07$. The states we shall be describing below, however, are characterized by $r_s \approx 1.5$. (Observe the experimental requirement of volume compressions approaching an order of magnitude: the convenience of the linear measure of density is entirely theoretical.)

Let the electrons in the system have a set of coordinates $\{\vec{r}_e\}$, and the protons or deuterons (which we refer to as ions) have the set $\{\vec{R}_i\}$. Then all of the phases that we will consider are described by a fundamental Hamiltonian

$$\hat{H} = \hat{H}(\{\vec{r}_e\}, \{\vec{R}_i\}) . \qquad (1)$$

All of the interactions in the system are strictly Coulombic. We consider only charge

neutral ensembles of fixed number so that the
thermodynamic functions are well defined. If
$v_c(k) = 4\pi e^2/k^2$ then for a macroscopic ensemble
of volume Ω, $\hat{H}$ takes the form

$$\hat{H} = \hat{T}_e + \frac{N}{2\Omega} \sum_{\vec{k} \neq 0} v_c(k)\{N^{-1}\hat{\rho}_e(\vec{k})\hat{\rho}_e(-\vec{k})-1\} \qquad (2a)$$

$$+ \hat{T}_i + \frac{N}{2\Omega} \sum_{\vec{k} \neq 0} v_c(k)\{N^{-1}\hat{\rho}_i(\vec{k})\hat{\rho}_i(-\vec{k})-1\} \qquad (2b)$$

$$- \frac{N}{\Omega} \sum_{\vec{k} \neq 0} v_c(k)\hat{\rho}_e(\vec{k})\hat{\rho}_e(-\vec{k}). \qquad (2c)$$

The first two terms (2a) of $\hat{H}$ represent a well-
known problem: they constitute the standard
Hamiltonian of N interacting electrons of mass
m_e and total kinetic energy $\hat{T}_e$ whose single
particle density operators have Fourier compo-
nents $\hat{\rho}_e(\vec{k})$. They move in a uniform compensating background at average charge density
(eN/Ω). Similarly (2b) represents the Hamiltonian of N protons or deuterons of mass m_i,
whose kinetic energy is $\hat{T}_i$ and whose density
components are $\hat{\rho}_i(\vec{k})$. They also move in a uniform compensating background with average
charge density $-eN/\Omega$. With appropriate boundary
conditions, (2a) and (2b) are themselves Hamiltonians for distinct and well-defined problems.
However, in the system under consideration they
are coupled, and coupled strongly, by the
attractive interaction described by (2c). In
the thermodynamic limit, which we assume, the
average of (2c) in states of fixed N is removed
by the requirements of charge neutrality.
Unlike the Hamiltonians that one might imagine
writing down for other simple metallic systems,
(2) is exact. There are no residual uncertainties stemming from the necessity of construct-
ing a pseudopotential, for example.

The Hamiltonian summarized in (2) has considerable symmetry.

3. NORMAL GROUND STATES

We are especially interested in ordered states
of the system described by (2). But first we
briefly describe the expected normal states and
their origin. Let Ψ^0 be the ground state wavefunction for the electron-proton or electron-
deuteron assembly. Then the determination of
the ground state energy requires us to evaluate

$$\langle\Psi^0(\{\vec{r}_e\},\{\vec{R}_i\})|\hat{H}(\{\vec{r}_e\},\{\vec{R}_i\})|\Psi^0(\{\vec{r}_e\},\{\vec{R}_i\})\rangle. \quad (3)$$

Since $(m_e/m_p)^{\frac{1}{4}} = 0.153$, and $(m_e/m_d)^{\frac{1}{4}} = 0.128$, it
is a reasonable starting point for such a goal
to factor the dependence on electronic and ionic
degrees of freedom in adiabatic form:

$$\Psi^0(\{\vec{r}_e\},\{\vec{R}_i\}) \doteq \Psi^{eo}_{\{\vec{R}_i\}}(\{\vec{r}_e\})\Psi^{io}(\{\vec{R}_i\}) \qquad (4)$$

where Ψ^{eo} is the ground state wavefunction for
the instantaneous configuration $\{\vec{R}_i\}$ of the
ionic system (whose ground state, in turn, is
described by Ψ^{io}). If the electron system
remains in its ground state or close to it then
it follows that the motion of the protons, or
deuterons, is described by the Hamiltonian

$$\hat{H}^i(\{\vec{R}_i\}) = \langle\Psi^{eo}_{\{\vec{R}_i\}}(\{\vec{r}_e\})|\hat{H}(\{\vec{r}_e\},\{\vec{R}_i\})|\Psi^{eo}_{\{\vec{R}_i\}}(\{\vec{r}_e\})\rangle.$$

$$(5)$$

The electronic degrees of freedom are clearly
integrated out, but the result of this procedure,
and hence of the resulting ionic motion, very
much depends on the nature of these electronic
states (whether itinerant or localized, for
example). The characteristics of these states
depend on density, as noted. Near $r_s = 3.1$
hydrogen and deuterium form well-defined
diatomic molecular distance $2d(r_s)$. (In principle, these units should be distinguished
according to the states of the ionic spins. The
matter of ortho-para differences will not concern us further in the low density phases.) A
physically acceptable choice for Ψ^{eo} for such
states is a product of (N/2) two-electron functions each localized about the mean molecular
coordinate. The electron functions in this
choice are thus identified by site and this
form for Ψ^{eo} therefore conflicts with the basic
requirements of overall antisymmetry. This
difficulty is overcome in practice by introducing into (5) phenomenological short range potentials which are generally taken as pairwise
functions of the relative intermolecular separations. At longer range, a product state based
on an assumption of little overlap leads, as is
well known, to fluctuating dipole forces in
lowest order. Thus when combined with the
Pauli repulsion (short range) terms we arrive
at the familiar picture of a system of interacting molecules with pairwise interactions (not
necessarily spherically symmetric) between them.
The ionic Hamiltonian can therefore be recast in
the form

$$\hat{H}^i = \hat{T}_m + \frac{1}{2} \sum_{m,m'} \Phi(\vec{R}_m - \vec{R}_{m'}) + \dots. \quad (6)$$

where m and m' denote molecular coordinates, and
Φ is a pair-potential.

If the density is such that the two-particle
functions seriously overlap, then terms beyond
simple pair interactions displayed in (6) will
become important, beginning with the dipole-
dipole-dipole term. At still higher densities
the overlap becomes so large that the very
starting point of the description, the assumption of well-defined localized two-electron
states, ceases to have any physically justifia-
ble validity. Under such conditions the density
of electrons near the boundaries of the Wigner-
Seitz cells has become an appreciable fraction

of the average density $\bar{\rho}_e$, and in these circumstances it is more appropriate to characterize the electron band states by band structure. In a very general sense the details of the electron band structure must still reflect the geometrical arrangement of the protons or deuterons. Thus, for example, we may think of a set of bands which continue to display at high density the basic diatomic ordering so characteristic of the low density system. For crystalline arrangements an integral number of bands are normally filled, but this paradigm of insulating configurations can itself revert to metallic behavior through the mechanism of band overlap. That such overlap can occur with diatomic ordering being preserved has been known for some time [3,4]. The corresponding conducting state is a forerunner of the more complete dissociative transition to a metal which, in crystals, for example, is expected to be represented by relatively simple structures (Bravais lattices are the simplest).

If metallization by band overlap is the correct mechanism, then the most interesting question concerns the density (and hence pressure) at which this occurs. To address this question, one requires a sequence of band structure calculations as a function of density for various separations 2d of the ions in the basis. Correspondingly one also requires a sequence of total energy calculations for both insulating and metallic phases. The metallic phases have been treated by a variety of methods including Wigner-Seitz techniques [5-9], localized orbital techniques [10-15], expansion methods [16-22] (in which (2c) (above) is treated as a perturbation) and density functional techniques [23]. The latter have the particular advantage that, in principle, they can treat metallic and molecular phases on the same footing with the errors of the method entering largely on a systematic basis. The combination of the results of this

method with those of band structure calculations [4] leads to a prediction for band overlap metallization at around 1.8 Mbar; complete dissociation in a static model does not appear to occur until much higher densities ($r_s \approx 1.1$), the corresponding pressure being quite dependent on the energies bound up in the ionic degrees of freedom and determined by (6). The motion of the ions is thus important. It is also extremely important in determining the possible states of long-range order of the metallic state once it is formed. This motion is again determined by an effective ionic Hamiltonian obtained by integrating out the electron degrees of freedom. For itinerant states appropriate to the metal the result is

$$\hat{H}^i = \hat{T}_i + E_0(r_s) + \frac{1}{2} \sum_{i,i'}' \tilde{\phi}(\vec{R}_i - \vec{R}_i') + \dots \quad (7)$$

where $E_0(r_s)$ is a term depending on volume arising from paramagnetic electron-gas energies, and $\tilde{\phi}$ is a screened pair-potential for the ions. We note that provided r_s is held fixed, the problem of solving for the motion of the ions is not essentially different from the equivalent problem in the insulating context.

4. ORDERED STATES

The description just given will determine the major contributions to the total energy of the system. The states we now discuss will involve ordering energies that are trivial in comparison. The states to be described are summarized in Table I.

If hydrogen and deuterium are both in crystalline states in their high density metallic manifestations then, with one possible exception, it is unlikely that the fact that one is a Fermion system while the other a Boson will lead to any marked physical distinctions. The

System	Normal States		Ordered States	
	Crystal	Liquid	Crystal	Liquid
Electrons	Fermi-Liquid	Fermi-Liquid	Super conductor	Super-conductor
Protons	Quantum -Crystal	Fermi-Liquid	(*)	Anisotropic-Superfluid
Deuterons	Quantum -Crystal	Bose-Liquid	(*)	Bose-Condensed Superfluid

Table I: States of order of electron, proton, and deuteron systems resulting from equation (6) and broken symmetry. The entries identified by (*) might well include magnetically ordered states.

exception concerns ground state vacancies [24] and defects that may occur in highly quantum systems. The spectrum of vacancy waves might then be quite different. However, the presumption that the metals are indeed crystalline, even in their ground states, is itself not at all well founded [1]. One can see that this might be the case by noting that a typical zone boundary phonon for the ions will have a frequency typical of an ionic plasma frequency, or an energy per ion of about

$$\hbar\omega_{p_i} \approx 2\sqrt{3}r_s^{-3/2}(m/m_i)^{\frac{1}{2}} \quad Ry. \quad (8)$$

For protons, for example, this amounts to 0.014 Ry at $r_s = 2.0$ and 0.04 Ry at $r_s = 1.0$. These energies are far in excess of the energy differences per ion (typically milliRydbergs) characteristic of different static structures [22]. Put another way, there is sufficient zero point energy in the ionic degrees of freedom to cause continuous rearrangement between a variety of common crystal structures. Though this argument is qualitatively correct, in practice it is necessary to treat electron and ionic degrees of freedom self-consistently [25]. Nevertheless, by direct simulation methods [1], it can be established that the conclusion to which one is being led, namely that liquid-like ground states may be energetically preferred, is in fact confirmed, at least over a specified range of densities. If the ground state is a liquid, however, then the quantum statistical differences between hydrogen and deuterium become extremely important.

4.1. Liquid Metallic Hydrogen

Even if the density is such that metallic hydrogen is a (quantum) crystal, the melting point is expected to be very low [1]. It has been pointed out that crystalline forms of metallic hydrogen are likely to be superconductors with high transition temperatures [26-28]. Consequently, for crystalline states we can also imagine that metallic hydrogen will be a superconductor at least up to its melting point. If we now assume that the lack of crystallinity is not inimical to superconductive pairing (and the existence of amorphous or glassy superconductors certainly gives support to this view) then we can conclude that metallic hydrogen will order, even as a liquid, into a superconducting state. The question is whether the transition temperature is lower or higher than in the corresponding crystal and whether the standard theory can be applied. The answer to the latter is that it can: to obtain T_c we need to solve the Eliashberg equations for the gap function and to find the temperature at which the gap is just suppressed by a pair breaking term. To carry out this procedure, the Eliashberg function $\alpha^2 F$ must be obtained from a description of density fluctuations which is more general than that used for crystals. In turn, this requires some knowledge of the frequency dependent density-density

response function $\chi(\vec{q},\omega)$ whose imaginary part appears explicitly [29] in the Eliashberg function. This can be obtained from memory function techniques, for example, but is most easily given by a generalization of the RPA like form:

$$\chi(q,\omega) = \frac{\chi^0(q,\omega)}{1-f(q,\omega)\chi^0(q,\omega)} \quad (9)$$

where $\chi^0(q,\omega)$ is the response function for non-interacting protons and $f(q,\omega)$ is the so-called polarization potential [30]. It can be obtained from the reasonable assertion that it is a continuous function with the appearance of a screened Coulomb interaction at long range, and taking a constant value at short range. The constant is fixed by the calculated sound speed as $q \to 0$. From (9) and from the electron-proton interaction $v_{ei}(q)$ we obtain the Eliashberg function according to

$$\alpha^2 F(\omega) = N(0) \int_0^{2k_F} dq(q/2k_F^2)|v_{ei}(q)^2\chi''(q,\omega)/\pi. \quad (10)$$

This is the required input for the gap function $\Delta(\omega)$: the solution of the Eliashberg equations [31] is relatively straightforward. The resulting transition temperature is $\sim 10^2\,°K$ and is a fairly strong function of density [29]. These temperatures do indeed exceed the melting temperature in certain density ranges so that the assumption that a liquid superconducting state exists appears to be substantiated.

It is worth noting that if the density is such that metallic hydrogen is in a liquid state, but the temperature is such that the system is normal, the properties of this 'normal' state are still rather remarkable. For it then constitutes a two-component Fermi liquid with one component (the electrons) in the high density limit and the other (the protons) in the low density limit. Both Fermi liquids have identical Fermi surfaces but very different Fermi temperatures ($T_{Fp} \sim 10^{-3} T_{Fe}$). From the standpoint of experimental detection of a (static) pressure generated phase of metallic hydrogen, it is also important to understand the properties of this phase. This can be achieved most easily through a straightforward generalization of Landau Fermi liquid theory to two (charged) components, both for equilibrium properties [32] and transport properties [33]. One of the most striking equilibrium properties is the appearance of a giant linear specific heat at low temperatures. It is a manifestation of the high density of states of the proton Fermi liquid and is in marked contrast to the normal T^3 phonon behavior expected when the system freezes.

The transport properties are no less dramatic. To study these one begins with a two-component generalization [33] of the Landau Silin transport equations describing the momentum, space, and time development of the electron and proton quasi-particle distribution functions. It is possible to obtain closed form expressions for the electrical conductivity, thermal conductivity κ, viscosity η, and spin diffusion coefficients. As $T \to 0$ all of these quantities diverge because of the phase-space considerations controlling the scattering. The electrical resistivity, in particular, vanishes as T^2, again in marked contrast to the solid where, in the absence of impurities, it would vanish as T^5. Most importantly, the transport coefficients of the liquid are found to be at least a factor $(m_e/m_p)^2$ smaller than in the corresponding solid metal (and in the case of the resistivity, the inverse of the conductivity, the factor can approach 10^8!).

Finally, in a fluid phase of metallic hydrogen, the Fermi temperature of the protons is $\sim (m_e/m_p)T_{Fe}$ where T_{Fe} is the Fermi temperature of the electrons. For $r_s \sim 1-2$, $T_{Fp} \sim 10^2 °K$. As noted above, there are effective (screened) interactions ϕ between the protons which have a repulsive region at short range, and an attractive and oscillatory region at long range [1]. Given this form of the interaction and with the example of He^3 before us, it is possible that at extremely low temperatures the protons may themselves be unstable toward pairing. The temperature at which such a protonic superfluid will form will be in the neighborhood of $10^{-3} T_{Fp}$. Whether the potential is sufficiently repulsive at short range to favor p-wave over s-wave pairing has yet to be determined.

4.2. Liquid Metallic Deuterium

A possible low temperature liquid metallic phase of deuterium represents a system with "mixed statistics," the electrons constituting a Fermi liquid, as before, but the deuterons belonging to the class of spin-1 Bosons. Unlike helium (spin-zero Bosons) the presence of the additional spin degree of freedom leads to a new branch in the quasi-particle spectrum even though there are no explicitly spin-dependent terms in the Hamiltonian (2). The dispersion of the branch follows a k^2 behavior at small k [35], and therefore differs in an essential way from the linear type of dispersions so characteristic of phonon-like quasi-particle excitations. There are thus 3 separate types of elementary excitations: the "phonons," the spin excitations, and electron-hole pairs. The system is quite unusual [36] and certainly contrasts markedly with, say, He^3-He^4 mixtures.

At low temperatures we may certainly expect, as with metallic hydrogen, that electron pairing can occur through the density fluctuations of the ions. The crucial difference is that $\chi(\vec{q},\omega)$ (see equation (9)) now reflects the Boson character of the ionic system. Again the approach is very similar: χ is constructed with the polarization-potential model [30] and the Eliashberg equations again solved. The resulting transition temperatures are somewhat lower than

in metallic hydrogen [37], but still in the neighborhood of $10^2 K$.

At even lower temperatures a potentially more interesting ordering can occur in liquid metallic deuterium. This is a superfluid transition associated with the Bose-condensation anticipated for the spin-1 system. At $r_s = 1.5$ an estimate based on non-interacting Bosons gives $T_{BC} \approx 30°K$ for this system. This suggests the rather intriguing possibility that we have a charged system in which the electrons display superconducting order while at the same time the deuterons display superfluid order. If it exists it is difficult to imagine a more remarkable state of condensed matter. At the very least it raises a number of interesting theoretical issues, for example, whether or not it is valid to describe the system with a two-fluid hydrodynamical picture, and whether or not persistent deuteron currents will actually exist, that is to say, whether combined superfluidity and superconductivity is stable.

The transport properties of the system are expected to be unusually rich in their possible temperature dependences. This is because in setting up the appropriate kinetic equations for the three types of quasi-particles it is recognized that their associated statistics have considerable bearing on the standard phase-space arguments determining the scattering characteristics. To give but one example, if the temperature exceeds both T_{BC} for the deuterons and T_C for the electrons, then the low temperature behavior of the electrical resistivity is expected to be linear in the temperature [36] rather than quadratic as is the case for the normal electron-proton fluid. The reason is that the phase space limitation on the scattering cross-section imposed by the Fermi-Dirac distribution of the protons is essentially removed when the protons are replaced by deuterons. A further consequence is that the magnitude of the low temperature resistivity of normal liquid metallic deuterium should greatly exceed that of liquid metallic hydrogen. As T falls below T_{BC} the deuterons will Bose condense and this transition should be manifested in the electrical transport properties, both in critical behavior near T_{BC} itself and in the form of the transport coefficients well below T_{BC} [36]. In particular, the manner in which electrons cause excitations between the condensate (presumably in one subspace of the spins) and the excitation gas (in the remaining subspaces) should be particularly interesting.

It was remarked above that the Hamiltonian (2) has considerable symmetry. In the macroscopic manifestations of (2) this symmetry is broken, the resulting states possessing physical properties that may be unique among the elements. The states we have discussed are probably only the simplest that can be imagined. There may be more complex possibilities involving magnetic order.

Acknowledgments. This work has been supported by NASA under Grant NSG 7487. I also wish to thank Drs. J. Oliva and J. Jaffe for many enlightening discussions throughout the continuing course of this work.

References

[1] K. K. Mon, N. W. Ashcroft, and G. V. Chester, Phys. Rev. B21, 2641 (1980).

[2] V. L. Ginzburg, "Key Problems of Physics and Astrophysics," MIR Publishers, Moscow (1978).

[3] D. E. Ramaker, L. Kumar, and F. E. Harris, Phys. Rev. Letts. 34, 812 (1975).

[4] C. Friedli and N. W. Ashcroft, Phys. Rev. B16, 662 (1977).

[5] E. Wigner and H. B. Huntington, J. Chem. Phys. 3, 764 (1935).

[6] R. Kronig, J. de Boer, and J. Korringa, Physica 12, 245 (1946).

[7] N. H. March, Physica 22, 311 (1956).

[8] W. C. DeMarcus, Astron. J. 63, 2 (1958).

[9] D. Styer and N. W. Ashcroft, to be published.

[10] A. A. Abrikosov, Astron. Zh. 31, 112 (1954).

[11] J. L. Calais, Arkiv. Fysik. 29, 255 (1965).

[12] F. E. Harris and H. J. Monkhorst, Solid State Communications 9, 1449 (1971).

[13] H. J. Monkhorst and J. Oddershede, Phys. Rev. Letts. 30, 797 (1973).

[14] A. K. McMahan, H. Beck and J. A. Krumhansl, Phys. Rev. A9, 1852 (1974).

[15] D. M. Wood and N. W. Ashcroft, to be published.

[16] W. J. Carr, Phys. Rev. 128, 120 (1962).

[17] A. Bellemans and M. Deleneer, Adv. Chem. Phys. 6, 85 (1964).

[18] G. F. Chapline, Jr., Phys. Rev. B6, 2067 (1972).

[19] E. G. Brovman, Yu. Kagan, and A. Holas, Sov. Phys. JETP 34, 1300 (1972); 35, 783 (1972).

[20] Yu. Kagan, V. V. Pushkarev and A. Holas, Sov. Phys. JETP 46, 511 (1977).

[21] L. Caron, Phys. Rev. B9, 5025 (1974).

[22] J. Hammerberg and N. W. Ashcroft, Phys. Rev. B9, 409 (1974).

[23] S. Chakravarty, J. Rose, D. M. Wood, and N. W. Ashcroft, Phys. Rev. BXX, XXXX (1981).

[24] A. F. Andreev and I. M. Lifshitz, Sov. Phys. JETP 29, 1107 (1969).

[25] D. M. Straus and N. W. Ashcroft, Phys. Rev. Letts. 38, 415 (1976).

[26] N. W. Ashcroft, Phys. Rev. Letts. 21, 1748 (1968).

[27] D. A. Papacoustanopoulos and B. M. Klein, Ferroelectrics 16, 307 (1977).

[28] J. M. Whitmore, J. P. Carbotte and R. C. Shukla, Can. J. Phys. 57, 1185 (1979).

[29] J. Jaffe and N. W. Ashcroft, Phys. Rev. B23, 6176 (1981).

[30] C. H. Aldrich, III and D. Pines, J. Low Temp. Phys. 32, 689 (1978).

[31] G. Bergmann and D. Rainer, Z. Phys. 263, 59 (1973).

[32] J. Oliva and N. W. Ashcroft, Phys. REv. B23, 6399 (1981).

[33] J. Oliva and N. W. Ashcroft, Phys. Rev. BXX, XXXX (1981).

[34] N. W. Ashcroft in *Modern Trends in the Theory of Condensed Matter*, edited by A. Pekalski and J. Przystawa (Springer, Berlin, 1979).

[35] B. I. Halperin, Phys. Rev. B11, 178 (1975).

[36] J. Oliva and N. W. Ashcroft, to be published.

[37] J. Jaffe and N. W. Ashcroft, to be published.

PHYSICS OF SOLIDS UNDER HIGH PRESSURE
J.S. Schilling, R.N. Shelton (editors)
© *North-Holland Publishing Company, 1981*

THE METALLIZATION OF SOME SIMPLE SYSTEMS*

M. Ross and A. K. McMahan

University of California, Lawrence Livermore National Laboratory
Livermore, California 94550, U.S.A.

We discuss the metallization of Xe, Ar, He, I_2, H_2, and N_2 in terms of some recent theoretical work and shock-wave experiments. New shock-wave data on liquid hydrogen and deuterium leads to a predicted pressure above 3 Mbar for the appearance of a monatomic metal phase. We expect CsI to become metallic near 0.8 Mbar.

1. INTRODUCTION

The metallization of inert gases and diatomic molecules has been a subject of continued activity. These endeavors combine the excitement of preparing exotic phases with tests of theoretical predictions. Although recent advances in the diamond anvil technique have opened the possibility of making metallization studies at near megabar pressures, most simple systems appear to require still higher compressions. Shock-wave experiments can achieve higher pressures and densities than static methods but with the added complication of simultaneously heating materials to very high temperatures. Our own interest in the subject of metallization began with the observation that shock compressed liquid xenon had an anomalously large compressibility which was apparently due to thermal excitation of electrons from the filled 5p valence band to an empty 5d-like conduction band.[1,2] Electron band calculations predicted that the 5p to 5d energy gap was decreasing with volume and would close at sufficiently high density leading to metallization of xenon. Thus the thermal electron excitations serve as probes of the band gap and anticipate phenomena occurring at higher density and low temperatures. Similarly the high temperatures achieved in shock-wave studies on liquid hydrogen have provided considerable information about the short-range repulsive forces between molecules permitting more reliable estimates of the molecular to monatomic transition.[3] In this paper we review some recent theoretical and experimental results.

2. XENON, ARGON AND HELIUM

Among the available inert gases, xenon is the prime candidate for metallization. One of the earliest predictions seems to have been made by Herzfeld in 1927.[4] He published a paper with the title, "On Atomic Properties which Make an Element a Metal", in which he predicted that of a closed shell atom or molecule would become a metal when the molar volume (V) became equal to, or less than its gas phase

molar refractivity (R). $R = 4\pi/3N\alpha$, where N is Avogadro's number and α is the atomic polarizability. He predicted xenon would become metallic at a volume of 10.2 $cm^3/$mol. Table I contains the Herzfeld predictions for the volume at metallization for a number of inert gases along with new results of augmented-plane-wave (APW) and linear-muffin-tin-orbital (LMTO) calculations for an fcc lattice. The simple Herzfeld model is in surprisingly good agreement with the more rigorous modern theory. Also included are predictions for some diatomic molecules.

Herzfeld's theory is based on the classical Lorentz oscillator model without damping, and assumes the system becomes metallic when the frequency of the oscillator placed in a dielectric medium approaches zero. A direct and illustrative derivation of this theory can be made by considering a simple classical atom in which a point positive charge is surrounded by a spherically symmetric negatively charged cloud of constant density extending to a radius r. This is the model of the atom as a perfectly conducting charged sphere. It can be shown from electrostatics that its polarizability is

$$\alpha = r^3,$$

and its molar refractivity is then

$$R = \frac{4}{3} \pi N r^3 .$$

R/N is also the volume of the isolated atom. We assume metallization occurs when the available volume, V, becomes equal to or less than R and the system of localized conducting spheres then becomes one large conductor. The model is limited to closed shell systems and the appropriate value of α is for the free atom or ion. The fact that α may vary in a substance under compression is not relevant. It is the free atom value that defines the size. The model should be applicable to predicting the metallization of closed shell molecules in which band overlap occurs and the molecule retains its structure, but should not be used for systems which become metallic by bond dissociation or rearrangement. The classical nature of the model suggests that it will be most accurate for atoms with large atomic charge.

The xenon band calculations were made for an fcc lattice using the APW method with Hedin-Lundqvist (HL) and Slater (S) exchange correlation potentials.[5] Although the Hedin-Lundqvist approximation is known to be better for calculating ground state properties such as cohesive energy and pressure, it does not accurately compute highly excited electronic states. The Slater exchange is believed more reliable for this purpose. However, as we approach the metallization volume we expect Hedin-Lundqvist to be correct for predicting gap closure. The calculated band gaps are shown in Fig. 1.

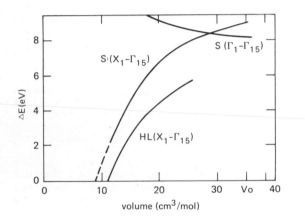

Figure 1: APW Conduction band gap versus volume. S and HL refer to calculations using Slater and Hedin-Lundqvist exchange-correlation potentials. The S results were extrapolated to · zero gap (dashed portion of curve).

At large volumes, the bottom of the empty conduction band (Γ_1 state) in xenon is of 6s character. The gaps in the figure are referred in energy from the top of the full 5p band (Γ_{15} state). Calculations made with the Slater exchange potential predicts the gap at normal solid volume, V = 34.7 cm^3/mol, to be 8.24 eV as compared with the experimental 9.28 eV. As volume is decreased, the bottom of the 5d band (X_1 state) becomes the lowest level in the conduction band, and is the state that first crosses the filled 5p band at which point xenon becomes metallic. Calculations using Hedin-Lundqvist and Slater exchange potentials are in approximate agreement as to the volume where this occurs, predicting 11 cm^3/mol (1.3 Mbar) and 9 cm^3/mol (~2 Mbar), respectively. On the basis of these theoretical results, we concluded that 1.3 Mbar is the pressure at which xenon is likely to become metallic.[2]

These results have been repeated by Ray et al.[6] who carried out APW and Gaussian type orbital calculations. They are in sharp disagreement with the experiments of Nelson and Ruoff[7] who report metallic conductivity at pressures above 0.33 Mbar and room temperature. However, Shiferl[8] compressed solid xenon to 0.44 Mbar using a diamond anvil, and was unable to visually detect any color change in the sample suggesting that the optical gap at that pressure is still above 4 eV in disagreement with Nelson and Ruoff. At 0.44 Mbar the optical gap predicted by calculations made using the Slater exchange is 4.5 eV and are consistent with Shiferl's observation.

Recently Hama and Matsui[9] have investigated the possibility that the bcc structure of xenon, and not the fcc, might be the stable phase at high pressure. They made APW calculations for both fcc and bcc xenon and found that up to 0.83 Mbar fcc has the lower energy, but bcc has the smaller band gap. At that pressure bcc becomes the stable form and is metallic. In a separate calculation these authors also found that including spin-orbit coupling further decreases the band gap lowering the transition pressure 0.66 Mbar, in closer agreement with Nelson and Ruoff. However, one should also include the larger relativistic effects due to mass-velocity and Darwin. The combination of all these relativistic effects should actually increase the gap. Nevertheless these authors have raised the interesting question of the extent to which structure influences the metallization conditions. Recently Young et al.[10] while studying the equation of state of helium made APW and LMTO band theory calculations for fcc, bcc, and hcp phases and found the relative stability of bcc and fcc to be similar to the xenon results of Hama and Matsui. However, they found that the hcp phase was the most stable in the region where otherwise there would have been an fcc-bcc transition. Thus the only structural high pressure phase transition in helium would be fcc to hcp. bcc is only stable in the nearly free electron high density limit. The work on helium is further discussed below. The hcp band gap in helium is slightly greater than for the fcc phase, and its metallization pressure is slightly higher. These results suggest that hcp calculations, including all relativistic corrections, should be made for xenon to test the possibility of an fcc to hcp transition below 0.83 Mbar. Figure 2 shows the calculated pressure versus volume for fcc and includes a speculated fcc-hcp transition. The existence of such a transition could be demonstrated by diamond anvil experiments.

Although the maximum compressions achieved with shock-wave experiments on liquid xenon have been insufficient for metallization, the high temperatures attained, up to 20000 K, are

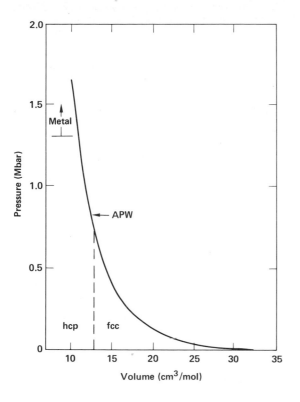

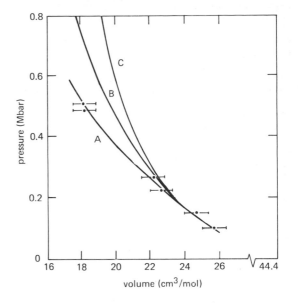

Figure 2: 0 K theoretical xenon isotherm and metallic transition with speculated hcp-fcc phase.

Figure 3: Xenon Hugoniot calculations and experimental data. The curves are theoretical results discussed in text.

sufficient to transfer a considerable number of electrons into the empty 5d level. The extent to which these thermal excitations modify the observed pressure-volume (Hugoniot) curve depends on the electron band gap and its volume derivatives. To see this a comparison of theoretical and experimental results for the liquid xenon Hugoniot are shown in Fig. 3. Curve A is a calculated xenon Hugoniot using the volume dependent band gap calculated with the Slater exchange and shown in Fig. 1. These results are in good agreement with experiment. Calculations that include no electronic excitation, the pure insulator case, are shown by the uppermost curve (C) and in curve B the calculations use only the normal solid xenon band gap and thus neglect the changes in electron partial pressures. By absorbing some of the shock energy, the excited electrons act as themal sinks, keeping the temperature down and lowering the pressure. The volume dependence of the energy gap also plays an important role. Because the 5p-5d gap is narrowing with decreasing volume, the pressure is lowered further. These electronic thermal processes lead to the observed softening of the experimental Hugoniot. Thus shock-

wave data can provide an estimate of the electron band gap under compression.

We have been not able to calculate a Hugoniot that agreed with the experimental shock-wave data and was also consistent with the gap closure reported by Nelson and Ruoff. Recently Nellis and Mitchell[10] extended the liquid xenon shock compression measurements to 1.3 Mbar and their results are in good agreement with the predictions of the present theory. The final temperatures are about 30000 K and approximately one electron/atom has been excited into the conduction band to form a cesium-like system.

We have recently made self consistent LMTO calculations for fcc argon using the Hedin-Lundqvist exchange-correlation potential. We find metallization to occur at 5.8 Mbar and a volume 4.45 cm^3/mol. The X_1 state at the bottom of the empty 3d conduction state crosses the Γ_{15} slate at the top of the filled 3p band and argon becomes metallic. These results are in Table I and are in good agreement with the Herzfeld theory.

This pressure is too high for current static methods. The only relevant experimental data for the metallization of argon is for the liquid which has been shock-compressed to 910 kbar.[12] At pressures up to 400 kbar electronic thermal excitations in argon are negligible and this substance remains a simple closed shell insulator fluid whose properties

are determined by the repulsive pair potential. Above this pressure the temperature is
greater than 12000 K and rising exponentially
with compression. In this higher temperature
regime the properties are increasingly dominated by electron thermal excitations from the
closed 3p shell into the condition band. The
sensitivity of the Hugoniots to the details of
electronic structure is demonstrated in Fig.
4, where the experimental Hugoniot data for
liquid argon are compared with theoretical
calculations. The upper curve INS (for insulator) was computed using an intermolecular
pair potential and does not include electronic
excitations. The two lower curves include
electronic thermal excitation calculated by
using an electron band gap for solid argon
computed using electron band theories. The
curves labeles WS and APW refer respectively
to Hugoniot calculations using the conduction
band gaps predicted by Wigner-Seitz and augmented plane wave methods.

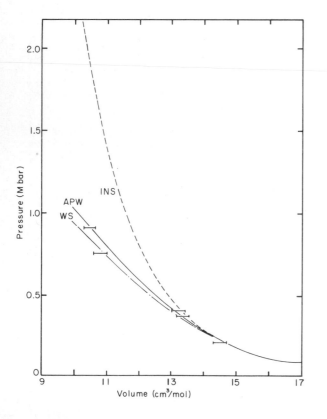

Figure 4: Argon Hugoniot calculations and
experimental data. The curves are
theoretical results discussed in text.

Helium and hydrogen are of great interest because of their simplicity, and they are also
the major components of the two largest
planets; Jupiter and Saturn. The determination

of helium's equation of state and metallization pressure has recently assumed new importance because of the possibility that the
hydrogen and helium mixture in Saturn may be
undergoing a phase separation. The highest
planetary pressures for which He is believed
to be under is about 45 Mbar in Jupiter's
interior.

We used the linear-muffin-tin-orbitals (LMTO)
method[13] to calculate the pressure and total
energy at zero temperature up to 250 Mbar.
Since the usual local-exchange-correlation
potentials may not be reliable for atoms with
as few electrons as He, we chose the $X\alpha$
potential with $\alpha = 0.6105$ in order to match
the T = 0 static lattice pressure. The Hedin-
Lundqvist potential yields pressures smaller
by about 40 kbar in this region, but gives
results within 9% and 4% of our $X\alpha$ pressures
near 10 and 100 Mbar, respectively.

Low pressure solid He is an insulator with a
full 1s band below a large energy gap. Under
extreme compression, helium becomes metallic
when the bottom of the 2p band (L_2' state
for fcc structure) drops below the Fermi
level. For the bcc, fcc, and hcp structures,
we find this occurs at pressures of 31.5, 97,
and 112 Mbar, respectively, corresponding to
volumes of 0.539, 0.310, and 0.291 cm^3/mole.
These results were obtained with the $X\alpha$
exchange-correlation potential as mentioned
above. Use of the Hedin-Lundqvist potential
increases the fcc transition pressure by only
2%. For a given volume the near neighbors are
closest in the bcc phase and as a result the
greater overlap of adjacent atomic orbitals
will lead to broader bands, and metallization
at the largest volume of the three. Herzfeld
predicted metallization to occur at 0.52
cm^3/mole, which is in close agreement with
the present bcc result.

Energy differences between the various lattice
types are shown in Fig. 5 (solid curves),
relative to the bcc structure. The quantity
actually plotted is $S \cdot (E - E_{bcc})$, where S
is the Wigner-Seitz radius in bohr units. As
can be seen in Fig. 5, fcc is the most stable
phase for S > 1.3 bohr, hcp is the most
stable for 0.4 < S < 1.3 bohr, and bcc is
the most stable for S < 0.4 bohr. The calculations were extended to sufficiently small
volumes to show that the stable bcc phase of
the one-component plasma is achieved. As the
hcp to bcc transition does not occur until
approximately 10^4 Mbar, it is clear that the
stable phase of T = 0 He will be close packed
throughout the range of planetary interest,
and that metallization will occur at 112
Mbar. This implies that planetary helium is
never metallic.

The dashed curves in Fig. 5 show for comparison the results of second-order free electron

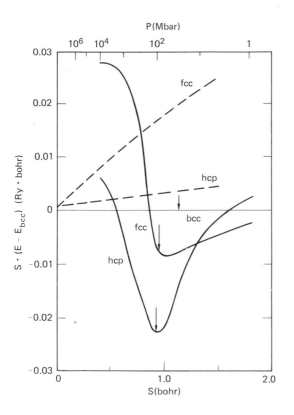

Figure 5: Energy differences $S \cdot (E(fcc) - E(bcc))$ and $S \cdot (E(hcp) - E(bcc))$ versus Wigner-Seitz radius S and pressure (upper axis). LMTO calculations (solid curves) and free electron perturbation theory (dashed curves). The arrows indicate metallization volumes.

perturbation theory.[14] This theory also referred to as screened Coulomb theory has been widely used as a model for planetary helium. The same $X\alpha$ exchange factor was used as in the LMTO calculations. The energy differences were multiplied by S so that the effect of the Madelung contribution to the total energy would show as the intercepts of these curves at $S = 0$. Thus it is the band structure term which dominates these curves. In the region $S < 0.5$ bohr the LMTO and perturbation theory results are in qualitative agreement as to the relative order of stability of the three lattices. For $S > 1$ bohr, the ls shell is fully intact, and so we expect the free electron perturbation theory to become inadequate.

3. IODINE, HYDROGEN AND NITROGEN

Diatomic molecules may enter a conducting or metallic state by dissociating into atoms, or by the overlap of an empty conduction band

with a filled valence band. Iodine plays a dual role in these studies. It is the only experimentally verified conversion of a diatomic molecule to a metal, and as a monatomic metal it behaves similar to xenon. Atomic iodine with one less electron than xenon has a nearly filled $5p^5$ shell. At equilibrium conditions the condensed phase is an insulating diatomic molecular solid. Due to significant anisotropy of the molecular solid, iodine gradually achieves metallic conductivity as the pressure is increased from 40 to 180 kbar, still retaining its diatomic structure. Finally at 210 kbar iodine transforms to a monatomic metal with a body-centered orthorhombic structure.[15] Herzfeld's theory predicts molecular iodine will turn metallic at 40 cm^3/mol, which is about 42 kbar, precisely where the resistivity begins to decrease. As we compress iodine further, the 5d level in monatomic iodine is expected to cross the top of 5p band just as in the case of xenon under compression. At $T = 0$ this is predicted to occur near 450 kbar. However since iodine has an unfilled 5p state the 5d will not cross the Fermi surface until much higher pressures. Under the high temperature condition achieved by shockwave experiments, (up to 33000 K for iodine) iodine is monatomic and electrons are excited into the d state with the resulting softening of the Hugoniot curve.[16] This can be seen in Fig. 6 where calculations including and excluding the 5d band are shown. Thus in iodine, as in xenon, the high temperature, high density monatomic properties are modified by the transfer with a decreased partial pressure.

The agreement at low pressure between molecular I_2 Hugoniot calculations and the experimental data, and at high pressures between monatomic iodine calculations and data is suggestive of a gradual transition from the molecular to monatomic phase, as is seen in x-ray crystallographic studies.

The metallization of hydrogen has been a subject of considerable theoretical and experimental interest. Although reports of experiments observing its metallization have appeared, these have not been widely accepted. The theoretically estimated necessary pressures are typically several megabar[3] and above the range of present day capabilities in static and shock-wave methods for this material. The probable absence of experimental results indicate it will be necessary to obtain better equations of state to predict the appearance of the metal phase.

In recent years there have been a number of static high pressure measurements on the thermodynamic properties of hydrogen and deuterium. These measurements include 4 K solid isotherms to 25 kbar[17] fluid isotherms to 300 K and 20 kbar,[18] and melting curves

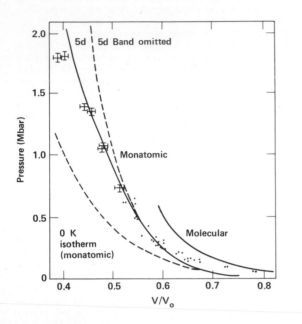

Figure 6: Iodine Hugoniot calculations of monatomic and molecular iodine compared with shock-wave data of Reference 16 and earlier work.

to 57 kbar at room temperature.[19] Recently shock-wave studies have been carried out on D_2 up to 800 kbars with final temperatures of about 7000 K, and on liquid H_2 to 100 kbars and 3250 K.[20] Although considerably more accurate than the earlier work on van Thiel, et al.,[21] the new, as yet unpublished, results are within the errors bars of the older data.

In principle, it should be possible to use the well established methods of quantum mechanics to calculate the forces acting between molecules and then to apply statistical mechanics to calculate thermodynamic properties. Fortunately, the calculation of thermodynamic properties with statistical mechanics has made great strides in the past 10 years and the available methods are sufficiently accurate. However, a satisfactory determination of the forces between molecules currently represents the chief uncertainty. An alternative to theoretical rigour, is to determine an effective intermolecular potential that, when used with proven statistical mechanical models, will reproduce the available experimental data. These models can then be used to compute thermodynamic properties. Some care must be taken when extrapolating to regions outside the range of the data from which the potential was determined.

The intermolecular potential we used is a modification of one first suggested by Silvera and Goldman.[22] They used a very accurate model for molecular solids, to determine a pair potential that fits the solid hydrogen and deuterium data up to 25 kbar. We subsequently used this potential with an accurate theory for the compressed fluid and were able to predict the experimental fluid isotherms from 0 to 20 kbar and 75 to 300 K as well as the melting curve up to 57 kbar. The calculation of the melting curves represents an even more severe test of the goodness of the theories because it is necessary to compute both free energies and pressures accurately in both phases. However, to correctly calculate the shock wave data it was necessary to modify the Silvera-Goldman potential at small intermolecular separations. Their potential had been fitted to low temperature data for solids below 25 kbar and does not extrapolate correctly beyond that region. Shock wave data represent a much higher temperature phenomenon where the important interactions occur at small intermolecular separations. This new potential, combined with our fluid theory, now fits H_2 shock data to 100 kbar and 3250 K, the D_2 shock data up to about 800 kbar and 7000 K, and the lower temperature static isotherms. The high temperatures generated in the shock process make it possible to determine the intermolecular potential as spacings comparable to those found in the molecular solid at densities near the metallic transition. We have recalculated the molecular to monatomic metal transition using a molecular equation of state based on our new intermolecular potential and metal equations of state calculated by the APW method[24] and third order perturbation theory.[25] Zero point properties were added. These theories predict metallic transitions at 3.1 Mbar (APW) and 3.6 Mbar (third order). The new results do not differ substantially from the earlier work.[3]

The other possibility exists that hydrogen may undergo the same gradual molecular-metallic transition observed in diatomic iodine. Some preliminary theoretical calculations for a solid at 0 K indicate that this may occur in hydrogen. Freidli and Ashcroft[26] using an approximate band theory predict a band crossing at 2.4 cm^3/mol (2.1 Mbar). The Herzfeld model predicts this will occur at 2 cm^3/mol, or about 3.4 Mbar, in the pressure range predicted for the metallic transition. The success of the Herzfeld model with iodine leaves us with some confidence that our predictions for hydrogen are reasonable. Thus we are lead to the possibility of a phase diagram in which molecular hydrogen first changed to an electrically conducting form of the molecule and then at some higher pressure into a monatomic metal. Some indication that this might occur could be drawn from the results of Sharma et al.[27] Using Raman spectroscopy these authors measured the molecular vibrational frequency under pressures of up to 600 kbar. They found that with increasing pressure the frequency increased as expected. But, above 300 kbar the frequency began decreasing. This could be interpreted as the gradual onset of intermolecular bonding.

Shock-wave data for oxygen and nitrogen have been compared to theoretical calculations made with a fluid theory using intermolecular pair potentials.[28] The data for oxygen shows no anamolous behavior and are in good agreement with the theory up to 0.9 Mbar. However, for nitrogen above 400 kbar, as shown in Fig. 7, the compressibility suddenly becomes unusually large, disagreeing with theory, and suggesting some fundamental change in the molecular structure. Diamond anvil studies on nitrogen to 410 kbar have found no evidence of metallization,[29] suggesting that the phenomena observed under shock loading has a thermal origin. Why this should happen in nitrogen but not oxygen is unclear, particularly since oxygen has a smaller dissociation energy and lower excited electronic states. Shown also in Fig. 7 is the theoretical 0 K isotherm for N_2

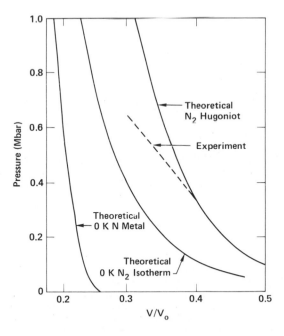

Figure 7: N_2 Hugoniot calculations, experimental shock data and 0 K isotherms of N_2 and monatomic N.

and a theoretical calculation of the 0 K isotherm of monatomic nitrogen made using the APW method with Hedin-Lundqvist exchange. Although no attempt was made to calculate the metallic

phase transition the results suggest the large volume change ($\sim$ 50%) is consistent with the interpretation that dissociation to a metal phase is occurring along the Hugoniot.

4. CESIUM IODIDE

CsI is isoelectronic with Xe and in the compressed state is expected to undergo similar changes in band structure. It has a band gap of 6.4 ev at 1 bar, compared to 9.3 ev for xenon, and is expected to become metallic at a lower pressure. The Herzfeld model predicts metallization at 24.7 cm^3/mol. This material has been the subject of two shock-wave studies[30,31] and a reduction of the data to a 0 K isotherm[31] results in a predicted transition pressure near 0.8 Mbar. The shock data suggests some anomaly near this pressure. Unfortunately, this occurs at the overlap of the two data sets and may be an artifact. In light of the success of the Herzfeld model, it would be useful if accurate band structure calculations were carried out to provide further guidance to possible diamond anvil studies.

Table I. Predicted Metallization Properties for Band Overlap

	$V_{Herzfeld}$ (cm³/mol)	$V_{Band Theory}$ (cm³/mol)	$P_{Band Theory}$ (Mbar)
He	0.5	0.3	112
Ar	4.2	4.5	5.8
Xe	10.2	11.0	1.3
H_2	2.0	(2.4)[a]	
I_2	40.0		
CsI	24.7		

[a]Reference 26.

REFERENCES

[1]   Ross, M., Phys. Rev. 171 (1978) 777.

[2]   Ross, M. and McMahan, A.K., Phys. Rev. B21 (1980) 1658.

[3]   Ross, M., J. Chem. Phys. 60 (1974) 3634.

[4]   Herzfeld, K.F., Phys. Rev. 29 (1927) 701.

[5]   Previous descriptions of the program used are given in Ross, M. and Johnson, K.W., Phys. Rev. B2 (1970) 4709, and in McMahan, A.K. and Ross, M., Phys. Rev. B15 (1977) 718.

[6]   Ray, A.K., Trickey, S.B., Weidman, R.S. and Kunz, A.B., Phys. Rev. Lett. 45 (1980) 933.

[7]   Nelson, D.A. and Ruoff, A.L., Phys. Rev. Lett. 42 (1979) 383.

[8]   Shiferl, D., personal communication (1981).

[9]   Hama, J. and Matsui, S., Solid State Comm. 37 (1981) 889.

[10] Young, D.A., McMahan, A.K. and Ross, M., Equation of state and melting curve of helium to very high pressure, Lawrence Livermore National Laboratory, UCRL-85788 (1981).

[11] Nellis, W.J. and Mitchell, A.C., personal communication (1981).

[12] Ross, M., Nellis, W.J. and Mitchell, A.C., Chem. Phys. Lett. 68 (1979) 532.

[13] The LMTO computer program used in this work is a modification of a program provided by Skriver, H.L. and is described in McMahan, A.K., Skriver, H.L. and Johansson, B., Phys. Rev. B, to be published.

[14] Friedli, C. and Ashcroft, N.W., Phys. Rev. B12 (1975) 5552.

[15] Takemura, K., Minomura, S., Shimomura, O. and Fujii, Y., Phys. Rev. Lett. 45 (1980) 1881.

[16] McMahan, A.K., Hord, B.L., and Ross, M., Phys. Rev. B15 (1977) 726.

[17] Anderson, M.S. and Swenson, C.A., Phys. Rev. B10 (1974) 5184.

[18] Mills, R.L., Liebenberg, D.H., Bronson, J.C. and Schmidt, L.C., J. Chem. Phys. 66 (1977) 3076.

[19] Mao, H.K. and Bell, P.M., Science 203 (1979) 1004; Mills, R.L., Liebenberg, D.H., Bronson, J.C. and Schmidt, L.C., Rev. Sci. Instrum. 51 (1980) 891; Diatschenko, V. and Chu, P., Science 212 (1981) 1393.

[20] Nellis, W.J. and Mitchell, A.C., to be published.

[21] van Thiel, et al., Physics of the Earth and Planetary Interiors 9 (1974) 54.

[22] Silvera, I.F. and Goldman, J.J., J. Chem. Phys. 69 (1978) 4209.

[23] Young, D.A. and Ross, M., J. Chem. Phys. 74 (1981) 6950.

[24] Ross, M. and McMahan, A.K., Phys. Rev. B13 (1976) 5154.

[25] Hammerberg, J. and Ashcroft, N.W., Phys. Rev. B9 (1974) 409.

[26] Friedli, C. and Ashcroft, N.W., Phys. Rev. B16 (1977) 662.

[27] Sharma, S.K., Mao, H.K. and Bell, P.M., Phys. Rev. Lett. 44 (1980) 886.

[28] Ross, M. and Ree, F.H., J. Chem. Phys. 73 (1980) 6146.

[29] Schwalbe, et al., Solid nitrogen at pressures up to 410 kbar, in Vodar, B. and Martean, P.H. (eds.), High Pressure Science and Technology 2 (Pergamon Press, Oxford, 1980).

[30] Altshuler, V.L., Pavlovskii, M.N., Kuleshova, L.V., and Simakov, G.V., Soviet Physics-Solid State, 5 (1963) 203.

[31] Pavlovskii, M.N., Vaschenko, V.Ya., and Simakov, G.V., Soviet Physics-Solid State, 7 (1965) 972.

*"Work performed under the auspices of the U.S. Department of Energy by the Lawrence Livermore Laboratory under contract number W-7405-ENG-48."

DISCLAIMER

PHYSICS OF SOLIDS UNDER HIGH PRESSURE
J.S. Schilling, R.N. Shelton (editors)
© *North-Holland Publishing Company, 1981*

SHOCK ANOMALY AND s-d TRANSITION IN HIGH PRESSURE LANTHANUM

A. K. McMahan

University of California, Lawrence Livermore National Laboratory
Livermore, California 94550, U.S.A.

H. L. Skriver

Research Establishment, Risö, DK-4000 Roskilde, Denmark
Nordita, Blegdamsvej 17, DK-2100, Copenhagen, Denmark

B. Johansson

Institute of Physics, University of Aarhus, DK-8000 Aarhus C, Denmark

We report linear-muffin-tin-orbital calculations of the band structure and pressure-volume isotherms for fcc La, both at zero and finite temperatures. The calculated bulk modulus shows a rapid stiffening in the range from 40-50% compression, due to termination of the 6s to 5d electronic transition. When combined with a simple Slater model analysis, these results yield a temperature dependent peak in the lattice Grüneisen parameter. Experimental confirmation of this peak is found in an anomalous stiffening seen in the shock compression data for La, and it may also have some bearing on the observed saturation of the superconducting transition temperature in La around 200 kbar.

1. INTRODUCTION

The concept of s-d transition was first introduced by Fermi,[1] and applied by Sternheimer[2] in trying to explain the 42 kbar isostructural transition in Cs.[3] In essence one considers the unoccupied 5d orbitals in Cs to be more compact spatially in comparison to the 6s orbitals, half of which are occupied at ambient conditions. On compression, the 6s orbitals overlap more strongly than do the 5d, causing their energies to rise faster. It becomes energetically favorable, then, for electrons to transfer from 6s to 5d states under compression. Modern band structure calculations[4,5] indicate that this process is not an abrupt collapse in size of the Cs atoms, as might directly explain the isostructural transition. It is rather of a continuous nature extending over a wide range in volume. This leads to an overall softening of the equation of state, i.e. pressure and bulk modulus increasing less rapidly with compression, then would have been the case in the absence of s-d transition. However, Grüneisen model calculations[5] based on the band structure results do predict the isostructural transition in Cs to arise indirectly from the s-d transition. The softening mechanism of the s-d transition has an enhanced effect on the lattice vibrational frequencies, and it is their contribution to the finite temperature equation of state which appears to induce the isostructural transition.

The purpose of this work is to report calculations which show that the effects of this same s-d transition are also apparent in La. A more detailed description of this work has recently appeared elsewhere.[6] In La, which has two more valence electrons than Cs, there is no isostructural transition. The softening effect is weaker in La because only a third of the valence charge participates in the s-d transition, while essentially all of it is involved for Cs. The interesting effect for La occurs at the end of the s-d transition, when all of the 6s electrons have been transferred to 5d states. At this point, the loss of the softening mechanism leads to an effective stiffening in the equation of state, which is particularly evident in shock compression data.[7-9] Figure 1 shows a plot of shock velocity, u_s, versus particle velocity, u_p for La, which

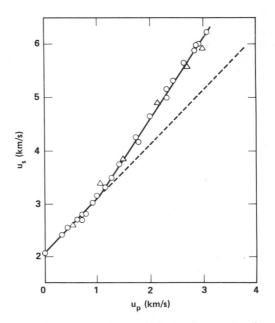

Figure 1: Experimental shock compression data for La. The data is from Refs. 8 and 9. The lines are drawn to emphasize the apparent slope change near u_p = 1 km/s.

are the two measured variables in shock compression. The apparent change of slope at u_p
= 1.07 km/s (P = 210 kbar) in Fig. 1 represents a rather dramatic stiffening in the La
equation of state which we propose is due
primarily to the termination of the s-d transition in this material. While this suggestion
is not new,[10] the present work offers the
first evidence from _ab initio_ calculations
that this, and not one of the several other
possible reasons, is the primary explanation
of the data.

Two other explanations have been proposed for
the apparent change of slope seen in Fig. 1,
overlap of the Xe cores,[8] and melting.[9]
Both explanations are in fact closely related
to s-d transition. Gust and Royce[8] were
able to correlate the volumes at which the
apparent slope changes occur, not only for La,
but also for other rare earths and group IIIA
elements having similar shock wave anomalies,
with overlap of the appropriate rare gas
cores. This suggests the stiffening is due to
core repulsion. However, band structure calculations show that significant s-d transition
is itself correlated with overlap of the rare
gas cores. This follows from orthogonality of
the wavefunctions. S and p valence bands
generally begin to rise with compression,
causing s-d transition if there are nearby d
bands, at about the same volume where s and p
core states begin to overlap. The present
calculations show effects of both core repulsion and s-d transition, but indicate that the
former is much too gradual to account for the
data in Fig. 1. The melting hypothesis[9]
presumes that the solid melts to a denser and
stiffer liquid phase. This can only occur if
the melting temperature is decreasing with
increasing pressure, which is a well known
characteristic of electronic transition. In
the case of Cs, for example, such behavior
occurs and is believed to result from the
different rates at which the s-d transition
takes place in the solid and liquid.[5,11] As
we are able to explain the essential behavior
seen in Fig. 1 without including liquid
corrections, and also get a normal melting
curve for La, the melting hypothesis does not
appear to be the correct explanation either.

2. RESULTS

Self-consistent linear-muffin-tin-orbital
(LMTO) calculations[12] were used to obtain
the zero and finite temperature pressure-
volume isotherms for fcc La. A region of
unusual stiffening is most apparent in a plot
of the pressure derivative, or bulk modulus,
shown in Fig. 2 for the zero temperature
case. The calculated bulk modulus (solid
curve) is seen to increase rather rapidly in
the region V/V_0 = 0.57 - 0.5, where V_0 is
the equilibrium volume of fcc La. This quantity is shown resolved in the figure (dashed

curves) into valence and Xe core contributions, and also a small muffin-tin correction
contribution. While the core overlap certainly contributes to the stiffening, it is the
more dramatic variation seen in the valence
contribution which is primarily responsible
for the structure seen in the total bulk
modulus. The maximum pressure is about 2 Mbar
for the volume range shown in Fig. 2, and the
experimental bulk modulus results of Syassen
and Holzapfel[13] (dotted curve) are included
for comparison.

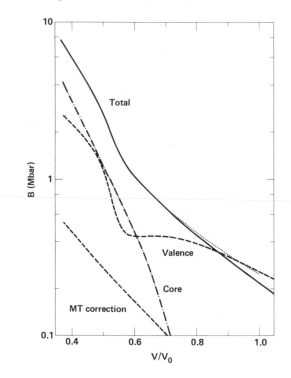

Figure 2: Bulk modulus vs relative volume
for T=0 La. The calculated results (solid
curve) are resolved into valence, core,
and muffin-tin correction contributions
(dashed curves). The room temperature
measurements (dotted curve) of Ref. 13 are
shown for comparison.

The behavior of the valence contribution to
the bulk modulus may be understood from a plot
of the unhybridized band edges shown in Fig.
3. As 6s levels rise above the Fermi energy,
$\varepsilon/\varepsilon_F = 1$, electrons must be transferred
from 6s to 5d states. This is just the s-d
transition.[14] Examination of the band
structure shows that the maximum density of 6s
states occurs at an energy which passes above
the Fermi level at V/V_0 = 0.57. At this
volume the s-d transition has achieved its
greatest rate, and with further compression
begins to diminish as there are successively
fewer 6s states to depopulate. The transition
ends at V/V_0 = 0.5 when the bottom of the 6s

band rises above the Fermi level. The effect of this process on the pressure may be crudely understood if one views the negative slope of each energy level in the figure as an effective pressure per electron. This viewpoint may be substantiated more formally in terms of the concepts of canonical band theory.[6],[15] Since the 4f and 5d levels in Fig. 3 have considerably smaller slopes than do 6s levels, it can be seen that conversion of a 6s electron to either a 5d or 4f state will reduce the pressure and the bulk modulus. According to the comments above about the rate of s-d transition, one can expect the greatest softening effect at $V/V_0 = 0.57$, and for this process to be exhausted by $V/V_0 = 0.5$. This is of course precisely the structure seen in the valence bulk modulus in Fig. 2.

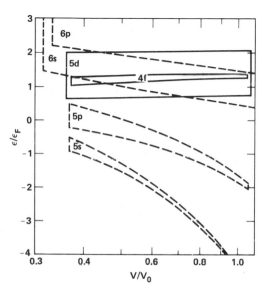

Figure 3: Unhybridized band edges vs relative volume for T=0 La. The energies are divided by the Fermi energy, ϵ_F, at each volume.

The lattice vibrational frequencies should undergo a region of rapid stiffening quite similar to that seen in Fig. 2. Within the Debye approximation one can calculate the lattice Grüneisen parameter, $\gamma = -(\partial \ln \omega_D / \partial \ln V)_T$, by relating the Debye frequency, ω_D, to the sound speed, and it in turn to the bulk modulus. This gives the Slater expression[16] for γ,

$$\gamma(V,T) = -\frac{1}{6} - \frac{1}{2} \left[\partial \ln B(V,T) / \partial \ln V \right]_T , \qquad (1)$$

where B should be the bulk modulus for a static lattice of atoms, i.e., without any lattice vibrational contribution. This is what the band structure calculations yield, and is the quantity plotted in Fig. 2 for the zero temperature case. At finite temperature

it can be argued that B should be the isothermal, static-lattice bulk modulus insofar as the short wavelength phonons, which dominate the equation of state, are concerned.[17] Results of Eq. (1) based on both the zero and finite temperature LMTO calculations are shown in Fig. 4 (solid curves). The calculations were performed in increments of 0.68 eV up to a maximum temperature of 3.4 eV. The dramatic peak in the zero temperature γ arises from the region of rapid stiffening in the bulk modulus seen in Fig. 2. It implies that the phonon frequencies themselves will undergo a similar region of rapid increase as the s-d transition ends. Note that the size of the peak in γ is considerably reduced with increasing temperature. This is to be expected from an electronic transition anomaly. At finite temperature one may consider the Fermi level to have a width of, say, $k_B T$. This causes the s-d transition to be smeared out over a much broader volume range than is the case at T = 0, which diminishes the rate at which the stiffening effect occurs, and thus the size of the peak in γ.

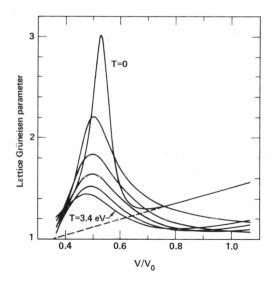

Figure 4: Lattice Grüneisen parameter, γ, vs relative volume. The temperature dependent, Slater model γ is shown at six temperatures ranging from 0 to 3.4 eV, in increments of 0.68 eV (solid curves). The temperature independent, linear γ (dashed curve) is used for a test calculation.

In order to calculate the shock compression or Hugoniot curve, a complete equation of state is needed, i.e. the lattice vibrational contribution to the pressure and energy must be added to the LMTO results. Within the Grüneisen model, one has

$$P(V,T) = P_{LMTO}(V,T) + 3k_B T\gamma(V,T)/V \qquad (2)$$

$$E(V,T) = E_{LMTO}(V,T) + 3k_B T(1 + \delta(V,T)), \quad (3)$$

where $\delta(V,T) = - (\partial \ln B/\partial \ln T)_V/2$, consistent with the model in Eq. (1). In shock compression, a shock wave of velocity u_s imparts a net velocity, the particle velocity u_p, to the material through which it passes. Conservation of mass, momentum and energy yield the Rankine-Hugoniot relations

$$P(V,T) - P_0 = \rho_0 u_s u_p \qquad (4)$$

$$V = V_0 (1 - u_p/u_s) \qquad (5)$$

$$E(V,T) - E_0 = \frac{1}{2}[P(V,T) + P_0](V_0 - V) \qquad (6)$$

Here V_0, P_0 and E_0 as well as the density ρ_0 are the initial conditions of the sample prior to passage of the shock wave. Given Eqs. (2) and (3), one solves Eq. (6) at each volume for the temperature along the Hugoniot in order to obtain the final state properties, V, $P(V,T_h)$ and $E(V,T_h)$. Eqs. (4) and (5) then given u_p and u_s.

The shock compression data for La offers direct evidence for the existence of a peak in the lattice Grüneisen parameter. This may be seen by comparison of the calculated Hugoniot (solid curve, Fig. 5) based on the present results for γ, with the data. The solid curve in Fig. 5 bends upward for $u_p > 1$ km/s in response to the rise in γ for $V/V_0 < 0.6$. Subsequently the solid curve begins to bend back downward at higher particle velocities in response to the decrease in γ on the small volume side of the peak. To see the magnitude of this effect, the calculation was repeated for the temperature independent, linear γ shown in Fig. 4 (dashed curve). Such linear behavior for γ is characteristic of most normal materials. The corresponding Hugoniot is the dotted curve in Fig. 5 and is shown for comparison. A 13% decrease in the size of the peak in γ would bring exact agreement of the solid curve with the data in the range near $u_p = 3$ km/s, whereas a factor of two reduction is needed to reach the dotted curve. Thus in spite of the crude Slater model estimate of γ used in this work, the shock data can be seen to clearly indicate a substantial peak in the lattice Grüneisen parameter for La.

The two intersecting dashed lines in Fig. 5 show the fit to the Hugoniot data proposed in Ref. 9, and on the basis of which it has been speculated that La undergoes a phase transition at the point where the lines cross. As can be seen, this fit fails to account for the very high pressure points at $u_p > 3.5$ km/s.

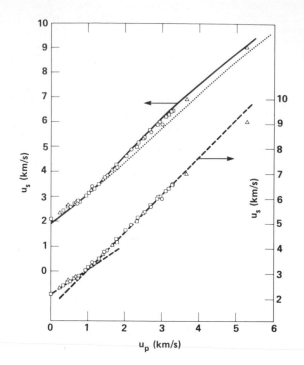

Figure 5: Comparison of theory and experiment for the shock compression curve of La. Theoretical calculations using the temperature dependent, Slater γ (solid curve) and the temperature independent linear γ (dotted curve) are compared to the data of Refs. 7-9. The two-line fit (dashed lines) of Ref. 9 is shown for comparison.

There is also no compelling evidence in the data for an abrupt change of slope at $u_p = 1.07$ km/s, as the present smooth curve (solid curve) fits the data equally well in the vicinity of 1 km/s. The slightly small $u_p = 0$ intercept of the solid curve follows from the calculated zero pressure bulk modulus being too small by 12%. As band theory generally only gives the normal density bulk modulus to within about 20% of experiment, such an offset is not unexpected. Using the Lindemann law[18] in conjunction with the results for γ in Fig. 4, we do estimate melting to occur in the vicinity of $u_p = 1$ km/s. Thus there may well be a change of slope due to melting. However, it is to be emphasized that the present calculations reproduce the essential anomalous features seen in the La shock data without including liquid corrections. This is consistent with the fact that it has not yet proved possible to detect melting in any metal solely from u_s-u_p Hugoniot data.

3. CONCLUSIONS AND FUTURE RESEARCH

The present work suggests that termination of the s-d electronic transition in La leads to a stiffening in the lattice vibrations for this material, which is indicated by a substantial peak in the calculated lattice Grüneisen parameter. It has been shown that an anomalous stiffening seen in the shock compression data for La can be explained by the existence of this peak. Since the initial rise in the low temperature γ in Fig. 4 occurs at about 200 kbar, the peak may also explain the saturation of the superconducting transition temperature which is observed in this pressure region.[19]

S-d transition is a fairly common phenomenon in metals. Most of the rare earths as well as other group IIIA elements (Sc, Y) exhibit essentially the same shock anomaly as discussed here for La.[7-9] As mentioned earlier, the 42.5 kbar isostructural transition in Cs can be attributed to the effects of s-d transition.[5] Neighboring Ba also exhibits an isostructural transition at 95 kbar,[20] and lighter alkalis and alkaline earths may well have such transitions at much higher pressures, all likely consequences of s-d transition. Clearly a thorough understanding of s-d transition and its consequences is important to an understanding of the high pressure behavior of many metals.

Because of the profound changes in electronic structure which accompany s-d transition, strictly electronic measurements such as de Haas-van Alphen are ideally suited to investigate this phenomenon. One can locate the various Lifshitz singularities which accompany the rise of the 6s band up through the 5d band. The final singularity, at the end of the s-d transition, occurs when the Γ_1 state rises above the Fermi level, which should occur in La at about 560 kbar ($V/V_0 = 0.5$) and in Cs at about 120 kbar ($V/V_0 = 0.25$).

Implicit in the present work is the fact that the lattice vibrational properties also offer a unique probe of systematic changes in the electronic structure. While the electronic structure determines the forces between atoms, the lattice vibrational properties probe (length) derivatives of these forces, and are therefore in some cases very sensitive indicators of changes in the electronic structure. Thus in both the present calculation for La and in previous work on Cs,[5] both based on Eq. (2), it has been the lattice vibrational contribution to the pressure, and specifically the Grüneisen parameter, which has lead to the important observable consequences in the equation of state. In the case of Cs, the structure in γ is sufficiently dramatic as to lead to an isostructural transition at room temperature. For the weaker s-d transition in

La it requires the very high temperatures induced by shock compression to sufficiently accentuate the γT term relative to the larger and more smoothly behaving electronic contribution to the pressure, in order to lead to significant effects in the equation of state.

Theoretical calculation of the lattice vibrational frequencies for non-simple metals, directly from first principles electron band theory, is an extremely difficult problem, but well worth pursuing. The elastic constants, phonon frequencies and their mode-Grüneisen parameters should all be rich in structure (as functions of volume) for materials undergoing electronic transition. Furthermore, calculation of the phonon frequencies is a necessary preliminary to calculations of resistivity and superconducting transition temperature, easily measured quantities which are well known for their extreme sensitivity to compression. Some theoretical progress has been achieved in this area: The tight binding approach developed by Varma, et al.[21] and the "frozen phonon" approach employed by Yin and Cohen[22] for Si, have both yielded frequencies in good agreement with experiment. Such calculations carried out for La and Cs under compression would be extremely valuable. Until this can be managed, however, there is still room for simple model calculations, such as the Slater model used in the present work to estimate the Grüneisen parameter. Another example would be the analysis of temperature and volume dependence of resistivity data, using some form of Bloch-Grüneisen theory,[23] in order to get an estimate of the volume dependence of an effective Debye frequency. Such analyses have been carried out for Cs,[24] and would be extremely useful for La in the region above 200 kbar in testing some of the predictions made in this work.

ACKNOWLEDGEMENTS

Part of this work was performed under the auspices of the U.S. Department of Energy by Lawrence Livermore National Laboratory under contract #W-7405-Eng-48. One of us (HLS) has been supported by the Danish Natural Science Research Council.

REFERENCES

[1]   Cited in Ref. 2.

[2]   Sternheimer, R., Phys. Rev. 78 (1950) 235.

[3]   See, e.g., McWhan, D.B., Parisot, G. and Bloch, D., J. Phys. F4 (1974) L69, and references therein.

[4]   McMahan, A.K., Phys. Rev. B17 (1978)
      1521; Louie, S.G. and Cohen, M.L., Phys.
      Rev. B10 (1974) 3237; Averill, F.W.,
      Phys. Rev. B4 (1971) 3315; and
      Yamashita, J. and Asano, S., J. Phys.
      Soc. Japan 29 (1970) 264.

[5]   Glotzel, D. and McMahan, A.K., Phys.
      Rev. B20 (1979) 3210.

[6]   McMahan, A.K., Skriver, H.L., and
      Johansson, B., Phys. Rev. B23 (1981)
      5016.

[7]   Al'tshuler, L.V., Bakanova, A.A., and
      Dudoladov, I.P., Zh. Eksp. Tero. Fiz. 53
      (1967) [Sov. Phys.-JETP 26, (1968) 1115].

[8]   Gust, W.H. and Royce, E.B., Phys. Rev.
      B8 (1973) 3595.

[9]   Carter, W.J., Fritz, J.N., Marsh, S.P.
      and McQueen, R.G., J. Phys. Chem. Solids
      36 (1975) 741.

[10] Al'tshuler, L.V. and Bakanova, A.A.,
      Usp. Fiz. Nauk. 96 (1968) 193 [Sov.
      Phys.-Uspeckhi 11 (1969) 678].

[11] Jayaraman, A., Newton, R.C. and
      McDonough, J.M., Phys. Rev. 159 (1967)
      527; Kennedy, G.C., Jayaraman, A. and
      Newton, R.C., Phys. Rev. 126 (1962) 1363.

[12] Andersen, O.K., Phys. Rev. B12 (1975)
      3060; Anderson, O.K. and Jepsen, O.,
      Physica 91B (1977) 317.

[13] Syassen, K. and Holzapfel, W.B., Solid
      State Commun. 16 (1975) 533.

[14] Because of hybridization between the 4f
      and 5d bands, the transition is at times
      as much s-f as s-d. The label s-d
      transition is still appropriate in that
      the electrons are transferring primarily
      from the 6s into the 5d bands, in spite
      of the growing f character of some of
      these states.

[15] Andersen, O.K., Madsen, J., Poulsen,
      U.K., Jepsen, O. and Kollar, J., Physica
      86-88b (1977) 249; Mackintosh, A.R. and
      Andersen, O.K., Electrons at the Fermi
      Surface, ed. by M. Springford (Cambridge
      University Press, Cambridge, 1980).

[16] Slater, J.C., Introduction to Chemical
      Physics (McGraw-Hill, New York, 1939)
      Ch. XIII and XIV; Landau, L.D. and
      Stanyukovich, K.P., Dokl. Akad. Nauk
      SSSR **46** (1945) 399.

[17] More, R.M., Lawrence Livermore National
      Laboratory Rept. UCRL-84379 (1980); R.
      Grover, private communication.

[18] Ziman, J.M., Principles of the Theory of
      Solids (Cambridge University, Cambridge,
      1964).

[19] Probst, C. and Wittig, J., Handbook on
      the Physics and Chemistry of Rare
      Earths, ed. by K.A. Gschneidner, Jr. and
      L. Eyring (North-Holland, Amsterdam,
      1978), V. I, Ch. 10, p. 749.

[20] Yoneda, A. and Endo, S., J. Appl. Phys.
      51 (1980) 3216.

[21] Varma, C.M. and Weber, W., Phys. Rev.
      B19 (1979) 6142.

[22] Yin, M.T. and Cohen, M.L., Phys. Rev.
      Lett. 12 (1980) 1044.

[23] Allen, P.B. and Butler, W.H., Physics
      Today 31 (1978) 44.

[24] McWhan, D.B. and Stevens, A.L., Solid
      State commun. 7 (1969) 301.

PHYSICS OF SOLIDS UNDER HIGH PRESSURE
J.S. Schilling, R.N. Shelton (editors)
© *North-Holland Publishing Company, 1981*

METALLIC MAGNETISM UNDER HIGH PRESSURE

E.P. Wohlfarth

Department of Mathematics, Imperial College, London
SW7 2BZ, England.

The pressure dependence of the Curie temperature, the saturation magnetization, the spin wave stiffness and the ferromagnetic susceptibility are discussed from the point of view of the itinerant electron model. The results are given in terms of simple yet sophisticated laws involving the concepts of single particle excitations, fine structure of the density of states curve, many body correlation effects, spin fluctuations and alloy concentration fluctuations, for all of which concepts high pressure research gives information. A wide variety of materials has been experimentally investigated to this end and the resuls are summarized in a Table.

1.   Introduction

The pressure dependence of the magnetic properties of metallic ferromagnets, such as the Curie temperature, the saturation magnetization and the spin wave stiffness, is of unusually wide physical interest. Since magnetism in these materials is a very complicated phenomenon a detailed knowledge of the dependence on volume of the various fundamental causes of this phenomenon has for the last 10 years or so helped considerably towards its satisfactory interpretation. At the beginning of this intense period of research we [1] reviewed the early experimental and theoretical situation at a previous high pressure meeting and it now seems appropriate to review the present status of the subject and the progress since 1969.   In this Introduction we now summarize the achievements brought about by the broadly based research in this field which demonstrates in a particularly gratifying way how much can be achieved by the interplay of usable theory and intelligently planned experiments:

(1) As will be shown, the measurable pressure coefficients are given on the basis of theoretical considerations in terms of simple yet richly endowed laws. These give, on the one hand, information of fundamental physical significance, and, on the other, an immediate qualitative or even quantitative interpretation of reasons for failures of these laws where such failures occur.

(2) The single particle excitations of metallic ferromagnetic materials are among the important elementary excitations which determine their physical properties. The resulting thermal effects are given in terms of the temperature scaled by the effective degeneracy temperature T_F which depends on the fine structure of the single particle density of states curve. This temperature may be a lot smaller than the Fermi energy given only by the particle density if the density of states curve has a complicated fine structure; this is always

the case for the d-electrons in transition metals. Pressure experiments have been used to estimate T_F on the basis of the simple laws referred to under (1).

(3) Theories of metallic ferromagnetism based on the Hartree-Fock approximation are known to be unreliable if not downright wrong. Several methods of estimating the required many body correlation corrections have been proposed, including a usable one by Kanamori [2]. This correction has also found its place in the simple laws governing pressure effects and its value may thus be estimated on this basis, at least in principle.

(4) Apart from the single particle excitations proposals have been made that spin fluctuations are also important elementary excitations in metallic ferromagnets [3]. Whereas the former are governed by Fermi statistics, the latter are governed by Bose statistics. Hence the simple laws will be different for the two cases and the possibility thus exists, in principle, of assessing the relative importance of these effects on the basis of pressure experiments.

(5) For several materials the simple laws have been observed to breakdown. Particularly detailed studies have been made of the pressure dependence of the Curie temperature. Here the law given by relation (8) is sometimes replaced by a law closer to relation (22). An analysis of this situation has led to the conclusion [4] that where there is a failure of (8) this is due to the non-homogeneous or spatially varying magnetization arising from alloy concentration fluctuations in the materials. The effect of these fluctuations is shown up particularly clearly in these pressure experiments as relations (8) and (22) are so utterly different, especially at low T_c's.

This brief review of the achievements of the successful interplay between theory and experiment shows what a wide range of ideas and phenomena has been covered in this way.   In the

following the various topics just summarized
will be described in greater detail. Attention
is focussed on transition metal and rare earth
materials since intermetallic uranium compounds
are being discussed elsewhere at this meeting
[5].

2. Pressure dependence of the Curie temperature; derivation of a simple law

The pressure derivative of the Curie temperature
T_c can be calculated on the basis of the
itinerant electron model of ferromagnetism by
using the following result [6]:

$$T_c^2 = T_F^2 (\bar{I}-1). \tag{1}$$

Here T_F was referred to above as the effective
degeneracy temperature depending on the fine
structure of the single particle density of
states curve [6]. Also

$$\bar{I} = I \, N(\varepsilon_F), \tag{2}$$

where I is the interaction between the itinerant
electrons after correcting for the many body
correlation effect [2] and $N(\varepsilon_F)$ the single
particle density of states at the Fermi energy
ε_F. To repeat, ε_F and T_F are not simply
related for the complicated d-bands of transition
metals. The pressure derivative of T_c is
calculated by assuming the following:

The value of I is related to that, I_b,
uncorrected for correlation, by the approximate
formula [2]

$$I/I_b = (1 + \gamma I_b/W)^{-1}, \tag{3}$$

where γ, a constant, and I_b the bare interaction, are taken to be independent of P and
the d-bandwidth W is given by Heine's formula
[7]

$$W \sim V^{-5/3}, \tag{4}$$

where $\omega = \Delta V/V = - \kappa P$ $\qquad$ (5)

is the volume strain due to a pressure P and
κ is the compressibility. Finally, T_F is
assumed to scale uniformly with W when pressure
is applied. On the basis of this set of
assumptions it follows that

$$T_c(P) \simeq T_c(0) (1-P/P_c)^\tau, \tag{6}$$

with $\tau = \frac{1}{2}$, $P_c = T_c^2(0)/2\alpha$, $\qquad$ (7)

and with α defined by (9). A more accurate
formula is the following, which in most
practical cases involves only little improvement over the approximate formula (6):

$$\frac{dT_c}{dP} = \frac{5}{3} \kappa T_c - \frac{\alpha}{T_c}, \tag{8}$$

where

$$\alpha = \frac{5}{6} \kappa \frac{I}{I_b} T_F^2. \tag{9}$$

The formulae (6) and (8) are "simple", i.e.
can be tested experimentally, but at the same

time sophisticated since they involve the concepts listed in §1. Thus many body correlation
effects as well as single particle fine structure
effects are combined in (9). It can be shown,
in addition [8], [3], that if spin fluctuation
effects are of supreme importance then (6) is
replaced by

$$T_c(P) \simeq T_c(0) (1-P/P_c)^\tau, \quad \tau = \frac{3}{4} \tag{6}*$$

so that measurements of one index τ should
suffice to settle this controversy.

The most important outcome of this simple analysis,
which reached this state of development soon
after [1] but has been used as a basis for
pressure experiments ever since, is the negative
pressure derivative of T_c given by (8) for
materials for which

$$T_c^2/T_F^2 < I/2I_b \tag{10}$$

(this inequality is amply fulfilled in most of
the substances listed in Table 1). This
phenomenon forms an important part of the invar
problem [9] whose other manifestations include
a large volume magnetostriction and thermal
expansion anomalies. As will be shown below
(§4) such a negative pressure effect is very
commonly observed (exceptions include metallic
nickel where the first term in (8) is operative),
but the dependence of dT_c/dP as $-T_c^{-1}$ given by
the second term in (8) is not always obtained
in high pressure research. The reasons for this
lie in alloy structure [4] as will be discussed
below. To prepare for this discussion we
present now a brief alternative formulation of
high pressure effects in metallic ferromagnets
obeying the itinerant electron model. This is
based on the Landau theory of phase transitions
for which the thermodynamic potential is
written as an expansion in a homgeneous small
magnetization M,

$$G = \frac{1}{2} AM^2 + \frac{1}{4} BM^4 + \frac{1}{2\kappa}\omega^2 - C\omega M^2 + P\omega. \tag{11}$$

Here A and B are Landau coefficients obtained,
using Fermi statistics, in the form

$$A = -\frac{1}{2\chi_0}(1-\frac{T^2}{T_c^2}), \quad B = \frac{1}{2\chi_0 M_0^2} (1+O(T^2)) \tag{12}$$

with χ_0 the ferromagnetic susceptibility at 0K,
M_0 the saturation magnetization at 0K, κ the
compressibility and there is no applied magnetic
field. The magnetoelastic coupling constant C
is basic to the present problems and it may be
obtained by minimizing G with respect to the
volume strain ω and substituting back to give

$$G = \frac{1}{2} A'M^2 + \frac{1}{4} B'M^4 + \text{const.}, \tag{13}$$

where $A' = A + 2\kappa CP$, $B' = B-2\kappa C^2$. $\qquad$ (14)

The Curie temperature is obtained by equating
A' to zero, giving, with (12),

$$\frac{dT_c}{dP} = - 2\kappa C\chi_0 T_c. \tag{15}$$

Comparing (8) and (15) gives, using also the formula for χ_O given in [6], i. e.

$$\chi_O = N(\varepsilon_F)\mu_B^2 T_F^2 / T_c^2 , \qquad (16)$$

that

$$C = \frac{5}{6} \frac{1}{N(\varepsilon_F)\mu_B^2} \left[\frac{1}{2} \frac{I}{I_b} - \frac{T_c^2}{T_F^2} \right]. \qquad (17)$$

The transformation B → B' was discussed in [10] to cover first order phase transitions. The form of (15) will be shown in §5 to be particularly useful when discussing pressure effects for heterogeneous magnetizations [4].

3. Other pressure effects

The saturation magnetization M_O is also in general pressure dependent. From (12) and (16) it follows that

$$M_O^2 = \frac{T_c^2}{2N(\varepsilon_F)\mu_B^2 T_F^2 B} , \qquad (18)$$

where B is the Landau coefficient in (11). Hence

$$\frac{d \ell n\, M_O}{dP} = \frac{d \ell n\, T_c}{dP} - \frac{5}{6} \kappa - \frac{1}{2} \frac{d \ell n\, B}{dP} . \qquad (19)$$

The two last terms on the right of (19) are both in general small for the materials listed in Table 1. Nevertheless, these small differences between the logarithmic derivatives of T_c and M_O have been observed by Tange et al.[11] for some binary alloys of nickel with Mn and Cr. Apart from these differences, the trend of the pressure variation is the same for both the Curie temperature and the saturation magnetization. At higher temperatures the ratio $\sigma*$ of the logarithmic derivative of $M \equiv M(T)$ to that of T_c is given by [12]

$$\sigma* = [1-(T/T_c)^2]^{-1} , \qquad (20)$$

and this result has been verified for two Fe-Ni invar alloys [12] (as well as for metallic nickel where the fit is remarkably good!).

The pressure derivative of the spin wave stiffness parameter D (defined by the ratio of the spin wave energy to the square of the wave vector) has been calculated and measured several times. Since the measurements are not as reliable as those for T_c and M_O it suffices to note that, according to [13], the logarithmic derivatives of D/a^2 (a, interatomic distance) and T_c are again almost equal, in good agreement with measurements [14] on Fe-Ni invar alloys and metallic nickel.

The pressure derivative of the ferromagnetic susceptibility χ_O is given by (16) as

$$\frac{d \ell n\, \chi_O}{dP} = -2 \frac{d \ell n\, T_c}{dP} + O(\kappa) , \qquad (21)$$

and this relationship has been verified approximately for Ni-Pt alloys [15]. It seems therefore that all the pressure derivatives characterizing metallic magnetism may be approximately reduced to that of the Curie temperature.

4. A summary of experimental data

We have analysed the pressure derivatives of T_c for a very wide range of metallic ferromagnets [4] and the results of this analysis are summarized in Table 1. In part A are shown materials obeying relation (8) with the positive first term often negligibly small. The values of the parameters τ, α, T_F and P_c have been deduced in some cases where the measurements were suitably extensive and their values are then given in the Table. To recall, τ is the index given in (6), α the parameter in (8) and (9), T_F the effective degeneracy temperature and P_c the critical pressure. It is seen that τ is close to $\frac{1}{2}$ in all cases but we can not yet claim complete certainty and thus superiority over the spin fluctuation value of $\iota = 3/4$, since the measurements are unfortunately not normally sufficiently accurate. The values of α span a wide range but it has been found [16] that, in accordance with (9), $\alpha/\kappa T_F^2$ varies much less between different systems. The values of T_F tend to agree with other estimates and with direct calculations [17]. Finally, the critical pressures P_c are frequently sufficiently low to be observable. Where several values are given in Table 1, more than one publication is involved; the references are all given in [4]. Very rarely has this analysis been applied to estimate I/I_b using (9). For Ni$_3$Al an analysis in terms of antistructure atoms [18] has given I/I_b between 0 and 1 which is reasonable. There is thus a future in high pressure research leading to much wider information on many body effects.

In part B of Table 1 are listed materials for which relation (8) does not apply. For most of these a relation of the form

$$dT_c/dP \sim -T_c \qquad (22)$$

is more closely obeyed but for a few a more cumbersome empirical relation of the form

$$dT_c/dP = aT_c - bT_c^2 \qquad (23)$$

is found to represent the data reasonably well. For most of the materials listed in part B there is other evidence (structural, magnetic, ...) that the alloys are heterogeneous in some sense. We have thus [4] extended our [19] Landau-Ginzburg approach to ferromagnetic materials with spatially varying $M \equiv M(\underline{r})$ to the present magneto-volume problem, and the results are briefly summarized in §5.

5. Pressure derivative of T_c for heterogeneous alloys

The Landau-Ginzburg model involves an expansion of the thermodynamic potential G in terms of a

TABLE 1

A.   Materials obeying $dT_c/dP = -\alpha/T_c$

Alloy System	τ	$\alpha (K^2/k\ bar)$	$T_F (K)$	$P_c (k\ bar)$
Ni-Pt	½	30		
Ni-Fe (far from M_s)	½	1600,1700,2050	1900	72
ZrZn$_2$	½	29		8.5, 18.0
Fe-Co-Ti (Fe rich)				
Mn-As-Sb	½	1760	1380	86
Fe-Ni-Mn	½	700		26.5
Cr-Te-Se	½			27.6
RE-Fe		840	~1000	
RE-Co		1950	~1000	
La-Co-Cu				
La-Co-Ni				
Amorph.				
Fe-B		3500, 1540		
Fe-B-Cr (high T_c)		2000		
Fe-Zr	½	1350		34

B.   Materials obeying $dT_c/dP \sim -T_c$ (or more complicated)

Ni-Cu	
Ni-Rh	
Ni-Pd	
Ni$_3$Al	0.63
Ni-Fe(near M_s)	~1
Sc$_3$In	-ve!
Fe-Co-Ti (Co rich)	
Cr-Te-Sb	
Fe-Ni-Cu	~1
Ti-Be-Cu	~1
RE-Ni	
Amorph.	
Fe-B-Cr (low T_c)	
Fe-Ni-B-P	
Fe-Ni-Zr	

small spatially varying magnetization $M(\underline{r})$ where, in (11), $A \rightarrow A(\underline{r})$, $B \rightarrow B(\underline{r})$ and an additional positive term $\frac{1}{2}C'|\nabla M(\underline{r})|^2$ occurs which limits the fluctuations of $M(\underline{r})$. The parameter C' determines the correlation length characterizing the heterogeneities of $M(\underline{r})$. By relating [4] the Fourier coefficients of the transform of $M(\underline{r})$ and of $\omega(r)$ to those γ_k of the transform of concentration $c(\underline{r})$ which occurs in the expressions of $A(\underline{r})$ and $B(\underline{r})$ and assuming a simple expression for γ_k it was found that the Curie temperature is again given by equating to zero A' in relation (14) but with

$$A \rightarrow \overline{A(\underline{r})}, \qquad (24)$$

where the bar denotes a spatial average. Hence from (12),

$$\frac{dT_c}{dP} = -2\kappa C \, \overline{\chi}_0 \, T_c, \qquad (15)*$$

and this extremely simple result must surely be very general, if only on dimensional grounds. It means that we are faced with a purely magnetic problem: How does the mean value of χ_0 of heterogeneous alloys vary with T_c on altering the average concentration? This is an extremely difficult problem in statistical mechanics and we are not aware of any generally useful explicit solutions.

It is reasonable to expect that for heterogeneous alloys the required relationship will be one involving a less rapid variation than that, $\chi_0 \sim T_c^{-2}$, for homogeneous alloys, i.e. relation (16). Using the Landau-Ginzburg formalism [20] to investigate this problem it was found that this less rapid variation occurred outside a range of concentrations close to critical which could not be investigated. In a simple mean field-CPA calculation of χ_0 for a "Stoner glass" Mathon [21] was able to obtain a susceptibility cusp at a temperature where $T_c = 0$, so that here χ_0 does not even diverge in this limit. Very similar results were obtained on the basis of a Landau-Ginzburg model applied to disordered alloys of the Pd-Ni type [22]. The presence or absence of a susceptibility divergence in heterogeneous alloys is thus still open to discussion.

The same uncertainty arises in considering experimental data since χ_0 is strictly speaking not a <u>high field</u> susceptibility. Most measurements of this quantity which have been reported have, however, not been obtained as is strictly necessary, i.e. by extrapolation to zero field with the aid of Arrott plots [6]. Some crude attempt [23] at an extrapolation was made for Cu-Ni alloys, part B of Table 1, to show some indication of a non-divergence of χ_0 at the critical concentration $(T_c \rightarrow 0)$. On the other hand, for Ni-Pt alloys, part A, Besnus et al [24] correctly obtained χ_0 from Arrott plots and found it to diverge as T_c^{-2}, in line with the homogeneous nature [15] of this alloys system.

The indication from imperfect theoretical and experimental sources is thus crudely that

$$\chi_0 \sim T_c^{-2} \quad , \quad \text{homogeneous}$$
$$\qquad (25)$$
$$\chi_0 \sim \chi_1 - \chi_2 T_c, \quad \text{very heterogeneous}$$

so that for these two extreme cases, using (15) with χ_0 the mean value of the susceptibility,

$$\frac{dT_c}{dP} \sim -T_c^{-1},$$
$$\qquad (26)$$
$$\frac{dT_c}{dP} \sim -\chi_1 T_c + \chi_2 T_c^2,$$

respectively. The qualitative agreement with most of the experimental results summarized in Table 1 should be taken to imply that the relationships (25) are as reasonable as can be expected in the absence of more reliable theoretical and experimental information. We hope that the present analysis will stimulate further work on this magnetic problem to which we have in (15)* reduced the magnetovolume problem of the pressure dependence of T_c.

6. Conclusions; future trends

The simple law (8) for the pressure dependence of the Curie temperature has been found to have a wide relevance to the many aspects of solid state physics to which we referred in the Introduction. Interest is aroused not only if the law is obeyed, as for the materials in part A of Table 1 but also where it is not obeyed, part B, since a clear indication of alloy structure is thus very sensitively provided. The analysis of the experiments summarized in the Table leads to the parameters τ, α, T_F and P_c which have the various fundamental meanings discussed in this article. The future of high pressure research in metallic magnetism is thus clearly to extend the list of materials studied in the systematic way described and to use the results of such studies to make further contacts with other branches of solid state physics.

REFERENCES

[1] Wohlfarth, E.P., Colloq. Int. CNRS 188 (1969) 363.

[2] Kanamori, J., Progr. Theor. Phys. 30 (1963) 275.

[3] Moriya, T., J. Mag. Mag. Mats. 14 (1979) 1.

[4] Wagner, D., and Wohlfarth, E.P., J. Phys. F. (1981).

[5] Franse, J.M.M. and Frings, P.H., this meeting (1981).

[6] Edwards, D.M. and Wohlfarth, E.P., Proc. Roy. Soc. A303 (1968) 127.

[7] Heine, V., Phys. Rev. 153 (1967) 673.

[8] Wohlfarth, E.P. Solid. St. Comm. 35 (1980) 797.

[9] Wohlfarth, E.P., Physics and Applications of Invar Alloys, Honda Memorial Series (Tokyo, Maruzen) 3 (1978) 327.

[10] Wohlfarth, E.P., J. Appl. Phys. 50 (1980) 7542.

[11] Tange, H. and Goto, M., J. Phys. Soc. Japan 49 (1980) 957; Tange, H., Yonei, T. and Goto, M. J. Phys. Soc. Japan 50 (1981) 454.

[12] Ponomarev, B.K. and Thiessen, V.G., Phys. Stat. Sol. (b) 104 (1981) 427.

[13] Mathon, J. and Wohlfarth, E.P., Phys. Lett. 34A (1971) 305.

[14] Gustafson, J.C. and Phillips, T.G., Phys. Lett. 29A (1969) 273.

[15] Alberts, H.L., Beille, J., Bloch, D. and Wohlfarth, E.P., Phys. Rev. B9 (1974) 2233.

[16] Wohlfarth, E.P., Procs. ICM Moscow 2 (1974) 28.

[17] Lipton, D. and Jacobs, R.L., J. Phys. C3 (1970) 389.

[18] Buis, N., Franse, J. and Brommer, P.E., Physica B (1981).

[19] Wagner, D. and Wohlfarth, E.P., J. Phys. F9 (1979) 717; J. Mag. Mag.Mats 15-18 (1980) 1345; Physica (1981).

[20] Yamada, H. and Wohlfarth, E.P., Phys. Stat. Sol.(b) 64 (1974) K71.

[21] Mathon, J., J. Phys. F8 (1978) 1790.

[22] Kato, T. and Mathon, J., J. Phys. F6 (1976) 1341.

[23] Sakakihara, M. Mollymoto, H., Okuda, K. and Date, M., J. Phys. Soc. Japan (1981); Acker,F. and Huguenin, R., Phys. Lett. 38A (1972) 343.

[24] Besnus, M.J. and Herr, A., Phys. Lett. 39A (1972) 83.

PHYSICS OF SOLIDS UNDER HIGH PRESSURE
J.S. Schilling, R.N. Shelton (editors)
© *North-Holland Publishing Company, 1981*

MAGNETIC PROPERTIES OF INTERMETALLIC URANIUM COMPOUNDS UNDER HIGH PRESSURES

J.J.M. Franse, P.H. Frings, F.R. de Boer and A. Menovsky,

Natuurkundig Laboratorium der Universiteit van Amsterdam,
Valckenierstraat 65, 1018 XE Amsterdam, The Netherlands

High pressure experiments are presented on the magnetic properties of various inter-metallic uranium compounds, mainly of the type UX_2 (X=Fe,Co,Ni,Pt,Ge,Ga). The general trend of pressure is to reduce the magnetic parameters with the exception of UGa_2 for which compound a positive pressure dependence of T_c is obtained. The high pressure data are combined with additional high magnetic field and specific heat studies for most of the investigated compounds. A detailed analysis of the magnetic properties of the Laves phase compounds UFe_2, UCo_2, UNi_2 and UAl_2 stresses the relevance of high pressure experiments on magnetism in metals.

1. INTRODUCTION

High pressure experiments on transition metals have proved to be of interest in establishing the type of ferromagnetic order in these materials. In general, large and negative pressure effects on the spontaneous magnetisation near zero temperature and on the Curie temperature are reported for weak itinerant electron ferromagnets [1]. Since it is often argued that the 5f electrons in the first half of the actinide series are of an itinerant nature [2], high pressure studies can be of help in investigating the magnetic behaviour of intermetallic uranium compounds.

High pressure studies on the magnetic properties of uranium compounds have been reported before for UPt, UAl_2 and UN. Huber et al. [3] observed a strong pressure dependence of the saturation magnetic moment per uranium atom for UPt at 4.2 K but did not find any shift of the Curie temperature. High pressure experiments on the spin fluctuation compound UAl_2 by Fournier and Beille [4] indicate a shift of the χ versus T curve under pressure towards lower χ with $\partial \ln \chi / \partial p \simeq -14$ Mbar^{-1} at 4 K. In studying the compound UN Fournier et al. [5] concluded from the experimental result $\partial \ln \mu_s / \partial p \simeq \partial \ln T_N / \partial p \simeq$ $\simeq -10$ Mbar^{-1}, where μ_s is the ordered magnetic moment at 4.2 K and where T_N is the Néel temperature, that this compound behaves like an itinerant electron antiferromagnet.

These high pressure studies show that the magnetic properties of uranium compounds can be strongly pressure dependent and that they are at least partly of an itinerant nature. In this paper we present additional information on the magnetic properties of various intermetallic compounds of uranium with gallium, germanium and platinum and some Laves phase compounds of uranium with iron, cobalt, nickel and aluminium. Most compounds will be discussed more extensively by comparing the high pressure data with high magnetic field and specific heat studies.

A common feature of many intermetallic uranium compounds is the strong dependence of magnetic properties on deviations from stoichiometry. The resistance minima and temperature dependent susceptibilities of UCo_x, with x close to 2, suggest that cobalt antistructure atoms give more magnetic character to these compounds [6]. The influence of non-stoichiometry has been studied for UFe_2 by Aldred [7] and by Zentko et al. [8] and for UPt by Huber et al. [3]. An interpretation of these results is not yet available. We observed that the presence of a second phase greatly influences the magnetic behaviour of UPt_2. In contrast to literature [11] we propose that UPt_2 is a paramagnetic material with a weakly temperature dependent susceptibility. The ferromagnetic order in UNi_2 is in our opinion due to a small fraction of local moments, probably situated on nickel atoms that occupy wrong sites in the hexagonal Laves phase structure.

Since preparation techniques of the uranium intermetallic compounds can have a large impact on the magnetic properties we shall present detailed information on sample quality, sample preparation and annealing procedures of the individual compounds.

2. EXPERIMENTAL

High pressure studies on the magnetic properties of the intermetallic uranium compounds were performed by an induction method using helium as the high pressure transmitting medium. Magnetisation measurements were carried out in magnetic fields up to 10 tesla with pressures up to 8 kbar. Typical sample sizes are: cylinders with a diameter of 6 mm and a length of 10 mm. Smaller pieces of the same batch were used for high magnetic field (up to 35 tesla) and susceptibility (up to 300K) measurements. Specific heat experiments were performed on some of the cylindrical samples.

The crystallographic data and the spacing between the uranium atoms of the investigated compounds were summarised in Table I. The uranium

TABLE I

Structure and U–U spacing of the investigated uranium compounds; data from |9,10,11|		
	Structure	U–U spacing (Å)
α–U	orthorhombic	2.75
UFe_2	$MgCu_2$	3.05
UCo_2	$MgCu_2$	3.03
UAl_2	$MgCu_2$	3.38
UNi_2	$MgZn_2$	3.0
UPt	CrB	3.61
UPt_2	distorted $InNi_2$	3.81
UGe_2	$ZrSi_2$	4.06
U_3Ge_4	orthorhombic	
UGa_2	AlB_2	4.01

TABLE II

Annealing temperatures and annealing times of the investigated intermetallic uranium compounds		
	Annealing temperature (^oC)	Annealing time (days)
UAl_2	800	10
$U(Fe,Co)_2$	950	7
UNi_2	–	–
UGa_2	800	10
UGe_2	950	7
U_3Ge_4	1000	7
UPt	900	10
UPt_2	1000	10

compounds were prepared by arc melting the appropriate amounts of the constituents in a titanium gettered argon atmosphere: uranium with a nominal purity of 99.8% (Al: 5 ppm; Fe: 5 ppm; Mn: 11 ppm; Cu: 55 ppm; Co,Ni: <10 ppm) and the other metals supplied by Johnson and Matthey – England (Fe,Co,Ni,Al,Pt,Ga, specpure) and Hobokon – Belgium (Ge,better than 6 N). All buttons were melted, turned over and remelted five times to increase sample homogeneity. The cylindrical samples were obtained by vacuum casting of the melt into a water-cooled copper crucible with the exception of UPt_2, UNi_2 and UGa_2 that were casted into an alumina crucible. Attempts to cast U_3Ge_4 were unsuccesful; this sample was taken from the button by spark erosion. The losses of material after melting and casting were negligibly small. All samples, with exception of UNi_2, were annealed in evacuated (2 x 10^{-6} Torr) sealed silica tubes at temperatures and for periods of time listed in Table II and subsequently furnace cooled.

To avoid contamination of the samples during the annealing procedure the samples were wrapped in tantalum foils; pieces of etched uranium served as a getter during the annealing. The $U(Fe_{1-x}Co_x)_2$ samples were water quenched. Part of the materials was crushed for X-ray studies. The Debye-Scherrer patterns showed that most of the materials were single phase and of the right structure. Only for UPt and UGa_2 additional weak lines of yet unidentified structures were found. X-ray patterns taken from the surface of the bulk samples always contain reflections of uranium oxides.

3.   RESULTS

A general survey of pressure experiments on the materials studied in the present work is given in Table III. With the exception of UGa_2 we observe negative pressure dependences of the Curie temperature for all ferromagnets. UGa_2 is known for its large magnetic anisotropy; its magnetic moment from experiments on a single crystalline sample amounts to 2.71 μ_B per uranium atom at 4.2 K. Although this large anisotropy points to localised magnetism, part of the magnetic moment could be due to conduction electron polarization \|12\|. Magnetisation measurements at low temperatures on polycrystalline samples are strongly influenced by this anisotropy. Nevertheless it is an interesting feature that the effect of pressure at 4.2 K is to reduce the magnetisation with a value for $\partial \ln \sigma/\partial p$ of -5 Mbar^{-1} in a field of 2.4 tesla.

Values for the spontaneous magnetic moment per uranium atom at 4.2 K in UPt, U_3Ge_4 and UGe_2 remain small and amount to 0.45 μ_B, 0.6 μ_B and 0.7 μ_B, respectively. For UPt this number is sample dependent and for different samples of the same melt values between 0.4 μ_B and 0.5 μ_B were obtained. The effects of pressure on the spontaneous magnetic moment at 4.2 K can be large, in particular for UPt where σ_o is reduced to almost zero at a pressure of approximately 10 kbar. It is a point of further study to investigate the effects of internal stresses on the value of the saturation moment of this com-

TABLE III

The effect of pressure on susceptibility (4.2 K), spontaneous magnetic moment (4.2 K) and Curie temperature of various ferromagnetic and paramagnetic intermetallic uranium compounds

	χ_o $\dfrac{m^3}{mol\ f.u.}$	σ_o $\dfrac{Am^2}{mol\ f.u.}$	T_c K	$\partial\chi_o/\partial p$ $\dfrac{m^3}{kbar\ mol\ f.u.}$	$\partial\sigma_o/\partial p$ $\dfrac{Am^2}{kbar\ mol\ f.u.}$	$\partial T_c/\partial p$ $\dfrac{K}{kbar}$	$\partial\ln X/\partial p$ Mbar^{-1}		
							$X=\chi_o$	$X=\sigma_o$	$X=T_c$
U(Fe$_{1-x}$Co$_x$)$_2$									
x = 0		6.2	160		−0.030±0.002	−0.52±0.04		−4.8	−3.2
0.1		4.7	129		−0.028±0.002	−0.63±0.05		−6.0	−4.8
0.2		3.0	87		−0.027±0.002	−0.8 ±0.2		−8.9	−9
0.3		0.76	24		−0.033±0.002			−44	−37
0.4	58x10^{-9}			−1.16 x10^{-9}			−20		
1	17x10^{-9}			−0.076x10^{-9}			−4.5		
UNi$_2$	14x10^{-9}	0.29	27	−0.15 x10^{-9}	−0.029±0.002	−0.18±0.05	−11	−10	−6.7
UPt		2.4-2.8	27			−0.5 ±0.1		≈−100	−18
UPt$_2$	65x10^{-9}			−0.52 x10^{-9}			−8		
U$_3$Ge$_4$		8.7	91		≈ 0	−0.24±0.05		≈ 0	−2.6
UGe$_2$		3.87	52		−0.026±0.002	−1.25±0.03		−6.8	−24
UGa$_2$			120			0.22±0.05			1.8
UAl$_2$	57x10^{-9}						−14*		

*From ref.4.

pound. The dramatic influence of pressure on the magnetisation curves of UPt is shown in fig.1. These curves can not easily be understood in the itinerant electron model as has previously been suggested for UPt |3|. In discussing the magnetic transition near 3 tesla in the curves shown in fig.1 one should also consider the effects, for instance, of the crystal field on the uranium moment and the possibilities of a more complex spin structure of the material. The magnetisation curves of UPt show a complex behaviour near T_c. The values for T_c and $\partial T_c/\partial p$ in Table III have been obtained from an analysis of σ^2 versus H/σ data at low magnetic fields.

In contrast with the Laves phase compounds of uranium with iron, cobalt and nickel, we observe quite different values for the parameters $\partial\ln\sigma_o/\partial p$ and $\partial\ln T_c/\partial p$ in UPt, U$_3$Ge$_4$ and UGe$_2$. This could be an indication of a more localised nature of ferromagnetism in these compounds. In order to further investigate the type of magnetism of the uranium-platinum and the uranium-germanium compounds we performed additional high magnetic field and susceptibility measurements on all these compounds and specific heat measurements on the uranium-platinum compounds only.

The specific heat data for UPt and UPt$_2$ are shown in fig.2. The data for UPt are in fair agreement with the work of Luengo et al. |13|. These authors analysed the difference in the

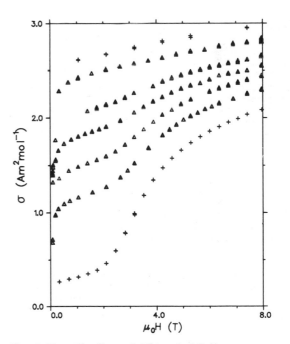

Fig. 1. Magnetisation of UPt at 4.2 K: (Δ) is at pressures of 4.5, 4.0, 3.7, 2.9 kbar and 1 bar; (+) is another sample at 7.7 kbar and 1 bar; (all pressure effects are negative).

TABLE IV

Specific heat and high magnetic field data for UPt and UPt$_2$; $C = \gamma T + \beta T^3 + DT^{3/2}$; $$S = \frac{\pi^2 k^2}{3\mu_o \mu_B^2} \frac{\chi}{\gamma} = 5.85 \times 10^6 \frac{\chi}{\gamma}$$					
	γ mJ/K^{2}mol f.u.	β mJ/K^{4}mol f.u.	D mJ/K$^{5/2}$ mol f.u.	χ_{HF} m^3/mol f.u.	S
UPt	110±5	1.4 ±0.1	6±2	47×10^{-9}	2.5
UPt$_2$	77±2	1.05±0.1	–	51×10^{-9}	3.8

C/T values between UPt and ThPt in terms of a spin wave contribution to the specific heat and did not pay much attention to the peak in C/T at approximately 19 K. Since this peak disappears in an applied magnetic field of 5 tesla, its magnetic origin is evident. The type of magnetic transition that is responsible for this peak is not yet clear. It is, however, another indication for the presence of localised magnetic moments. Below 10 K we analysed the specific heat data with the expression $C/T = \gamma + \beta T^2 + DT^{1/2}$, where the last term corresponds with a spin wave contribution to the specific heat. Values for the parameters γ, β and D are presented in Table IV. The large γ value of UPt is in agreement with the XPS spectra for UPt |11| that give evidence to a large contribution of 5f electrons to the density of states at the Fermi level. In an applied magnetic field of 5 tesla a reduction of the γ value of UPt with 10 percent is observed. The γ value for UPt$_2$ is of comparable magnitude; for this compound a small upturn in the C/T versus T^2 plot, that is partly suppressed by a magnetic field of 5 tesla, is found below 4 K. The specific heat results for UPt and UPt$_2$ can be combined with the differential susceptibility data at 4.2 K in order to derive values for the Stoner enhancement factor S. From the fact that the Stoner parameter reaches quite reasonable values we conclude that the differential susceptibility of UPt and UPt$_2$ is governed by the band electrons.

High magnetic field data for UPt, UPt$_2$, U$_3$Ge$_4$ and UGe$_2$, from which the differential susceptibility values in Table IV have been taken, are shown in fig.3.

Susceptibility data in the temperature interval $T_c < T < 300$ K are presented in fig.4 in a plot of χ^{-1} versus T for UPt and UGe$_2$. The strongly curved plots suggest a description of χ by a sum of a nearly temperature independent susceptibility and a Curie-Weiss term. A different interpretation of these susceptibility data can be

given, however, with effective uranium moments that are temperature dependent due to crystal field effects or to an admixture of higher J multiplets. More detailed information on the uranium-platinum and the uranium-germanium compounds will be published later.

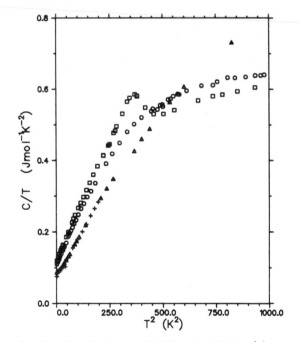

Fig. 2. Specific heat of UPt: at 0 tesla (□), at 5 tesla (○); and of UPt$_2$: 0 tesla (△) and 5 tesla (+).

The data in Table III show that the quantities $\partial \ln \sigma_o/\partial p$ and $\partial \ln T_c/\partial p$ are of equal sign and equal magnitude for the ferromagnetic Laves phase compounds. In the series U(Fe$_{1-x}$Co$_x$)$_2$ the

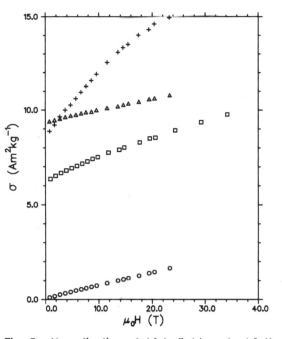

Fig. 3. Magnetisation at high fields and 4.2 K of:
(□) UPt, (○) UPt₂, (△) UGe₂ and
(+) U₃Ge₄.

values for these parameters tend to diverge at low T_c values. These observations point to itinerant magnetism in the cubic Laves phase compounds |14|. Although UNi_2 and $U(Fe_{0.7}Co_{0.3})_2$ have quite similar T_c values, the corresponding shifts of T_c with pressure differ largely. The small value for the pressure derivative of T_c for UNi_2 provides arguments to ascribe the ferromagnetic order in this compound to the presence of a relatively small number of localised magnetic moments that interact via the susceptible UNi_2 matrix. For that reason we also present in Table III values for the matrix susceptibility of UNi_2 and its pressure dependence. The intrinsic matrix susceptibility of UCo_2 is of the same magnitude as that of UNi_2. Still, localised moments in a low concentration as for UNi_2 are not likely to be formed in UCo_2. This conclusion can be drawn from an analysis of the specific heat in zero magnetic field and in an applied field of 5 tesla. Specific heat data for UFe_2 and UCo_2 are presented in fig.5 in a plot of C/T versus T^2. For both compounds no effect of the external field on the specific heat is observed. In particular, the upturn below 7 K that is found for UCo_2 is not affected by the magnetic field. A high temperature (T > 8 K) analysis results in the following values for the parameters γ and β in the expression $C = \gamma T + \beta T^3$: γ = 55 mJ/K²mol and β = 1.4 mJ/K⁴mol for UFe_2; γ = 10 mJ/K²mol and β = 1.1 mJ/K⁴mol for UCo_2. These values must be considered with caution. An analysis below 8 K would result in different values for these parameters, in particular for UCo_2.

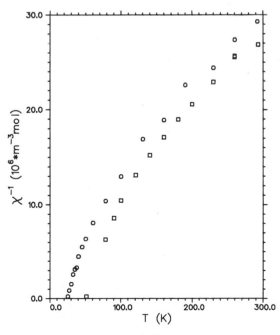

Fig. 4. Inverse susceptibility of UGe₂ (□) and UPt (○).

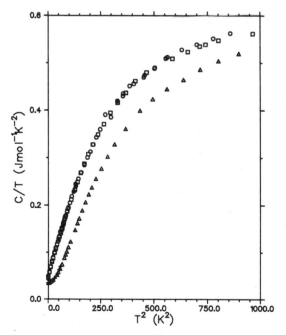

Fig. 5. Specific heat of UFe₂: at 0 tesla (□), at 5 tesla (○); and of UCo₂: at 0 tesla (△) and 5 tesla (+).

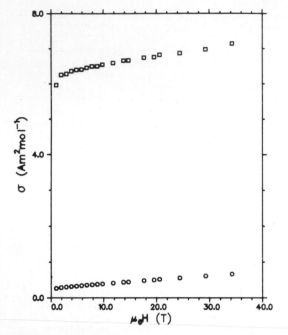

Fig. 6a. Magnetisation at high fields and 4.2 K of :
(□) UFe$_2$ and (○) UNi$_2$.

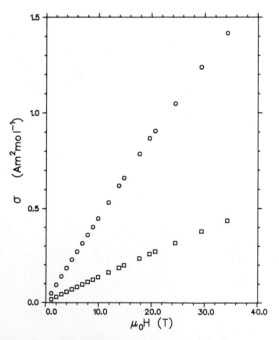

Fig. 6b. Magnetisation at high fields and 4.2 K of
(□) UCo$_2$ and (○) UAl$_2$.

High magnetic field data for all four Laves phase compounds are shown in figs.6a and 6b. For the ferromagnetic compounds UFe$_2$ and UNi$_2$ only a weak field dependence of the differential susceptibility is observed at higher fields. The paramagnetic compounds UCo$_2$ and UAl$_2$ show a decreasing differential susceptibility with increasing magnetic field. It is not only in these high magnetic fields that UAl$_2$ and UCo$_2$ show similar characteristics. In UAl$_2$ too, an upturn in C/T versus T^2 has been found which is also not affected by a magnetic field of 4.3 tesla |15|.

In the next section we shall go into a more detailed discussion of these Laves phase uranium compounds.

4.   DISCUSSION

A detailed inspection of high pressure, high magnetic field, susceptibility and specific heat data for the Laves phase uranium compounds reveals some interesting features that could be of general interest for the theory of magnetism in metals. In this discussion the relevance of high pressure experiments on the magnetic properties of these uranium compounds will be stressed.

UFe$_2$.

An accurate and extensive study of the magnetic properties of this compound has been performed by Aldred |7| on a single crystalline sample. The magnetic anisotropy in this compound is relatively small (see also ref.16), the temperature dependence of the magnetic moment is partly due to Stoner excitations and the critical exponents near the magnetic phase transition are close to values reported for pure 3d metals. Polarised neutron experiments at 4.2 K revealed magnetic moments at the iron and uranium sites of 0.59 μ_B and 0.06 μ_B, respectively |17|. These facts suggest that the magnetism of UFe$_2$ is similar to that of the 3d elements and that it can be treated in an itinerant electron picture. The value for the electronic specific heat coefficient γ and the results for the logarithmic pressure derivatives of the spontaneous magnetic moment at 4.2 K and of the Curie temperature support this suggestion.

By a closer inspection of the magnetisation data near T$_c$ we shall stress the importance of high pressure experiments in the study of magnetism in metals.

Magnetisation data in a small temperature region around T$_c$ are presented in fig.7 in a plot of σ^2 versus H/σ. The curves represent a description of the data with the expression |18|:

$$(H/\sigma)^{1/\gamma} = B_o(T-T_c)/T_c + B_1\sigma^{1/\beta}, \qquad (1)$$

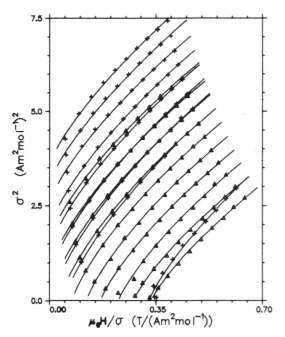

Fig. 7. Magnetisation of UFe_2 around T_c at 5kbar ($\triangle$) and 1 bar (+), solid lines represent eq.(1). Temp. are: 154.7 156.0 157.0 158.0 159.0 160.0 161.0 162.0 163.0 and 164.0 K at 5 kbar and 154.0 155.0 156.0 157.0 158.0 159.0 160.0 161.0 and 166.15 at 1 bar.

TABLE V

Pressure derivatives of the magnetic parameters of UFe_2 at 4.2 K and near T_c		$\partial \ln X / \partial p$ $Mbar^{-1}$
σ_o (Am^2/mol)	6.22 ± 0.02	-4.8 ± 0.5
T_c (K)	160.0 ± 0.1	-3.2 ± 0.3
m_o (Am^2/mol)	8.44 ± 0.08	-8.5 ± 1.3
h_o (T)	$162 \quad \underline{+}10$	$-25 \underline{+} 6$
m_o/σ_o	1.36 ± 0.02	

where values for the critical exponents γ and β of 1.26 and 0.45, respectively, were taken from the work by Aldred [7]. A plot of the various results for $B_o(T-T_c)/T_c$ and B_1 versus T yields values for T_c, B_o and B_1 at pressures of 1 bar and 5 kbar. The quantities B_o and B_1 can be combined in order to derive values for the parameters m_o and h_o that enter the theory of critical phenomena: $m_o = (B_o/B_1)^\beta$ and $h_o = B_o m_o$. In a molecular field type of description in which the local magnetic moments are assumed to be temperature independent the value of m_o should be equal to $f\sigma_o$ with $f = 1.732$ for $S = 1/2$ [18]. Values for the parameters T_c, m_o, h_o and f and for the logarithmic pressure derivatives of σ_o, T_c, m_o and h_o, where σ_o is the spontaneous magnetic moment at 4.2 K, are given in Table V. The results for m_o and $\partial \ln m_o/\partial p$ provide interesting information on the decrease with temperature of the magnetic moment in an itinerant electron system. This information may contribute to the actual question of the persistence of magnetic moments above T_c in an itinerant electron ferromagnet [19]. The numerical value of the factor f in the expression $m_o = f\sigma_o$, however, is dependent on the type of interaction between the magnetic moments [18]. Any conclusion on the decrease of the magnetic moment between zero temperature and the Curie temperature from the ratio of m_o/σ_o rests heavily on the type of interaction that is assumed so that no conclusive answers can be drawn from this ratio. For that reason we believe that the information on the relative

pressure dependence of m_o is superior to the value of m_o itself. The value for $\partial \ln m_o/\partial p$ in Table V turns out to be almost two times larger than the result for $\partial \ln \sigma_o/\partial p$ at 4.2 K. This means that the relative pressure dependence of the local magnetic moment σ_ℓ (local in the sense as it has been introduced by Shiga [20], for instance) is about two times larger at T_c than at 4.2 K. In a first attempt to understand this result we shall discuss it in the Stoner model in which σ_ℓ^2 (H=O) = $-A/B$ where A and B are defined by the expression:

$$H/\sigma_\ell = A + B\sigma_\ell^2 . \qquad (2)$$

The quantities A and B in (2) can be expressed in elementary band parameters [21].

Neglecting the temperature and pressure dependences of the parameter B we can write for the pressure dependence of $\sigma_\ell^2(o) = \sigma_\ell^2(H=o)$:

$$\partial \sigma_\ell^2(o)/\partial p = -2\kappa C/B \qquad (3)$$

with C, the magnetovolume parameter, defined as $C = -\frac{1}{2}\partial A/\partial \ln V$ and with κ the compressibility. By taking the magnetovolume parameter independent of temperature we obtain

$$(\partial \ln \sigma_\ell(o,T)/\partial p)(\partial \ln \sigma_\ell(o,o)/\partial p)^{-1} = \sigma_\ell^2(o,o)/\sigma_\ell^2(0,T) \qquad (4)$$

In this way we are able to derive the temperature dependence of the local magnetic moment from the relative pressure derivatives of the magnetic moments at different temperatures. By writing $m_o = f\sigma_\ell(o,T_c)$, $\sigma_\ell(o,o) = \sigma_\ell$ and $\partial \ln m_o/\partial p = \partial \ln \sigma_\ell(o,T_c)/\partial p$ we calculate from the data of Table V: $\sigma_\ell^2(o,T_c) = (0.55 + 0.1) \sigma_\ell^2(o,o)$ and $f = 1.83 \pm 0.2$. A closer inspection of the pressure dependence of all relevant band parameters of UFe_2 in the way it has been carried out for Ni_3Al by Brommer [22], is required before a firm conclusion can be drawn from these experiments. This simplified treatment, however, may serve as a demonstration of the importance of high pressure experiments in the study of metallic magnetism.

UNi$_2$

This compound is known from the literature to order ferromagnetically at approximately 30 K with a spontaneous magnetic moment at 4.2 K between 0.13 μ_B |23| and 0.07 μ_B |24| per formula unit. In this paper we present a description of the magnetic properties of UNi$_2$ in terms of a low concentration of local magnetic moments in a susceptible matrix. This description emerges from an analysis of high pressure and high magnetic field experiments on this compound.

The pressure dependence of the magnetisation of UNi$_2$ at 4.2 K, shown in fig.8, can be described by two contributions: a decrease of the spontaneous magnetic moment (σ_o) with pressure and a decrease of the differential susceptibility (χ_{HF}) at higher field strengths with pressure. For both contributions the relative pressure effects are of the same order of magnitude (Table III). This result will be used for evaluating the magnetisation as the sum of a localised part, that is saturated at 4.2 K in magnetic fields of a few tesla; and a matrix part σ_m that is induced by the sum of an internal and the external magnetic field:

$$\sigma = \sigma_1 + \sigma_m \qquad (5)$$

with

$$\sigma_m = \chi_{HF}(H_i + H_u) \qquad (6)$$

By taking the localised moment σ_1 and the field experienced by the local moment on the matrix, H_i, independent of pressure we derive the relation:

$$\partial \ln \sigma_o / \partial p = (\chi_{HF} H_i / \sigma_o) \, \partial \ln \chi_{HF} / \partial p \qquad (7),$$

where σ_o is the spontaneous magnetic moment at zero temperature. Inserting the experimental results for the logarithmic pressure derivatives given in Table III we find: $\chi_{HF} H_i = 0.91 \sigma_o$, $H_i = (18\pm2)T$ and $\sigma_1 = (0.03\pm0.03)$ Am^2/mol. With these data for H_i and σ_1 in mind we shall consider the results of high magnetic field experiments on UNi$_2$. A plot of σ^2 versus H/σ shows an upturn that is characteristic for the presence of local magnetic moments in a susceptible matrix, as was discussed for the Ti(Fe,Co) system by Brommer et al. |25|. Assuming that the matrix magnetisation σ_m follows the relation

$$H/\sigma_m = A + B\sigma_m^2 \qquad (8)$$

with $H = H_i + H_u$, we reanalyse the magnetisation data for UNi$_2$ in a plot of $(\sigma - \sigma_1)^2$ versus $(H_i + H_u)/(\sigma - \sigma_1)$, see fig.9. In this way linear Arrott plots result, although the linearity itself is not a sharp criterion. Susceptibility data below 300 K are shown in fig.10 for UNi$_2$ and UCo$_2$. The data for UNi$_2$ above T_c can be described by the expression

$$\chi(T) = \chi_o(1 - \alpha_1 T^2 + \alpha_2 T^4) + C/(T - \theta) \qquad (9)$$

with $\chi_o = 1.2 \times 10^{-9}$ m^3/mol, $\alpha_1 = 1.4 \times 10^{-6}$K^{-2}, $\alpha_2 = 9 \times 10^{-12}$ K^{-4}, $C = 9.43 \times 10^{-8}$ Km^3/mol and $\theta = 27$ K.

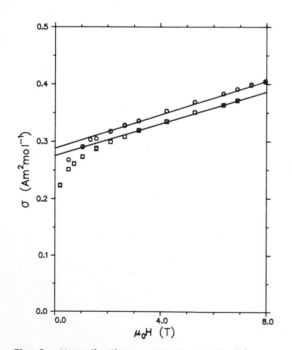

Fig. 8. Magnetisation at 4.2 K of UNi$_2$ (□) 4.4 kbar and (○) 1 bar. Solid lines represent eq. 5 and 6.

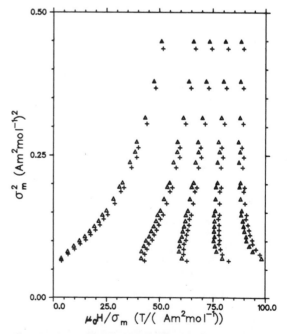

Fig. 9. Magnetisation of UNi$_2$ at 4.2 K, plot conform eq.8; (△): σ_1 is 0.00, (+): σ_1 is 0.01, $\mu_0 H_i$ is 0. 10. 15. 20. and 25 tesla.

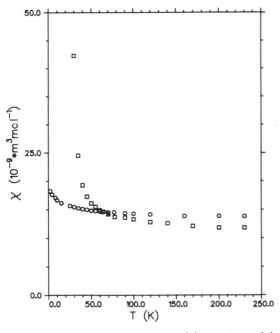

Fig. 10. Susceptibility of UNi$_2$ (□) and UCo$_2$ (o).

The specific heat of UAl$_2$ was succesfully ana-
lysed by Trainor et al. |27| with the expression
from spin fluctuation theory:

$$C/T = \gamma m^*/m + \beta T^2 + (\alpha\gamma/T_{SF}^2)T^2 \ln T/T_{SF}, \quad (10)$$

where m* is the enhanced mass, T$_{SF}$ a characteristic temperature and α a parameter containing
the Coulomb interaction, the density of states
at the Fermi level and the Stoner enhancement
factor.

In fig.11 we present the specific heat of UCo$_2$
below 6.3 K in zero field and in an applied
field of 5 T in a plot of C/T versus T^2. The
experimental data have been fitted to the expression:

$$C/T = A + BT^2 + DT^2 \ln T. \quad (11)$$

Numerical results for the parameters A, B and
D are given in Table VI, together with values
for these parameters reported by Trainor et al
for UAl$_2$. By comparing these numerical data with
the above given expression (10) for the specific
heat in the spin fluctuation theory we find with
β = 1.16 mJ/K^4 a value for T$_{SF}$ of 40 K and a
value for αm/m* of 20. A value of 25 K for T$_{SF}$,
however, with β = 0.94 mJ/K^4 mol and
αm/m* = 7.8 is equally well acceptable.

Writing for the effective moment μ$_{eff}$ = pμ$_B$ and
for the saturation magnetic moment μ$_s$ = qμ$_B$
we calculate N$_1$p^2/N$_A$ = 0.060 and
N$_1$q/N$_A$ = 0.054±0.005, where N$_1$ is the number
of local moments per mol UNi$_2$, N$_A$ is Avogadro's
number and σ$_1$ = N$_1$μ$_s$. With p ≈ q ≈ 1 we find
N$_1$/N$_A$ ≈ 0.05.

We tentatively ascribe these local moments to
defects or to a partial disordering of the
hexagonal Laves phase structure. A fraction of
2x10^{-2} of the nickel atoms, occupying wrong
lattice sites could explain, for instance, the
observed phenomena.

The values for ∂ ln T$_c$/∂p and ∂ ln χ$_o$/∂p, that
are of nearly equal magnitude for UNi$_2$, can be
understood with Shimizu's description of ferromagnetism of dilute alloys with localised
moments |26|.

UCo$_2$ and UAl$_2$

As we mentioned in the preceeding section there
is an upturn in the plot of C/T versus T^2 for
UCo$_2$ and UAl$_2$ that is not affected by a magnetic
field of approximately 5 tesla. This upturn in
C/T versus T^2 corresponds for UCo$_2$ with a peak
in C versus T at approximately 4 K. If this
peak were due to local magnetic moments in an
internal magnetic field we would estimate this
internal field to be of the order of 4 tesla.
Since the position of the peak does not change
when a magnetic field of 5 T is applied we conclude that this upturn can not be ascribed to
local magnetic moments.

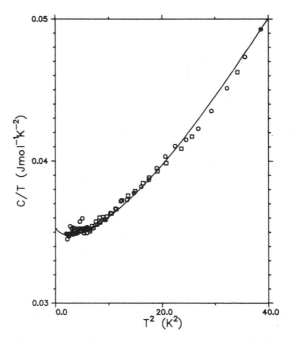

Fig. 11. Specific heat of UCo$_2$: at 0 tesla (□),
and at 5 tesla (O); solid line represents eq. 11.

TABLE VI

Specific heat data for the compounds UAl$_2$ and UCo$_2$ below 6 K; $C = AT+BT^3+DT^3\ln T$; C in mJ/K mol f.u. T in K		
	UCo$_2$	UAl$_2$*
A	35.4 ± 1	143
B	-0.42 ± 0.01	-4.38
D	0.43 ± 0.01	1.94

* Data from ref.26.

To discuss the field dependence of the specific heat of UCo$_2$ we follow Brodsky and Trainor |15| and make use of the expression given by Béal-Monod et al. |28|:

$$\delta C/T \cong 0.1 \ (\mu H/kT_{SF})^2(S/\ln S)C/T \qquad (12)$$

With a value for S/ln S of approximately 3 (corresponding with a Stoner enhancement factor between 2 and 5), a value for μ_0H of 5 T and a value for T_{SF} between 25 K and 40 K we calculate a relative change in C/T of the order of 1 percent or less, in agreement with experiment.

Our result for $\gamma m^*/m$ (= 35.4 mJ/K^2 mol) indicates that γ of UCo$_2$ is at least two times smaller than the γ value of UAl$_2$ for which a value of 70 mJ/K^2 mol has been reported |27|. This decrease of γ going from UAl$_2$ to UCo$_2$ is in qualitative agreement with the photoemission study on UAl$_2$ and UCo$_2$ by Naegele et al. |29,30|.

Susceptibility measurements on UCo$_2$ in the temperature interval 1.5 K - 300 K (fig.10) show an upturn in the low temperature region. This upturn, that essentially differs from a Curie-Weiss behaviour, was ascribed before to the onset of magnetic ordering |23| or to a local enhancement on wrong site Co atoms |6|. We want to stress the similarity between the UAl$_2$ and the UCo$_2$ data. The relative increase of the susceptibility below T_{SF} is of comparable magnitude for UAl$_2$ and UCo$_2$.

The high magnetic field data on UAl$_2$ and UCo$_2$ of fig. 6 show a decrease of the differential susceptibility with increasing field, see fig. 12. For UAl$_2$ we observe that the relative decrease of the differential susceptibility is comparable with the relative decrease of the susceptibility at reaching T_{SF} |4,15|. Apparently the enhancement of the susceptibility can be suppressed by a magnetic field of which the magnitude is given by the relation $\mu H \simeq kT_{SF}$. In UCo$_2$ roughly the same feature is present.

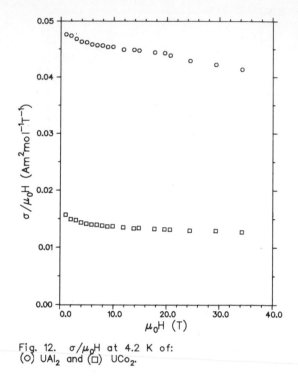

Fig. 12. σ/μ_0H at 4.2 K of: (O) UAl$_2$ and (□) UCo$_2$.

Finally we discuss the reduction of the susceptibility of UAl$_2$ and UCo$_2$ under pressure. Since the enhancement by spin fluctuations is of comparable magnitude for UAl$_2$ and UCo$_2$ we assume that the difference in the $\partial\ln\chi_0/\partial p$ values given in Table III is due to a difference in the enhanced Pauli susceptibilities of the two compounds. For the pressure dependence of the enhanced susceptibility χ we write |22|:

$$\partial\ln\chi/\partial p = -\kappa\lambda\Gamma(S-1) - \kappa\Gamma \qquad (13)$$

where $\Gamma = \partial\ln N(E_F)/\partial\ln V$ and $\Gamma\lambda = \partial\ln IN(E_F)/\partial\ln V$, with I the effective electron-electron interaction.

With values for the Stoner enhancement factor S of 10 and for the compressibility of 1.65 Mbar^{-1} |4| we derive for UAl$_2$ values for $\Gamma\lambda$ and Γ of 0.6 and 3.6, respectively.

Taking for the compressibility of UCo$_2$ the value for UAl$_2$ and for the Stoner factor of UCo$_2$ a value between 3.5 and 7 we obtain values for $\lambda\Gamma$ between 0.12 and 0.3 and a value for Γ of 2.3. The difference in the Γ values for UAl$_2$ and UCo$_2$ is in qualitative agreement with the different amount of d and f character of the density of states at the Fermi level.

ACKNOWLEDGEMENTS

The authors like to thank Dr. P.E. Brommer for helpful discussions. The experimental assistance of C.J.M. Eijkel is gratefully acknowledged.

REFERENCES

|1| Wohlfarth, E.P., this conference (1981).

|2| Johansson, B., Skriver, H.L., Mårtensson, N., Andersen, O.K. and Glötzel, O., Physica 102 B (1980) 12-21.

|3| Huber, J.G., Maple, M.B. and Wohlleben, D., J.Magn.Magn.Mat. 1 (1975) 58-63.

|4| Fournier, J.M. and Beille, J., J. de Phys. C4 (1979) 145-146.

|5| Fournier, J.M., Beille, J., Boeuf, A., Vettier, C. and Wedgwood, A., Physica 102 B (1980) 282-284.

|6| Hrebik, J. and Coles, B.R., Physica 86-88B (1977) 169-170.

|7| Aldred, A.T., J.Magn.Magn.Mat. 10 (1979) 42-52.

|8| Zentko, A., Diko, P., Miskuf, J., Svec, T., Kosturiak, A. and Drab, J., J.Magn.Magn. Mat. 13 (1979) 310-312.

|9| Trzebiatowski, W., Ferromagnetic Materials (North-Holland, Amsterdam, 1980).

|10| The Actinides: Electronic Structure and Related Properties (Academic Press, New York, 1974).

|11| Schneider, W.D. and Laubschat, C., Phys.Rev. B23 (1981) 997-1005.

|12| Andreev, A.V., Belov, K.P., Deryagin, A.V., Kazei, Z.A., Levitin, R.Z., Menovsky, A., Popov, Yu.F. and Silant'ev, V.I., Sov.Phys. JETP 48 (1978) 1187-1193.

|13| Luengo, C.A., Maple, M.B. and Huber, J.G., J.Magn.Magn.Mat. 3 (1976) 305-308.

|14| Franse, J.J.M., J.Magn.Magn.Mat. 10 (1979) 259-264.

|15| Brodsky, M.B. and Trainor, R.J., Physica 91B (1977) 271-280.

|16| Popov, Yu.F., Levitin, R.Z., Zelený, M., Deryagin, A.V. and Andreev, A.V., Sov.Phys. JETP 51 (1980) 1223-1226.

|17| Lander, G.H., Aldred, A.T., Dunlap, B.D. and Shenoy, G.K., Physica 86-88B (1977) 152-154.

|18| Kouvel, J.S. and Comley, J.B., Phys.Rev. Letters 20 (1968) 1237-1239.

|19| Panel discussion on Itinerant Electron Magnetism, Inst.Phys.Conf.Ser. 55 (1981) 669-687.

|20| Shiga, M., Inst.Phys.Conf.Ser. 55 (1981) 241-250.

|21| Edwards, D.M. and Wohlfarth, E.P., Proc. Roy.Soc. A303 (1968) 127-137.

|22| Brommer, P.E., to be published in Physica.

|23| Lam, D.J. and Aldred, A.T., AIP Conf.Ser. 24 (1975) 349-350.

|24| Sechovsky, V., Smetana, Z., Hilscher, G., Gratz, E. and Sassik, H., Physica 102 B (1980) 277-281.

|25| Brommer, P.E., Franse, J.J.M. and Hölscher, H., Inst.Phys.Conf.Ser. 55 (1981) 279-282.

|26| Shimizu, M., Rep. Prog. Phys. 44 (1981) 329-409.

|27| Trainor, R.J., Brodsky, M.B. and Isaacs, L.L., AIP Proc. 24 (1974) 220-221.

|28| Béal-Monod, M.T., Shang-Keng Ma, and Fredkin, D.R., Phys.Rev.Letters 20 (1968) 929-932.

|29| Naegele, J.R., Manes, L., Spirlet, J.C., Pellegrini, L. and Fournier, J.M., Physica 102 B (1980) 122-125.

|30| Naegele, J.R., Manes, L., Spirlet, J.C. and Fournier, J.M., Applications of Surface Science 4 (1980) 510-517.

PHYSICS OF SOLIDS UNDER HIGH PRESSURE
J.S. Schilling, R.N. Shelton (editors)
© *North-Holland Publishing Company, 1981*

ELECTRONIC STRUCTURE OF BULK SOLIDS AND ARTIFICIAL
MATERIALS AT POSITIVE AND NEGATIVE PRESSURES

A. J. Freeman

Department of Physics and Astronomy
Northwestern University
Evanston, Illinois 60201, U.S.A.

We discuss the application of local density functional theory to the determination of
the electronic structure of selected examples of bulk solids and some new artificial
materials when subjected to positive and negative pressures and for both hydrostatic
and non-hydrostatic pressures.

1. INTRODUCTION

It is widely recognized that experiments at
high pressures still represent one of the most
difficult and challenging areas of research in
condensed matter physics. By contrast, the
life of the theorist is much simpler since he
can produce high pressures in a trivial fashion
simply by changing lattice parameters. With
the change of this parameter, the theorist can
make both positive and, as I will emphasize,
negative pressures. He can simulate hydrostatic pressures (a simple thing to do because it
retains the symmetry) and nonhydrostatic pressures (which add to the complexity of the
calculation).

2. METHODOLOGY AND APPROACH

In our work we use the local density and local
spin density functional theory approach of
Hohenberg, Kohn and Sham [1,2] which has been
shown to yield accurate solutions of the ground
state electronic band structure and properties
of solids. The calculations are carried out
self-consistently in order to account for
important charge transfer effects. Since these
charge transfer effects become greater at high
pressures, the need for self-consistency in the
theoretical calculations becomes more apparent.
As a result of self-consistency, highly accurate solutions of the Schrödinger or Dirac
equations can be obtained which are independent
of any assumptions about initial configurations
and/or starting potentials. The remaining
assumption, and indeed the only empirical input
(other than the atomic number of the atoms) is
the observed lattice parameter. This last need
will be obviated when total energy calculations
(in progress) can be done at reasonable cost
and speed.

One reason for the present happy state of the
theory lies in the development of several simplified energy band schemes which can be shown
to yield very good results and which overcome a
number of the limitations inherent in the
traditional energy band methods when treating
complex materials. These methods known as the
linear muffin-tin orbital (LMTO) method [3] and
the linearized augmented plane wave (LAPW)
method [4] are "linearized" versions of the KKR

and APW schemes, respectively. Their major
virtues are that they avoid some of the computational complexities and high costs of treating complex (many atoms per unit cell systems)
inherent in the regular plane wave based methods. These linearized methods retain
relatively high accuracy and suffer little,
if any, loss in computational speed compared
to pseudopotential methods.

3. PROPERTIES OF BULK SOLIDS UNDER PRESSURE

As is well known, the importance of pressure
as a parameter is that it can often greatly
affect the properties of a material in a non-
destructive reversible way by changing its
lattice parameter and not otherwise affecting
its basic structure. Here we cite several
examples of the role of pressure on magnetism/
superconductivity. In general, positive
pressures promote superconductivity and
destroy ferromagnetism and negative pressures
do the opposite. We shall see this effect in
this and the next section.

A. f-electron Bonding, Electronic Structure, and Phase Transitions in La and Ce

Under pressure La and Ce show effects due to
varying degrees of 4f localization, much the
same as occurs in the actinide series where
the 5f states change from itinerant to localized with increasing atomic number. These
effects are shown very clearly in the self-
consistent LAPW calculations carried out by
Pickett, Freeman and Koelling, [5,6]. In
both metals the 4f occupation number, Q_f, is
found theoretically [5,6] to increase only
slightly with pressure: from 0.05 to 0.10
under a 25% decrease in atomic volume in La
and from 0.88 to 0.94 under a 15% decrease in
atomic volume in Ce. This increase in Q_f is
the result of significant broadening of the
4f bands under pressure, as a result of both
nearest neighbor f-f overlap and sd-f hybridization. The 4f band broadening slightly more
than compensates the raising of the center of
the 4f band under pressure which tends to
decrease Q_f. For the γ and α phases of Ce,
the 4f band center is found to lie at 0.5 (0.7)
eV above E_F in γ- (resp. α) - Ce, with an
effective half-band width of 0.7 (1.0) eV. In

La, the 4f bands are slightly wider but lie 2.5 eV above E_F resulting in the small value of Q_f.

Early speculation on the γ-α (isostructural) fcc-fcc transition centered on a proposed promotion of the 4f electron (or fraction thereof) into the itinerant sd bands as the 4f level rises under pressure. This interpretation, although apparently confirmed by model Hamiltonian calculations [7], was questioned by Kmetko and Hill [8] on the basis of band structure calculations and by Johansson [9] in terms of empirical promotion energies. The present results are consistent with the proposal of Johansson that the γ-α transition is a localized-to-itinerant transition of the 4f state ("Mott transition"), with negligible change in 4f occupation. Such a transition, which is usually metal-to-antiferromagnetic insulator, but not so here due to the background metallic sd bands, is usually discussed in terms of relative magnitudes of an intra-atomic correlation energy U and an effective bandwidth W, with the ratio U/W varying through a critical value versus pressure (or "density"). We have also made some estimates of U and W and find it is reasonable to conclude that U/W is at least 50% larger in the γ-phase than in the α-phase and thar variation of this ratio through its critical value could easily occur at the γ-α transition [6].

One of the strongest arguments for the promotional model of the γ-α transition was that the differences in atomic volume could be accounted for in terms of the extra bounding due to $\sim$0.5 more sd electrons in the α-phase: γ-Ce was said to have a "valency" of $\sim$3.1 while α-Ce had a "valence" of 3.6. Study of the band structure and non-spherical charge density in detail [6] indicates the f-f overlap in α-Ce to contribute to binding. However, this contribution is weaker and qualitatively different from metallic sd bonding in Ce; for example, f-f σ-bonding seems to be nearly absent. Although f-f bonding in α-Ce has not been put on a quantitative basis, it is likely that it can account for the "increased bonding" (smaller atomic volume) in α-Ce. The concept of "valency" loses its quantitative significance in this picture, with the bonding in Ce becoming analogous to that in the 5f series.

Turning now to the case of La, there are indications at low T, from resistivity and the superconducting T_c, that La undergoes structural transformations at 25 and 53 kbar [10]. The intermediate phase is likely to be fcc (as it is in a wide pressure range at 300 K) and it has been proposed that the 53 kbar transition is isostructural [10]. The 25 kbar "transition" may represent the point at which the dhcp phase, which coexists with the fcc phase at ambient pressure in most samples, is finally "squeezed out" in favor of fcc La.

We have carried out self-consistent calculations of the band structure of fcc La corresponding to

atomic volumes at ambient pressure, 50 and 120 kbar [5]. It may be significant that at each of these pressures the Fermi surface is found to have a different connectivity. In the range 50-230 kbar, a saddle point in band 2 along the Γ-K direction crosses the Fermi level. Such an occurrence, leads, in principle, to a Lifshitz-Dagens lattice instability [11], which arises from a singular derivative in the electronic free energy. The coupling of the electronic system to the lattice gives rise to a structural transition which may be isostructural. Since at 50 kbar $T_c \simeq 10$ K, evidently the electron-lattice coupling is strong, which makes the interpretation of this transition in terms of a Lifshitz-Dagens instability even more attractive.

The 25 kbar transition presents a dilemma; this could be fcc-dhcp. However, there is some evidence from the La_xCe_{1-x} alloy system [12] that this transition is at least partially isostructural (fcc-fcc). This latter possibility, with the change in Fermi surface topology noted earlier, seems to suggest another Lifshitz-Dagens instability. However calculation of the generalized susceptiblity [5] for each lattice constant shows very large peaks to grow, as the pressure is reduced from 50 kbar, at just those wave vectors necessary to distort the fcc lattice into a dhcp lattice. Thus, the present results lend some support to both of these interpretations of the 25 kbar transition. Clearly more experimental work to determine the phase changes in La under pressure would be desirable.

Finally, a detailed comparison was carried out between the theoretical predictions and a large number of experiments. Because of space limitations we refer the reader to the original papers [5,6].

B. Magnetism/Superconductivity in C15 Compounds

The cubic Laves or C15 compounds of type $MgCu_2$ exhibit a number of interesting phenomena including 1) the much studied weak ferromagnetism in $ZrZn_2$ and 2) correlations between high superconducting transition temperatures T_c and lattice instability in ZrV_2 and HfV_2 which are well known in the more familiar A15 and rock salt transition metal compounds. Interest in superconductivity and itinerant magnetism in the C15's has recently been revitalized following the report of itinerant antiferromagnetism in $TiBe_2$ by Matthias et al. [13]. This discovery followed the expectation raised by the theoretical proposal of Enz and Matthias [14] that the well-known weak itinerant ferromagnetism of $ZrZn_2$ arises from inhibited p-wave pairing and that above the critical pressure at which the magnetism is destroyed, p-state superconductivity would appear in $ZrZn_2$ and in $TiBe_2$ would coexist with any antiferromagnetic ordering.

We have carried out detailed self-consistent LMTO energy band studies of $TiBe_2$ and $ZrZn_2$ at ambient and high pressures in their paramagnetic states [15]. To aid in understanding the magnetism/superconductivity in these C15 compounds, the high T_c superconductor ZrV_2 and the ultra-low T_c superconductor YAl_2 were also considered. Results obtained include total and partial (by atom type and ℓ-value) density of states, important Fermi surface nesting, Stoner-like parameters to describe the conditions for magnetism, electron-phonon coupling parameter, λ, and T_c values and their behavior under pressure. From the calculated results we have obtained a good qualitative understanding of magnetism/superconductivity in these materials. Furthermore, on the basis of the pressure calculations, we have predicted the possibility of superconductivity in $ZrZn_2$ with application of pressure without invoking p-state pairing. Most recently in a similar study for $HfZn_2$, we find that the application of reasonable pressures may result in superconductivity in this system as well. These theoretical studies show that both $ZrZn_2$ and $TiBe_2$ are more strongly exchange-enhanced systems than is Pd (and like Pd have strongly nested jungle-gym Fermi surfaces). Further, the calculations for the paramagnetic state show Stoner factors which are greater than unity and hence that ferromagnetism is favored.

The later report [16] of ferromagnetism in $TiBe_{1.8}Cu_{0.2}$ and the measurement of its neutron magnetic form factor gave further impetus to undertaking local spin-density calculations to study the ferromagnetic state for this interesting material. We have obtained results of self-consistent spin-polarized band studies of $TiBe_2$ and $ZrZn_2$ in an assumed ferromagnetic state [16]. From these calculations we are able to discuss and assess itinerant ferromagnetism in $ZrZn_2$ and $TiBe_{1.8}Cu_{0.2}$ using the calculated magnetic moments, spin magnetization densities and neutron magnetic form factors and their comparison with experiments on these materials. It would be now interesting to carry out these spin-polarized calculations with application of high pressures to study theoretically the destruction of magnetism with application of pressure.

4. Artificial Materials and the Effect of Negative Pressures

A. General Considerations

Just as external pressure changes the environment experienced by an atom in a bulk solid, a changed environment produced by nature or by an experimentalist in the laboratory can simulate the effects of pressure including, as we shall see, ultra-high negative pressures not otherwise achievable. To place these ideas in proper perspective, let me first describe the fascinating challenge which lies in modern submicron technology and physical phenomena on the microstructure scale [17].

One of the most important developing trends in the last decade lies in the preparation of synthetic structures on the submicron level, which will permit, in the near future, new scientific phenomena to be investigated and novel device applications to be made on artificial materials not found in nature. It is clear, even now, that the range and variety of new phenomena that can be studied, pose an exciting challenge. Much of the incentive for these technological efforts lies in the universal recognition of the enormous economic potential inherent in large scale applications of micro-electronics.

Examples on the microstructure domain include: a) small particles, b) thin film junctions (sandwiches) of dissimilar metals which are either magnetic or superconducting, etc., and c) modulated structures (both crystalline and amorphous) consisting of alternating layers of materials A and B, i.e., ABABA....

In addition to the technological reasons for focusing attention on submicron problems, there are interesting phenomena to study which are of fundamental physical interest in themselves. Attention spreads from bulk properties to surfaces and the next natural extention is to obtain a better fundamental understanding of interfacial and reduced dimensionality phenomena. Thus one needs to focus on the possible breakdown of traditional concepts when treating some of the microstructure systems mentioned above. One must particularly ask whether concepts such as energy bands, effective mass and transport are applicable.

Materials prepared on the microstructure domain represent systems with changed atomic environments -- really modifications of their atomic arrangements. Thus, for ferromagnetic thin films, the surface layers have their magnetic moments substantially altered: in Ni(001), the moment is reduced to 044 μ_B but is not magnetically "dead" [18] in agreement with electron capture measurements [19]; by contrast Fe(001) has an increased magnetic moment [20], in agreement with Mössbauer effect measurements [21]. Recently, the modification of environment as in the NiCu modulated structures of Thaler, et al. [22] show a large change in the ferromagnetic resonance and interpreted to be due to an increase in magnetization. Theoretical calculations [23] have shown that the magnetization of the Ni atoms is actually decreased -- in agreement with later magnetization [24] and neutron experiments [25]. More recently, we have studied CuNi modulated structures consisting of eight layers modulated along [100] and varying the Cu/Ni concentrations from 7/1 to 4/4. [26] These studies give more detailed information about the electronic structure and magnetism for differing numbers of Ni layers (1-4) in different surroundings of Cu layers. In particular, the interface moment increases from 0.14 μ_B for the single

Ni layer, to 0.4 for the 4 Ni layers, i.e.,
reduced magnetic moments but no total quenching
of the Ni magnetization.

B. Enhanced Magnetism of Au/Pd/Au Sandwiches

The exciting promise of artificial materials
lies in the possibility of producing to speci-
fication new materials with desired properties.
An example is the enhanced magnetization of Pd
in an Au/Pd/Au sandwich observed experimentally
by Brodsky and Freeman [27].

Arguing that changes in the properties of
materials are often caused by the application
of high pressures which decrease the lattice
parameter, a_0, but do not otherwise affect their
basic structure and that the common use of
alloy methods to increase a_0 is not satisfactory
for studying intrinsic magnetism/superconducti-
vity in Pd, we set out and prepared thin films
of Pd (presumably with expanded lattice parameters) sandwiched between thicker Au films.
We hoped to obtain large increases in the
Stoner factor and hence to increase the possibility of either producing ferromagnetism or
p-wave pairing superconductivity in pure Pd.
Pd metal was chosen because of its incipient
ferromagnetic nature and the large effects of
spin fluctuations. With its high density of
states at the Fermi energy, and large Stoner
exchange enhancement ($S \sim 10$), it is easily
polarized by dilute magnetic impurities leading
to giant magnetic moments and often to ferromagnetism at relatively high temperatures.
More recently, interest in p-wave pairing
superconductivity has focused on Pd as a most
likely candidate because it has the largest
value of S for pure metals.

The expansion of the Pd lattice parameter in
the sandwich environment can be understood as
follows. During deposition of metal A on a
single crystal of an isostructural metal B, the
initial layers of A form with an a_0 between
those of A and B. As the A layer thickens, its
a_0 relaxes toward the normal value; thus there
is a limit to the thickness of the modified a_0.
The large misfit between the Pd and Au lattice
constants, 4.72% leads to an expanded a_0 (Pd),
calculated to be $\sim$2.5% and to relaxation within
the several layers. In order to stretch as
many Pd layers as possible, the samples were
prepared as Au/Pd/Au sandwiches so as to lock
the expanded Pd lattice from both sides. Since
each layer of Pd atoms has a different a_0 and
tetragonal distortion, the experimental data
result from an averaging process.

The measurements showed the first successful
observation of exceptionally large increases in
the magnetic susceptibility, χ, of pure Pd in
these sandwiches at low T. The films do not
show magnetic order ($T \gtrsim 2.2K$) or superconductivity ($T \lesssim 0.05K$). From the results of band
calculations on Pd at negative pressure, and
the measured χ at low T, we derived very large

values of S ($\sim$350-25,000) for the sandwiches.
Since large S values in the modified paramagnon
model of Levin and Valls [28] for superfluid
He_3 yield sizable values of T_c from p-wave
pairing superconductivity in the absence of
magnetic ordering, we suggested the possibility
of superconductivity in similar expanded Pd
systems but in thin sandwiches which would
have a lower proximity effect [27].

This sandwich technique holds great promise
for modifying properties of other systems and
is now being pursued further by experiments on
metal film sandwiches of Au/Cr/Au and Au/V/Au,
[29].

C. Effects of Tetragonal Distortions:
Stretching or Negative Pressures

In both the CuNi modulated structures and the
Au/Pd/Au sandwiched discussed above, the
local environment of the Ni or the Pd can be
considered to result from a negative pressure
in the form of a tetragonally distorted stretching of the interatomic distances. Thus the
local environment greatly affects the properties of the system subjected to this negative
pressure.

Several recent theoretical studies have
focussed on the change in its properties when
Ni is subjected to external hydrostatic pressure [30]. The local environment of Ni in
the compositionally modulated films is quite
different from that in fcc Ni metal: the
structure of Ni-layers between Cu-layers in a
coherent modulation consists of a fcc lattice
with the [111] direction perpendicular to the
layers. For example, for a 50-50 CuNi composition, the lattice constants in the planes
may be approximated by the average of the Cu
and Ni fcc lattice constants; in the [111]
direction the Ni atomic spacing will be compressed by approximately the same amount (δ)
as the stretching which occurs in the other
two directions. Thus the Ni structure is a
tetragonal distortion of the fcc lattice along
the [111] direction.

To understand the effect of tetragonal stretching on the properties of Ni, we determined the
electronic structure of paramagnetic and
ferromagnetic fcc Ni and tetragonal Ni by means
of self-consistent semi-relativistic linear
muffin-tin orbital (LMTO) energy band studies
[31]. The self-consistent fcc calculations
were performed using 89-k-points in the
irreducible Brillouin zone (IBZ) while the
tetragonal calculations used 45-k-points
in the hexagonal IBZ. For tetragonal Ni,
we used $\delta = |a_{tetr} - a_{fcc}| /a_{fcc} = 0.0125$

(a_{fcc} is the lattice constant equal to 6.6488

au) which is a reasonable value for Ni layers
between Cu layers in a coherent modulated
structure. (This gives a volume increase of

the tetragonal Ni compared to the fcc by about 1.5%). The tetragonal unit cell is chosen as a hexagonal unit cell with the c-axis along [111], so that $c/a = \sqrt{3}(0.975)$ where c and a (=6.7318 au) are the hexagonal lattice parameters. This unit cell contains 3 atoms.

The paramagnetic studies used the local density exchange-correlation potential of Hedin, et al. [32] and the ferromagnetic ones used the exchange-correlation potential of Gunnarsson, et al. [33]. The calculations for ferromagnetic fcc Ni yield magnetic moments, magnetization density, hyperfine fields and Fermi surface areas which are in good agreement with experiment and with other theoretical calculations such as those of Wang and Callaway [34]. Our results for tetragonal Ni with a structure equal to that of Ni in coherent modulated CuNi structures show that the effect of d-band narrowing due to the stretching is weaker than the DOS broadening caused by the band splitting in the tetragonal symmetry. In particular, the DOS values at E_F are slightly lower than for fcc Ni, which makes a Stoner splitting of the bands less effective and, as shown from our spin-polarized results, the magnetic moment decreases. The reduction in magnetic moment is also reflected in the reduction of the calculated hyperfine field by about the same amount, 15-20%. Thus, in layers of Ni, in CuNi (111) modulated structures which are thin enough to be coherent but yet thick enough not to be influenced by the proximity of Cu layers, we expect from our calculations a reduction of the magnetic moments compared to pure fcc Ni in agreement with the results for the CuNi modulated structures actually calculated [23].

In order to understand the behavior of Pd in the Au/Pd/Au environment, we undertook spin polarized calculations of fcc Pd and tetragonally stretched Pd as described above for Ni. In these studies, an external magnetic field is first applied and then removed in later stages of the self-consistency procedure. In order to study the possible onset of magnetism in tetragonally stretched Pd, we had to do the comparison calculations for fcc Pd at ambient pressure so as to be able to attribute any differences in properties to the "stretching". An important first result of these studies was the discovery of metamagnetism in Pd in high magnetic fields.

For some time now itinerant metamagnetism in high magnetic fields has been a subject of intense interest since its prediction by Wolfarth [35] for Pd and other strongly ex-changed enhanced metals. By itinerant metamagnetism, one means a magnetic field induced first-order phase transition from the paramagnetic to ferromagnetic state. While metamagnetic behavior has been inferred or observed for several systems, no first-principles theoretical calculations have been reported in support of such observations.

We have reported the first ab initio theoretical study of the electronic structure, magnetic susceptibility, and magnetization of Pd in the presence of external magnetic fields which demonstrates the onset of a metamagnetic transition. [36] We find that the calculated magnetic susceptibility, χ, and magnetization, m, show an almost step-like increase at a critical field which indicates the onset of a metamagnetic phase transition. The result also demonstrates a strong hysteresis effect of the magnetization with magnetic field. Starting with the magnetized state above the metamagnetic transition and gradually reducing the strength of the field, the self-consistent results show m to be fairly stable at about 0.19 μ_B/Pd until a low field value is reached. Below this value of the supporting field, the system cannot sustain a net magnetization. The appearance of hysteresis in the observed magnetism of $Co(S_xSe_{1-x})_2$ was cited by Wolfarth as evidence for a metamagnetic transition [37].

Stimulated by this result and the enhanced χ observed for the Au/Pd/Au sandwiches [27], we have also considered conditions that may cause the metamagnetic transition to occur at lower external fields and, thus to, make its observation likely. Hence, we have also studied the metamagnetic behavior of tetragonally "stretched" Pd as in the Au/Pd/Au sandwiches. As a model for the stretched Pd in the Au/Pd/Au sandwich, we used a tetragonally distorted fcc unit cell that matched onto the Au lattice in the xy plane in a manner similar to the coherent modulated structures of Thaler, et al. [22]. The calculated moments for stretched Pd are found to be ~0.1 μ_B larger than those for fcc Pd at the lowest fields so that at a 0.2 mRy field the moment is already 0.13 μ_B. This leads to a susceptibility which is ~40 times larger than that calculated for fcc Pd. Thus tetragonally stretched Pd is either metamagnetic in very low fields or ferromagnetic at zero field. This would be the first time that a paramagnetic metal like Pd was made into a ferromagnet and provides strong support for the experimental efforts with microstructures to produce in the laboratory materials "engineered" to have desirable properties not found in nature.

I am grateful to my colleagues who participated in the work referred to in this paper. This work was supported by the National Science Foundation (Grant No. DMR 77-23776) and the Air Force Office of Scientific Research (Grant No. 81-0024).

[1] P. Hohenberg and W. Kohn, Phys. Rev. 136, (1964) 864.

[2] W. Kohn and L.J. Sham, Phys. Rev. A 140, (1965) 1133.

[3] O.K. Andersen, Phys. Rev. B 15, (1975) 3060.

[4] D.D. Koelling and G. Arbman, J. Phys. F. 5, (1975) 2041.

[5] W.E. Pickett, A.J. Freeman and D.D. Koelling, Phys. Rev. B 22, (1981) 2695.

[6] W.E. Pickett, A.J. Freeman and D.D. Loelling, Phys. Rev. B 23, (1981) 1266.

[7] B. Coqblin and A. Blandin, Advan. Phys. 17 (1968) 281; R. Ramirez and L.M. Falicov, Phys. Rev. B3 (1971) 1255.

[8] D.D. Koelling and A.J. Freeman, Phys. Rev. B2 (1970) 290; A.J. Freeman and D.D. Koelling, Ch. 2, Vol. 1, The Actinides: Electronic Structure and Related Properties, A.J. Freeman and J.J. Darby, Jr., eds. (Academic Press, New York, 1974); E.A. Kmetko and H.H. Hill, in Plutonium 1970 and Other Actinides, ed. W.N. Miner (AIME, New York, 1970) p. 223.

[9] B. Johansson, Phil. Mag. 30 (1974) 469; B. Johansson, J. Phys. F7 (1977) 877.

[10] H. Balster and J. Wittig, J. Low Temp. Phys. 21 (1975) 377.

[11] I.M. Lifshitz, Sov. Phys. JETP 11 (1960) 1130; L. Dagens, J. Phys. Lett. 37 (1976) L37; L. Dagens, J. Phys. F8 (1978) 2093.

[12] E. King and I.R. Harris, J. Less-Common Metals 27 (1972) 51.

[13] B.T. Matthias, et al. Phys. Lett. A 69 (1978) 221.

[14] C.P. Enz and B.T. Matthias, Science 201 (1978) 828.

[15] T. Jarlborg and A.J. Freeman, Phys. Rev. B 22 (1980) 2332.

[16] T. Jarlborg, A.J. Freeman and D.D. Koelling (to be published).

[17] See discussion by A.J. Freeman, J. Mag. Magn. Matls. 15-18, (1980) 1591.

[18] C.S. Wang and A.J. Freeman, Phys. Rev. B 19, 793 (1979); J. Mag. Magn. Matls. 15-18 (1980) 869.

[19] S. Eichner, C. Rau and R. Sizmann, J. Mag. Magn. Matls. 6, (1977) 204; C. Rau, Bull. Am. Phys. Soc. 25, (1980) 234.

[20] C.S. Wang and A.J. Freeman, J. Mag. Magn. Matls. 15-18 (1980) 869 and to be published.

[21] J. Tyson, A.H. Owers and J.C. Walker, J. Appl. Phys. 52, (1981) 2487.

[22] B.T. Thaler, J.B. Ketterson and J.E. Hilliard, Phys. Rev. Lett. 41 (1978) 336.

[23] T. Jarlborg and A.J. Freeman, Phys. Rev. Lett. 45, (1980) 653.

[24] E.M. Gyorgy et al. Phys. Rev. Lett. 45, (1980) 57.

[25] G.P. Felcher, et al. J. Mag. Magn. Matls. 21, (1980) L198.

[26] T. Jarlborg and A.J. Freeman (to be published).

[27] M.B. Brodsky and A.J. Freeman, Phys. Rev. Lett. 45, (1980) 133.

[28] K. Levin and O.T. Valls, Phys. Rev. B 17, (1978) 191.

[29] M.B. Brodsky, J. Appl. Phys. 52 (1981) 1665 and M.B. Brodsky and A.J. Freeman (to be published).

[30] P.C. Riedi, Phys. Rev. B 20 (1979) 2203; J.F. Janak, Phys. Rev. B 20 (1979) 2206.

[31] T. Jarlborg and A.J. Freeman, J. Mag. Magn. Matls. 22 (1980) 6.

[32] L. Hedin, B.I. Lundqvist and S. Lundqvist, Solid State Commun. 9 (1971) 537.

[33] O. Gunnarson, B.I. Lundqvist and J.W. Wilkins, Phys. Rev. B 10 (1974) 1319; O. Gunnarson and B.I. Lundqvist, Phys. Rev. B 13 (1976) 4274.

[34] C.S. Wang and J. Callaway, Phys. Rev. B 15 (1977) 298.

[35] E.P. Wohlfarth and P. Rhodes, Philos. Mag. 7 (1926) 298.

[36] T. Jarlborg and A.J. Freeman, Phys. Rev. B 23 (1981) 3577.

[37] E.P. Wohlfarth, J. Mag. Magn. Matls. 20, (1980) 77.

PHYSICS OF SOLIDS UNDER HIGH PRESSURE
J.S. Schilling, R.N. Shelton (editors)
© *North-Holland Publishing Company, 1981*

ORGANIC SUPERCONDUCTORS

Meir Weger

Hebrew University, Jerusalem,
Israel

In a one-dimensional metal, electron-phonon coupling gives rise to superconductivity when the electron-phonon coupling constant is not too large. A large coupling constant gives rise to an insulating Peierls state. In organic metals, pressure has the effect of reducing the electron-phonon coupling constant $\lambda = I^2 n(E_F)/K$, where $K = M\omega^2$ is the lattice force constant, because it makes the lattice rigid and increases K by about 20% for each 1% change in lattice constant. Therefore it brings about the transition from the peierls state to the superconducting state. This behavior was observed recently in $(TMTSF)_2PF_6$ by Bechgaard and Jerome. The temperature of the crossover is related to the phonon frequency and varies from $\omega/50$ to $\omega/20$ for strong forward scattering. Tunneling between the chains must overcome the Coulomb coupling to observe superconductivity. The umklapp interaction in this commensurate system gives rise to a SDW state. This umklapp is due to a novel two-phonon process, and is therefore easily suppressed by pressure, and also by a disordered anion lattice.

1.   INTRODUCTION

The subject of one-dimensional superconductivity was started by Frohlich (1) even before the advent of the BCS theory. Frohlich pointed out that a Peierls state, with a gap, is invariant under translation (for an incommensurate system) and there is no mechanism for energy loss of individual electrons, and thus should give rise to a resistance-less state.

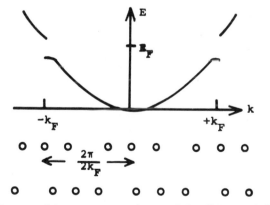

By now this state was observed in $NbSe_3$ and is called "sliding CDW" state (2), but it is not superconducting. Following Frohlich's work it was observed (3) that in materials of the A15 class, like Nb_3Sn, the Nb atoms are arranged in chains and the high superconductivity transition temperature was attributed to this one-dimensional structure.

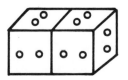

Both softening of the $2k_F$ phonons (due to an incipient Peierls transition(3)) and the one-dimensional $1/\sqrt{E}$ divergence at band edges (4) were invoked.(5).

At about the same time, Little(6) suggested room temperature superconductivity of organic polymers, due to an exciton mechanism.(7).

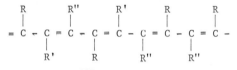

2.   GORKOV THEORY.

Following these suggestions, Bychkov, Gorkov and Dzyaloshinsky (8) investigated the Peierls and BCS transitions Tp, T_c of one dimensional metals in some depth. These transitions can be described as divergences of Bubble

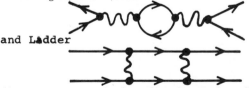

and Ladder

diagrams, respectively. They found that these divergences occur simultaneously, i.e. $T_p = T_c$. The calculation was performed for an *unretarded* interaction. Thus we cannot ignore the Peierls transition when we investigate superconductivity (and vice versa) even in the first approximation. Moreover, in addition to bubble and ladder diagrams, we also have to consider "Parquet" diagrams like:

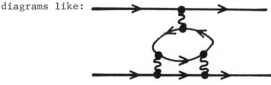

which are combinations of bubble and ladder diagrams, since these diverge just as strongly as the transition $T=T_P=T_c$ is approached. Gorkov et al succeeded to solve this extremely difficult problem for an unretarded interaction.

3. RETARDED INTERACTION

Gutfreund, Horovitz and Weger tried to apply the BGD theory to the A15's; this required taking into account the retardation due to the phonons, whose frequency is lower than the electronic bandwidth. The BGD formalism is extremely complicated, and up to now we are not aware of any work that succeeded to incorporate retardation into the Parquet-diagram formalism. However, in the A15's three-dimensional coupling due to tunneling of electrons between chains, is strong. When $t_{\perp} > k_B T_c$ ($t_{\perp}$ is the transverse transfer integral), the Parquet diagrams diverge less strongly than the ladder and bubble diagrams, and we can ignore them (9). This simplifies the problem considerably, since the Peierls and BCS channels become decoupled. Because of this simplification, the Eliashberg equations can be used, and thus the retarded nature of the interaction can be properly taken into account.

The solution of the Eliashberg equations is still complicated, since both Peierls and BCS channels diverge simultaneously (for a nested Fermi surface), but the solution was obtained by Horovitz (10).

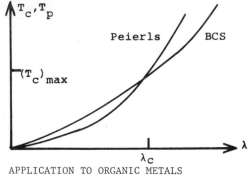

4. APPLICATION TO ORGANIC METALS

At this time, TTF-TCNQ was discovered to possess a metallic electrical conductivity, and a Peierls transition at 52 K (11). Thus, the theory was immediately applied to this new, exciting material. According to the GHW theory, for small values of the electron-phonon coupling constant λ, the BCS channel wins; $T_P < T_c$. For large values of λ, the Peierls channel wins; $T_P > T_c$ (12). At the crossover T_c has its maximum value, $(T_c)_{max}$. This temperature is related to the phonon frequency; $(T_c)_{max} \simeq \omega/50$ for a "normal" situation, while $(T_c)_{max} \simeq \omega/20$ for very strong forward scattering of the phonons, as exists in organic metals because of the large size of the molecules (13).

In TTF-TCNQ, λ is somewhat larger than the crossover value λ_c. So, in order to obtain superconductivity, λ must be reduced, and this can easily be

achieved by applying pressure since

$$\lambda = <I^2> n(E_F)/(M\omega^2)$$

where ω is the phonon freuency. Applying pressure increases ω (14) and since the material is soft, the effect is very large. A preesure of 1 GPa reduces λ by a factor of 2, approximately. The pressure dependence of ω was found to account for the large pressure dependence of the resistivity (15) extremely well (16).

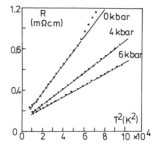

Resistance plotted vs. T^2 at various pressures

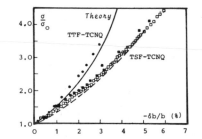

Thus, the GHW theory made the following predictions:
(1)   A pressure of order 1 GPa may cause crossover from the Peierls state to the BCS state.
(2)   At the crossover, $(T_c)_{max} \simeq \omega/20$.
(3)   Passing the crossover, T_c falls rapidly, due to the continued decrease of λ.

Based on these predictions, the Orsay project was started (17).

5. "g"-OLOGY

Gorkov theory was extended for *unretarded* interaction using the very powerful theory of the renormalization group (18), and certain exact solutions of model hamiltonians (19). The results can be expressed in form of a phase diagram, in terms of two parameters – the backward scattering amplitude g_1 and the forward scattering amplitude g_2 ("g-ology").

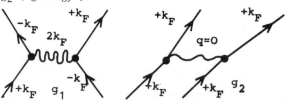

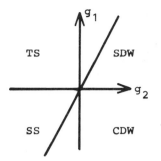

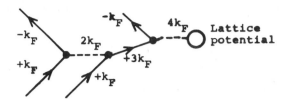

There are four main phases: Singlet (BCS) superconductivity, SS; Triplet superconductivity (TS);
Peierls phase (CDW); and an antiferromagnetic
(SDW) phase; each occupying the appropriate part
of the phase diagram. Surprisingly, the rather
different theoretical approaches yield essentially
the same phase diagram. This phase diagram (with
a few more phases at large coupling constants)
was obtained in mean field theory by Horovitz(20).

The energies of all four phases are very close
together, so very minor perturbations can cause
changeover from one phase to another. Such
perturbations are interchain Coulomb coupling
(21); umklapp scattering (22), etc.
*The near degeneracy of SS, TS, CDW, and SDW
states is the salient property of one dimensional
metals.*

6.   SUPERCONDUCTIVITY OF $(TMTSF)_2PF_6$

The work of Bechgaard and Jerome on $(TMTSF)_2PF_6$
observing superconductivity at 1 GPa and 1.1 K
(22) confirmed all the predictions of the GHW

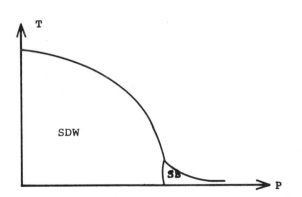

theory; the crossover occurs near 1 GPa; since
$\omega \simeq 35$ K, T_c is about $\omega/20$; and T_c drops
rapidly with pressure (23). However, the insulat-
ing state is a SDW state rather than a CDW one
(24). This result was not predicted by the GHW
theory, although the near degeneracy of the four
states did not cause this result to be surprising.

The cause for this is as follows. $(TMTSF)_2PF_6$ is
commensurate; the band is 1/4 filled. As a result,
umklapp processes are possible. In a "normal"
umklapp process, the lattice potential has a
component with momentum $4k_F$, and this component

gives rise to a transition in which *both*
electrons change their momentum from $+k_F$ to $-k_F$.
These processes were introduced into the theory
of 1-D metals by Klemm and Gutfreund (22) and
are represented by a coupling constant g_3. They
favor the CDW and SDW phases over the superconducting phases. Thus a state in the g_1-g_2 plane
at a point represented by the star, will be

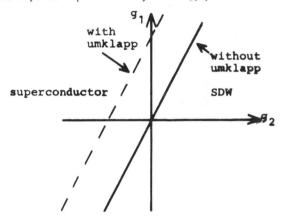

superconducting without umklapp, and a SDW state
with umklapp.

7.   PRESSURE DEPENDENT UMKLAPP

The symmetry of the $(TMTSF)_2PF_6$ chains is such
(25) that there is an approximate glide plane
symmetry, and therefore the component of the

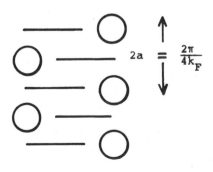

lattice potential *along the chain* with $4k_F$
momentum is very small. Therefore the g_3 term is
very small. This is because the electronic motion
is one-dimensional along the chain, so they do
not see the $4k_F$ component of the lattice potential.

However, the *phonons* are three dimensional, and do not necessarily move along the chain. Therefore they see the $4k_F$ potential due to the anions and are reflected by it, giving rise to a novel type of phonon-mediated umklapp process (26):

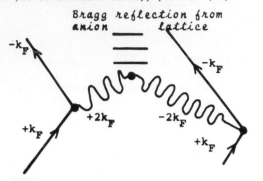

Since this process requires <u>two</u> phonons, $+2k_F$ and $-2k_F$, it depends more strongly on the phonon frequency than the first-order process, $\lambda_3 \propto 1/\omega^4$ while $\lambda_1, \lambda_2 \propto 1/\omega^2$. Therefore this umklapp is strongly suppressed by pressure, in agreement with experiment. Also, in a disordered lattice, like the ClO_4 lattice in $(TMTSF)_2ClO_4$, there is no umklapp and the superconducting state is observed at zero pressure (27). The phase diagram, including the retarded nature of the interaction, looks like (26):

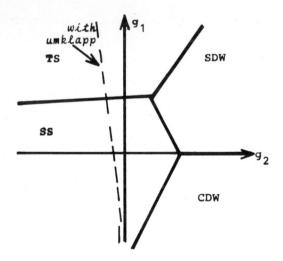

The full line shows the phase diagram without umklapp, and the dashed line shows the effect of umklapp. The point marked by a star is a SDW state with umklapp, and superconducting without umklapp. The Coulomb repulsion may exceed somewhat the phonon-mediated attraction, as in many "ordinary" metals, but superconductivity wins because of the retarded nature of the attractive interaction, as in many "normal" metals. (μ^* is considerably smaller than μ, and while $\mu > \lambda$, we have $\mu^* < \lambda$).

8.   CONCLUSION

We see the cardinal role played by pressure in this research, as a means to vary the electron-phonon coupling constant λ and bring it to the right range for superconductivity.

REFERENCES

1. H. Frohlich, Proc. Roy. Soc. A223(1954)296.
2. P. Monceau, Invited paper, LT 16 Conf.
3. M. Weger, Rev. Mod. Phys. 36(1964)175
4. J. Labbe & J. Friedel, J. Physique 27(1965)153
5. M. Weger & I.B. Goldberg, Solid State Physics 28(1973)1. M. Weger, J. Less Com. Met.62(1978) 39.
6. W.A. Little, Phys. Rev. 134(1964)A1416.
7. H. Gutfreund & W.A. Little in "Highly Conduct. One Dimensional Solids"J.T. Devreese R.P.Evrard V.F. Van Doren Edts. Plenum New York 1979.
8. A. Bychkov, L.P. Gorkov, I.E. Dzyaloshinsky Sov. Phys. JETP 23(1966)489.
9. B. Horovitz H. Gutfreund M. Weger Phys. Rev B12(1975)3174.
10. B. Horovitz, Phys. Rev. B16(1977)3943.
11. L.B. Coleman M.J. Cohen D.J. Sandeman F.G. Yamagishi A.F. Garito A.J. Heeger Solid State Commun. 12(1973)1125.
12. B. Horovitz H. Gutfreund M. Weger J. Phys. C. 7(1974)383.
13. M. Kaveh, M. Weger, H. Gutfreund Solid State Commun. 31(1979)83.
14. M. Nicol M. Vernon J.T. Woo J. Chem Phys. 63(1975)1992.
15. M. Weger, M. Kaveh, H. Gutfreund J. Chem. Phys. 71(1979)3916; Solid State Comm. 32 (1979)323.
16. H. Gutfreund, M. Weger, M. Kaveh Solid State Commun. 27(1978)53.
17. D. Jerome & M. Weger, "Chemistry & Physics of One Dimensional Metals" H.J. Keller Ed. NATO ASI Series B25 Plenum Press New York 1977.
18. N. Menyhard & J. Solyom, J. Low Temp. Phys. 12(1973)529.
19. A. Luther & V.J. Emery Phys. Rev. Lett. 33 (1974)589.
20. B. Horovitz Solid State Comm. 18(1976)445.
21. R.A. Klemm & H. Gutfreund Phys. Rev. B14 (1976)1086.
22. H. Gutfreund & R.A. Klemm, Phys. Rev. B14 (1976) 1073.
23. D. Jerome, A. Mazaud, M. Ribault, K. Bechgaard J. Physique Lett. 41(1980)L95.
24. H.J. Pedersen, J.C. Scott, K. Bechgaard, Solid State Commun. 35(1980)207.
25. K. Bechgaard, C.S. Jacobsen, K. Mortensen, H.J. Pedersen, N. Thorup Solid State Comm. 33(1980)1119.
26. B. Horovitz, H. Gutfreund, M. Weger, Solid State Commun. 39(1981)541.
27. K. Bechgaard, K. Carneiro, M. Olsen, F.B. Rasmussen, C.B. Jacobsen, Phys. Rev. Lett.

PHYSICS OF SOLIDS UNDER HIGH PRESSURE
J.S. Schilling, R.N. Shelton (editors)
© North-Holland Publishing Company, 1981

HIGH PRESSURE EFFECTS ON METAL-SEMICONDUCTOR PEIERLS TRANSITION : MEM-$(TCNQ)_2$

D. Bloch[+], J. Voiron[+], J. Kommandeur[++] and C. Vettier[+++]

[+]Laboratoire Louis Néel, C.N.R.S., 166X, 38042 Grenoble-cédex, France
[++]Laboratory for Physical Chemistry, University of Groningen, the Netherlands
[+++]Institut Laue-Langevin, 156X, 38042 Grenoble-cédex, France

The donnor-acceptor compound MEM^+-$(TCNQ)_2^-$ has an electronic-Peierls transition temperature T_{EP}, at 335 K which separates the metallic and semiconducting temperature ranges. High pressure neutron scattering gives the pressure dependence of the crystallographic axis and electrical conductivity experiments indicate a rapid decrease of T_{EP} with pressure. At P = 3.5 kbar, T_{EP} is found to be around 155 K.

The MEM (N-methyl-N ethyl-morpholinium : $C_7H_{16}NO$) molecule combines with TCNQ (tetra-cyanoquinodimethane $C_{12}N_4H_4$) molecule through a charge transfer mechanism to form $(MEM)^+ (TCNQ)_2^-$. The structure consists of planar TCNQ molecules forming columns parallel to the <001> axis, with sheets parallel to the (010) plane of the triclinic structure (1). Above 335 K, the TCNQ molecules are stacked almost uniformly along the <001> axis (2). At T = 335 K a dimerization occurs along the <001> axis (1), followed, at 19 K, by a quadrimerization (3). The three different phases have different transport and magnetic properties. There is an electronic-Peierls transition, with a metallic regime for T > T_{EP} and a semiconducting one for T < T_{EP}, at T_{EP} = 335 K. The magnetic susceptibility, in the metallic regime is of the exchange-enhanced Pauli type ; in the semiconducting range it has the characteristic features of one-dimensional antiferromagnetic chains. There is a spin-Peierls transition between the magnetic high temperature and the non magnetic low temperature regime (4,5) at T_{SP} = 19 K. This behaviour can be understood by considering the number of electrons per unit cell : 1/2 in the metallic, uniform Pauli regime, 1, with $\sim 1 \mu_B$ magnetic moment, in the magnetic semiconductor electron-Peierls dimerized range, and 2 with a singlet ground state in the semiconducting non magnetic quadrimerized spin-Peierls state.

EXPERIMENTAL TECHNIQUES

Neutron scattering experiments (figure 1) have been performed using a clamped type alumina pressure cell (6). A fully deuterated single crystal is mounted in the alumina cell together with a small NaCl single crystal for measurement purposes. Inactive fluorinert (C_6F_6) was used as a pressure transmitting medium. Experiments were performed on the two-axis spectrometer D1A at the Laue-Langevin Institute. High resolution was achieved by using of a long wave length, λ = 2.988 Å from a germanium [113] monochromator. Independent measurements were obtained for different scattering planes.

Electrical conductivity was measured on needle shape single crystals, using silver paint to insure electrical contact. The sample was located in a maraging steel high pressure cell capable to maintain a 10 kbar helium gas pressure. The temperature of the high pressure cell could be varied using a temperature regulated gaseous helium flow and a platinum resistor was mounted inside the high pressure cell, for temperature measurements. Insulated electrical inlets allow for temperature and resistance measurements.

EXPERIMENTAL RESULTS

First we have studied the pressure dependence of the crystallographic parameters of deuterated MEM-$(TCNQ)_2$ at 300 K. The room temperature values were found to be nearly identical to those previously determined for hydrogenated MEM-$(TCNQ)_2$ (2). In order to obtain the six parameters a, b, c, α, β, γ of the triclinic cell from two-axis measurements two orientations of the sample were used. Measurements were performed in the (0, k, 1) and the (h, k, 0) scattering planes. The six parameters of the triclinic cell, as well as the volume of the cell, as obtained after refinement procedures, are given figure 2 (a to g). The parameters b, c,

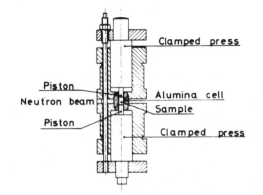

Figure 1 : Clamped alumina cell for high pressure neutron scattering (6).

D. Bloch et al.

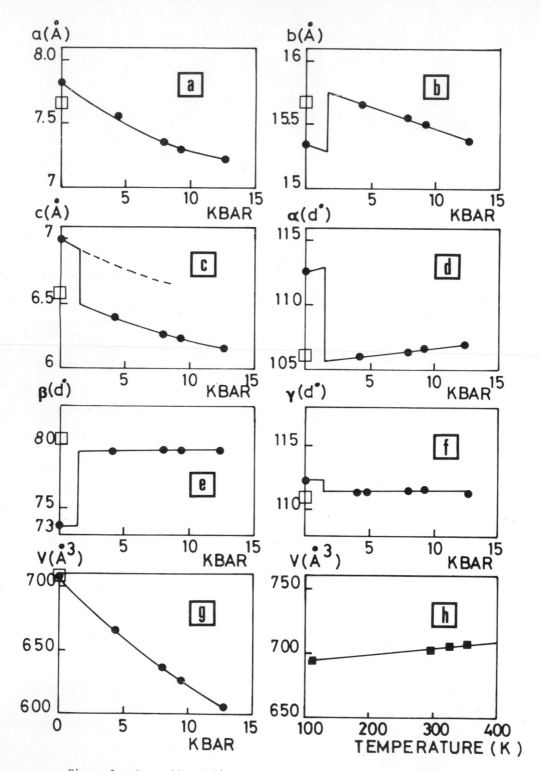

Figure 2 : Crystallographic parameters of triclinic MEM-(TCNQ)$_2$.
a to g : pressure dependence at 300 K (deuterated sample),
h : thermal dependence at atmospheric pressure. The atmospheric pressure data (2) (□),
figure a to g, are taken in the high temperature uniform phase at 348 K.

α, β, γ present clearly a discontinuity below
4 kbar. The value for the critical pressure,
1.5 kbar, is determined from electrical resistivity experiments (described in the next
section).

No volume discontinuity can be detected in the
thermal dependence of V as determined at
atmospheric pressure (figure 2-h) nor in the
pressure dependence of V as determined at room
temperature (figure 2-g). In order to characterize the phase above ∿ 1.5 kbar we have
represented, on the same diagram, the crystallographic parameters of MEM-(TCNQ)$_2$ at 348 K,
that is above the electronic-Peierls transition
temperature, in the uniform state. The room
temperature high pressure phase clearly appears
as identical to the high temperature uniform
phase.

To get a more extensive knowledge of the phase
diagram we have measured, at various constant
pressure or temperature, the temperature or
pressure dependence of the electrical conductivity of MEM-(TCNQ)$_2$. The metal insulator transition was observed around 337 K at atmospheric
pressure (figure 3). It rapidly decreases with
pressure (figure 4). The metal insulator transition is 155 K for a pressure of 3450 bars.
At low temperature the metal-insulator transition is always accompanied with a breaking of
the sample. Thus a new sample is needed for each
set of experiment. The c-axis dependence of the
electronic Peierls transition is given in figure
4-b where both have been taken into account the
thermal and pressure dependences of the stacking
c-axis.

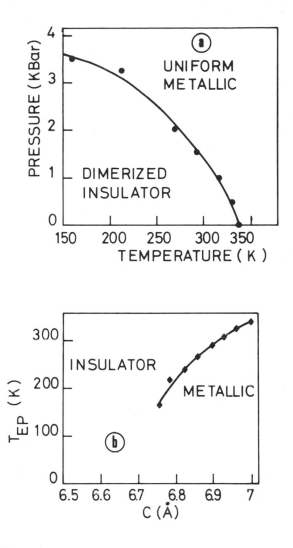

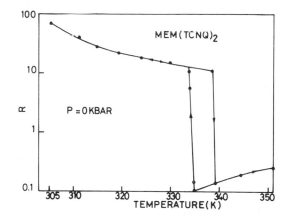

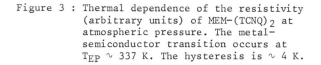

Figure 3 : Thermal dependence of the resistivity
(arbitrary units) of MEM-(TCNQ)$_2$ at
atmospheric pressure. The metal-
semiconductor transition occurs at
T_{EP} ∿ 337 K. The hysteresis is ∿ 4 K.

Figure 4 : a - Temperature-pressure phase diagram
for MEM-(TCNQ)$_2$.
b - Temperature-crystallographic c-axis
phase diagram for MEM-(TCNQ)$_2$.

The compressibility has been taken as temperature
independent, with a value equal to that determined at 300 K. Note the c-dependence of T_{EP} is
not linear. Its slope, at atmospheric pressure is

$$\frac{d \log T_{EP}}{d \log c} \sim 10.$$

CONCLUSION

MEM-(TCNQ)$_2$ has a very large compressibility. A
pressure as low as 7 kbar is sufficient to reduce
its volume by 10 %, and thus is to be compared,

for instance, with the 250 kbar needed to obtain a similar volume reduction in transition metals and oxides.

Some details of the phase diagram of MEM-(TCNQ)$_2$ are given in this preliminary report but other experiments are required to extend this phase diagram to lower temperatures and higher pressures. The low temperature region of the metallic high pressure phase has to be characterized and the pressure dependence of the spin-Peierls transition to be determined. These experiments are now in progress.

REFERENCES

(1) A. Bosch and B. Van Bodegom,
     Acta Cryst., B33, 3013 (1977).
(2) B. Van Bodegom,
     Thesis, Groningen (1979).
(3) B. Van Bodegom, B.C. Larson and H.A. Mook,
     to be published.
(4) S. Haizinga, J. Kommandeur, G.A. Sawatzky,
     B.T. Thole, K. Kopinga, W.J.M. de Jonge
     and J. Roos,
     Phys. Rev., B19, 4723 (1979).
(5) D. Bloch, J. Voiron, J.W. Bray, I.S. Jacobs,
     J.C. Bonner and J. Kommandeur,
     Phys. Letters, 82A, 21 (1981).
(6) D. Bloch, J. Paureau, J. Voiron and
     G. Parisot,
     Rev. Scient. Instr., 47, 296 (1976).

PHYSICS OF SOLIDS UNDER HIGH PRESSURE
J.S. Schilling, R.N. Shelton (editors)
© *North-Holland Publishing Company, 1981*

RECENT HIGH PRESSURE FERMI SURFACE STUDIES ON $(TMTSF)_2PF_6$ AND ReO_3[*]

J. E. Schirber

Sandia National Laboratories[*]
Albuquerque, New Mexico 87185 USA

Hydrostatic pressure coupled with temperatures near absolute zero has proven to be an extremely important environment in the study of the Fermi surface and electronic structure of metals. This environment allows critical testing of theoretical models for the electronic structure and allows access to regions of the phase diagram of the material where interesting and desirable properties may be isolated and studied. Two recent examples of the latter type of study are reviewed involving the "compressibility collapse transition" in ReO_3 and quantum oscillatory behavior at high-magnetic fields in $(TMTSF)_2PF_6$.

Introduction

The use of high pressure in the study of the Fermi surfaces of metals was really in its infancy 16 years ago when the First International Conference on Physics at High Pressure was held in Tucson. The only technique available for direct measurement of the Fermi surface (FS) was high-field magnetoresistance and the attendant Shubnikov-deHaas oscillations so studies were limited primarily to semi-metals and very small sheets of the Fermi surface. The intervening time period saw most elemental metals and a large number of metallic compounds prepared in high-quality single-crystal form. Direct FS data have been obtained on many of these and pressure measurements of the cross sectional areas performed on a substantial fraction.[1-3]

This surge of activity stemmed from concurrent development of sensitive measurement techniques with the availability of long mean-free-path crystals. Of particular utility to pressure studies of the FS was the perfection of the field modulation de Haas-van Alphen (dHvA) method[4] which allows measurement of extremal cross sectional areas of samples under pressure without having to attach electrical leads. In fact in many cases it is unnecessary to bring any leads whatsoever into the high-pressure environment which obviously simplifies the experiments substantially.

The need for long-electron mean-free paths in order to see the quantum effects such as the dHvA oscillations puts severe restrictions upon the pressure technique. All pressure media are solid in the low-temperature regime necessary for direct examination of the FS except He below 25 bar. In certain situations it is actually possible to measure FS cross section derivatives directly with precision with only 25 bar. This happy situation has made it possible to critically test the isobaric nature of the more conventionally employed solid He medium. The solid He pressure technique is now quite widely employed and, if used with sufficient care, will result in purely, from an experimental point of view, hydrostatic conditions on even the most fragile and anisotropic materials. The pressure technique has been discussed in detail.[5]

The combination of high-quality samples, easily obtainable high fields (~ 100 kG) via superconducting solenoids, the field modulation dHvA scheme and the solid He pressure technique have resulted in a large body of experimental information on the effect of interatomic spacing on the FS and band structure of metals. These data fall into two broad categories as far as their use is concerned. The first and most obvious use for such data is to provide critical tests for the various model descriptions for a given material. The philosophy is to use all the normal volume information available to arrive at the best possible description of the electronic structure of the material. The band calculation is then redone at a slightly different interatomic spacing and if the description is physically significant it must predict the results of the direct measurements of the FS under pressure. This type of approach has been very useful in refining potentials and in showing importance of self-consistent and relativistic treatments and similar effects in the calculations.

The second category of pressure studies involves the investigation of phase changes and the use of pressure to reversibly reach interesting portions of the phase diagram of materials. Examples are metal/insulator or normal/superconducting phase boundaries. The pressure environment is particularly suitable for such studies because purely hydrostatic conditions usually allow fully reversible and non-destructive traversals of the boundary. This is often difficult or impossible with non-hydrostatic pressures or other variables such as changes in electron/atom ratios via alloying.

In the limited space it will not be possible to discuss studies of the first category of critical test of the band models, but reviews do exist in the literature.[1-3] The second category of pressure-induced electron transitions, the simplest case of which involves changes of the Fermi surface topology with no attendant volume change,[6] also will not be reviewed as again this information is available.[7] However two recent studies of this general type will be discussed. The first deals with the understanding of a novel phase transition in ReO$_3$ which we[8] have called the "compressibility collapse transition." This designation has more aliterative appeal than accuracy in that the extremely unusual feature is that the compressibility _increases_ by nearly an order of magnitude discontinuously at the transition. We have studied the electronic structure implications of this and the compressibility divergence itself is the subject of the following paper by Dr. Batlogg.

The second example that will be covered involves obtaining direct information about the FS via the observation[9] of the oscillatory phenomena in the high-field magnetoresistance of the new organic superconductor[10]bis-tetra-methyltetraselenafulvalenium hexafluorophosphate,(TMTSF)$_2$PF$_6$, (hereafter designated TSP).

ReO$_3$

The bright red perovskite-structured compound ReO$_3$ has been the subject of much study for about a dozen years. It was one of the first compounds to show very large quantum effects. This was particularly important because it is not a semi-metal and indeed has a room temperature conductivity only a factor of five less than Cu so it demonstrated that compounds could be fabricated with mean-free paths comparable to elemental metals. Various anomalies in the lattice properties[8] of ReO$_3$ and in FS studies under pressure by Razavi,Altounian and Datars[11] indicated some kind of phase transition was occurring although none could be detected as a function of temperature at normal volume. We have shown rather convincing evidence[12] that ReO$_3$ undergoes a second-order phase transition from the simple cubic perovskite to a body-centered cubic Bravais lattice (Im3) with a doubled lattice constant. Figure 1 shows the central (110) and (100) cross sections for this simple cubic ReO$_3$ Fermi surface which consists of the three sheets shown in Fig. 2, all of which are Γ centered electron pockets. The shift with pressure in phase of dHvA oscillations due to the smallest sheet of the simple cubic FS, the α(100), is shown in Fig. 3. The slope of this shift is simply related to the pressure derivative of the cross sectional area of the particular FS

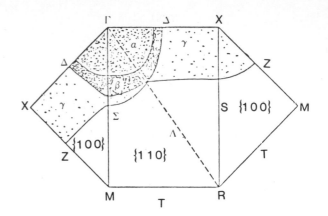

Figure 1: Central {110} and {100} cross sections of the normal volume simple-cubic Fermi surface of ReO$_3$.

sheet involved. The data, which consititute four excursions through the transition, indicate a discontinuous increase in the compressibility of about a factor of seven. We have made direct measurements of the compressibility at room temperature in a diamond cell in the 20-100 kbar regime and under hydrostatic conditions to 4 kbar which are consistent with this interpretation. Extensive NMR measurements have been carried out[13] using the [187]Re resonance. Quadrupolar effects as the transition is crossed indicate that there is a single Re site but that its cubic symmetry is lowered. The first-order quadrupole effects were used to map out the transition line from 4 K to 300 K as shown in Fig. 4.

These results are consistent with the above-mentioned change from simple cubic to the Im3 structure. The latter structure can be thought of as a kinking of the mutually perpendicular Re-O-Re chains at the O sites in such a way as to remain body-centered cubic with a doubled lattice parameter. This has the very attractive feature that one can envision an increased compressibility because of the ability to "hinge" at the O rather than compress the O-Re bond. There is a complication introduced by this picture in that the FS was (as in Fig. 3) observed through the transition with no change except in the rate of its variation with pressure. This is in spite of a proposed change from simple cubic to body-centered cubic and a doubled lattice parameter which results in a much smaller Brillouin zone.

The FS for the proposed high-pressure structure is shown in Fig. 5. Several of the sheets such as the α and β electron spheroids and cross sections of the γ jungle gym (see Fig. 2) are identical because they fall within the boundaries of the smaller Brillouin zone. There are however new sheets[14] centered at H in both the second and third zones and a much different

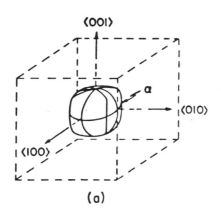

(a)

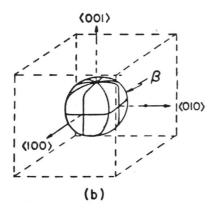

(b)

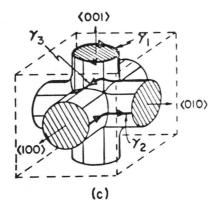

(c)

Figure 2: Three-dimensional sketch of the normal volume simple-cubic Fermi surface of ReO$_3$.

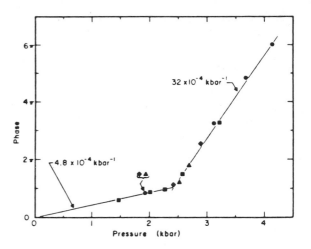

Figure 3: Shift in phase vs pressure for the α frequency of the Fermi surface of ReO$_3$ for B parallel to [110] and of magnitude equal to 100 kG.

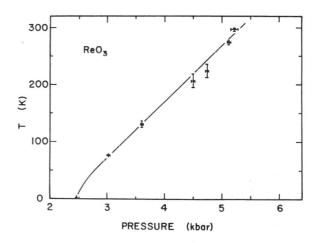

Figure 4: Temperature-pressure dependence of the compressibility collapse transition in ReO$_3$.

first zone hole sheet (h$_1$) than found in the simple cubic case. A three-dimensional sketch of h$_1$ is shown in Fig. 6. Recently we have succeeded in observing the $\bar{\gamma}_2$ and γ_4 orbits[14,15] shown in Fig. 6 and in this observation is the key for understanding why the FS is seemingly unchanged across the transition. In Fig. 7 we show the amplitude of the $\bar{\gamma}_2$ oscillations vs reciprocal field. These plots are typically monotonic with a slope proportional to the Dingle temperature (or the scattering) of electrons on the particular orbit in question. Here we observe a maximum in the amplitude which is characteristic of magnetic breakdown. Carriers are tunnelling the energy

gaps at the Brillouin zone faces of the Im3
structure at very low fields and thus finding
themselves back on the simple cubic Fermi surface. Therefore except for these two very low
frequencies which can be observed at low fields,
one is always at higher magnetic fields than
breakdown so the original FS is observed.

Thus the picture for the "compressibility collapse trnsition" in ReO_3 seems fairly well established except for direct crystallographic
verification of the high-pressure structure.
Time-of-flight powder neutron studies are under
way at the WNR facility at Los Alamos National
Laboratory. With this information it will be
of interest to study the nature and critical
behavior of this unusual transition in more
detail.

$(TMTSF)_2PF_6$

Superconductivity was demonstrated[10] in the organic metal $(TMTSF)_2PF_6$ only a year and a half
ago prompting an enormous flurry of activity
and speculation as to the nature of the effects
in these very anisotropic charge-transfer materials. We find superconductivity above ~ 6.5
kbar with a very steep boundary between metal
and non-metal. Jerome and coworkers have a
similar shaped T-P diagram but with the metal/
non-metal boundary about 2 kbar higher. It is
not known at this time how much of this dispar-
ity stems from sample or pressure generation
technique differences. We have taken transverse high-field magnetoresistance data[9] on
single crystals of TSP with current along the
a (or stacking) axis in fields up to 100 kG.
Pressures to ~ 9 kbar were generated by careful
isobaric freezing of He.

Magnetoresistance vs field plots for the applied
field along the b an c axes are shown in Fig. 8.
The oscillating behavior discernable in the c
axis data is accentuated by field modulation
third harmonic detection as shown in Fig. 9.
These oscillations are periodic in reciprocal
field and have a frequency of 7.6×10^5 G. The
temperature dependence of the amplitude gives
an effective mass ratio value of roughly 1.1
and the field dependence of the data gives a
Dingle temperature of about 3 K.

This magnetoresistance anisotropy, coupled with
the closed orbits for H‖c indicates that we
are dealing with a compensated material (equal
electron and hole populations) consisting of cylindrical sheets of FS with their axes oriented
along the c axis of the crystal.

A striking feature of the data is the sudden
turn-on of the oscillatory behavior near 70 kG.
This onset is closely related to a sharp anomaly[16] in the methyl group proton resonance spin-
lattice relaxation time. These T_1 measurements
were performed with a high-frequency,low-power,
phase-coherent NMR spectrometer useful to 500
MHz. This anomaly shown in Fig. 10 is a

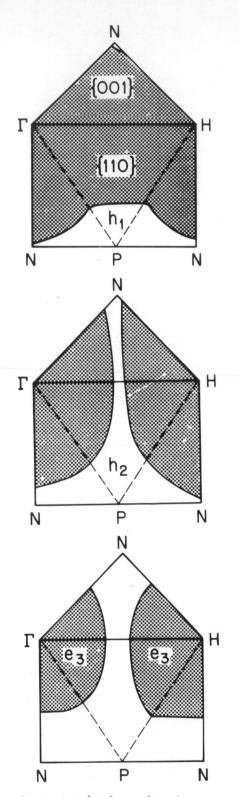

Figure 5: Central {110} and {100 } cross
sections of the proposed high-
pressure phase ReO_3 Fermi surface.

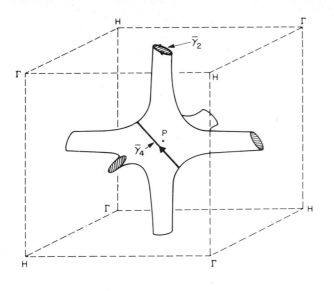

Figure 6: Three-dimensional sketch of the h$_1$
sheet of the high-pressure-phase
ReO$_3$ Fermi surface.

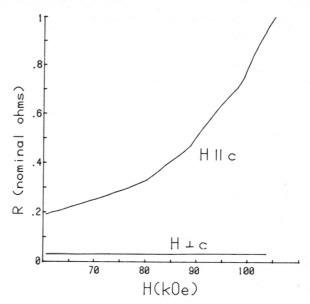

Figure 8: Nominal resistive a-axis signal
from a single crystal vs magnetic
field at T = 1.1 K and P = 7.4 kbar
for fields oriented roughly parallel
and perpendicular to the crystal c
axis.

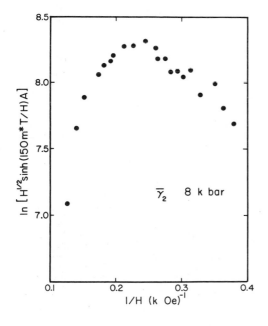

Figure 7: Dingle plot for the γ$_2$ frequency of
ReO$_3$ at ~ 8 kbar for H∥[100] at 1.1 K.

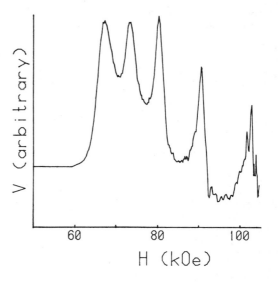

Figure 9: Signal at 150 Hz with constant-ampli-
tude 50-Hz field modulation vs magnetic field (raw data).   T = 1.1 K,
P = 6.9 kbar.

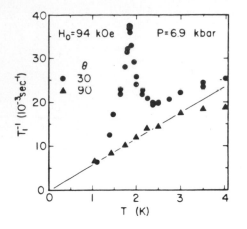

Figure 10: Methyl-group proton relaxation rate
          vs temperature for two field orienta-
          tions in $(TMTSF)_2PF_6$.

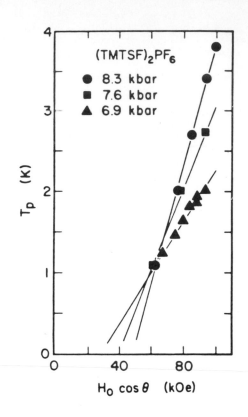

Figure 11: Field dependence of the transition
          temperature at several pressures.

strong function of the projection of the field
along the c axis and pressure as shown in Fig.
11. In both Figs. 10 and 11, θ refers to the
angle in the bc plane away from the c axis.
(The structure of this compound is triclinic
so referal to crystal axes is only approximate.)
We have thus found a transition in the metallic
state at high fields above which the material
is compensated and the FS is two dimensional
with closed orbits for H∥c.

This finding is in substantial disagreement with
the viewpoint that there is one electron per
unit cell and that the material is one dimensional. We do not at this point understand the
nature of this transition so it is clear that
much further study is warranted to link up these
findings at high fields with the low-field
superconducting behavior.

Conclusions

These two examples of pressure studies of the
Fermi surface illustrate the importance of low-
temperature hydrostatic pressure as a reversible parameter for accessing interesting portions of the phase diagram. In both examples
it would be difficult to envision a comparably
useful means to get into the phases of interest
without damaging the material and/or increasing
the mean-free path of the carriers and thus
obviating direct observation of the intriguing
quantum mechanical effects. In both examples
as yet not fully understood phases are being
probed so further study is warranted, but it
is clear that hydrostatic pressures, probably
only achievable in solid He, will be required.

Acknowledgements

This paper constitutes a brief review of work
involving important contributions of a large
number of people. The ReO_3 studies involved
L. F. Mattheiss, B. Morosin, L. J. Azevedo,
A. Narath, D. L. Overmyer and R. L. White. The
organic superconductor studies were carried
out with J. F. Kwak, L. J. Azevedo, R. L.
Greene and E. M. Engler.

*This work was supported by the U. S. Depart-
ment of Energy under Contract DE-AC04-76-
DP00789. †A U. S. Department of Energy facility.

References

[1] Brandt, N. B., Itskevich, E. S. and Minina,
    N. Ya., Usp. Fiz. Nauk. 104 (1971) 459
    [Soviet Phys.--Uspekki 14 (1972), 438].
[2] Schirber, J. E., Materials Under Pressure,
    Tokyo (1974) p. 141.
[3] Schirber, J. E., in High Pressure and Low
    Tempurature Physics, ed. C. W. Chu and J. A.
    Woollam (Plenum, New York, 1978) p. 55.
[4] Stark, R. W. and Windmiller, L. R., Cryo-
    genics 8 (1968) 272.
[5] Schirber, J. E., Cryogenics 10 (1970) 418.

[6] Lifshitz, I. M., Zh. Eksp. Teor. Fiz. 38 (1960) 1569 [Sov. Phys.--JETP 11 (1960) 1130].

[7] Schirber, J. E., in High Pressure Science and Technology, Vol. 1, ed. K. D. Timmerhaus and M. S. Barber (Plenum, New York, 1979), p. 130.

[8] Schirber, J. E. and Morosin, B., Phys. Rev. Lett. 42 (1979) 1485 and references therein.

[9] Kwak, J. R., Schirber, J. E., Greene, R. L. and Engler, E. M., Phys. Rev. Lett. 46 (1981) 1296.

[10] Jerome, D., Mazaud, A., Ribault, M., and Bechgaard, K., J. Phys. Lett. 41 (1980) L-95.

[11] Razavi, R. S., Altounian, Z., and Datars, W. R., Solid State Commun. 28 (1978) 217.

[12] Schirber, J. E., Azevedo, L. J., Narath, A., Morosin, B., Matls. Sci. and Engr. 49 (1981) 7.

[13] Schirber, J. E., Azevedo, L. J., and Narath, A., Phys. Rev. B 20 (1979) 4746.

[14] Schirber, J. E. and Mattheiss, L. F., Phys. Rev. B

[15] Schirber, J. E. and Overmyer, D. L., Solid State Commun. 35 (1980) 389.

[16] Azevedo, L. J., Schirber, J. E., Greene, R. L. and Engler, E. M., Proc. XVIth Intl. Conf. on Low Temp. Phys.,Los Angeles, CA, 8/19-25/81.

PHYSICS OF SOLIDS UNDER HIGH PRESSURE
J.S. Schilling, R.N. Shelton (editors)
© *North-Holland Publishing Company, 1981*

ABNORMAL COMPRESSIBILITY DIVERGENCE IN ReO_3 UNDER HIGH PRESSURE

B. Batlogg, R. G. Maines, and M. Greenblatt

Bell Laboratories, Murray Hill, NJ 07974 USA

High-resolution measurements of the pressure-volume relationship in ReO_3 up to 30 kbar using a highly sensitive strain gauge technique reveal unusual softening of the crystal, as the transition pressure of 5 kbar is approached. The compressibility diverges at the transition point. The additional volume strain in the high pressure phase is proportional to the 2/3 power of the pressure. The results are discussed with respect to the problem of formulating a microscopic model for ReO_3 in particular, and for perovskites in general.

1. INTRODUCTION

The structural instability of the perovskite structure and the associated phenomena such as ferroelectricity and phonon softening are well known. ReO_3 may be considered as belonging to the cubic perovskite structure, with the A sites empty. Contrary to most perovskite-type compounds, no distortion due to the rotation of the oxygen octahedra has been found in ReO_3, at least at atmospheric pressure. Recently, however, Schirber and Morosin [1] found a transition at high pressure ($\sim$ 5 kbar at room temperature) into a new and more compressible phase, which they termed a "compressibility collapse" transition and they attributed this to a buckling of the Re-O bonds. Any understanding of such an unusual phase change requires information on the crystal structure of the high pressure phase as well as its electronic structure. However a detailed knowledge of the pressure dependence of the volume could be revealing.

In this paper we report the results of high-resolution measurements of the pressure-volume relationship and of the electrical resistivity up to 30 kbar. The new findings are: (1) an exponentially increasing softening as the transition is approached from low pressures; (2) a divergence of the compressibility at the transition point; and (3) a decrease of the volume proportional to the 2/3 power of the pressure above the transition.

2. THE PRESSURE-VOLUME RELATIONSHIP

The samples investigated are single crystals grown by iodine vapor transport and are up to $6 \times 6 \times 4$ mm^3 in size. An improved strain gage technique was employed to monitor the length change of the sample under pressure.

Four strain gages are glued in pairs onto the ReO_3 crystal and a silicon standard, and then connected to a full bridge. Even for moderate

excitation currents (1 mA), relative length changes of 2 ppm can be resolved and the dynamic range spans 4.3 decades. In combination with the Teflon-cell technique to generate up to 40 kbar hydrostatic pressure with a resolution of $\sim$ 3 bars on extremely slow compression or decompression, this constitutes, as far as we know, the most sensitive technique to measure the pressure-volume relationship. The experimental details will be given elsewhere [2].

Figure 1 shows the relative volume up to 30 kbar. The initial compressibility is $5.0 \pm 0.1 \times 10^{-7}$ bar^{-1}, in good agreement with the ultrasonic value of $5.12 \pm 0.18 \times 10^{-7}$ bar^{-1} [3,4].

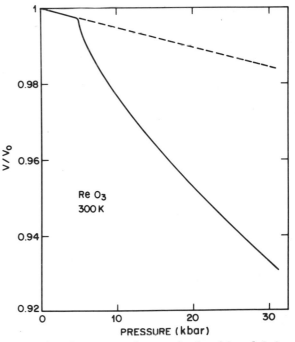

Figure 1: Pressure-volume relationship of ReO_3 at room temperature. The dashed line is the extrapolation of the low pressure behavior.

It may be noted here that in the particular case of ReO_3, the isothermal and adiabatic values are the same, for the thermal expansivity is found to be very small ($\lesssim 10^{-6}$ K^{-1}) (5). Just prior to the transition at 5 kbar, the volume decreases faster than linearly; in fact an additional <u>exponential</u> decrease is observed. This precursor effect amounts to 16% of the expected normal strain at the transition, and grows by a factor of 10 every 2.3 kbar. At the transition point, the slope of the volume curve is infinite but gradually decreases at higher pressures. The drastic change in compressibility is best illustrated by noting that ReO_3 at atmospheric pressure is almost (80%) as hard as sapphire but at 7 kbar is as soft as NaCl.

The dashed line in Fig. 1 is an extrapolation of the low pressure region and represents the expected volume strain in the absence of the transition. The excess volume reduction in the high pressure phase is plotted in Fig. 2 on a log-log scale versus pressure measured from the transition. Obviously a simple power law holds over the entire range of the measurement. Within the given accuracy and as observed in all five independent experiments, the exponent is 0.667 ± 0.005, close enough to be called 2/3.

3. THE ELECTRICAL RESISTANCE

ReO_3 is known as a very good electrical conductor with a resistivity of 10-15 $\mu\Omega$cm at room temperature. This is due to the seventh electron from Re which occupies the lowest part of a wide p-d like conduction band (6-8). Under pressure the resistivity decreases monotonically. The phase transition is reflected in a change of slope and the overall behavior is reminescent of the p-V curve. At 15 kbar, the resistance is reduced to $\sim 55\%$ of its zero pressure value.

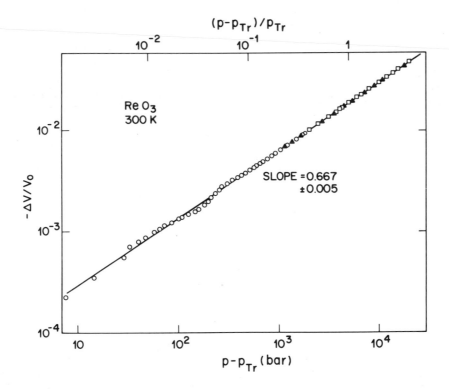

Figure 2: The pressure dependence of the excess volume change in the high pressure phase. It corresponds to the difference between the full and the broken line in Fig. 1.

4. DISCUSSION

As mentioned earlier, distortions of the ideal perovskite structure are well known. For ReO_3 clear indications of an incipient lattice instability can be seen in the elastic constants deduced from ultrasonic measurements (3,4). The magnitude of C_{11} is typical of cubic transition metal oxides and is a measure of the resistance to a change of the Re-O bond length. In contrast to C_{11}, C_{12} is only $\sim 1\%$ of C_{11} and even the sign is not clear. Also C_{44} is exceptionally small, only $\sim 12\%$ of C_{11}. These macroscopic quantities reflect the freedom of the oxygen to move in the plane perpendicular to the Re-O-Re bonding direction. Therefore it is not surprising that moderate external pressures of a few kbar suffice to induce structural distortions.

Realizing that C_{11} is the dominant elastic modulus, the observed divergence of the compressibility κ leads to the conclusion that C_{11} is also going to zero, since $\kappa = 3/(C_{11}+2C_{12})$. This situation is different from other cases, where only a particular combination of elastic moduli vanishes. Consequently we infer the complete softening at the zone center of the LA phonon branch in the [100] and of one of the TA branches in the [110] direction.

Of significance is the 2/3 power law connecting strain and stress in the high pressure phase. Several experimental evidences (not a structural analysis) led Schirber and Mattheiss (9) to propose a cubic T_h^5(Im3) symmetry for ReO_3 at high pressure. In this the unit cell parameter is doubled while the Re positions are unchanged. The oxygen octahedra are tilted equally about each of the cubic axes. However a definitive structural study remains to be done. Symmetry considerations would allow the transition from the low pressure Pm3m symmetry into Im3 to be continuous. Accordingly we find no evidence for an abrupt volume change at the transition.

In analogy to other perovskite compounds, the tilt angle ϕ may be regarded as the order parameter, although this choice needs confirmation by later experiments. As the rigid oxygen octahedra are rotated by ϕ, the volume reduces proportional to ϕ^2. This holds true also for other symmetries then Im3. The experimental result $(\Delta V/V) \propto (p-p_{Tr})^{2/3}$ then translates to $\phi \propto (p-p_{Tr})^{1/3}$, and this power law is obeyed with the same exponent up to the highest pressures, where ϕ exceeds 10°. When the closely related compounds $SrTiO_3$ and $LaAlO_3$ are cooled below the transition temperature, the critical exponent of ϕ changes from 1/3 to the mean field value of 1/2 (10) for $\phi < 1°$.

The change of the Gibbs free energy can be calculated for the high pressure phase from Fig. 1. It is proportional to the 5/2 power of $(\Delta V/V_o)$, or equivalently to ϕ^5. The exponent of 5/2 is reminescent of the total energy of electrons in a parabolic band, which is proportional to $E_F^{5/2}$. A closer numerical study, however, convinces us that we are not dealing with a simple Lifshitz transition where a new band is shifting linearly with V below the Fermi level E_F. One would also expect a thermal broadening of the transition, at variance with the experiment.

The difference between ReO_3 and $SrTiO_3$ lies in the fact that in the former the presence of the conduction electrons will certainly play an important part, at least by their screening of the long-range forces and in the different occupation of the A sites. One specific situation can be anticipated: part of the Fermi surface of ReO_3 with Im3 symmetry consists of a "jungle gym" (9). The calipers of the arms are small but will vary as the strength of the superlattice potential grows. The associated Kohn anomalies at small wave vectors will influence the elastic properties. Such irregularities in the phonon dispersion curves have been found in the isostructural Na_xWO_3 (11), which are known to have the same jungle-gym topoloty of the Fermi surface as ReO_3. In addition to the influence of the conduction electrons, a major question still concerns the general instability of the perovskite structure. Bilz and coworkers (12,13) pointed to the importance of the polarizability of the oxygen ion. They calculated it in a simplified model for the hybridization of the O 2p with the transition metal d states. The most recent observation of a temperature dependent charge transfer from p to d in $LiTaO_3$ (14) is fully in accordance with such a mechanism. It would be very interesting to extend these ideas to ReO_3 and start with the detailed knowledge of the band structure as given by Mattheiss, who found strong p-d hybridization (6).

The explanations set forth in the preceding paragraphs are meant to point out the complexities and challenges inherent in formulating a microscopic model for ReO_3 as well as for perovskites. ReO_3 is a unique test case because the structural stability is affected by the occupation of the A sites (at least via the Madelung potential) and the number of conduction electrons. Only in ReO_3 are the A sites empty and the conduction band is populated.

Investigations of Li-doped ReO_3 and Na_xWO_3 are under way in order to sort out the various contributions experimentally.

ACKNOWLEDGEMENTS

The authors are very grateful to A. Jayaraman, C. M. Varma, M. Y. Azbel and L. F. Mattheiss for helpful discussions.

REFERENCES

(1) Schirber, J. E. and Morosin, B., Phys. Rev. Lett. 42 (1979) 1485.

(2) Batlogg, B. **and** Maines, R. G., to be published.

(3) Tsuda, N., Sumino, Y., Ohno, I., and Akahane, T., J. Phys. Soc. Japan 41 (1976) 1153.

(4) Pearsall, T. P. and Coldren, L. A., Solid State Commun. 18 (1976) 1093.

(5) Matsuno, N., Yoshimi, M., Ohtake, S., Akahane, T., and Tsuda, N., J. Phys. Soc. Japan 45 (1978) 1542.

(6) Mattheiss, L. F., Phys. Rev. 181 (1969) 987.

(7) King, C. N., Kirsch, H. C., and Geballe, T. H., Solid State Commun. 9 (1971) 907.

(8) Pearsall, T. P. and Lee, C. A., Phys. Rev. B 10 (1974) 2190.

(9) Schirber, J. E. and Mattheiss, L. F., Phys. Rev. B (1981) (in print).

(10) Müller, K. A. and Berlinger, W., Phys. Rev. Lett. 26 (1971) 13.

(11) Kamitakahara, W. A., Harmon, B. N., Taylor, J. G., Kopp, L., Shanks, H. R., and Roth, J., Phys. Rev. Lett. 36 (1976) 1393.

(12) Bilz, H., Bussmann, A., Benedek, G., Büttner, H., and Strauch, D., Ferroelectrics 25 (1980) 339.

(13) Bussmann, A., Bilz, H., Roenspiess, R., and Schwarz, K., Ferroelectrics 25 (1980) 343.

(14) Lönert, M., Kaindl, G., Wortmann, G., and Salomon, D., Phys. Rev. Lett. 47 (1981) 195.

PHYSICS OF SOLIDS UNDER HIGH PRESSURE
J.S. Schilling, R.N. Shelton (editors)
© *North-Holland Publishing Company, 1981*

THE EVOLUTION OF THE CHARGE DENSITY WAVE STATE OF 2H-TaSe$_2$

D. B. McWhan and R. M. Fleming

Bell Laboratories, Murray Hill, NJ 07974 USA

The charge density wave, CDW, structures in 2H-TaSe$_2$ have been studied using high resolution X-ray diffraction as well as neutron scattering techniques, and the temperature-pressure phase diagram determined up to a pressure of 4.4 GPa. The phase diagram is more consistent with theoretical models of Rice which include inter-layer coupling than with strictly 2D models of Bak et al.

Charge density waves, CDW, have been observed in a number of compounds such as transition metal dichalcogenides and trichalcogenides and organic conductors. The CDW and the associated periodic lattice modulation may be incommensurate with the underlying lattice under certain conditions and commensurate under other conditions. There has been considerable theoretical and experimental interest in the phase diagrams of these and related systems such as rare gases physisorbed on surfaces (1). The earliest theories for the commensurate-incommensurate (CI) transition suggested that the incommensurate phase consisted of commensurate regions separated by narrow regions in which the phase or amplitude of the incommensurate superlattice varied rapidly (2). In the case of physisorbed gases, these regions are misfit dislocations and the CI transition is determined by the energy needed to introduce misfit dislocations or discommensurations, DC (3). One area of current research involves trying to determine how sharp the DC are in real systems. In two-dimensional systems, the concept of DC leads to the idea of domains and to whether the domains exist as regular arrays with the symmetry of the lattice or to linear (striped) arrays (4) or to disordered arrays of domains (5). 2H-TaSe$_2$ is the first system in which the striped phase has been observed in high resolution X-ray scattering experiments (6), and recently the striped domain pattern has been observed directly by lattice imaging techniques in the electron microscope (7). 2H-TaSe$_2$ is a layered material and the ratio of the linear compressibilities perpendicular and parallel to the layers is ~ 5 (8). This large anisotropy means that by using the pressure variable the relative strengths of the inter- and intra-layer coupling terms in the free energy can be changed substantially. The temperature-pressure phase diagram has been determined by a combination of resistivity (9); X-ray (10) and neutron (11) scattering experiments and these results are compared below with Landau theories for the phase diagram (12). During the course of these experiments it became apparent that small uniaxial stresses lead to domain orientation and substantially change the hysteresis in one of

the transitions. X-ray measurements were made on samples of 2H-TaSe$_2$ under uniaxial tension, and some preliminary results are given below.

The structure of 2H-TaSe$_2$ consists of layers of Ta atoms at the center of trigonal prisms formed by six Se atoms. The hexagonal unit cell contains two such Se-Ta-Se layers which are related by a center of symmetry. The early neutron scattering results (13) showed that on cooling there was a second order transition at T = 122 K from the normal phase to an incommensurate CDW state with a wave vector of $\vec{q} = (1-\delta)\vec{a}^*/3$ where $\delta \approx 0.03$ at onset (Fig. 1). With further decrease in temperature δ initially decreases rapidly, but then has an unusual nearly level region at $\delta \approx 0.007$. A second transition occurred at T = 90 K at which the wave vector locked in at $\vec{q} = a^*/3$, i.e. $\delta \equiv 0$. Two other important features of the neutron diffraction results were the observation of additional superlattice reflections around the main Bragg reflections at $\vec{q}_1 - \vec{q}_2$ and at $2\vec{q}$. The first of these established that the CDW was a triple $\vec{q}$ state, i.e. it resulted from the coherent addition of three CDW; one along each of the three symmetry equivalent directions in the hexagonal plane. The appearance of a reasonably strong reflection at $2\vec{q}$ indicated that the CDW was not a simple sinewave, and this was the first evidence for the existence of discommensurations. More recent thermal expansion measurements suggested a third phase transition at T = 112 K (14). High resolution X-ray scattering results established that on warming a transition takes place from the commensurate phase to a striped phase in which one of the wave vectors remains commensurate and the other two become incommensurate as illustrated in the lower right of Fig. 2 (6). Because of the finite size of the X-ray beam an average of all possible orientations is observed and two reflections corresponding to both the commensurate and incommensurate reflections are observed in scans along [h00] through the superlattice reflections around a Bragg peak at $\vec{q} \approx a^*/3$. (The two incommensurate reflections are not resolved because of the relaxed vertical resolution of the diffractometer.) By studying the distribution of reflections around $\vec{q}_1 - \vec{q}_2$ the structure shown in Fig. 2 was uniquely determined. With increasing temperature the two reflections from the striped phase evolve in position as shown in Fig. 1 and at T = 112 K there is a first order transition to a CDW phase with hexagonal symmetry, i.e. three equal incommensurate CDW as shown in the upper right of Fig. 2. On cooling in the striped phase the lock-in transition was found to be continuous with hysteresis. Similarly the high resolution X-ray measurements show that the lock-in transition from the hexagonal phase during the initial cooling cycle was continuous.

The early Landau models for the C-I transition assumed a complex order parameter in which only the phase is allowed to vary (3).

$$\psi(\vec{r}) = A \, e^{i[\vec{q}_c \cdot \vec{r} + \phi(\vec{r})]} .$$

If the CDW were a simple sinewave then $\phi(\vec{r}) = \vec{\delta} \cdot \vec{r}$ and the incommensurate wave vector is given by $\vec{q}_I = (1-\delta)\vec{q}_c = (1-\delta)\vec{a}*/3$. In the discommensuration model the phase is constant over large regions and then the phase changes rapidly over a narrow region by $2\pi/3$. This is equivalent to commensurate regions of length L, where $L = 2\pi/3\delta$, separated by domain walls. The domain wall patterns in real space and the corresponding phases of the order parameter in each domain are shown for hexagonal and striped phases on the left of Fig. 2.

The single $\vec{q}$ model of Frank and van der Merwe considered a string of atoms connected by springs of restoring force K subjected to an

incommensurate sinusoidal potential of amplitude V (2). This model leads to a difference in free energy between the commensurate and incommensurate phases in terms of L of the form

$$F(L) = \frac{8}{3L}(KV)^{1/2}(1+4e^{-L/\xi}) - q\frac{2\pi K}{3L}$$

where $\xi = \frac{1}{3}\left(\frac{K}{V}\right)^{1/2}$ is a measure of the width of the discommensuration and q is a measure of the incommensurability of the potential. The incommensurate state becomes stable when F < 0, and there is a continuous C-I transition as $q \rightarrow q_c = \frac{4}{\pi}(KV)^{1/2}$. In the discommensuration limit Bak et al (4) considered the triple $\vec{q}$ state of 2H-TaSe$_2$ and pointed out that there would be a competition between the repulsion of parallel DC and a DC crossing energy, Y. This leads to a difference in free energy of:

$$F_{HEX} - F_{STRIPE} = \frac{8}{3L}(KV)^{1/2}(e^{-2L/\xi} - e^{-L/\xi}) + Y/L^2 .$$

If Y is positive then in the limit of $L \rightarrow \infty$, i.e. close to the I-C transition, the striped domain phase is favored. With decreasing L, a striped to hexagonal transition may occur for the appropriate range of the parameters K, V, and Y. In a Landau theory the free energy is expanded in powers of the order parameter and one assumes that one or more of the different parameters varies monotonically with temperature and pressure. The usual assumption is to hold q and K constant and assume that V varies linearly as $(T_o - T)^{1/2}$ where T_o is the temperature at which the

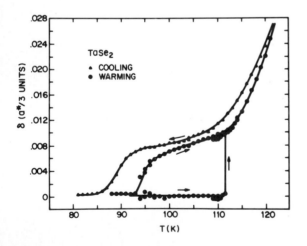

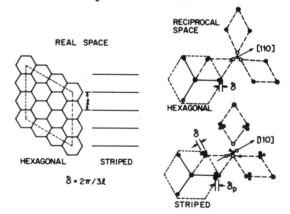

Figure 1: Incommensurability as a function of temperature at P = 1 atm. On cooling the CDW remains in the fully incommensurate hexagonal phase until lock-in at 85 K. On warming the striped CDW phase is seen in the range 93-112 K and a striped to hexagonal transition occurs at 112 K (Ref. 6).

Figure 2: Comparison of the distribution of satellites around the main Bragg peaks of 2H-TaSe$_2$ in reciprocal space for the striped to hexagonal domain phases and the corresponding array of domain walls in real space.

CDW appears, i.e. T = 122 K for 2H-TaSe$_2$. As V increases with decreasing T, $q_c = 4/\pi(K\tilde{V})^{1/2}$ becomes greater than q and a lock-in transition occurs. A measure of the incommensurability at onset is q and direct X-ray measurements show that q increases linearly with pressure and that q has increased over 60% by a pressure of P = 4.4 GPa (10). This simple model predicts that the temperature of the lock-in transition should decrease with increasing pressure relative to the onset transition. Similarly if one considers the model of Bak et al (4) for the striped to hexagonal transition, δ increases and L decreases with increasing temperature (Fig. 1) and at some point the hexagonal phase will be favored. With increasing pressure the value of L at a given temperature decreases because δ_0 increases. Therefore one might expect a large decrease in the temperature of the striped-hexagonal transition with increasing pressure.

A tentative temperature-pressure phase diagram for 2H-TaSe$_2$ has been determined from a combination of resistivity (9), X-ray (10), and neutron (11) scattering measurements, and it is shown in Fig. 3. The solid lines for the onset and lock-in transitions up to P = 1.8 GPa are taken from the paper of Chu et al (9) and are based on resistivity measurements. The filled circles are from neutron scattering measurements (11) and the rest of the points are from X-ray measurements (10). Further measurements will be required to determine the phase boundary between the striped and the reentrant commensurate phase. The existence of the boundary is established although questions of whether the apparent suppression of the commensurate phase around 1.5-2.5 GPA represents

an equilibrium phase boundary remain to be explored. It is clear that the upper two phase boundaries are not sensitive functions of pressure whereas the lock-in transition is. The initial decrease in the lock-in transition is consistent with the predictions of the model given above but no simple monotonic variation of the parameters with pressure will produce the observed reentrant behavior. Similarly the insensitivity of the striped-hexagonal transition to pressure would require a fortuitous variation of Y to match the observed variation of L at the transition with pressure. Rice (12) has shown that this failure of the simple model results from the fact that in 2H-TaSe$_2$ there are three CDW in each plane and two planes per unit cell. As all these have complex order parameters, the temperature-pressure diagram is determined by the competition between the intra- and inter-layer phasing of the different CDW. There are several possible commensurate states, which are local minima of the free energy, but which differ in the phase difference of the order parameter between the layers. In this model the only effect of pressure is to change the inter-layer term which leads to a switch of the absolute minimum from one phase value to another. Initially, pressure reduces the stabilization energy of the commensurate phase, but at higher pressures the stabilization energy increases. This accounts qualitatively for the minimum in the temperature of the lock-in transition. An extension of this model also accounts for the striped-hexagonal transition. The striped state has an additional degree of freedom since the three components of the order parameter are no longer related by hexagonal symmetry. As a result the stabilization energy of the striped state is less sensitive to the phase competition and its onset is less pressure sensitive. By a suitable choice of the coefficients, Rice can qualitatively reproduce the complete pressure temperature phase diagram.

One of the interesting features of the striped phase is the observed hysteresis both in the striped-commensurate transition and in the striped to hexagonal transition. The former appears to be continuous but shows hysteresis which is uncharacteristic of a second order transition. The latter is first order, but it is only seen on warming out of the commensurate phase. In the course of the neutron scattering measurements to determine the striped-hexagonal phase boundary it became apparent that small uniaxial stresses along a* led to domain orientation and also removed almost all of the hysteresis in the striped-hexagonal transition so that it was observed both on warming and cooling (11). In order to show this effect directly, we have made some preliminary measurements on 2H-TaSe$_2$ subjected to varying amounts of tension in the basal plane perpendicular to $\vec{a}$*. (The softness of the crystals precluded applying a compressive stress along $\vec{a}$*.) The earlier X-ray measurements had been made in a diamond anvil device mounted in a

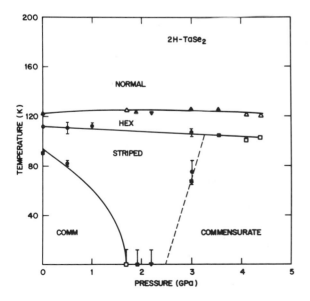

Figure 3: Tentative temperature-pressure phase diagram for 2H-TaSe$_2$ (Ref. 11).

closed cycle refrigerator in such a way that the cell could be rotated in situ approximately around c* (10). A small tension device was mounted in place of the diamond cell as indicated in the top right of Fig. 4. Scans were taken through the CDW reflections near 5/3 along both $\vec{a}*$ and $\vec{b}*$ and the results are shown in Fig. 4. Along $\vec{a}*$ a commensurate peak with a small shoulder corresponding to an incommensurate peak is observed. Along b* the reverse is observed and in addition the 2δ reflection expected along the incommensurate direction is clearly seen. The corresponding reciprocal lattice is shown at the lower right of Fig. 4, and it shows that the domain walls in the striped phase are preferentially aligned normal to the axis which is being stretched.

The scan along b* in Fig. 4 also gives information that is relevant to the question of the sharpness and regularity of the dis-commensurations. An upper limit can be set for the intensity of the 4δ reflection of less than 10% of the 2δ reflection. In the absence of thermal or static disorder the intensity of the 4δ reflection is calculated to be comparable to the 2δ reflection in the model of phase modulation with $L/\xi \sim 12$ but the data are not in agreement with this calculated value. This result and the recent electron microscope studies (7) indicate that the effects of disorder will have to be included.

In conclusion, further studies will be needed to establish the details of the C-I transition above P = 2.5 GPa. Also it would be interesting to try to observe the changes in the phase of the order parameter predicted by the model of

Rice. However the technical difficulties of balancing the required instrumental resolution against the required signal to noise suggest that these experiments will necessitate the use of a synchrotron source.

We thank J. D. Axe, F. J. DiSalvo, D. E. Moncton, and T. M. Rice for helpful discussions and A. L. Stevens for technical assistance.

REFERENCES

[1] Sinha, S. K. (ed.), <u>Ordering in Two Dimensions</u> (North-Holland, Amsterdam, 1980).
[2] Frank, F. C. and van der Merwe, J. H., Proc. R. Soc. London A 198 (1949) 205.
[3] McMillan, W. L., Phys. Rev. B 14 (1976) 1496.
[4] Bak, P., Mukamel, D., Villain, J., and Wentowska, K., Phys. Rev. B 19 (1979) 1610.
[5] Villain, J., in Riste, T (ed.), <u>Ordering in Fluctuating Systems</u> (Plenum, New York, 1980), p. 221.
[6] Fleming, R. M., Moncton, D. E., McWhan, D. B. and DiSalvo, F. J., Phys. Rev. Lett. 45 (1980) 576.
[7] Chen, C. H., Gibson, J. M., and Fleming, R. M., to be published.
[8] Feldman, J. L., Vold, C. L., Skelton, E. F., Yu, S. C., and Spain, I. L., Phys. Rev. B 18 (1978) 5820.
[9] Chu, C. W., Testardi, L. R., DiSalvo, F. J., and Moncton, D. E., Phys. Rev. B 14 (1976) 464.
[10] McWhan, D. B., Fleming, R. M., Moncton, D. E. and DiSalvo, F. J., Phys. Rev. Lett. 45 (1980) 269.
[11] McWhan, D. B., Axe, J. D., and Youngblood, R., Phys. Rev. B (to be published).
[12] Rice, T. M., Phys. Rev. B 23 (1981) 2413 and to be published.
[13] Moncton, D. E., Axe, J. D., and DiSalvo, F. J., Phys. Rev. B 16 (1977) 801.
[14] Steinitz, M. O. and Grunzweig-Genossar, J., Solid State Comm. 29 (1979) 519.

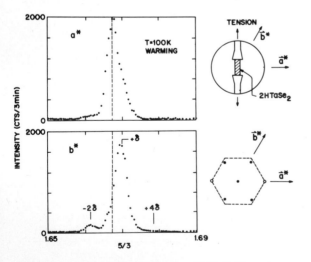

Figure 4: Scans through the (5/3,0,0) and (0,5/3,0) primary superlattice positions for a crystal of $2H-TaSe_2$ in the striped phase under tension as shown in the upper right. The observed domain orientation in reciprocal space is illustrated in the lower right.

PHYSICS OF SOLIDS UNDER HIGH PRESSURE
J.S. Schilling, R.N. Shelton (editors)
© *North-Holland Publishing Company, 1981*

HIGH PRESSURE STUDIES OF GRAPHITE INTERCALATION COMPOUNDS

Roy Clarke

Department of Physics
The University of Michigan
Ann Arbor MI 48109,   USA

The use of high pressure in the study of graphite intercalation compounds (GIC's) is
reviewed. X-ray diffuse scattering measurements on the higher stage alkali-metal
GIC's show that the stage sequence is unstable to applied pressure. Above approximately
3 kbar, stage 2 potassium-graphite transforms reversibly to a pure stage 3 stacking
sequence. Accompanying this transition is a 2x2 superlattice ordering within the
intercalant layer. Experiments on intercalated single crystal and pyrolytic graphites
reveal interesting sample dependence arising probably from defects in the graphite
host. Anomalies observed recently in pressure-dependent electrical transport
measurements of donor and acceptor compounds may have structural origins.

1.   INTRODUCTION

Graphite intercalation compounds (GIC's) are
currently the subject of intense research activity: they display a remarkably rich variety of
phase transition phenomena (1,2) and are also of
potential importance for various energy-related
technologies.(3) The most fascinating aspect of
intercalation compounds, of key importance to
both these areas, is the property of 'staging'(4)
i.e. the formation of long-range sequences of
intercalant and host layers. For example, in the
case of GIC's, layers of electropositive or
electronegative intercalant species may be chemically inserted into the Van der Waals gap
between the planes of carbon atoms of the
graphite host structure (Figure 1). A stacking
sequence in which an intercalant layer altern-
ates with n planes of carbon atoms is referred to
as 'stage n'(Figure 2).

Historically, as a result of the seminal work of
Bridgman, Drickamer and others, the application
of pressure, and in particular, hydrostatic
pressure, has become a remarkably fruitful
technique for exploring the properties of solids
and for testing theoretical models of those
properties. In this connection, recent experimental (5,6,7) and theoretical (8) works suggest
that applied pressure is an important thermodynamic variable for the stability of the staging
arrangements in GIC's, in addition to the more
commonly considered parameters of temperature
and chemical potential.

This contribution will review the use of high
pressure techniques for the study of graphite
intercalation compounds, principally the heavy
alkali-metal GIC's which are the best characterized systems to date. I will be concerned mostly
with X-ray diffuse scattering experiments since
these have provided the most direct and powerful
insight into the structural behavior in
question.

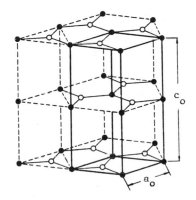

Figure 1 : Crystal structure of graphite,
a_o = 2.46 Å, c_o = 6.70 Å.

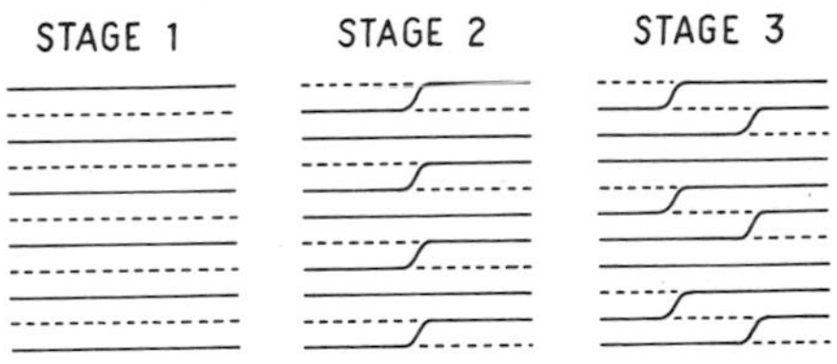

STAGE 1   STAGE 2   STAGE 3

Figure 2 : Staging in GIC's. Solid lines
represent carbon layers and broken
lines, intercalant layers. Domains
in stages n > 1 are macroscopic.

2.   EXPERIMENTAL TECHNIQUES

2.1   X-ray Diffuse Scattering

Depending on the scattering geometry employed
(Figure 3) three types of structural inform-

ation can be gained:

i) (00ℓ) : diffraction vector parallel to $\underline{c}$-axis.
    Intercalant layer $\underline{c}$-axis spacings can be
    determined from position of peaks.

ii) (hk0) : diffraction vector in basal plane.
    This arrangement gives intralayer spacings
    and symmetries.

iii)(hkℓ) : diffraction vector oblique to $\underline{c}$-axis.
    Positions of peaks and systematic absences
    give sequencing of the layers e.g. ABABA...
    for pure graphite; A/AB/BC/CA..... (9) for
    stage 2 alkali metal GIC's. (A,B refer to
    carbon layer positions and '/' to the
    intercalant layer.)

For all three geometries, the inverse widths of
the diffraction profiles are a measure of the
coherence length or range of structural order in
the particular direction chosen. (10)

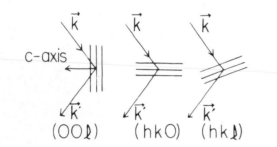

Figure 3 : X-ray scattering geometries.

Experiments have been carried out on two types
of intercalated graphite: highly oriented pyrolytic graphite (HOPG), a polycrystalline form
in which only the $\underline{c}$-axes of the crystallites are
well aligned (to within~0.5°), and natural
single crystals of graphite. The latter samples
are preferable for most structural studies in
that the diffraction data from HOPG samples are
powder averaged and, apart from special directions like (00ℓ), the data are difficult to un-
tangle. One important reason for using two types
of starting material is to study the effects of
sample defects (stacking faults, dislocations,
etc.) which are known to be important for the
staging mechanism. (11)

The X-ray equipment used for these studies
consisted of an automated four-circle diffractometer utilizing graphite-monochromated Mo Kα
radiation from a 12 kW rotating-anode generator.

2.2.   Diamond Anvil cell

Figure 4 shows schematically the apparatus used
for X-ray diffraction studies at pressures up to
12 kbar. The minature pressure cell, after the
design of Merrill and Bassett (12), is gasketed,
with heavy mineral oil as a hydrostatic

pressure transmitting medium. Special precautions were taken to ensure chemical stability
during the experiment because of the highly
reactive nature of the intercalated samples. If
the mineral oil was initially vacuum degassed,
it was found that samples were not degraded in
this medium for several weeks. Pressures were
measured immediately before and after each run
using the ruby R_1 fluorescence technique. (13)

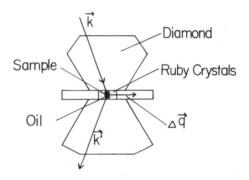

Figure 4 : A schematic representation of the
           gasketed diamond anvil apparatus.

3.   RESULTS AND DISCUSSION

3.1   Pressure-induced staging transition

As mentioned above, the stability of the staging
arrangement is expected to be sensitive to
applied pressure. More specifically, interactions arising from elastic deformations of the
host, caused by the presence of intercalant ions,
have been proposed (8) as a mechanism to explain
the formation and equilibrium of 'pure' stages
in dilute GIC's. In this model the elastic deformations are approximated by point dipoles
(Figure 5). Thus intercalant ions in the same
plane would be attracted to each other and would
tend to coagulate into 'islands'. Similarly, the
intercalant islands in neighboring layers would
repel each other, the lowest free energy state
being that of a pure-stage configuration. This
model is clearly very crude considering the
complexity of real systems and it is likely that
electrostatic effects will also play an important role. (14) However, this simple picture
is very useful for the interpretation of the
pressure experiments on GIC's.

Figure 6 shows a sequence of 00ℓ diffraction
profiles as a function of increasing pressure
in an HOPG sample of stage 2 potassium graphite.
At a pressure of approximately 2.5 kbar (Figure
6b) several new peaks appear and with increasing
pressure these can be seen to grow at the expense of the stage 2 peaks. At pressures of
approximately 6.5 kbar (Figure 6e) and above,
the 00ℓ profile can be identified as that of a

pure stage 3 stacking sequence. Coexistence of stage 2 and stage 3 sequences (see Figure 7a)

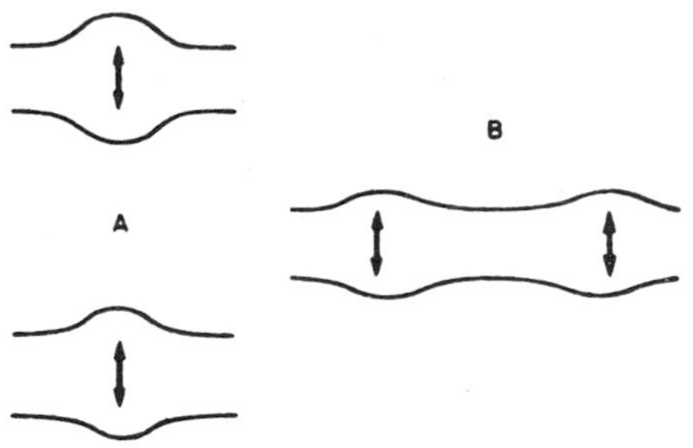

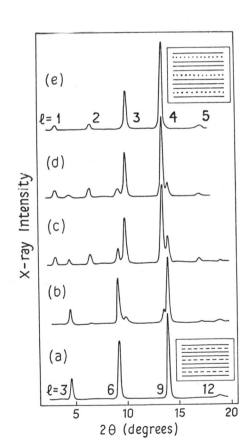

Figure 5 : Pairs of elastic dipoles (arrows) in an infinite anisotropic medium. The dipoles in configuration A repel while those in configuration attract. (from Safran and Hamann Ref.8)

Figure 6 : Room-temperature 00ℓ profiles for

HOPG sample of Stage 2 potassium-graphite: (a) ambient pressure, (b) 2.5 kbar, (c) 4.0 kbar, (d) 5.5 kbar, and (e) 6.5 kbar. Insets: (a) Stage 2 stacking (intercalant planes indicated by broken lines) (e) Stage 3 stacking.

marks the staging transition as being of first order. The transition is hysteretic and reversible. Thus the intercalant is not being squeezed out of the sample but is redistributing within the host layers to form a stage 3 structure. This observation (5) is perhaps the strongest evidence to date in support of the domain model of intercalation of Daumas and Hérold (15) in which every interlayer space contains an equal amount of intercalant. The intercalant ions are distributed inhomogeneously (as islands) rather than uniformly as in the simple 'empty layer' configurations depicted in the insets of Figure 6. Figure 7b shows schematically how the intercalant ions and carbon layers may be reconfigured according to this model as the sample evolves from stage 2 to stage 3 under pressure.

3.2   Defects

The dynamics of staging is a fascinating topic but is still largely unexplored. Our experiments indicate that if the sample is kept at fixed pressure in the mixed-stage coexistence region (2 - 6 kbar for HOPG samples), the structure will evolve towards the pure stage 3 configuration after a few days. This is consistent with the prediction (8) mentioned above that the mixed stage state is metastable. A further interesting feature of the pressure-induced staging transition is that it depends strongly on sample quality. For example, the stage 2-stage 3 coexistence region extends up to ∽ 12 kbar for single crystal samples. (∼ 6 kbar) is thought to be due to a higher density of basal plane dislocations (16) in the HOPG case which would facilitate slippage of carbon layers relative to each other, a mechanism necessary for the change of stage (see Figure 7b). Also of relevance here are the results shown in Figure 7(a) indicating sample annealing effects upon pressure cycling.

The main section of Figure 7 shows how such layer slippage disrupts the stacking sequence in the coexistence region: the very broad c* profile shows that the sequence is almost completely incoherent in this region and only when the pure stage is formed does the stacking become well defined again.

3.3   Changes in intralayer structure during staging transition

The structure of the potassium layer in the ambient pressure and temperature phase of stage 2 GIC is disordered. (17) The nature of this disorder is somewhat controversial and is not

yet well understood. Evidently the structure is
determined by a subtle balance of intercalant-
intercalant and intercalant-carbon interactions.
The nominal <u>layer</u> stoichiometry is KC_{12} for all
stages <u>n</u> $\geqslant$ 2. (4) Thus a transition from stage 2

Experiments on other alkali metal GIC's by
Wada (7) confirm that the staging instability
under pressure is pervasive in such compounds.
In particular, the in-plane 2x2 ordering is
always observed at high pressures but unless the

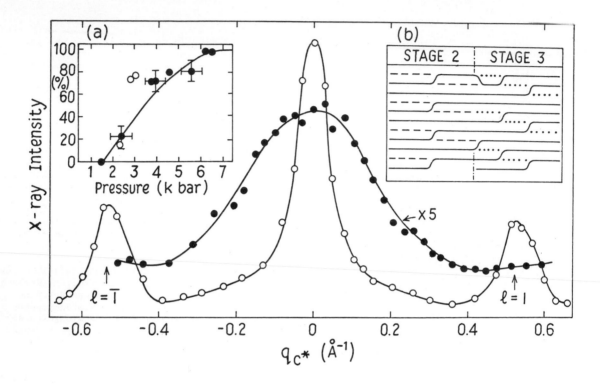

Figure 7 : X-ray intensity as a function of diffraction
wavevector, q_c*, along the 10ℓ reciprocal-
lattice row   of single crystal KC_{24} at
4 kbar (closed circles) and 11 kbar
(open circles). (The 10ℓ scan at ambient
pressure is indistinguishable from background.) Insets: (a) Percentage of Stage-3
phase in an HOPG sample of KC_{24}, as a function
of pressure. First run, closed circles; second
run, open circles. (b) Domain model for
graphite layers during staging transition.

to stage 3, assuming no expulsion of the intercalant from the sample, implies an increase in
the local areal density of potassium ions by a
factor of 3/2 in order to maintain the overall
stoichiometry. i.e. the local layer stoichiometry becomes KC_8 in the high-pressure stage 3
phase. This is reminiscent of the 2x2 superlattice layer structure observed in stage 1 K, Rb,
and Cs intercalates. (18) Indeed, hk0 diffraction patterns (Figure 8) clearly reveal the half
order spots associated with such a superlattice
at higher pressures. Thus the pressure-induced
staging transition also involves an ordering of
the intercalant <u>ions</u> into a structure with dis-
crete symmetry and long-range order. This
represents a most unusual freezing transition!

bulk stoichiometry is favorable (e.g. stage 2→3,
stage 4→6, etc) an unstable mixture of stages
is usually formed under pressure.

3.4   Compressibilities

During the course of our staging experiments,
data on the <u>c</u>-axis compressions could also be
extracted. Our results for pristine graphite
(Figure 9 and Table I) agree well with bulk
moduli obtained from other techniques. In
Figure 9 it can be seen that stage 1 KC_8 is
somewhat "harder" than pristine graphite and
linear behavior is observed for both materials
up to at least 10 kbar. Recent measurements reported by Wada (7) show that CsC_8 has a much

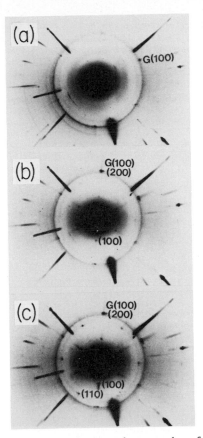

Figure 8 : X-ray oscillation photographs of KC_{24}
single crystal at (a) 1 kbar,
(b) 4 kbar, and (c) 11 kbar. The
intense streaks are X-ray reflections
from the diamond anvils and the diffraction ring is from the gasket.

HOPG (o) and stage 1, KC_8 prepared from
single crystals (△) and HOPG (▲) are
shown. C(p) denotes the lattice constant
at pressure, p. The straight lines are
least-squares best fits.

Table I : The Bulk Modulus $C_{33} \approx 1/k_c$ of Graphite

Method	C_{33} (10^{11} dyn/cm^2)
X-ray (this work)	3.66 ± 0.12
Neutron (Ref. 19)	3.71 ± 0.05
Ultrasonic + static compression (Ref. 20)	3.65 ± 0.10

lower compressibility than KC_8, RbC_8 or pristine
graphite, indicating a strengthening of the int-
ercalant-host interaction with increasing size
of intercalant species.

In the stage 2 GIC's (Figure 10) the effects of
the staging transition can be seen as a break in
the slope of the c-axis compression curve at
~4 kbar. The "softness" of the stage 2 compounds can be attributed to the staging instability. In contrast, the stage 3 samples appear
to be much less compressible than pristine
graphite (see Table II). This measurement,
however, may be unreliable because of the
difficulty of identifying diffraction peaks from
mixed-stage 3 and 4 phases at high pressures. (7)

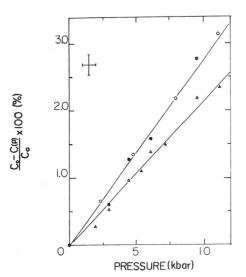

Figure 9 : The percentage of c-axis lattice
contraction vs. hydrostatic pressure.
Pristine single crystal graphite (●),

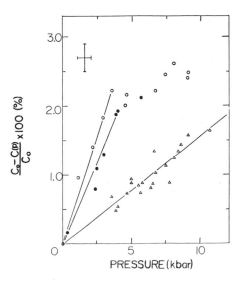

Figure 10 : The percentage of c-axis lattice

contraction of KC_{24} vs. hydrostatic pressure. Stage 2, single crystal (o), HOPG (●), and stage 3 single crystal (△) and HOPG (▲) samples were used.

Table II : Spacing between intercalant layers, $c_0{}'$, compressibility, k_c, and least squares coefficient of determination (R^2 from Figures 9 and 10) for KC_8 and KC_{24}. From Ref. 6

	$c_0{}'$	k_c $(\times 10^{-12}cm^2/dyn)$	R^2
(graphite)	3.353±0.003	2.73±0.09	0.990
KC_8	5.355±0.004	2.13±0.09	0.986
KC_{24} stage 2 [a,b]	8.670±0.020	5.00±0.80	0.861
stage 2 [c,b]	8.706±0.009	4.77±0.39	0.968
stage 2 [d,b]	8.676±0.009	6.25±0.47	0.983
stage 3 [e]	11.960±0.020	1.60±0.20	0.814

Notes: a) HOPG and single crystal data points were used with a single linear fit.
b) Data points below 4 kbar only
c) HOPG sample
d) Single crystal sample
e) $c_0{}'$ extrapolated to zero pressure

4. OTHER TECHNIQUES

4.1 Electrical Transport

In no case so far has a pressure-induced structural instability been observed in a saturated stage 1 compound. However, recent transport measurements by Fuerst et al. (21) do appear to show a c-axis resistivity anomaly in KC_8 at approximately 15 kbar, a somewhat larger pressure than was reached in the X-ray measurements. There is as yet no diffraction information to correlate with this anomaly.

In connection with the basal plane transport properties, Uher (22) has noted that the pressure induced staging transition could lead to a significant enhancement of the in-plane electrical conductivity. The enhancement, estimated at a factor of 3, is proposed to arise from:
a) an increase in the local intercalant density from MC_{12} to MC_8; hence increasing the conduction electron density, and
b) simultaneously achieving higher mobilities associated with the higher stage structure (23)

4.2. Light Scattering

The effects of the pressure-induced staging transition have recently been observed in a spectacular way using Raman scattering. (7) Figure 10 shows the spectrum of stage 4 RbC_{48} associated with the Raman active layer-shearing mode E_{2g} which appears at $\omega \approx 1581$ cm^{-1} in pure graphite. Because there are two different environments for the carbon layer in the intercalation compound for stages $\underline{n} \geqslant 3$ (carbon layer next to an intercalant layer and carbon layer bounded by two other carbon layers), this mode appears as a doublet.(25) The relative intensities of the two components change according to the stage: as higher stages are approached the 'interior' graphitic component dominates. This behavior is beautifully illustrated by the data of Wada reproduced in Figure 10 with increasing pressure. The lattice dynamical effects have been correlated with structural changes using X-rays as described above.

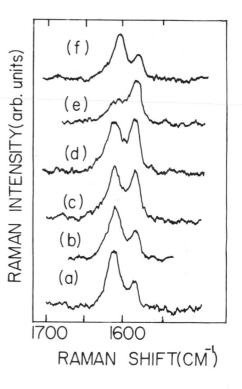

Figure 11 : Raman spectra of RbC_{48} taken at (a) 2.2 kbar, (b) 3.3 kbar, (c) 5.3 kbar, (d) 8.8 kbar, (e) 9.8 kbar and (f) ambient pressure. The spectra were taken consecutively. After Wada (7).

5. OTHER GRAPHITE INTERCALATION COMPOUNDS

In an attempt to induce staging instabilities in other GIC's we have recently begun to investigate two acceptor compounds: $SbCl_5$-graphite (26) and HNO_3-graphite (27).

Unfortunately, the full structures of these acceptor compounds are not known at this time but the staging in each of the materials is very well defined. Various stages of the $SbCl_5$ compound have been pressurized to 12 kbar, so far without any sign of an instability. During the course of our experiments we were able to identify the layer structure of $SbCl_5$-graphite (stages 2-7) as a closed-packed $\sqrt{7} \times \sqrt{7}$ superlattice.(28) Evidently the molecules (which incidentally may be disproportionation products such as $SbCl_3$ and $SbCl_6^-$ in the GIC (29) are strongly bound to the host layers and much higher pressures may be required to induce an instability.

The second acceptor compound we have studied, however, is more promising in this respect. It is reported to have a disordered arrangement of molecules (again, probably not HNO_3 but NO_3^-) at ambient temperatures and pressures. Presumably this would allow for a redistribution of the intercalant molecules at reasonably low pressures. There is some evidence, from resistivity measurements,(30) that a phase transition of some kind occurs at pressures of approximately 6 kbar in stage 3 and 4 HNO_3-graphite but again there is little structural information available. The situation here is complicated by the fact that two chemical modifications of HNO_3-graphite are known to exist: a bilayer configuration (27) with a staggered molecular arrangement, and a single layer 'residual' form. (31) We have found that the bilayer form at ambient temperature evolves slowly (over a few days) to the residual form. The final structure is usually found to be a stable pure stage, in agreement with the elastic deformation model (8) briefly discussed above. Our pressure measurements on these compounds are at a preliminary stage but we can observe partially reversible structural changes. We need to perform a set of experiments at reduced temperatures (but above the ordering temperature T = 253 K) in order to separate out pressure induced effects from the time-dependent sample evolution which is thermally activated.

6. FUTURE DIRECTIONS

Clearly high pressure is an important variable for the study of staged compounds such as GIC's. In the donor compounds structural instabilities are observed under rather modest pressures. Both staging and in-plane rearrangements are taking place. It would be interesting to extend these experiments to higher pressures to look for possible re-orderings of the registered structures such as the stage 1 MC_8 case and the $\sqrt{7} \times \sqrt{7}$ superlattice. Such experiments may shed more light on the role of incommensurability and discommensurations in GIC's.(32)

Recent evidence (21,30) from transport measurements indicates that there are significant pressure effects on the c-axis resistivity of GIC's. It would be very instructive to also probe the basal-plane resistivity in which case the pressure effects should be even more

pronounced. Speculatively, if such instabilities could be induced in the acceptor compounds it may be possible to enhance their conductivities beyond the already very high values that have been observed, for example in AsF_5-graphite under ambient conditions.(33)

Perhaps the most interesting avenue for basic research on GIC's that high pressure affords is the provision of a means to study the dynamics of the staging mechanism under well-controlled conditions of chemical potential and temperature. Given the advances that have been made in high pressure techniques over the past few years, together with qualitative improvements in instrumentation for structural studies (particularly the advent of synchrotron radiation), it is likely that a combination of these techniques will contribute greatly to our understanding of this fascinating class of materials.

ACKNOWLEDGEMENTS

Much of the work described herein was carried out in fruitful collaborations with N. Wada and S.A. Solin, with support from the NSF Materials Research Laboratory of the University of Chicago. I thank H. Homma for his contributions to the research on graphite nitrates and chlorides, and Ctirad Uher for several stimulating discussions. The more recent stages of this research at the University of Michigan were supported by the NSF (grant DMR-8001655) and the Dow Chemical Co. Foundation.

REFERENCES

[1] For recent reviews, see Solin, S.A., Adv. in Chem. Phys. (in press).
[2] Dresselhaus, M.S., and Dresselhaus, G., Adv. in Phys. (in press).
[3] Ebert, L.B., Annual Rev. Mater. Sci. 6 (1976) 181.
[4] Rüdorff, W., and Schulze, E., Z. Anorg. Allg. Chem. 277 (1954) 156.
[5] Clarke, R., Wada, N., and Solin, S.A., Phys. Rev. Lett. 44 (1980) 1616.
[6] Wada, N., Clarke, R., and Solin, S.A., Synthetic Metals 2 (1980) 27.
[7] Wada, N., Phys. Rev. B (in press).
[8] Safran, S.A., and Hamann, D.R., Phys. Rev. Lett. 42 (1979) 1410.
[9] Parry, G.S., and Nixon, D.E., Nature 216 (1967) 909.
[10] Guinier, A., X-ray Diffraction (Freeman, San Francisco, 1963).
[11] Hooley, J.G., Mater. Sci. Eng. 31 (1977) 17.
[12] Merrill, L., and Bassett, W.A., Rev. Sci. Instrum. 45 (1974) 290.
[13] Piermarini, G.J., Block, S., and Barnett, J.D., J. Appl. Phys. 44 (1973) 5377.
[14] Safran, S.A., DiSalvo, F.J., Haddon, R.C., Waszczak, J.V., and Fischer, J.E., Physica 99B (1980) 494.

[15] Daumas, N., and Hérold, A., C.R. Acad. Sci. 268 (1969) 373.

[16] Soule, D.E., and Nezbeda C.W., J. Appl. Phys. 39 (1968) 5122.

[17] See Ref. 9 and also Zabel, H., Moss, S.C., Caswell, N., and Solin, S.A., Phys. Rev. Lett. 43 (1979) 2022.

[18] See Ref. 4 and also Parry, G.S., Mater. Sci. Eng. 31 (1977) 99.

[19] Blakslee, O.L., Proctor, D.G., Seldin, E.J., Spence, G.B., and Weng, T., J. Appl. Phys. 41 (1970) 3373.

[20] Nye, J.F., Physical Properties of Crystals (Clarendon, Oxford, 1972) p.131.

[21] Fuerst, C.D., Moses, D., Denenstein, A., Heeger, A.J., and Fischer, J.E., Bull. Amer. Phys. Soc. 26 (1981) 451.

[22] Uher, C., unpublished.

[23] Onn, D.G., Foley, G.M.T., and Fischer, J.E., Phys. Rev. B19 (1979) 6474.

[24] Nemanich, R.J., Lucovsky, G., and Solin, S.A., in Balkanski, M. (ed.) Proc. of Intl. Conf. on Lattice Dynamics (Flammarion, Paris, 1978) p.619.

[25] Nemanich, R.J., Solin, S.A., and Guérard, D., Phys. Rev. B16 (1977) 2965.

[26] Mélin, J., and Hérold, A., C.R., Acad. Sci. 269 (1969) 877.

[27] Rüdorff, W., Z. Phys. Chem. B45 (1939) 42.

[28] Clarke, R., and Homma, H., Bull. Amer. Phys. Soc. 26 (1981) 452.

[29] Eklund, P.C., and Boolchand, P. (private communication).

[30] Iye, Y., Takahashi, O., Tanuma, S., Tsuji, K., and Minomura, S. in Proc. of Yamada Conference on Intercalation Compounds, 1980.

[31] Fusellier, H., Mélin, J., and Hérold, A., Mater. Sci. Eng. 31 (1977) 91.

[32] Clarke, R., Gray, J.N., Homma, H., and Winokur, M.J., (to be published).

[33] Vogel, F.L., J. Mater. Sci. 12 (1977) 982.

PHYSICS OF SOLIDS UNDER HIGH PRESSURE
J.S. Schilling, R.N. Shelton (editors)
© *North-Holland Publishing Company, 1981*

SOME RECENT RESULTS IN METAL HYDROGEN SYSTEMS IN THE HIGH PRESSURE REGION

Stanislaw Filipek, Bogdan Baranowski and Marek Klukowski

Institute of Physical Chemistry, Polish Academy of Science
Warsaw, Poland

Properties of the Ni-Fe-H system under high pressure of gaseous hydrogen are discussed.

Influence of the high pressure of gaseous hydrogen on thermoelectric power and electric resistivity of several metallic glasses has been determined. Conclusions concerning the hydrogen absorption in these glasses are presented.

1. INTRODUCTION

The Ni-Fe-H belongs to the most intensively investigated metal-hydrogen systems because of its very important role in modern technology as well as for its interesting electronic, magnetic and structural properties.

In the beginning of the 70's the hydrogenation of Ni-Fe alloys has been undertaken by electrochemical methods [1]. The introduction of high pressure devices gave the possibility to study phase equillibria and properties of hydrogenated phases in well defined thermodynamic conditions and, in that way, to enlarge our knowledge about Ni-Fe-H system.

The measurements of electrical resistivity and thermoelectric power of Ni-Fe alloys as function of hydrogen pressure in isothermic conditions, together with X-ray diffraction data, cleared some structural and electronic properties of Ni-Fe-H system [2-4]. The Ponyatovski's high pressure group studied T - P phase diagram of Ni-H system in the large pressure and temperature range (630 K, 18 kbar) [5], phase diagram of Ni-D and Ni-H [6] (375°C, 20 kbar) and critical phenomena in Ni-Fe-H system in the high pressure of gaseous hydrogen [7]. They achieved also the synthesis of iron hydride from elements (condition of hydride formation: $p(H_2)$ = 45 kbar, t = 250°C) [8].

In this paper, the results of experiments, realised with different high pressure techniques and concerned with the influence of hydrogen on Ni-Fe system, are presented. Moreover, taking into account the increasing interest about absorption of hydrogen in metallic glasses (for example it was found that hydrogen absorption in some metallic glasses is higher than in corresponding crystallic phase [9]) and experimental result, demonstrating that order-disorder transition in Ni-Fe system plays an important role in hydrogen absorption, the investigation of nickel and Ni-Fe derivative metallic glasses was also included. The interesting behaviour of nickel derivative glass under pressure of hydrogen encouraged us to make additional measurements on palladium silicon glass.

2. EXPERIMENTAL

The electrical resistivity and thermoelectric power were measured in the manner described in previous papers [3,4]. High pressure high temperature measurements were performed on the apparatus supplied with an internal alumina microheater. The absorption and desorption isotherms were determined on high-pressure apparatus shown in Fig. 1.

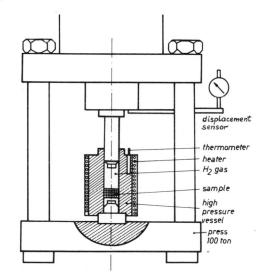

Figure 1: Apparatus for direct determination of hydrogen absorption and desorption isotherms.

The measurements started from 1 kbar. After each change of pressure the piston was moved in order to compensate the pressure changes during absorption or desorption time. When absorption or desorption was finished, the procedure was repeated at another pressure level. For low-temperature measurements a standard helium cryostat has been used. All specimens were made from 5 - 25 micrometer foils, from which the pieces 1 x 10 mm were cut for electrical properties measurements. The similar foils, but in quantity 5 - 15 g were used for direct absorption and desorption curves determinations.

3. RESULTS AND DISCUSSION

3.1. The Ni-Fe-H system

In the behaviour of Ni-Fe alloys in contact with
hydrogen under very high pressure several characteristic ranges, determined by iron concentration, can be distinguished. Description of these
ranges is presented below.

a) In the range of 0 - 18 at % Fe the hydride
phases are formed. The formation of hydrides occurs at higher pressures than their decomposition.
Both formation and decomposition pressures increase
with iron concentration. Simultaneously, the difference between these pressures decreases and
disappears near Ni-Fe 18 % composition. Formation
and decomposition of hydride phases is followed
by rapid changes of such properties like electric
resistivity and thermoelectric power. Typical
changes of these properties as function of hydrogen pressure are presented on Fig. 2 and Fig. 3.

The conclusions concerning hydride phases formation
in this concentration range were confirmed not
only by changes of the electric properties, but
also by X-ray diffraction analysis and by direct
determination of hydrogen absorption and desorption isotherms. These isotherms, received for
Ni-Fe 2,09 % and Ni-Fe 5,7 % alloys, are presented
on Fig. 4 and Fig. 5. Hydrogen concentration
values determined from these isotherms are in good
agreement with the data from mass-spectrometric
analysis of samples removed from high pressure
vessel after cooling down to about -60° C.

Our measurements of the electrical resistivity
as a function of pressure of hydrogen at elevated temperatures correspond to the data from
Ponyatovski's paper[7] concerning critical phenomena in Ni-Fe-H system. The Figure 6 presents,
as an example, the changes of electrical resistivities of Ni, Ni-Fe 5,7 %, Ni-Fe 9,6 % and
Ni-Fe 17,2 % samples versus hydrogen pressure at
175°C.

It is worthwhile to point out, that at 25°C, the
thermoelectric power of alloys containing more
than 2 % Fe reach almost the same value when the
hydride phase is formed (thermoelectric powers of
these alloys without hydrogen depend strongly on
iron concentration). However, the electrical resistivities of the hydride phases are higher for
alloys containing larger iron concentration
(Ni-Fe 16,2 % after hydrogenation has about a six
times higher resistivity than the same alloy but
without hydrogen).

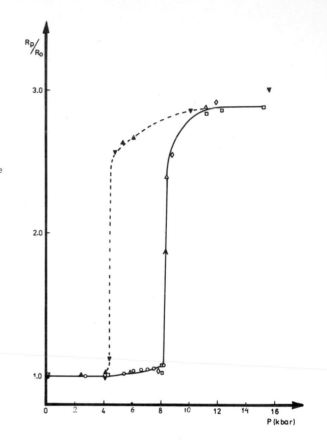

(continuous line - pressure increase, dotted
line - pressure decrease)

R_p - resistance of sample at pressure p.

R_o - resistance of sample at atmospheric pressure.

Figure 2: Electrical resistivity of Ni-Fe 4,2 %
as function of hydrogen pressure at
25°C

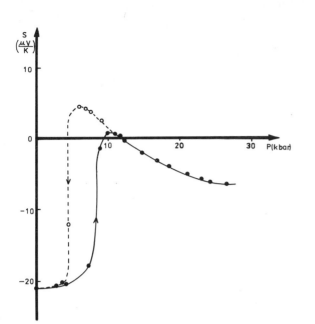

Our low temperature measurements have demonstrated that at elevated hydrogen concentrations in nickel the residual resistivity was only about 50 % higher than value received for pure nickel. On the other hand, even a small addition of iron into nickel caused a remarkable increase of the residual resistivity of the hydride phase formed, compared to the hydrogen free metal. The decrease of the thermal part of resistivity, just like in the case of nickel hydride formation, takes place only for alloys containing up to about 1 % of iron. At higher iron concentrations the increase of both parts, residual and thermal, is observed. The results of low temperature measurements for hydride phases obtained under high pressure of hydrogen are presented in Table 1.

Figure 3: Thermoelectric power of Ni-Fe 4,2 %
as function of hydrogen pressure at
25°C.

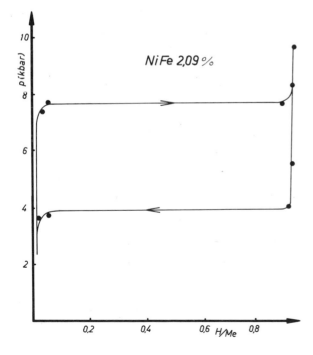

Figure 4: Absorption and desorption isotherms
of hydrogen in Ni-Fe 2,09 at % alloy
at 25°C.

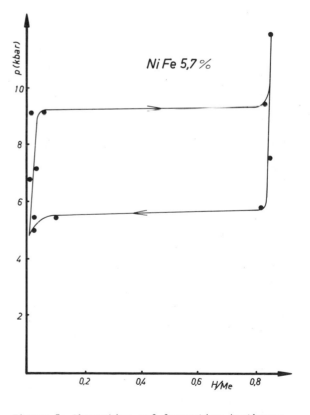

Figure 5: Absorption and desorption isotherms
of hydrogen in Ni-Fe 5,7 % alloy at
25°C.

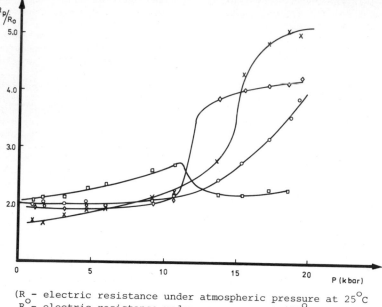

(R$_o$ - electric resistance under atmospheric pressure at 25°C
R$_p$ - electric resistance under pressure p at 175°C)

Figure 6: The dependences of electrical resistivities of nickel (□), Ni-Fe 5,7 % (◊), Ni-Fe 9,6 % (✗),
and Ni-Fe 17,2 % (o) on hydrogen pressure at 175°C.

Table 1. Results of low temperature measurements of hydride phases synthetised under high pressure of
hydrogen.

Composition of alloy	Residual resistivity of alloy without hydrogen ($\mu\Omega$ cm)	Alloys after hydrogenation		
		Residual resistivity ($\mu\Omega$ cm)	Increase of thermal part of resistivity ($\mu\Omega$ cm)[1]	Hydrogen concentration (H/Ni)
Pure nickel	0,263	0,312	−0,826	0,94
Ni-Fe 0,23 %	0,537	2,433	−0,822	0,94
Ni-Fe 0,50 %	0,700	4.392	−0,905	0,94
Ni-Fe 4,22 %	3,664	19,950	+1,257	0,90
Ni-Fe 9,60 %	5,032	28,650	+8,874	0,78

[1] The differences between thermal part of resistivity of hydrogenated and hydrogen free
alloy at 130 K.

b) The range 18 - 38 % Fe: The hysterisis of hydrogen absorption-desorption process in nickel-iron alloys disappears from about 18 % of iron. However, even in the Ni-Fe 21,4 % alloy the creation of hydride phase has been observed. Remarkable changes of the electric resistivity and the thermoelectric power, as well as mass-spectrometric analysis give good arguments for strong hydrogen absorption in alloys belonging to this range of concentration.

Near the concentration of Ni$_3$Fe disorder-order transition occurs. The change of the electric resistivity of the Ni-Fe 30,2 % alloy in disordered and ordered state is shown in Fig. 7. The ordered superstructure was obtained after annealing during one month in 460°C. From Fig. 7 it is evident that for ordered sample the resistance-pressure relation is shifted towards higher pressures. Such a shift testifies that ordering in Ni-Fe system restrains the hydrogen absorption. A different behaviour was observed in Pd-Fe system [10]. The curves R$_p$/R$_o$=f(PH$_2$) for that system in ordered and disordered state are presented on Fig. 8.

The influence of ordering process on the behaviour of Ni-Fe-H system under high hydrogen pressure can be attributed to two effects:

(i) Decrease of holes in the d-band of the alloy indicated by the reduction of electronic specific heat caused by ordering process[11].

(ii) Decrease of the lattice parameter caused by the ordering process[12].

Supposing, that during the ordering process the number of holes in the d-band diminishes and that electrons from absorbed hydrogen fills the empty states, one can expect that the ordering process should play negative role in hydrogenation. It is worthwhile to point out that up to the Ni-Fe 30 % composition, the introduction of iron decreases the specific heat in disordered alloys and in the same time decreases the amount of hydrogen absorbed in the alloy. If the same tendency is valid for ordering process, the experimental results seem to be quite reasonable in spite of rather tough approximation like the band model. Besides the band model factors the negative influence of ordering on hydrogen absorption can be attributed to contraction of the lattice during ordered structure formation.

The explanation of the difference between behaviour of ordered Ni-Fe and Pd-Fe alloys gives Switendick[13] basing on calculation results actually performed. These calculations show the different character and range of Pd-H and Ni-H interaction in both kinds of alloys. Contrary to Ni-Fe, in Pd-Fe alloys the interaction between Pd and H atoms should have rather localised character.

c) <u>Ni-Fe alloys containing 38 - 60 % Fe</u>: The typical changes of electrical resistivity and thermoelectrical power for alloys from this concentration range are shown in Fig. 9 and Fig. 10. The shape of these curves is similar to greek lambda letter. The pressure coordinates of maxima points of both these relations are almost equal for all alloys belonging to this range of iron concentrations. Besides this, in spite of quite different initial values, the values of thermoelectric powers in maxima points are approximately equal for all of these alloys.

Such a change of electrical properties with hydrogen pressure (or with hydrogen activity) could be attributed to phase transition of the

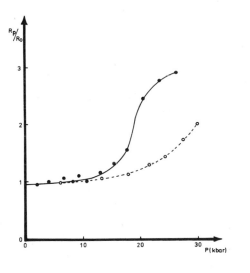

o - ordered sample
● - disordered sample

Figure 7: Influence of Ni$_3$Fe ordering in Ni-Fe 30,2 % alloy on the resistivity changes as function of hydrogen pressure.

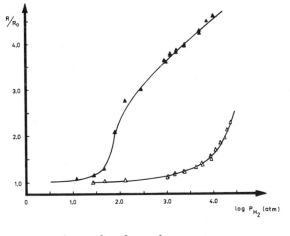

▲ - ordered sample
△ - disordered sample

Figure 8: Influence of Pd$_3$Fe ordering on the resistivity changes as function of hydrogen pressure[10].

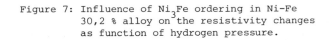

second order. Soomura[14] in a paper concerning
Mössbauer effect in hydrogenated Ni-Fe 55 %
alloy showed the strong interaction between iron
atoms and hydrogen dissolved. This interaction
decreases remarkably the magnetic moment of iron
and results in a partial filling of the 3d holes
by electrons introduced by the hydrogen dissolved.
As a consequence of the appearance of such a
phase in which 3d holes are filled up, the decrease of electric resistivity can be expected.
Therefore, the first, rising part of the curves:

$$R_p/R_o = f(p_{H_2})$$

can be interpreted by the formation, in the first
step, hydrogen solid solution which is inhomogenous and have strong local distortions. Continued hydrogenation causes that at the critical
hydrogen activity (which would correspond to
the hydrogen pressure in maximum point of resistivity and thermoelectric power relations)
a continuous phase transition is starting. The

homogenous distribution of the large amount of
hydrogen in this phase, as well as a strong hy-
drogen-iron interaction cause the decrease of the
electric resistivity of samples which is observed in the second part of the curves discussed.

3.2. Metallic glasses $(NiFe)_{78}X_{22}(X=B+Si)$

The first step of this study was the investigation of a metallic glass not containing iron but
only nickel and silicon as glass-making elements.
In the Fig. 11 the thermoelectric power change
with hydrogen pressure is presented. From this
figure it can be seen that in the case of inert
medium pressure effect on thermoelectric power
of the $Ni_{78}X_{22}$ is to be neglected. However, when
hydrogen is used as pressure transmitting medium, the interesting changes of thermopower are
observed. These changes start from about 5 kbar,
thus at lower pressure than in the case of pure
nickel. Between 5 and 10 kbar relatively rapid
change is observed. Afterwards the change rate be-

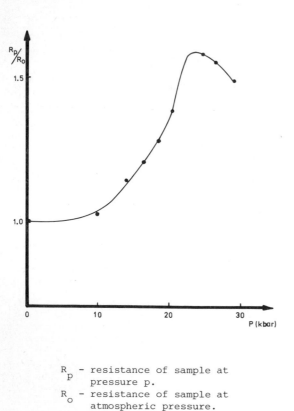

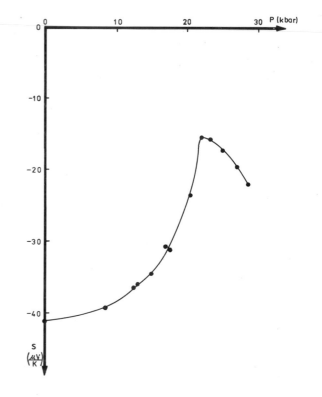

R_p - resistance of sample at
     pressure p.
R_o - resistance of sample at
     atmospheric pressure.

Fig. 9: Electrical resistivity of Ni-Fe 48,5 %
        as function of hydrogen pressure at
        25°C.

Figure 10: Thermoelectric power of Ni-Fe 48,5 %
           as function of hydrogen pressure at
           25°C.

came smaller but still detectable. During the pressure reduction the change of the thermo-electric power does not follow the curve obtained during pressure increase. The hysteresis is relatively large. Even at 1 kbar the thermoelectric power is still different from the initial value.

These results point out that in comparison with crystalline nickel the absorption of hydrogen in glassy nickel starts at lower hydrogen activity and the stability of hydrogenated phase is higher. It is hard to say now if such behaviour is caused by the glassy state or by the presence of silicon and boron. This could be cleared up by observing the tendency after introducing these elements into crystalline nickel. However, considering the above behaviour of $Ni_{78}X_{22}$ glass, one could expect that replacing nickel by some amount of iron only a limited shift of the hydrogen absorption should be observed, similar to the crystalline Ni-Fe alloys. It is therefore surprising

that in the $Ni_{624}Fe_{15,6}X_{22}$ and $Ni_{39}Fe_{39}X_{22}$ metallic glasses the changes measured in hydrogen are almost the same as in inert pressure transmitting medium up to about 20 kbar. Also the electric resistivity measurement do not show any possibility of hydrogen absorption in these glasses. However, the influence of hydrogen on magnetoelastic behaviour in $Ni_{40}Fe_{40}X_{20}$ glass, which has approximately the same composition as our's $Ni_{39}Fe_{39}X_{22}$, has been published[15]. It can therefore be concluded that iron addition into glassy nickel have stronger dumping effect for hydrogenation than in crystalline state or that absorption occurs but does not cause significant changes of the electrical properties measured.

3.3. $Pd_{83}Si_{17}$ metallic glass

The interesting result obtained for glassy nickel encouraged us to study the glassy $Pd_{83}Si_{17}$ alloy which should exhibit hydrogen absorption at much moderate pressures.

In the case of $Pd_{81}Si_{19}$ glass Ehrlih[16] found out that hydrogen absorption causes increase of electric resistivity. The maximum hydrogen concentration achieved by this author using electrochemical method, was H/Me = 0,075. It was therefore reasonable to expect that measurement of electric resistivity and other properties can give information about hydrogen absorption in this metallic glass. The measurements of electric resistivity and thermoelectric power of $Pd_{83}Si_{17}$ glass has been performed and the obtained results are presented on Fig. 12 and Fig. 13. From the shape of these curves it can be concluded that the influence of pressure transmitted by inert medium is very small for glassy $Pd_{83}Si_{17}$ in the pressure range investigated. On the other hand, when hydrogen is used as pressure transmitting medium, the remarkable changes of both of these electrical properties are observed even at moderate pressures.

Almost the total change of resistivity and thermoelectric power occurs up to 5 kbar. Later changes are relatively small. Also in that system the hysteresis, which in general is typical for first order phase transitions, has been observed.

The analysis of samples after hydrogenation at 1 kbar resulted in concentration $H/Pd_{83}Si_{17}$ = 0,14, so almost two times higher than Ehrlih result. It could be expected that at higher pressures the sample can absorb much more hydrogen.

The investigations of these systems are continued.

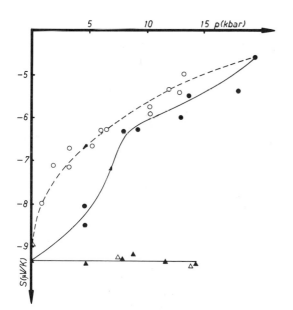

● - increase of pressure)
○ - decrease of pressure) in H_2

△▲ - increase and decrease of pressure (in inert medium)

Figure 11: Thermoelectric power of metallic glass $Ni_{78}X_{22}$(X=B+Si) as function of pressure at t = 25°C.

S. Filipek et al.

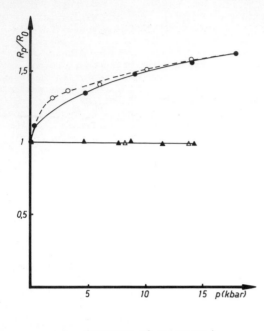

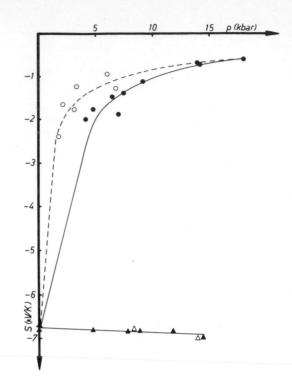

● – increase of pressure)
○ – decrease of pressure) in H_2

△▲ – increase and decrease of
    pressure (in inert medium)

Figure 12: Electrical resistance of glassy
Pd$_{83}$Si$_{17}$ as function of pressure
transmitted by hydrogen or by inert
medium at 25°C.

● – increase of pressure)
○ – decrease of pressure) in H_2

△▲ – increase and decrease of pressure
    (in inert medium)

Figure 13: Thermoelectric power of glassy
Pd$_{83}$Si$_{17}$ as function of pressure
transmitted by hydrogen or by inert
medium at t = 25°C.

ACKNOWLEDGEMENT

Thanks are due to Dr. A. Calka for providing
samples of metallic glasses.

REFERENCES

1. For example: M.L. Wayman, G.C. Smith,
   (1971)   J. Phys. Chem. Solids 32, 103
2) B. Baranowski, S. Filipek, Roczniki Chem. 45,
   1353 (1971)
3. B. Baranowski, S. Filipek, Roczniki Chem. 47,
   2165 (1973)
4. S. Filipek, B. Baranowski, Roczniki Chem. 49,
   1149 (1975)
5. E.G. Ponyatovskij et. al., Dokl. Akad. Nauk
   SSSR, 229 (2), 391 (1976)
6. V.E. Antonov et al., Dokl. Akad. Nauk SSSR,
   233 (6) 1114 (1977)
7. E.G. Ponyatovskij et al., Dokl. Akad. Nauk
   SSSR, 230 (3), 649 (1976)
8. V.E. Antonov et al., Dokl. Akad. Nauk SSSR,
   Techn. Fiz. 252 (6) 1384 (1980)

9. A.J. Maeland, L.E. Tanner, G.G. Libowitz,
   J. Less-Common Met. 74, 279 (1980)
10. T.B. Flanagan, S. Majchrzak, B. Baranowski,
    Phil. Mag. 25, 435 /1972)
11. T.G. Kollie, J.O. Scarborough, D.L. McElroy,
    Phys. Rev. B.2, 2831 (1970)
12. P. Leech, C. Sykes, Phil. Mag. 27, 742 (1939)
13. A.C. Switendick, private communication
14. T. Soomura, Ph. D. Thesis, Osaka University
    (1976)
15. B.S. Berry, W.C. Pritchet, J. Appl. Phys. 52
    (3) 1865 (1981)
16. A.C. Ehrlih, D.J. Gillespie, Report of NRI
    Progress, December 1974

PHYSICS OF SOLIDS UNDER HIGH PRESSURE
J.S. Schilling, R.N. Shelton (editors)
© North-Holland Publishing Company, 1981

PRESSURE DEPENDENCE OF THE ELECTRICAL RESISTIVITY OF SOME METALLIC GLASSES

G. Fritsch[*], J. Willer, A. Wildermuth, E. Lüscher

Physik-Department, Technische Universität München, D-8046 Garching, W.Germany
[*] ZWE Physik, Hochschule der Bundeswehr München, D-8014 Neubiberg, W.Germany

We have measured the electrical resistance of metallic glasses as a function of temperature between 1.5 and 300 K under pressures up to 13 GPa (130 kbar). Depending on the system, the resistivity is found either to increase ($Mg_{70}Zn_{30}$) or to decrease ($Pd_{30}Zr_{70}$, $Cu_{40}Zr_{60}$, $Ti_{50}Be_{40}Zr_{10}$). This behaviour is in contrast to the negative temperature coefficient of resistivity which all the amorphous alloys considered here exhibit. Our results are discussed within a simple semiempirical model.

1. INTRODUCTION

It has been shown [1,2] that the temperature dependence of the electrical resistivity of nonmagnetic amorphous alloys can be successfully described by the Ziman equation. This expression is based on the nearly free electron model (NFE-model) in the weak scattering limit. However, the mean free path of the electrons, as derived from the measured absolute value of the resistivity, turns out to be of the order of the inter-atomic distances. Therefore, it is difficult to understand the success of such a model. In order to get further information on this problem we would like to analyse the volume (pressure) dependence of the electrical resistivity ϱ. Data are obtained by varying the pressure up to 13 GPa. It will be shown that the simple Ziman formulation is sufficient to describe at least qualitatively the variation of ϱ with pressure of the alloys examined. Such a conclusion seems to be in accordance with the statement by Cochrane et al. [3] for the metglas 2204 ($Ti_{50}Be_{40}Zr_{10}$). Amorphous alloys containing metalloids like B, P or Si behave quite differently [3,4,5].

In order to compare theory and experiment, the compressibilities $\varkappa$ are needed. These numbers are partly available directly [3], partly may be derived from Young's modulus and Poisson's ratio [6]. In case of MgZn $\varkappa$ has been calculated from a concentration weighted average of the element values [7], since the densities are governed by a similar law [8]. However, no pressure dependence of $\varkappa$ has been included. Apart from the fact that it is not known, it should be small.

2. SOME THEORETICAL CONSIDERATIONS

In transition metal systems the resistivity in the Ziman formulation can be expressed as [2]:

$$\varrho = \frac{30\,\pi^3\hbar^3}{me^2k_F^2E_f\Omega} \cdot \sin^2[\eta_2(E_F)] \cdot S(2k_F), \qquad (1)$$

where k_F and E_F are the Fermi-wave vector and energy, respectively and Ω denotes the atomic volume. The quantity $\eta_2(E_F)$ describes the phase-shift of the electronic d-wave function in a scattering event. The other symbols have the usual meaning.

If one considers the T-dependence of ϱ, the only relevent contribution comes from the static structure factor, since the other quantities depend only weakly on T via the thermal volume expansion. This effect may be safely neglected. Such a conclusion is no longer valid when varying the pressure, since all factors in equ. (1) depend on volume V.

Whereas the T-dependence has been discussed widely in literature [1,2], the V-dependence has been not. Therefore, we would like to present this analysis for small relative volume changes: $\Delta V/V \ll 1$.

According to Nagel [9] the structure factor at $2k_F$ $S(2k_F)$ can be written to a good approximation as:

$$S(2k_F) = 1+[S_0(2k_F)-1]\exp[-2\{W(T,V)-W(0)\}] \qquad (2)$$

Here $S_0(2k_F)$ is the structure factor at T = 0 and $W(T,V) - W(0)$ denotes the Debye-Waller factor. For the latter we assume:

$$W(T,V) - W(0) = \frac{12\,\hbar^2k_F^2(V)}{2Mk_B\,\theta^2(V)} \cdot F(T), \qquad (3)$$

where F(T) contains T as a leading term plus higher powers in T. M is the atomic mass and $\theta(V)$ the Debye-temperature. θ can be defined by fitting the specific heat to a Debye-spectrum. In the crystalline case equ. (3) is the approximation for $T \gtrsim \theta$. However, we expect it to be valid also at lower temperatures $T < \theta$, because quantum effects in the phonon spectrum may be smeared out by the amorphous structure.

Taking equs. (1) to (3) and considering only V-dependent terms, yields within the NFE-model:

$$\varrho(V,T) = \varrho_0(V) + \varrho_T(V,T). \qquad (4)$$

Here, we have used the abbreviations:

$$\rho_0(V) = A V^{1/3} \sin^2 \eta_F(V) \cdot S_0(2k_F[V] \,) \tag{5}$$

and

$$\rho_T(V,T) = -A V^{1/3} \sin^2 \eta_F(V) \cdot [S_0(2k_F(V))-1]$$

$$\cdot B \left(\frac{k_F(V)/k_{F_0}}{\theta(V)/\theta_0}\right)^2 \cdot F(T) \; , \tag{6}$$

where the quantities A and B are constants.

$$A = \frac{30 \pi^3 \hbar^3 V_0^{-1/3}}{me^2(k_{F_0}^2 E_{F_0} \Omega_0)} \quad \text{and} \quad B = \frac{12 \hbar^2 k_{F_0}^2}{M k_B \theta_0^2} \; . \tag{7}$$

The index zero refers to a constant T or P, usually taken as T = 0 and P = 0. The Debye-Waller factor has been expanded in a power series. Only the first two terms have been taken into account, since $2[W(T,V)-W(0)]$ is smaller than about $5 \cdot 10^{-2}$.

Assuming $\Delta V/V$ to be small, we get from equ.(5):

$$\rho_0(V) \approx \rho_{00} \{1 + [1/3 + 2\Delta] \cdot \Delta V/V\} \quad , \tag{8}$$

where

$$\rho_{00} = A V_0^{1/3} \sin^2 \eta_F(V_0) \cdot S_0(2k_F) \tag{9}$$

and

$$\Delta = 2 [ctg \, \eta_F(V_0)] \cdot V_0 [d \, \eta_F(V)/dV]_{V_0} \; . \tag{10}$$

Since the first peak in S(k) at kp should roughly coincide with $2k_F$ and k_F as well as kp scale in the same way with V, $S_0(2k_F)$ should be independent from volume. Hence ρ_{00} is a constant.

Using the definition of the compressibility $\mathscr{æ}$, we can write:

$$\rho_0(P) = \rho_{00} [1 - S_0 \cdot P] \tag{11}$$

with

$$S_0 = [1/3 + 2\Delta] \cdot \mathscr{æ} \; . \tag{12}$$

The parameter Δ contains information on the scattering process. For d-electrons it can be estimated by assuming [10]

$$ctg \, \eta_F \sim E_F/V_s \quad , \tag{13}$$

where V_s is the scattering potential. If V_s does not depend on V, Δ is calculated within the NFE-model:

$$\Delta = 4/3 \cos^2 \eta_F \leq 4/3 \; . \tag{14}$$

A similar procedure for $\rho_T(V,T)$ yields from equ. (6):

$$\rho_T(V,T) \approx - \rho_{00} [S_0(2k_F)-1] \cdot B$$

$$\{1+[2\Delta+2\gamma_G-1/3] \cdot \Delta V/V\} \cdot F(T). \tag{15}$$

In order to derive equ. (15), we have used:

$$k_F(V)/k_{F_0} \approx 1-1/3 \cdot \Delta V/V \text{ and } \theta(V)/\theta_0 \approx 1-\gamma_G \cdot \Delta V/V.$$

The quantity γ_G is the Grüneisenparameter, defined as usually:

$$\gamma_G = -d \ln \theta(V)/d \ln V.$$

With the help of the compressibility we finally get:

$$\rho_T(V,T) \approx - \rho_{T0} [1-S_T \cdot P] \cdot F(T), \tag{16}$$

where

$$\rho_{T0} = \rho_{00} [S_0(2k_F)-1] \cdot B \tag{17}$$

and

$$S_T = [2(\Delta+\gamma_G)-1/3] \cdot \mathscr{æ} = S_0 + 2\mathscr{æ}(\gamma_G-1/3) \tag{18}$$

Finally, if we assume for F(T) the approximation

$$F(T) = T + D_0 \cdot T^n \tag{19}$$

with D_0 and n constants we get:

$$\rho(P,T) = \rho_{00}[1-S_0 P]-\rho_{T0}[1-S_T P]\cdot[T+D_0 T^n] \; . \tag{20}$$

Equ. (20) constitutes the basic equation for the pressure and temperature variation of the electrical resistivity ρ within the NFE-model.

The following conclusions can be derived directly from equ. (20):

$$\rho_{T0}/\rho_{00} = [S_0(2k_F) - 1] \cdot B \; . \tag{21}$$

The analysis of the data delivers the parameters Δ, γ_G and $S_0(2k_F)$. Since $\Delta \leq 4/3$ and γ_G as well as $S_0(2k_F)$ should be numbers around one, a test of the model is possible.

For binary or ternary alloys [11] equ. (1) has to be extended in order to incorporate the partial structure factors and the various pseudopotentials. However, for a qualitative analysis we would like to ignore this complication and replace the amorphous alloy by an artificial "element" based on average quantities.

3. EXPERIMENTAL RESULTS

We like to compare the results from the amorphous alloys $Mg_{70}Zn_{30}$, $Ti_{50}Be_{40}Zr_{10}$, $Pd_{30}Zr_{70}$ and $Cu_{40}Zr_{60}$ with the simple model outlined in Sect. 2. The data on the PdZr and CuZr alloys are published elsewhere [12]. Our results on MgZn and TiBeZr (metglas 2204) are given in Figures 1, 2 and 3 resp.

A least square fit to the data according to

$$\rho(P,T) = \rho_0(P) + \rho_1(P) \cdot T + \rho_2(P) \cdot T^{3/2} \tag{22}$$

proved to be very successfull, as can be seen from Figs. 1 and 2. By introducing the notations

$$\rho_i(P) = \rho_{i0}(1 - S_i P) \tag{23}$$

for i = 0 to 2 the results, given in Table I can be derived.

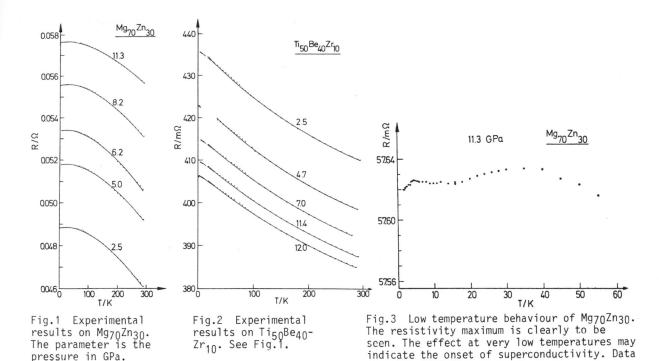

Fig.1 Experimental results on $Mg_{70}Zn_{30}$. The parameter is the pressure in GPa.

Fig.2 Experimental results on $Ti_{50}Be_{40}Zr_{10}$. See Fig.1.

Fig.3 Low temperature behaviour of $Mg_{70}Zn_{30}$. The resistivity maximum is clearly to be seen. The effect at very low temperatures may indicate the onset of superconductivity. Data from the 11.3 GPa run.

Table I   Results from a fit of equs.(22), (23) to the data

Alloy	$Pd_{30}Zr_{70}$	$Cu_{40}Zr_{60}$	$Ti_{50}Be_{40}Zr_{10}$	$Mg_{70}Zn_{30}$
$\rho_{00}/\mu\Omega cm$	148 ± 15	180 ± 20	247 ± 25	43 ± 4
$\rho_{10}/n\Omega cmK^{-1}$	$-(57 \pm 2)$	$-(60 \pm 10)$	$-(103 \pm 3)$	9 ± 2
$\rho_{20}/n\Omega cmK^{-3/2}$	1.7 ± 0.1	1.8 ± 0.1	2.8 ± 0.2	$-(1.2 \pm 0.2)$
$S_0/GPa^{-1}\cdot 10^2$	1.05 ± 0.07	1.3 ± 0.1	0.9 ± 0.1	$-(1.1 \pm 0.1)$
$S_1/GPa^{-1}\cdot 10^2$	3.2 ± 0.1	2.9 ± 1.2	2.7 ± 0.2	7 ± 2
$S_2/GPa^{-1}\cdot 10^2$	3.1 ± 0.4	3.0 ± 0.7	3.6 ± 0.3	6 ± 2
$\varkappa/m^3 J^{-1}\cdot 10^{11}$	1.0 [7]	1.0 [6]	1.02 [6]	2.5 [7]

In view of the data, the $\rho_i(P)$ are given in Figs 4 and 5, only a linear pressure dependence has been assumed.

4. DISCUSSION

By comparing equ. (22) with equ. (20) we find the indentifications:

$$\rho_{10} = -\rho_{T0} \ , \quad \rho_{20} = -D_0\rho_{T0} \text{ and } D_0 = \rho_{20}/\rho_{10}. \quad (24)$$

In addition we have

$$S_1 = S_2 = S_T \text{ as long as } D_0 \neq D_0(P) \quad (25)$$

The parameter Δ can be calculated from equ.(12).

$$\Delta = (S_0/\varkappa - 1/3)/2 \quad (26)$$

and from equ. (18) we get for the Grüneisen-parameter γ_G:

$$\gamma_G = 1/3 + (S_1-S_0)/(2\varkappa) \quad (27)$$

Finally, equ. (21) yields:

$$S_0(2k_F) = 1 - (\rho_{10}/\rho_{00})/B \ . \quad (28)$$

The numbers, derived from equs.(26) to (28) together with the Fermi vector k_F and the Debye temperature θ are summarized in Table II.

Since S_1 is almost equal S_2 for PdZr, CuZr and MgZn the quantity D_0 is independent of pressure in these cases. This is, however, not true for TiBeZr.

The d-electron alloys show a parameter Δ in accordance with equ. (14). Hence, the NFE-model is not in contradiction with theory. The MgZn alloy which contains only such p-electrons exhibits $\Delta < 0$. This is a consequence of

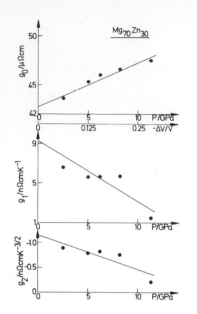

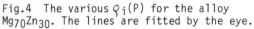

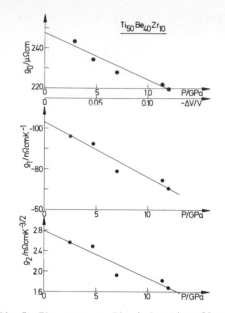

Fig.4 The various $\varrho_i(P)$ for the alloy Mg$_{70}$Zn$_{30}$. The lines are fitted by the eye.

Fig.5 The same as Fig.4 for the alloy Ti$_{50}$Be$_{40}$Zr$_{10}$.

Table II Comparison between experiment and model

Alloy	Pd$_{30}$Zr$_{70}$	Cu$_{40}$Zr$_{60}$	Ti$_{50}$Be$_{40}$Zr$_{10}$	Mg$_{70}$Zn$_{30}$
Δ	0.4 + 0.1	0.5 + 0.1	0.3 + 0.1	-(0.4 + 0.1)
γ_G	1.4 + 0.1	1.2 + 0.4	1.2 + 0.1	1.9 + 0.4
$S_0(2k_F)$	2 + 0.1	1.8 + 0.2	3.5 + 0.3	0.4 + 0.2
$D_0/K^{-1/2}$	-(0.03+0.003)	-(0.03+0.006)	-(0.027+0.003)	-(0.13+0.04)
k_{F0}/nm^{-1}	14 [8]	14 [15]	13.8 [22]	14.1 [14]
θ_0/K	180 [13]	180 [13]	425 [22]	295 [14]

of the increase of the resistivity with pressure.

The Grüneisenparameters γ_G are between one and two in all cases. It should be mentioned that an analysis of the pressure dependence of the superconducting transition temperature T_c in case of PdZr and CuZr [13] yields similar values.

Also the magnitude of $S_0(2k_F)$, governing the ratio of the constant to the linear T-dependend part is surprisingly close to the expected value, at least for the Zr-alloys.

The negative sign of D_0 indicates that the $T^{3/2}$ term has always the opposite sign to the linear one. Whereas in the Zr-alloys the linear term is always negative due to the dominance of the elastic scattering, it is just the other way around for MgZn. This is not surprinsing, since MgZn has no relevant d-electrons. Therefore, the assumption that the main contribution to the resistivity comes from the region around $2k_F$ is no longer valid. We must have an appreciable contribution from inelastic "phonon"-scattering, which makes the linear term in T positive. This fact is also reflected in the number for $S_0(2k_F)$ which is smaller than one. Hence, the quantities $S_0(2k_F)$ and Δ are averaged over the whole range of k, from 0 to $2k_F$. Nevertheless, the description of the resistivity by the model outlined above seems to remain valid for averaged quantities. The special behaviour of MgZn is also the reason for the small maximum in $\varrho(T)$ (see Fig.3) at low temperatures. Here the positive linear term dominates, but at higher temperatures it is overwhelmed by the negative $T^{3/2}$-term.

We will conclude the discussion with two remarks. First, the deviation of $\varrho(T)$ from the extrapolated fit-curves at the lowest temperatures is due to superconductivity. As can be shown [16], superconducting fluctuations above T_c yield a negative curvature in the $\varrho(T)$ curves. If these fluctuations are quenched, the resistivity continues to rise further. Secondly, our conclusions are based on data up to T = 300 K. It is not our claim, that $\varrho(T)$ may be described by equ. (22) up to the crystallization temperature. Higher powers of T can play an ever increasing role.

5. CONCLUSION

The pressure dependence of the electrical resistivity of amorphous alloys, containing no metalloids, can be described qualitatively by the NFE-model using the Ziman formulation. This statement supports earlier conclusions derived from the temperature dependence of the resistivity [2] and of the thermopower [17]. This analysis does not work definitely for the Fe-B-alloys [5]. Those alloys should be quite different in structure, if one assumes for the moment that the NFE-model does not break down. Magnetic contribution apart from other effects should only influence the $T^{3/2}$-term and higher powers in T [19]. Recent e^+-annihilation experiments [19] and determinations of the partial structure factors [20] arrive at similar conclusions. The question why the Debye-Waller factor should contain a $T^{3/2}$-term is yet unsolved. There are certain experimental hints, that this might be possible [21].

ACKNOWLEDGEMENT

We would like to thank Prof. M. Kalvius for giving us the opportunity to use his apparatus. We appreciate very much the help of Prof. H.-J. Güntherodt, Prof. J. Hafner and Prof. W. Triftshäuser with the samples. Financial support by the DFG (Lu 109/21) is acknowledged.

LITERATURE

1 Glassy Metals I, Topics in Applied Physics Vol.46, Eds. H.-J. Güntherodt and H. Beck, Springer Verlag, Berlin, Heidelberg, New York (1981)

2 S.R. Nagel, Phys.Rev. B16 (1977) 1694

3 R.W. Cochrane, J.O. Strom-Olsen et al. Solid State Com. 35 (1980) 199

4 D. Lazarus, Sol.State Com. 32 (1979) 175

5 G. Fritsch, J. Willer, H. Schink, E. Lüscher, to be published

6 H.S. Chen, J.Appl.Phys. 49 (1978) 462

7 K.A. Gschneidner, Jr. in Solid State Phys. Vol.16, Eds. F.Seitz, D.Turnbull, Academic Press New York (1964)

8 G.R. Gruzalski, J.A. Gerber, D.J. Sellmyer, Phys.Rev. B19 (1979) 3469

9 S.R. Nagel, Phys.Rev.Lett. 41 (1978) 990

10 K. Fischer, phys.stat.sol. (b), 46 (1971)11

11 R. Evans, D.A. Greenwood and P. Lloyd, Phys.Lett. 35A (1971) 57

12 J. Willer, G. Fritsch, E. Lüscher, to appear in J.Non-Cryst.Solids

13 J. Willer, G. Fritsch, E. Lüscher, Appl. Phys.Lett. 36 (1980) 859

14 U. Mizutani, T. Misoguchi, J.Phys.F11 (1981) 1385

15 H.S.Chen, Y.Waseda,phys.stat.sol.51(79)593

16 J. Willer, G. Fritsch, E. Lüscher, to be published

17 S.R. Nagel, Phys.Rev.Lett. 41 (1978) 990

18 R. Richter, W. Wolf, F. Goedsche, phys. stat.sol. b95 (1979) 473

19 W. Triftshäuser, private communication

20 S. Steeb, private communication

21 T.J. Bastow, Phys.Lett. 60A (1977) 487 K. Nishiyama, D. Riegel in Hyperfine Interactions 4, p.490, North Holland Publ.Co. Amsterdam(1978)

22 R. Clarke, S.R. Nagel, R.L. Hitterman, M.H. Mueller, Sol.State Com. 36 (1980) 751

PHYSICS OF SOLIDS UNDER HIGH PRESSURE
J.S. Schilling, R.N. Shelton (editors)
© *North-Holland Publishing Company, 1981*

ELECTRONIC STRUCTURE OF THE ACTINIDE METALS

Börje Johansson

Institute of Physics
University of Aarhus
8000 Århus
Denmark

Hans L. Skriver
Research Establishment, Risö
4000 Roskilde
Denmark

and

Ole Krogh Andersen
Max-Planck-Institut für Festkörperforschung
7000 Stuttgart 80
Federal Republic of Germany

We discuss some aspects of the electronic structure of the actinide metals, both at ambient conditions and under compressions.

1. INTRODUCTION

The lanthanide and actinide elements are obtained by the gradual filling of the 4f and 5f shells, respectively. Since for the lanthanides the 4f shell is located deeply within the atom, it follows that the outer valence regions of the different lanthanide atoms are very similar to each other. In the metallic state the open 4f electron shell retains an integral occupation number and gives rise to a well-defined localized magnetic moment.[1,2] Thus, there are two types of electronic states in the rare-earth metals:[3] a) a normal transition-metal-like conduction band state of (sd) type, and b) a localized f state. Since the valence (bonding) (sd) electron density distributions will be very similar to each other for the various lanthanides, one expects only a gradation of the bonding characteristics when one proceeds through the series. The number of (sd) valence electrons is three except for europium and ytterbium, where only two electrons participate in the (sd) bonding. For these two anomalous metals the third valence electron goes into the f shell and forms a localized $f^7(^8S)$ and $f^{14}(^1S)$ configuration,[4] respectively.

One must, however, be cautious in attempting to apply the same type of picture for the 5f electron orbitals in the actinide series of elements, because although the 5f shell is fairly localized it forms a more extensive part of the actinide atoms than does the 4f shell in the lanthanide atoms.[5,6] As we shall see, this is of uttermost importance for the electronic structure and bonding properties of the earlier actinide metals. In the latter part of the 5f series, however, the usual shell contraction has diminished the spatial extension of the 5f shell to the extent that it becomes comparable to that of the 4f shell for the lanthanide elements.[7] For heavier actinides we might therefore expect a normal rare-earth-like behaviour where the solids are becoming analogous to the lanthanides. Still, this will only be partially true since a predominance of divalent metals is expected in the end of the actinide series.[8,9] However, in as far as one only implies a localized f-electron behaviour, the heavier actinide metals can be said to form a second rare-earth series.

The fact that the 5f orbitals in the actinides are relatively more extended than the 4f orbitals in the lanthanides originates from their orthogonality to the 4f core. This orthogonality can

effectively be looked upon as a repulsive contribution to the atomic potential experienced by
the 5f electrons. For the same reason a similar
and perhaps more familiar difference in orbital
extensions occurs between the 4d and 3d orbitals
in the 4d and 3d transition series or between
the 3p and 2p orbitals in the third and second
period. Thus the expansion of the 5f shell relative to the 4f shell is a general feature of
an $(n+1)m$ shell relative to the corresponding
nm shell in the Periodic System.

That there is a distinct difference between the
electronic structure of the 5f and 4f series of
metals is very well illustrated by the behaviour
of the metallic radius. In Fig. 1 we have plotted the experimental metallic radii for the two
series.[10,11] The smooth variation in the rare-
earth series (except for Eu and Yb due to their
divalent state) has no counterpart among the actinides. Instead, the actinide behaviour has a

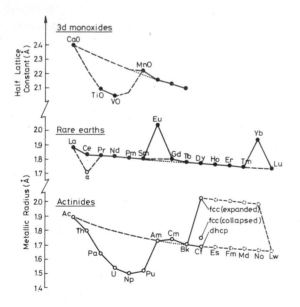

<u>Fig. 1</u> – Metallic radii for the lanthanides
          and actinide elements. For the 3d
          monoxides half the lattice constant
          is plotted (NaCl-structure).

great similarity to the variation of the metallic radius for the 4d (or 5d) transition series,
at least for the earlier elements in the 5f

series.[11] This similarity was originally taken
as a support for the picture of thorium, protactinium, uranium and neptunium as forming a
<u>6d transition series</u>, having no occupation of
the 5f shell and thus showing an electronic
structure of tetra-, penta-, hexa- and heptava-
lent transition metals,[12] respectively. Only
when arriving at the element plutonium some occupation of the 5f shell had to be introduced
and the 5f electron(s) was then assumed to be
of non-bonding localized rare-earth type. However, this "valence picture" was not universal-
ly accepted. For example the exceptional crystal
structures found in uranium, neptunium and plutonium led Friedel[13] to propose that there is a
substantial covalent (df) bonding present in
these metals. Later, non-relativistic energy-
band calculations by Hill and Kmetko[5] and relativistic calculations by Freeman and Koelling[6]
gave good evidence for that the 5f electrons are
behaving quite differently in the actinides
than the 4f electrons in the lanthanides. The
continued efforts by Freeman and Koelling,[14]
which culminated in a band-structure calculation for the α-uranium structure,[15] have given
clear indication for a metallic 5f electron behaviour for the earlier actinides. Quite a number of other independent investigations have
pointed in the same direction. A most direct
evidence of 5f bonding was obtained from an
analysis of the experimental cohesive energies
of the earlier actinide metals.[9] Also the behaviour of the room temperature entropy of the
actinide metals was shown to be very different
than that of other series of elements in the
Periodic Table.[16] Especially, no magnetic contribution to the entropy was found in the analysis. In addition, direct susceptibility measurements for the earlier actinide metals have
shown the absence of local magnetic moments.[17-19]
A very direct demonstration of the presence of
5f electrons in uranium metal has been obtained
from its valence band photoelectron spectrum.[20,21] Also the measured binding energies
of the core electrons can only be understood if

there is a substantial occupation of the 5f level.[22,23]

The rather intensive experimental and theoretical work on the actinides during the last decade has clearly shown that they form a series of elements that is absolutely unique among the Elements. Indeed, from the work that we somewhat schematically will review below,[24-26] it has become evident that the bonding in the elements Pa-Pu is dominated by the 5f electrons. Thus, at ambient conditions these actinides are the only elements in the Periodic System, where f electrons are the main bonding electrons, a circumstance which can be directly correlated with the abnormal crystal structures found in these elements. The 5f electrons, in fact, here play very much the same role as the d electrons for the transition metals. Since the 5f electrons prefer very short bonds,[26] one can at least qualitatively understand the general occurrence of rather exotic and distorted structures in the (P,T) phase diagrams of these elements. A quantitative understanding of the origin of these structures is, however, still lacking and it will certainly take some time before, for example, the details of the extremely complicated phase diagram of plutonium[27] will be clarified. Nevertheless, from the picture of an f-electron dominated bonding a realistic basis for further developments is now available.

Arriving at americium in the actinide series, a very drastic change of properties occurs. This is for example clearly demonstrated by its equilibrium atomic volume (Fig. 1). The crystal structure of americium is dhcp[28] (double hexagonal close-packed), a structure which otherwise is found only among the lighter lanthanide metals. Already this fact gives a very strong evidence for that in this particular actinide element a normal rare-earth-like behaviour is reappearing in the series.[4] As a trivalent metal, it will have an $5f^6$, $J = 0$, non-magnetic ground state. This led to the prediction of superconductivity,[4] which later was found experimental-

ly.[29] This in combination with its non-magnetic susceptibility nicely confirms that the rare-earth type of behaviour starts with the americium metal in the actinide series.

Since the 5f electrons are localized in americium but itinerant in the preceding element plutonium, one can look upon this change as a Mott transition[7,30] (as a function of atomic number). A similar change is well-known among the 3d monoxides (compare Fig. 1 and Refs. 11 and 31). Therefore, it was suggested that high pressure would convert the localized 5f electrons in americium into an itinerant state.[32] This expectation was based on the simple fact that a decrease of volume would compensate for the 5f orbital contraction. Calculations showed that this should occur already at about 100 kbar.[25] This now seems to have obtained at least partial experimental support, which will be more explicitly discussed below.

In the present contribution we will first concentrate ourselves to a review of the main features of the present understanding of the electronic structure of the actinide metals at zero pressure. For the heavier rare-earth-like actinides we will thus consider the question about which metallic valence state will be the stable one, but also briefly indicate what effect high pressure should have on the electronic structure of these materials. For the earlier actinides we will discuss calculations where the deep influence on the bonding of the 5f electrons is clearly exhibited. Indeed, this metallic 5f bonding accounts quite well for the initial variation of the atomic volume in the actinide series. Furthermore, we will present calculations where the 5f localization is shown to take place between plutonium and americium. As expected, the main reason for this dramatic change is found to be due to the decreased 5f band width in americium as compared with plutonium. From this it becomes immediately evident that under high pressure quite spectacular changes would be expected for americium, since compression would give rise to an increased 5f band width and thus facilitate the possibility of a delocalization of the atomic-like $5f^6$ configuration. An investigation of the behaviour of the americium metal under pressure therefore becomes a central question in connection with the present conference.

The obvious next question to pose is: What will happen with the elements following americium in the series, namely curium and berkelium, when subjected to compression? Firstly, it should be emphasized that when americium, curium and berkelium are in the localized $5f^n$ state, they will behave very similarly to what is expected of a trivalent rare-earth-like metal.[4] Thus, as long as the pressure does not exceed the critical one for 5f delocalization, we have from the structural point of view no reason to expect anything new in comparison with the well-known behaviour of compressed, trivalent lanthanide metals. As will be pointed out below, there might, however, be a very interesting exception to this for the berkelium metal due to a possible change from trivalent to tetravalent behaviour before the 5f delocalization comes into operation! The mentioned deep relationship between the trivalent lanthanides and actinides was emphasized in Ref. 4, where it was argued that americium should be most similar to praseodymium in the lanthanide series. This connection has obtained further support from recent high pressure studies.

From the discussion that will follow about the actinide electronic structure, it should hopefully become clear that high pressure studies will continue to be of uttermost importance for the further elucidation of the physical properties and the fundamental electronic structure of these unique materials. It has to be kept in mind, however, that only a very few specially equipped laboratories can handle the health hazards involved when dealing with these generally highly radioactive actinide materials. However, advances in the theoretical[30,33-35] and experimental studies[36-38] of the earlier rare-earth metals point to that itinerant f states can also be found for these elements. Here, no special precautions in the handling are necessary, except those required due to their strong chemical reactivity. Thus, itinerant f states and f-electron delocalizations are phenomena available for study in any high-pressure laboratory.

2. VALENCE STATE IN THE HEAVIER ACTINIDE METALS

As mentioned in the Introduction, for the heavier actinides we expect a rare-earth-like behaviour with localized magnetic 5f configurations. For the lanthanides it has become well estbalished that the localized 4f electrons have no influence on the bonding of these metals.[4,39] Therefore, as far as the cohesion of these elements is concerned, the 4f electrons can be regarded as part of the ionic core. The very close interrelation between the electronic structure of the lanthanide metals has for example been amply demonstrated by a construction of a generalized structural phase diagram,[4] where all the trivalent metals have been included. Thus, for the individual elements the typical crystallographic structures (like the hcp→Sm-type→dhcp→fcc structure sequence[40]) are exhibited independent of the number of 4f electrons in the element. Also alloys between different lanthanide metals show the same crystal structure sequence as a function of the alloying concentration. It is of special significance that this structure sequence is found in the binary phase diagram between lanthanum and yttrium, i.e., between two elements possessing no f electrons. This clearly demonstrates the non-influence of the 4f electrons on the crystal structures. From this follows that the heavier actinides should be expected to behave in essentially the same way. Among the lanthanides, europium and ytterbium are of exceptional character due to their divalent metallic state. The question then arises whether or not there will be similar exceptional metals among the heavier actinides.

As it turns out, this question can be studied quite accurately but still in rather simple terms.[4,9] That this is possible is due to the inert character of the localized f electrons. The procedure to be used will heavily rely on regularities shown in the behaviour of the cohesive energy among the Elements. In Fig. 2 we have depicted some representative trends for the cohesive energy[41] of the elements in the left part of the Periodic Table. For divalent metals (like barium and strontium) the energy gained in the

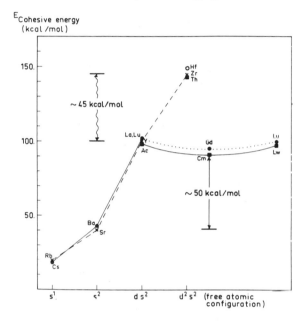

Fig. 2 — Characteristic behaviour of the cohesive energy for elements to the left in the Periodic System.

solid relative to the free atom is consistently close to 40 kcal/mol. For trivalent metals (like lanthanum and yttrium) the corresponding quantity is about 100 kcal/mol. Similarly, tetravalent metals (like zirconium and hafnium) all have a cohesive energy of about 145 kcal/mol. For the rare-earths with non-bonding f electrons, one therefore expects a cohesive energy close to 100 kcal/mol relative to the trivalent atomic state $(f^n ds^2)$. However, a

slight decrease can be noticed from the experimental value for gadolinium $(E_{coh}(Gd) = 95$ kcal/mol). (Note that gadolinium as a free atom has the $4f^7 5d6s^2$ configuration.) This decrease is due to the multiplet coupling between the open 5d and 4f shell electrons in atomic gadolinium but which is absent in the metallic state.[39] A similar but still relatively small coupling between the 4f and 5d shells will also be present for the other trivalent lanthanide atoms. (The details of this effect have been treated in Ref. 39, where it was explicitly shown that the bonding energies of the lanthanides are independent of the filling of the 4f shell.) To account for this purely atomic effect in an average way we ascribe to the lanthanides a cohesive energy of 95 kcal/mol relative to the atomic $f^n ds^2$ state. Similarly, the measured cohesive energy for the actinide element curium is about 90 kcal/mol,[42] a value which we will take as representative for the cohesive energies of the trivalent actinide metals relative to their trivalent $5f^n 6d7s^2$ atomic configuration.

The important thing to notice here is that the cohesive energy for the trivalent rare-earths (relative to the trivalent atomic configuration) is about 55 kcal/mol larger than for the divalent metals. This number appears to be about 50 kcal/mol for the actinides. The corresponding difference between trivalent and tetravalent metals is close to 45 kcal/mol (Fig. 2).

From these general numbers it is now easy to construct a direct energy comparison between the various possible valence states in the metallic state of the f element in question. This is illustrated in Fig. 3. Imagine that we start out from the divalent f^{n+1} state of the metal and then atomize the metal to the divalent atomic state $(f^{n+1} s^2)$. In this process we lose the energy A (≈ 40 kcal/mol). After this we excite the free divalent atoms into the trivalent atomic configuration $(f^n ds^2)$. Then, finally, we bring these prepared atoms together to form the trivalent metal, thereby gaining the cohesive

energy A + 55 (lanthanides) or A + 50 (actinides).
From this it now becomes a trivial matter to derive the preferred valence state of the condensed phase. In Fig. 4 we have plotted the

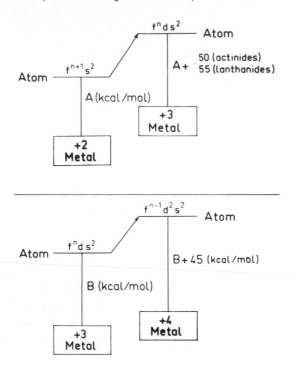

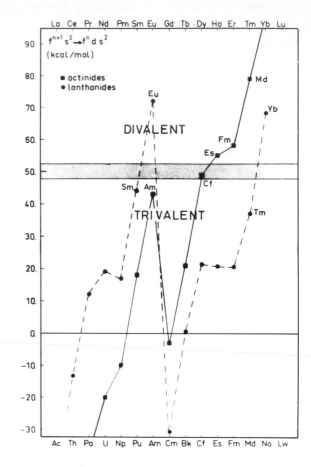

Fig. 3 – Energy comparison between the divalent and trivalent metallic
states (upper part) and trivalent
and tetravalent metallic states
(lower part) for the f elements.

atomic excitation energy $f^{n+1}s^2 \to f^n ds^2$ for the
lanthanides[43] and actinides[44,45] and we have also included the critical value for a valence
change between the di- and trivalent metallic
states. (For clarity, only the critical value
for the actinides is shown.)

First of all we note from Fig. 4 that europium
and ytterbium should be the only divalent metals
among the lanthanides. This finding agrees nicely
with well-established experimental facts. For the
actinides we find that the "twin" element to europium, i.e., americium, should be trivalent.
Also this is in agreement with the experimental
experience. It might here be remarked upon that
the common view on the reason for why europium

Fig. 4 –   Energy comparison between the
atomic excitation energy,
$f^{n+1}s^2 \to f^n ds^2$, and the critical
value separating the divalent
and trivalent metallic states.

and ytterbium are divalent is that it is due to
the strong preference of forming a half-filled
and filled 4f shell, respectively. From Fig. 4
we see that this is indeed true. However, the
present analysis gives a quantification of the
competition between the divalent and trivalent
metallic states. Thus the important thing is the
magnitude of the energy separation between the
f^{n+1} and $f^n d$ configurations, namely whether this
quantity in the atom exceeds the critical value
55 kcal/mol or not. In the actinides we find
from the atomic excitation energies a similar
preference of the half-filled shell in americium
(compare Fig. 4), but here the effect is not

sufficiently pronounced to maintain a divalent state in the condensed phase! Thus, the somewhat smaller energy separation between the f and d shell in americium as compared with europium is the fundamental reason for why americium in its condensed phase has a totally different appearance than its rare-earth analogue, europium.

For the heavier actinides we see from Fig. 4 that not only nobelium (the analogue of ytterbium) but also mendelevium and probably fermium and einsteinium should be divalent metals. Unfortunately, the extremely limited amount available of these elements has so far prevented a study of No, Md and Fm in their elemental metallic states (for Es see below). An extremely interesting case is californium, which according to Fig. 4 is found to be situated just on the borderline between divalent and trivalent metallic behaviour. The experimental study in Ref. 46 seems to confirm its close neighbourhood to a valence instability. Depending on the preparation conditions, various forms of the californium metal were obtained showing divalent, intermediate valent and trivalent behaviour (cf. Fig. 1). These experiments were, however, performed on a microgram scale, whereas the more recent heat of vapourization experiments were made on considerably larger samples.[47] These latter experiments showed that californium is a trivalent metal. It is, nevertheless, of great interest to point out that the samarium metal, which is trivalent in the bulk, has been found to have a divalent surface.[48-50] The close neighbourhood of californium to a divalent state then implies that the surface of californium must be divalent, which to some extent might explain the microscale experimental results in Ref. 46. Recently, microtechniques were used to prepare thin films of einsteinium.[51] Electron diffraction studies showed that the crystal structure was fcc (like in ytterbium) and that the lattice constant was typical for a divalent metal. This nicely confirms the energy comparison in Fig. 4. From the experience with californium one should, however, be cautious before

concluding that also <u>bulk</u> einsteinium is divalent, although it seems most likely. In this connection it should be brought to attention that even high pressure experiments have been attempted for californium,[52] but the presently available results seem to indicate that the samples employed might have been contaminated. Still, this serves to show that it is not unrealistic to start considerations about what can be achieved from high pressure experimental studies of heavy actinide materials. Thus, if bulk einsteinium and fermium turns out to be divalent at ambient conditions, it then directly follows from Fig. 4 that only very modest compressions should be required to induce a transformation to the trivalent state in these two elements. On the other hand, for mendelevium, and especially for nobelium, considerable compressions would be required before a valence change could be pressure induced.

From Fig. 4 it is also clear that curium and berkelium will be trivalent metals. The experimental finding that all the elements Am-Bk crystallize in the dhcp structure[10] constitutes a strong confirmation of this expected trivalent behaviour.

The main difference between the lanthanides and actinides is thus the steeper variation of the f-d separation through the actinide series. This first of all makes americium trivalent but also implies that a <u>series</u> of elements in the end of the actinide series will be divalent. Indeed, the preference of the divalent state in nobelium is so strong that among its halide compounds probably the only trivalent compound that will form is the trifluoride.[32]

The energy difference between the trivalent and tetravalent metallic states (again assuming nonbonding f electrons) can be investigated in the same way as above for the divalent-trivalent case. The pertinent energy comparison is made in Fig. 5. For the lanthanides it is obvious that none of them can be a tetravalent metal (not even cerium[30]). For the actinides we first of

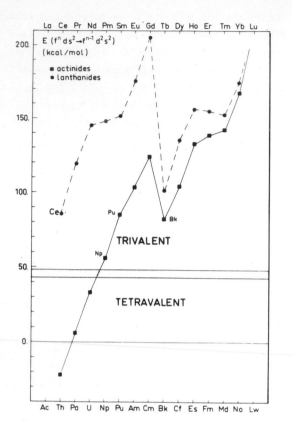

Fig. 5 – Energy comparison between the
atomic excitation energy $f^n ds^2 \rightarrow$
$f^{n-1}d^2s^2$, and the critical value
separating the trivalent and
tetravalent metallic states.

all find that americium, curium and berkelium
all prefer the trivalent state confirming the
lanthanide character of these metals. However,
it is of special significance to note that neither neptunium nor plutonium could be tetravalent metals. Thus, the present valence picture
shows that a trivalent state is to be expected
for both these metals (compare also Fig. 4). But
from the plot of the metallic radius of the actinide series, Fig. 1, it is obvious that neither plutonium nor neptunium can be simple trivalent (non-bonding f) metals. Therefore, something in our applied scheme must have gone wrong.
Certainly the energy comparison by itself must
be correct, but it should be recalled that the
method required non-bonding f electrons. The
found inconsistency clearly points to a more in-

volved and active participation of the 5f electrons in the cohesive properties of neptunium
and plutonium. This important aspect will be
treated in some detail in the next section.

3. ITINERANT 5f ELECTRONS

In recent years it has become increasingly
clear[53],[54] that a number of ground-state properties of metals can be described accurately using
a one-particle picture[55] and a local approximation for the exchange-correlation energy functional.[56] Starting from this reduction of the
original many-body problem to a self-consistent
band-structure problem, further computational
and conceptual simplifications may be introduced.
Thus for example the so-called Linear Muffin-Tin
Orbitals (LMTO)[57] method has been applied successfully to calculate various bulk properties
of the transition metals.[54],[58] From the given
description of the actinides the expectation of
itinerant 5f electrons immediately suggests that
this method should also be applicable to at least
the earlier actinide elements. From the expectation of a 5f localization in the series, one
should also be aware about that one here en-
counters a situation, where the band theory is
pushed to its limits of applicability. However,
in order to simulate the possibility of localization of the 5f electrons, a ferromagnetic spin
polarization is allowed for in the calculations.
The theoretical calculations to be reviewed below were all performed for the fcc crystal structure, were fully hybridized and included s, p, d
and f partial waves. Moreover, the relativistic
shifts were included, the spin-orbit coupling
was neglected, and the atomic radon cores were
frozen. The electronic pressure was calculated
using the relation given by Pettifor.[59] From
this the zero pressure atomic volume was calculated. In Fig. 6 we compare the so calculated
equilibrium atomic volumes with those found experimentally in the most dense crystal structures.

From this figure we note that the agreement between theory and experiment is very satisfactory for thorium, protactinium and uranium. It is only when we arrive at neptunium and plutonium that we observe significant deviations between the calculated and experimental equilibrium volumes. The paramagnetic (non-polarized) calculation for americium shows a serious discrepancy

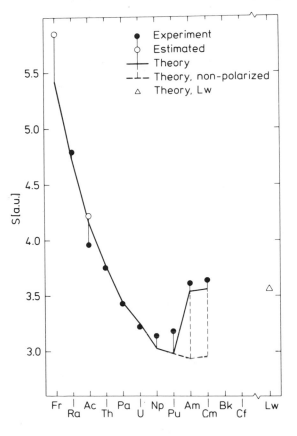

Fig. 6 - Comparison between the experimental and the calculated Wigner-Seitz radius, S, for the actinide metals. The estimated value for Ac was obtained from Ref. 12.

with the experimental value. The found deviations for neptunium and plutonium might be due to correlation effects not included in the present one-particle scheme and it is significant that the binding is overestimated. Both neptunium and plutonium are known to behave anomalously in most of their physical properties and are close to being magnetic,[18] which is indicative of strong f-electron correlations. Another

possibility is that the presently applied local spin density approximation for exchange and correlation might give rise to a slight error in the energy position of the f-band relative to the d-band. Finally, the omission of the spin-orbit interaction might also be of some significance for the f-electron bonding.

In agreement with experiment, the present calculations give no spin-polarized solution for neptunium and plutonium at zero pressure, but for somewhat expanded volumes (negative pressure) a net magnetization is calculated. Arriving at americium we find that the completely 5f spin-polarized solution exists at normal pressure and has the lowest energy. In Fig. 6 this is seen to have a dramatic effect on the equilibrium volume for americium, indicating an almost non-bonding character of the 5f electrons. Furthermore, we observe that allowing for this polarization the calculated atomic volume agrees well with the experimental value. Similar results are obtained for curium.

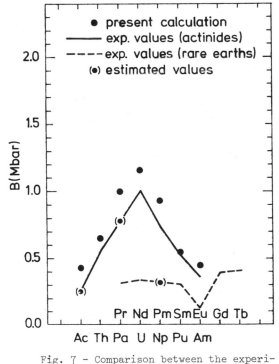

Fig. 7 - Comparison between the experimental and calculated bulk modulus for the actinide metals.

In Fig. 7 we compare the calculated bulk modulus with experiment.[60] Also here a good general agreement is obtained. In particular, the spin-polarized calculations for americium give a bulk modulus close to the experimental value, a value which is also close to typical for the rare-earth metals. The finding that both the experimental equilibrium volume and the bulk modulus are quite well reproduced by the calculations means that the equation-of-state of the actinides is fairly well described. (At this point it should be stressed that there are no free parameters available in the theory.) Still it is true that the calculations are performed for the simple fcc structure, while in reality much more complex crystal structures are met with in these actinides. Nevertheless, the course features of the PV-relation are quite well accounted for and the calculated electronic structure should provide a realistic basis for future investigations of the fundamental origin of the anomalous crystal structures.

In the LMTO-method the partial contribution to the pressure might readily be investigated. The results of such an analysis are shown in Fig. 8.

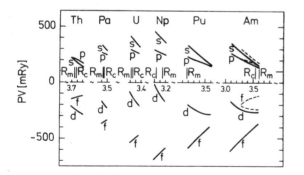

Fig. 8 – Partial contributions to the pressure from the s,p,d and f partial waves. The dashed lines represent the spin-polarized calculation for Am. The measured and calculated Wigner-Seitz radii are denoted by R_m and R_c, respectively.

First we note that even in thorium there seems

to be some significant contribution to the bonding from the 5f states. However, it should be kept in mind that this does not necessarily imply that the f orbitals are occupied and instead is due to the fact that the partial-wave analysis of the tails of the s,p, and d orbitals will give rise to a small f-partial-wave contribution. Going to the following elements in the series, the contribution to the binding increases strongly as a result of the progressive occupation of the 5f orbitals. Due to the narrowing of the 5f band this binding contribution seems to saturate between neptunium and plutonium. The unpolarized calculations for americium and curium (Fig. 6) also show the effect of reaching the half-filled band situation, namely that of a maximal use of the available bonding part of the f band. For the elements beyond curium, the anti-bonding part of the f band becomes occupied and a parabolic-like type of variation of the atomic volume should develop as it does for the d transition elements. In Fig. 8 it should be especially noticed that while for the unpolarized calculation in americium there is an appreciable f contribution to the binding, the fully spin-polarized f-state has essentially a non-bonding character. From this point of view the totally polarized solution represents correctly the localized non-bonding f^6 state in americium.

The calculated state density shows essentially no occupation of the f band in thorium. For the next, element, protactinium, the f band intersects the Fermi energy and the f occupation number, n_f is close to 1. Thus, protactinium is the first element in the actinide series where a 5f-band electron actively takes part in the bonding. Therefore, the filling of the 5f level in the actinide series does not start in the element following actinium, in contrast to the lanthanides, where the 4f level is becoming occupied in cerium, the element after lanthanum (compare also with Fig. 5). In uranium the calculated value of n_f is somewhat larger than 2 and for neptunium and plutonium the n_f occupation numbers are about 4 and 5, respectively. For the spin-polarized solution in americium we obtain the occupation numbers $n_{f\uparrow} \simeq 6$ and $n_{f\downarrow} = 0$. If the corresponding spin moment is combined with the orbital moment, the total moment is equal to zero.

The calculated f-band width (at the experimental equilibrium volume) is comparatively large in actinium, being about 5 eV. In the elements thorium-neptunium it is about 3 eV, and for plutonium somewhat larger than 2 eV. In americium it is less than 1 eV. The origin of this contraction of the 5f band width is of course the usual orbital contraction in a series of elements.

At this point it might be useful to investigate in some detail the origin of the f-band contribution to the bonding. The partial f pressure, which in Fig. 8 is seen to be negative, i.e., attractive, can be approximately represented by the following expression[61] (where s is the Wigner-Seitz radius, i.e., $4\pi s^3/3 = V =$ the atomic volume),

$$(3PV)_f = - n \left[\frac{\delta C}{\delta \ln s} + (\bar{E} - C) \frac{\delta \ln W_f}{\delta \ln s} \right] . \qquad (1)$$

Here δ denotes the change with a rigid atomic-sphere potential and the volume dependence of the f-band centre, C, has been separated from that of the f-band energies, E, measured relative to the centre. The quantity W_f is the band-width and $\bar{E}$ is the centre of gravity of the occupied part of the f band. Finally, n is the number of f electrons. To derive an estimate of this expression we approximate $\bar{E} - C$ by $-\frac{1}{2}(1 - \frac{n}{14})W_f$, a relation which holds exactly for a rectangular shape of the f-projected density of states. Furthermore, we have used the expression given in Ref. 61 to estimate that $\delta C/\delta \ln s$ is essentially zero and $\delta \ln W_f/\delta \ln s \simeq - (2\ell+1) = -7$. The self-consistent band calculations show that through the volume range of interest the f-occupation number n_f remains rather constant. With these simplifications we obtain the approximate

relation

$$(3PV)_f \simeq - 3.5 \, n \, (1 - \frac{n}{14}) W_f . \qquad (2)$$

From the calculated partial pressures in Fig. 8 we first of all note that the contribution is dominated by the f electrons. Furthermore, we see from relation (2) that this contribution is maximal for the half-filled shell and indeed corresponds directly to what is known for the d transition metals. Thus we find that the 5f electrons play the same dominating role for the bonding of the actinide metals as do the d electrons for the bonding of the d transition metals. This clearly demonstrates that the earlier actinide metals constitute a unique series of elements in the Periodic Table, where f electrons play the active and dominating role in the bonding characteristics.

4. THE 5f DELOCALIZATION IN AMERICIUM

From the results in the previous section it becomes rather obvious that something rather dramatic ought to happen when americium is subjected to high compressions, since thereby the 5f band-width will become affected. Before we discuss the calculated results for this situation, it might be useful to consider the basis for the effect of the polarization on the metallic binding. For a ferromagnetic spin polarization of the f band, $m = n_\uparrow - n_\downarrow$, we divide the approximate relation for the f-band partial pressure in eq. (2) into a spin-up and spin-down contribution. Thus we write

$$(3PV)_f^{pol} \approx - 3.5 \left[n_\uparrow(1 - \frac{1}{7} n_\uparrow) + n_\downarrow(1 - \frac{1}{7} n_\downarrow) \right] W_f$$

$$= - 3.5 \left[n(1 - \frac{1}{14} n) - \frac{1}{14} m^2 \right] W_f . \qquad (3)$$

This clearly shows that spin polarization will reduce the binding contribution of the f electrons. For a fully-polarized state, $m = n$ ($n \leq 7$), we then find

$$(3PV)_f^{fully \, pol} = - 3.5 \, n(1 - \frac{n}{7}) W_f . \qquad (4)$$

For a nearly half-filled f band ($n \simeq 7$) the spin

polarization therefore removes the f pressure almost completely, meaning that for such a situation the calculation will closely mimic the properties of a non-bonding f state. Thus the fact
that the 5f localization takes place in the element americium having six f electrons means
that our fully spin-polarized calculations will
quite well reproduce the localized character of
the 5f electrons in this element. (Such a fortu-
nate situation is not met with in the rare-earth
elements cerium and praseodymium, where although
the calculated ferromagnetic spin polarization
is found to take place at approximately the correct volume,[34,35] there is still in the calculations a substantial contribution left from the
f-partial pressure even for the fully polarized
case, so that the corresponding calculated pressure-volume relation does not account so well
for the experimental data.) For americium it has
furthermore been found that the spin-polarized
local density approximation accounts quite well
for the energetics of the $5f^6$ atomic configuration.[23] This means that a calculation of the 5f
delocalization in americium should be expected
to be fairly accurate.

In Fig 9 we show the calculated behaviour of the
quantity 3PV as a function of the Wigner-Seitz
radius. The curve labelled 'spd' refers to the
contribution to the pressure from the spd electrons in americium, the 'non'-curve represents
the results from a non-polarized calculation. The
difference between these two curves shows the
itinerant f contribution to the pressure. As a
function of the atomic volume, the spin polarization m is calculated to vary in a stepwise manner
around s = 2.98 a.u. (see lower part of Fig. 9).
The corresponding pressure-volume curve is denoted by 'Pol' in Fig, 9, and is found to exhibit
a van der Waals-like behaviour. The critical
pressure for the first-order transition is obtained by means of the familiar Maxwell equal-
area construction. From this we find that the
fully polarized solution goes discontinuously
over into the non-polarized solution. As already

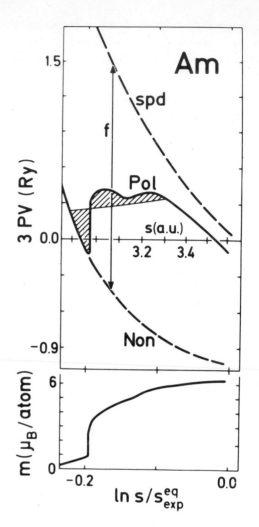

Fig. 9 – Calculated partial and total
pressures for americium (see
text for explanation). In the
lower part the calculated spin-
polarization, m, is shown as a
function of the Wigner-Seitz radius, s, on a logarithmic scale.

mentioned, the calculated equilibrium volume and
bulk modulus agree quite well with experiment.
To obtain this good agreement it should again be
stressed that it is important that the f band is

nearly half full (n≃6.4) and that the spin po-larization is complete. Furthermore, 6.4 is suf-ficiently close to 6 so that, in the high volume, high-spin state the 5f electrons may be consi-dered as being essentially in the local $5f^6$ con-figuration, for which Hund's rules give S = 3, L = 3 and J = 0. This agrees with the observed lack of magnetic moments in the americium metal as well as its superconductivity property.

The calculated pressure-volume relation for ame-ricium is shown in Fig. 10. The most noticeable feature from this is of course the dramatic volume decrease accompanying the 5f delocaliza-tion process (Mott transition). Before we com-pare with presently available experimental data, we refer back to the found deviations between the calculated and experimental equilibrium volumes for neptunium and plutonium. For the latter metal the deviation was about 15%. As al-ready mentioned, this points to a failure in our description of the itinerant f state for these elements, a failure which seems to increase with the atomic number. Therefore, for americium the experimental discontinuity in the atomic volume should be expected to be much reduced in comparison with our calculated change. It should also be kept in mind that for the f-band metals (like plutonium) the temperature has quite an important effect on the atomic volume. There-fore, data taken at room temperature would tend to further decrease the volume change accompa-nying the 5f delocalization.

In Fig. 10 we have also included the presently available experimental pressure-volume data. As can be seen, for low pressures the calculated compressibility agrees very well with experi-ment. The dhcp to fcc crystallographic change observed at about 50 kbar[62] will be discussed in the following section. Around 110 kbar an-other phase transition has been reported.[62] Re-cently, the crystal structure for the new phase above 110 kbar has been analyzed and found to be the so-called α"-uranium structure.[63] This structure has otherwise been identified in cer-

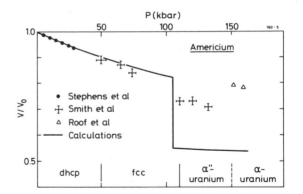

Fig. 10 – Theoretical and experimental pressure-volume relations. For the experimental and theore-tical points V_0 is equal to the experimental and theoreti-cal equilibrium volume, res-pectively. This choice of nor-malization is made to clearly display the goodness of the calculated compressibility.

tain quenched U-alloys (like 10 at.% Mo). From Fig. 10 it is clear that the experimental volume change is much smaller than the one calculated. Some possible reasons for this discrepancy were listed above and further investigations are needed to clarify the situation. Above 160 kbar americium has been reported to attain the ortho-rombic α-uranium structure.[64] From the experi-mental pressure-volume points in Fig. 10 it is, however, clear that there is an inconsistency between the results reported by Smith et al. and those by Roof et al. Thus, further experimental studies of the high-pressure phases are also ne-cessary.

Despite the apparent disagreement between the two experimental groups, it is in our opinion very significant that typical light actinide metal structures are found for compressed ameri-cium. This heavily supports the picture that a 5f delocalization has taken place. In this con-nection it is also significant to point out that cerium in its α'-phase has the α-uranium structure,[65] strongly suggesting an itinerant 4f band in this metal. The high-pressure transition in praseodymium[37] ($\sim$200 kbar) is likely to be of the same origin and one should be prepared to find some of the light actinide-type structures also for this metal at high compressions. From the crystallographic data on the actinide metals it seems possible to phenomenologilly infer that when f electrons participate in the bonding (i.e., when itinerant), and especially so when they dominate the bonding, they give rise to quite distorted structures. The reason why α-cerium, although having an itinerant f band, still has the simple fcc crystal structure might then be due to that in this phase the f-electron binding is not yet the dominating part of the bonding properties. It is apparently only at higher compressions that the f electrons in cerium become the decisive factor as far as the crystal structure is concerned.

5. THE dhcp-fcc TRANSITION IN AMERICIUM. COMPRESSED CURIUM AND BERKELIUM METALS.

The observed dhcp to fcc structural change in compressed americium[62] is of special interest since the same crystallographic transition is also known to occur under pressure for the light rare-earth elements lanthanum, praseodymium and neodymium. This gives further evidence for the rare-earth-like behaviour of the americium metal. In Ref. 4 a simple parameter F ($\sim R_{WS}/R_I$, where R_I is a measure of the radial extension of the trivalent ion and R_{WS} the Wigner-Seitz radius) was introduced to correlate the crystal structure sequence for the lanthanide elements both as a function of atomic number and compression. The corresponding value of F for americium was used to place this actinide element relative to the lanthanides in a position equivalent to praseodymium. Experimentally, the room-temperature dhcp→fcc transition has been found to take place at 40 and 50 kbar for praseodymium and neodymium,[27] respectively. Therefore, the found transition pressure of 50 ± 10 kbar in ameri-

cium[63] fits very well with this effective position of americium relative to the lanthanide series. From the same picture negative pressures would be required for the dhcp-fcc transition in actinium. This is in good agreement with the observed fcc structure in this element.

By means of a canonical (unhybridized) band theory, a direct relation between the crystal structure sequence in the rare-earths and the d-occupation number, n_d, was calculated by Duthie and Pettifor.[66] From a fully hybridized calculation for lanthanum the critical n_d number for the dhcp to fcc transition was recently empirically determined to be 1.95.[35] For actinium the equilibrium value of n_d is calculated to be 1.85,[26] which would then, in disagreement with experiment, imply a dhcp structure. Similarly, the corresponding n_d number for americium at equilibrium is calculated to be about 1.7, which leads to that a substantially higher pressure than the observed 50 kbar is needed before it would attain the critical value of 1.95 for its n_d occupation. Still it is true that the number of d electrons increases with pressure in americium and from this point of view the Duthie-Pettifor theory accounts at least qualitatively for the observed structural change. Obviously, the effect of hybridization adds some complications to the analysis made by Duthie and Pettifor, and the detailed explanation of the lanthanide structure sequence may be somewhat more involved than implied by just the d-occupation number.

From the analysis in Ref. 4 of the superconductivity for the non-magnetic rare-earth elements one would in americium expect a rather rapid increase of T_C with compression. At the 5f delocalization transition, however, the superconductivity should become suppressed (partially or totally), due to the relatively narrow 5f band. An experimental study of this behaviour would be most interesting and could provide valuable further insight into the understanding of superconductivity among the class IIIA metals (Sc, Y

La, Ac, and Lu).

From the generalized phase diagram in Ref. 4 it follows that both curium and berkelium should transform from the dhcp structure into the fcc crystal structure under pressure. The required pressures for this to happen should be somewhat higher than that found for americium, and higher in berkelium than in curium.

From the 5f delocalization in americium the next, quite obvious question to consider is when the corresponding transition should take place for curium and berkelium. The same type of calculations as above for americium suggests a critical pressure of about 450 kbar in curium[26] and a still higher value in berkelium. Thus, it may take some time before the 5f delocalization will become experimentally studied under well-controlled conditions in these elements. However, there is another possibility which makes the berkelium metal of special interest, namely that the 5f delocalization might be preceded by a valence change from three to four. At zero pressure the localized $5f^8$ configuration in berkelium is expected to be situated at about 1.5 eV below the Fermi energy.[67] A reasonable (but still probably a somewhat generous) estimate gives that this energy difference might decrease with pressure by about 10 meV/kbar. Thus, somewhere above 150-200 kbar a valence change from trivalent into tetravalent behaviour might be possible. At these high pressures trivalent berkelium should have the fcc crystal structure. Since in the tetravalent state (especially under pressure) the crystal structure of berkelium is likely to be the same as in thorium, i.e. fcc, the expected valence transition should be of isostructural type (which would facilitate the formation of an intermediate valence state over a certain pressure range). From this picture we would as a function of pressure first obtain a decrease of the number of f electrons from eight to seven. However, when we finally reach the 5f delocalization transition the number of f electrons would

then _increase_ to about eight. Therefore, a very unusual and peculiar f occupation-number dependence should take place in berkelium with pressure. In the case of curium the $5f^7$ configuration is energetically quite far away from the Fermi energy[67] so that a valence transition should here be excluded. One might then also ask if such a valence change would be possible for americium. However, a trivalent to tetravalent transition should here only be expected for pressures beyond 200 kbar (assuming localized 5f electrons), so that for this element the 5f delocalization precedes the valence change (which then of course loses any meaning), just the opposite to what is expected for berkelium.

6. SHORT SUMMARY AND OUTLOOK

From the fundamental electronic structure point of view, the most fascinating aspect of the actinide elements is the dual role played by the 5f electrons; in the earlier elements they are itinerant (band) electrons, while for the later elements they form localized non-bonding magnetic states. Compression of the first element showing 5f localization, americium, should be expected to give rise to a Mott transition into an itinerant 5f state at a relatively low pressure. Such a remarkable behaviour is of course a great challenge to high pressure physicists. Indeed, such experiments have already begun and one may hope that some inconsistencies in the present data analysis will soon be remedied. It would be of special significance if the experiments on americium could be extended into the low temperature region, where for example a study of the pressure dependence of the superconductivity transition temperature should be particularly valuable. For the following elements, curium and berkelium, the pressures required for the 5f delocalization seem to be somewhat too high for presently available equipments. Still, it would be of interest if the dhcp→fcc transition could be confirmed. In berkelium there might in addition take place an isostructural valence transition from three to four for reasonably low pressures. If the element following californium, i.e. einsteinium, would form in a trivalent metallic state this would be expected to be in the Sm-type crystal structure.[4] Unfortunately, this possibility of finding another elemental metal possessing this unique structure is likely to be prevented due to that einsteinium will form as a divalent metal. If this turns out to be the case, there is, however, still some chance that the Sm-structure could be obtained in compressed einsteinium.

From these examples it is clear that high pressure experiments will continue to play a key-role in future studies of the basic properties of the actinide metals. At the same time, high pressure experiments on cerium and praseodymium are likely to provide further insight into the understanding of itinerant f states. Thus, for example, it is known that cerium under pressure shows a close relationship to the actinide metals.[30] A similar behaviour would be expected for praseodymium at high pressure. Historically, however, the phase diagram of cerium was first interpreted as a trivalent-tetravalent valence change, a picture which still seems to dominate the literature. Therefore, high pressure studies of americium are particularly valuable since they would help to clarify various aspects of the f delocalization phenomenon.

Above we have limited ourselves to a discussion about the pure actinide elements and their electronic structure. Turning to actinide compounds, intermetallics and alloys, an almost unlimited number of various behaviours can be imagined. As we have noticed, for the pure metals the interatomic 5f orbital overlap plays a particularly significant role for their electronic structure, and it is likely to continue to do so also in more complex systems. Indeed, this seems to be the main reason for the success of the Hill plot.[68] Since high pressure is the cleanest way by which one can manipulate this overlap, it is clear that this experimental technique will play a particularly central role in the future investigations of actinide materials.

REFERENCES:

[1]   Gschneider, K.A., Jr., *Rare Earth Alloys* (Van Nostrand, Princeton, N.J., 1961).

[2]   Taylor, K.N.R. and Darby, M.I., *The Rare Earths* (Wiley, New York, 1972).

[3]   Freeman, A.J. in *Magnetic Properties of Rare Earth Metals*, ed. by R.J. Elliot (Plenum, London, 1972).

[4]   Johansson, B. and Rosengren, A., Phys.Rev. B 11 (1975) 2836.

[5]   Kmetko, E.A. and Hill, H.H., *Plutonium 70*, ed.: W.N. Miner (AIME, New York, 1970) p. 233.

[6]   Freeman, A.J. and Koelling, D.D., *The Actinides; Electronic Structure and Related Properties*, ed. by A.J. Freeman and J.B. Darby, Jr. (Academic Press, New York, 1974), p. 51.

[7]   Johansson, B., Phys.Rev. B 11 (1975) 2740.

[8]   Nugent, L.J., Burnett, J.L. and Morss, L.R., J.Chem.Thermodynam. 5 (1973) 665.

[9]   Johansson, B. and Rosengren, A., Phys.Rev. B 11 (1975) 1367 and Johansson, B., *Proceedings of the 2nd Int.Conf. on the Electronic Structure of the Actinides*, eds.: J. Mulak, W. Suski and R. Troc (Ossolineum, Wroclaw, 1977), p. 49.

[10] Lee, J.A. and Waldron, M.B., Contemp.Phys. 13 (1972) 113.

[11] Johansson, B., Phys.Rev. B 12 (1977) 5890.

[12] See, e.g., Zachariasen, W., J.Inorg.Nucl. Chem. 35 (1973) 3487.

[13] Friedel, J., J.Phys.Chem.Solids 1 (1956) 175.

[14] Freeman, A.J., Inst.Phys.Conf.Ser. No. 37 (1978) 120 and Koelling, D. J.Phys. (Paris) C 4 (1979) 117.

[15] Freeman, A.J., Koelling, D.D. and Watson Yang, T.J., J.Phys. (Paris) C 4 (1979) 134.

[16] Ward, J.W. and Hill, H.H., *Heavy Element Properties*, eds.: W. Müller and H. Blank (North Holland, Amsterdam, 1976) p. 65.

[17] Nellis, W.J., and Brodsky, M.B., *The Actinides; Electronic Structure and Related Properties*, eds.: A.J. Freeman and J.B. Darby, Jr. (Academic, New York 1974),Vol.II,265.

[18] Brodsky, M.B., Rep.Prog.Phys. 41 (1978) 1547.

[19] Blaise, A. and Fournier, J.M., Sol.State Commun. 10 (1972) 141.

[20] Veal, B.W. and Lam, D.J., Phys.Rev. B 10 (1974) 4902.

[21] Baer, Y. Physica 102 B (1980) 104.

[22] Johansson, B. and Mårtensson, N., Phys.Rev. B 21 (1980) 4427.

[23] Johansson, B., Skriver, H.L., Mårtensson, N., Andersen, O.K. and Glötzel, D., Physica 102 B (1980) 12.

[24] Skriver, H.L., Andersen, O.K. and Johansson, B., Phys.Rev.Letters 41 (1978) 42.

[25] Skriver, H.L., Andersen, O.K. and Johansson, B., Phys.Rev. Letters 44 (1980) 1230.

[26] Skriver, H.L., Andersen, O.K. and Johansson, B., (to be published).

[27] Donohue, J., *The Structures of the Elements* (Wiley, N.Y., 1974).

[28] McWhan, D.B., Cunningham, B.B. and Wallman, J.C., J.Inorg.Nucl.Chem. 24 (1962) 1025.

[29] Smith, J.L. and Haire, R.G., Science 200 (1978) 535.

[30] Johansson, B., Philos.Mag. 30 (1974) 469.

[31] Andersen, O.K., Skriver, H.L., Nohl, H. and Johansson, B., Pure & Appl.Chem., Vol. 52 (1979) 93.

[32] Johansson, B., J.Phys.Chem.Solids 39 (1978) 467.

[33] Johansson, B., FOA 4 Rapport C4588-A2 (1974).

[34] Glötzel, D., J.Phys. F 8 (1978) L 163.

[35] Skriver, H.L., (present conference).

[36] Kornstädt, U., Lässer, R. and Lengeler, B., Phys.Rev. B 21 (1980) 1898 and Glötzel, D. and Podloucky, R., Physica 102 B (1980) 348.

[37] Wittig, J., Z.Physik B 38 (1980) 11.

[38] Platau, A. and Karlsson, S.-E., Phys.Rev. B 18 (1978) 3820.

[39] Johansson, B., (to be published).

[40] Jayaraman, A. and Sherwood, R.C., Phys. Rev. 134 (1964) A 691.

[41] Brewer, L., LBL Report No. 3720 (1975).

[42] Ward, J.W., Ohse, R.W. and Reul, R., J.Chem. Phys. 62 (1975) 2366.

[43] Martin, W.C., Zalubas, R. and Hagan, L., *Atomic Energy Levels – The Rare-Earth Elements*, Nat.Bur.Stand.(US) Spec.Publ. 60 (1978).

[44] Brewer, L., J.Opt.Soc.Am. 61 (1971) 1101.

[45] Vander Sluis, K.L. amd Nugent, L.J., Phys. Rev. A 6 (1972) 86.

[46] Haire, R.G. and Asprey, L.B., Inorg.Nucl. Chem.Lett. 12 (1976) 73 and Noé, M. and Peterson, J.R., *Proc. 5th Int.Conf. on Plutonium and other Actinides*, eds.: H. Blank and R. Lindner (North Holland, Amsterdam, 1976).

[47] Ward, J.W., Kleinschmidt, P.D. and Haire, R.G., J.Phys. (Paris) C 4 (1979) 233.

[48] Wertheim, G.K. and Creselius, G., Phys.Rev. Letters 40 (1978) 813.

[49] Allen, J.W., Johansson, L.I., Bauer, R.S., Lindau, I. and Hagström, S.B.M., Phys.Rev. Letters 41 (1978) 1499.

[50] Johansson, B., Phys.Rev. B 19 (1979) 6615.

[51] Haire, R.G. and Baybarz, R.D., J.Phys. (Paris) C 4 (1979)101.

[52] Burns, J.H. and Peterson, J.R., Int.Phys. Conf.Ser. No. 37 (1978) 52.

[53] Moruzzi, V.L., Williams, A.R. and Janak, J.F., Phys.Rev. B 15 (1977) 2854.

[54] Poulsen, U.K., Kollar, J. and Andersen, O.K. J.Phys. F 6 (1976) 587; Andersen, O.K., Madsen, J., Poulsen, U.K., Jepsen, O. and Kollar, J., Physica 86-88 B (1977) 249.

[55] Kohn, W. and Sham, L.J., Phys.Rev. 140 (1965) A 1135.

[56] Hedin, L. and Lundqvist, B.I., J.Phys. C 4 (1971) 2064.; von Barth, U. and Hedin, L., J.Phys. C 5 (1972) 1629.

[57] Andersen, O.K., Solid State Commun. 13
     (1973) 133; idem, Phys.Rev. B 12 (1975) 3060
     and Andersen, O.K. and Jepsen, O., Physica
     91 B (1977) 317.

[58] Glötzel, D. and Andersen, O.K. (to be pub-
     lished.

[59] Pettifor, D.G., Commun.Phys. 1 (1976) 141.

[60] Stephens, D.R., Stromberg, H.D. and Lilley,
     E.M., J.Phys.Chem.Solids 29 (1968) 815.

[61] Mackintosh, A.R. and Andersen, O.K., *Elec-
     trons at the Fermi Surface*, ed. by M.
     Springford (Cambridge Univ.Press, England,
     1979).

[62] Akella, J., Johnson, Q., Thayer, W. and
     Schock, R.N., J.Less Common Met. 68 (1979)
     95.

[63] Smith, G.S., Akella, J., Reichlin, R.,
     Johnson, Q., Schock, R.N. and Swab, M.,
     (to be published).

[64] Roof, R.B., Haire, R.G., Shiferl, D.,
     Schwalbe, L., Kmetko, E.A. and Smith, J.L.,
     Science 207 (1980) 1353.

[65] Ellinger, F.H. and Zachariasen, W.H., Phys.
     Rev. Letters 32 (1974) 773.

[66] Duthie, J.C. and Pettifor, D.G., Phys.Rev.
     Letters 38 (1977) 564.

[67] Johansson, B. (to be published).

[68] Smith, J.L., Physica 102 B (1980) 22.

PHYSICS OF SOLIDS UNDER HIGH PRESSURE
J.S. Schilling, R.N. Shelton (editors)
© *North-Holland Publishing Company, 1981*

FIRST PRINCIPLES CALCULATION OF THE ELECTRONIC PRESSURE: APPLICATION TO

TRANSITION METALS, SEMICONDUCTORS AND CERIUM

Dieter Glötzel

Max-Planck-Institut für Festkörperforschung
D-7000 Stuttgart 80, Federal Republic of Germany

In the last decade, self-consistent local density band structure calculations have become an important tool for understanding and predicting cohesive and magnetic ground state properties of crystals. The stability of solids with respect to hydrostatic compression is conveniently understood in terms of the electronic pressure. We discuss Linear muffin-tin orbitals calculations of the pressure-volume relation for transition metals, semiconductors and cerium, with particular emphasis on the γ-α-transition.

1. INTRODUCTION

Experimental and theoretical high pressure research are complementary in so far as the independent variable in experiments is pressure, whereas in electronic structure calculations it is atomic volume or lattice constant. Therefore if we want to compare some electronic property measured under pressure with a theoretical calculation at reduced lattice constant, we need either the experimental pV-isotherm (not known for too many systems) or some theoretical estimate of the hydrostatic pressure. As is obvious from the increasing literature and several contributions at this conference, there exist by now theoretical pV-relations for many solids, which exhibit a remarkable accuracy: e.g. equilibrium lattice constants have been determined for most elemental metals within a few percent and the bulk moduli within 10 to 20 % from their experimental values [1-3]. These calculations are essentially parameterfree, the only input being the lattice structure and the nuclear numbers of the constituent atoms. They are all based on the local density expression for the ground state energy of an inhomogeneous electron system:

$$E_{tot} = \sum_{i,\sigma} <\psi_{i\sigma}|-\nabla^2|\psi_{i\sigma}> + \int\rho(r)[\tfrac{1}{2}\Phi\{\rho(r)\}+V_N]$$

$$+ E_{NN} + \int\rho(r)\varepsilon_{xc}\{\rho(r),m(r)\} \qquad (1)$$

The four terms are in turn: the kinetic energy of the independent electrons, the Coulomb interaction of the electrons with each other and the nuclei, the nuclear-nuclear repulsion, and finally the exchange-correlation energy, which depends on the local electron density
$\rho(r) = \sum|\psi_{i\sigma}(r)|^2$ and local spin density
$m(r) = \sum|\psi_{i\uparrow}|^2 - \sum|\psi_{i\downarrow}|^2$. The functional form of ε_{xc} can be either of the simple Xα-type:
$\rho\varepsilon_{xc} \sim \rho_\uparrow^{4/3} + \rho_\downarrow^{4/3}$ or of the more elaborate local spin density form, which takes into account

many-body effects of the homogeneous electron gas [4]. Stationarity of Equ. (1) with respect to electron and spin density yields an effective one particle Schrödinger equation, which has to be solved self-consistently. The effective potential is $V_{eff}^\sigma = \Phi+V_N+\mu_{xc}^\sigma$, where $\mu_{xc}^\sigma = \dfrac{\delta}{\delta\rho_\sigma}(\rho\varepsilon_{xc})$ is the exchange correlation potential. The stationary point of the functional (1) is the total energy of the system and the pressure may be expressed in terms of the virial of the electrons:

$$-\frac{\delta E_{tot}}{\delta\ln s} = 3pV = 2T + U \qquad (2)$$

where V is the volume of the unit cell,
$s = (3V/4\pi)^{1/3}$ is the Wigner-Seitz radius, T the kinetic energy and U the potential energy. The first pressure calculations for solids (alkali metals, noble gas solids, Al) appeared about ten years ago [5]. They were based on APW-Xα calculations, and for the pressure the authors used either the virial (Equ. (2)) or numerical differentiation of Equ. (1). From a numerical point of view both of these methods are inconvenient, since they involve the subtraction of large numbers which arise from the core electron contributions. In 1971 Liberman [6] managed to eliminate the core electrons from the problem by transforming (2) into a surface integral over the Wigner-Seitz cell, where by definition the core wave functions vanish. All these pressure calculations were quite expensive in computer time and applications of the virial or the Liberman expression remained rather scarce until 1976, when Pettifor [7] combined the latter with the atomic sphere approximation (ASA) of Andersen [8]. Starting with a cellular expansion of the Bloch functions:

$$\psi_{\underline{k}} = \sum_{\ell,m} B_{\ell,m}^{\underline{k}} Y_{\ell,m}(\underline{\hat{r}}) \phi_\ell(E_{\underline{k}},|r|) \qquad (3)$$

with expansion coefficients $B_{\ell m}$, spherical harmonics $Y_{\ell m}$ and normalized radial functions ϕ_ℓ, he evaluated the Liberman expression over the atomic sphere instead of the polyhedron and found for a monatomic solid:

$$(3pV) = \sum_{\underline{k},\ell} n_\ell^{\underline{k}} \, s\phi_\ell^2(E_{\underline{k}},s) \quad \times \tag{4}$$

$$\{ (D_\ell(E_{\underline{k}},s)-\ell)(D_\ell(E_{\underline{k}},s)+\ell+1) + (E_{\underline{k}}-\varepsilon_{xc}(\rho(s)))s^2 \}$$

Here, $n_\ell^{\underline{k}} = \sum_m |B_{\ell m}^{\underline{k}}|^2$ is the population of angular momentum ℓ in state $\psi_{\underline{k}}$, and

$D_\ell(E,s) = s\phi_\ell'(E,s)/\phi_\ell(E,s)$ is the logarithmic derivative of the radial function at the sphere boundary. Equ. (4) represents just the sum of the deformation potentials of all the occupied states for a change of lattice constant, but keeping the rigid self-consistent potential, which is modified only at the sphere boundary from the exchange-correlation potential μ_{xc} to the exchange correlation energy density ε_{xc}. This point of view has been emphasized by Mackintosh and Andersen [9], who proved the following more general theorem. The first order change in total energy of an atomic arrangement with respect to a displacement of the nuclei is given by the first order change of the sum of one particle energies with the rigidly displaced self-consistent potentials plus the change in Coulomb energy due to the rigidly displaced charges. In the special case of a monatomic solid in the ASA this Coulomb term vanishes and we are left with Pettifors formula Equ. (4). In the remainder of the paper we shall first discuss the electronic pressure in connection with the Linear muffin-tin orbitals (LMTO) method of Andersen [8] and present selected results for all transition metals. Then we compare several recent theoretical results on the cohesive properties of Si, Ge and diamond, and finally we discuss the $\gamma-\alpha$ transition of cerium in the light of spin-polarized band and cohesive energy calculations.

2. LMTO-ASA METHOD AND ELECTRONIC PRESSURE

The newly developed fast linear methods of band structure calculation [8] have contributed a lot to our present understanding of magnetic and cohesive properties of simple solids as well as compounds [9-11]. We focus here not so much on the computational aspects but rather on the interpretative aspects of the LMTO-ASA method. The atomic sphere approximation (ASA) to the LMTO method is equivalent to the secular equation of the Greens's function method:

$$\det | S_{\ell'm',\ell m}^{\underline{k}} - P_\ell(E)\,\delta_{\ell'm',\ell m} | = 0 \tag{5}$$

where the muffin-tin radius of non-overlapping spheres has been replaced by the Wigner-Seitz radius s and the kinetic energy in the interstitial region has been put equal to zero. $S_{\ell'm',\ell m}^{\underline{k}}$ are the hermitean structure constants independent of energy and lattice spacing, and the potential function $p_\ell(E)$ is the analogue of the cotangent of the scattering phase shift in the Green's function method. The energy dependence of the potential functions $p_\ell(E)$ can be accurately parametrized in terms of the expansion coefficients of a Taylor series of the normalized radial function $\phi_\ell(E_\nu,s)$ around a fixed but arbitrary energy E_ν. The details of this procedure are described in Ref. [8]; here we need only some derived potential parameters, in particular for each angular momentum ℓ the center of the ℓ-band C_ℓ, the band mass μ_ℓ and the distortion parameter γ_ℓ. If one set of ℓ-bands dominates the band structure, like the d-bands in transition metals, Andersen [8] has shown that it is a good approximation to neglect the hybridization of the d-states with the other angular momenta. This then leads to the concept of canonical ℓ-bands, characteristic of a given lattice structure, which have to be scaled with the non-linear potential function $p_\ell(E)$ characteristic of the atomic species. The canonical d-bands $S_d(\underline{k})$ are the eigenvalues of the d-d block of the structure constants matrix and the actual (unhybridized) d-bands are then to second order in $(E-E_\nu)$ given by

$$E_d(\underline{k}) = C_d + (\mu_d s^2)^{-1/2} S_d(\underline{k})[1-\gamma_d S_d(\underline{k})]^{-1}(\mu_d s^2)^{-1/2} \tag{6}$$

This means that the center of the canonical d-band is shifted to C_d, the bandwidth is scaled by $\mu_d s^2$, and finally the d-band is distorted by the function in brackets. Only recently it was recognized that the form of Equ. (6) can be carried over to the fully hybridized case (also to order $(E-E_\nu)^2$). The hybridized energy bands are then the eigenvalues of the following Hamiltonian matrix:

$$\underline{H} = \underline{C} + (\underline{\mu}s^2)^{-1/2}\,\underline{S}\,[\underline{1}-\underline{\gamma S}]^{-1}\,(\underline{\mu}s^2)^{-1/2} \tag{7}$$

where $\underline{C}$, $\underline{\mu}$ and $\underline{\gamma}$ are to be understood as diagonal matrices with the corresponding potential parameters C_ℓ, μ_ℓ and γ_ℓ on the diagonal, and the matrix in brackets has to be inverted. This more elegant formulation of the LMTO-ASA method is discussed in Ref. [12].

Since the structure constants matrix $\underline{S}$ does not depend on the lattice constant, the change in the eigenvalues of the Hamiltonian (7) under compression is solely determined by the change of the potential parameters. The expectation value of the derivative of Equ. (7) yields the LMTO-ASA pressure formula (equivalent to Pettifor's formula Equ. (4)):

$$3pV = \sum_{\underline{k},\ell} n_\ell^{\underline{k}} \{ -\frac{\delta C_\ell}{\delta \ell n s} + (E_{\underline{k}}-C_\ell)\frac{\delta \ell n(\mu_\ell s^2)}{\delta \ell n s}$$

$$- (E_{\underline{k}}-C_\ell)^2 \mu_\ell s^2 \frac{\delta \gamma_\ell}{\delta \ell n s} \} \tag{8}$$

Following the analysis of Andersen et al. [11], we now want to discuss the leading terms of Equ. (8). The first term is proportional to

$2(C_\ell - \varepsilon_{xc})s^2/\mu_\ell$ and therefore repulsive for the valence bands of transition metals (see Fig. 1).

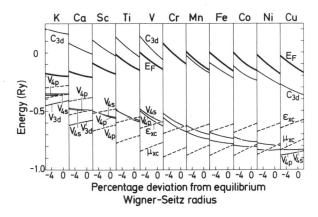

Fig. 1: Band energies in Ry for the 3d series K to Cu as function of atomic radius. The heavy line (marked E_F) is the Fermi level, C_{3d} is the center of gravity of the 3d-band, and V_{4s} and V_{4p} are the square-well pseudopotentials [8]. V_{4s} coincides with the bottom of the 4s-band. The exchange-correlation energy density at the sphere boundary ε_{xc} and the corresponding potential μ_{xc} are indicated by dashed lines.

$$\frac{\delta \ln(\mu_\ell s^2)}{\delta \ln s} = 2\ell+1 + 2/\mu_\ell + \ldots \qquad (9)$$

in the second term of Equ. (8) describes the band broadening under compression of the solid. Because of the prefactor $(E_k - C_\ell)$ this term is bonding (anti-bonding) for states below (above) the center of gravity. This term is dominating for the d-pressure in transition metals; it may be evaluated using a rectangular density of states, which then yields the well known parabolic behaviour for pure d- or f-bands:

$$(3pV)_{bond} = \left\{ \begin{array}{l} -\dfrac{\Delta_d}{4} n_d(10-n_d) \\[2mm] -\dfrac{\Delta_f}{4} n_f(14-n_f) \end{array} \right. \qquad (10)$$

where n_d and n_f are the occupation numbers in the d- or f-band and $\Delta_d = 25/(\mu_d s^2)$ and $\Delta_f = 40/(\mu_f s^2)$ are the corresponding band widths for close packed metals. The ferromagnetic pressure due to repopulation in the d- or f-band is then given by

$$(3pV)_{mag} = \frac{\Delta}{4} m^2 \qquad (11)$$

where m is the spin moment per atom. The parabolic trend of Equ. 10 is demonstrated

in Fig. 2, which displays the partial angular momentum pressures obtained from fully hybridized LMTO-ASA calculations for the 3d metals K to Cu.

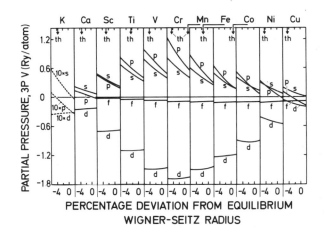

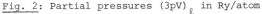

Fig. 2: Partial pressures $(3pV)_\ell$ in Ry/atom for the 3d series K to Cu. The arrows on the top line indicate the theoretical equilibrium (p(s) = 0) for the non magnetic ASA calculations.

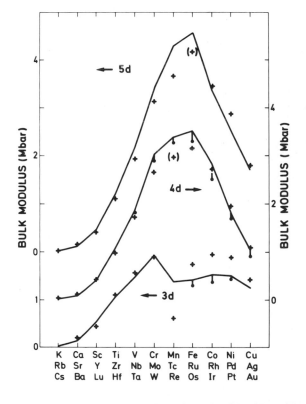

Fig. 3: Bulk moduli in Mbar for the 3d, 4d and 5d series. Solid line: theory; + : expt.

On the other hand, the electronic pressure of the s-p electrons is dominated by the first term of Equ. (8), which because of core orthogonalization is strongly repulsive. The balance between bonding d and repulsive s-p pressure stabilizes the transition metals and since the d-pressure only weakly depends on volume, the bulk moduli $B = -dp/d\ln V$ are mostly determined by the s-p pressure. In Fig. 3 we compare the bulk moduli of the 3d, 4d and 5d metals, calculated at the observed equilibrium lattice constants, with the experimental values (crosses). The solid circles indicate for the 3d series the results of non magnetic calculations, whereas for the 4d series they show the size of the muffin-tin Coulomb corrections (see e.g. Ref. [13]).

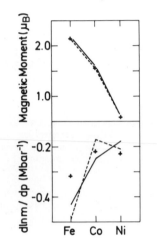

Fig. 4: Saturation magnetization and its pressure derivative for Fe, Co and Ni. Dashed line: calculation of Ref. [1]; solid line: this work; + : experiment.

The local spin density approximation is known to predict correctly the occurence of ferromagnetism in the 3d metals Fe, Co and Ni [1,14-16]. In Fig. 4 we compare theoretical predictions [1,2] of the saturation magnetization and its pressure derivative with experiment. The recently determined experimental pressure coefficient for Ni [17] agrees better with both calculations than the older, less precise value quoted in Ref. [1]. The agreement of the pressure coefficients is quite satisfactory, and part of the deviations from the experimental values can be attributed to the unknown change of the electronic g-factor.

3. SEMICONDUCTORS

The LMTO-ASA method was originally designed to study the electronic structure of close packed metals and compounds. In order to apply this method to more open structures, like the diamond structure, one has to improve the space filling, obtained by non-overlapping spheres around the atomic sites, by introducing additional interstitial spheres, in which again charge density

and potential are spherically averaged. This amounts then for e.g. Si in the diamond structure to a two component compound calculation, where the proper Madelung (point charge) contribution has to be included in the potential and the electronic pressure (see e.g. Ref. [11]). Employing this technique we found for Si, Ge and diamond in Ref. [18] not only good agreement of the calculated cohesive properties with experiment, but also close agreement of the band energies with virtually exact linearized APW calculations [19]. In the following table we compare our LMTO-ASA results on the cohesive properties of the elemental semiconductors with experiment and recent theoretical calculations.

	E_{coh} (eV)	a_o (a.u.)	B (kbar)
Silicon			
Expt.	4.7	10.26	988
LMTO [18]	4.8	10.22	980
LCAO [20]	4.8	10.37	870
Pseudo [21]	4.1	10.42	1230
" [22]	5.2	10.28	940
" [23]	4.6	10.32	990
Germanium			
Expt.	3.9	10.70	772
LMTO [18]	3.6	10.78	660
Diamond			
Expt.	7.6	6.73	4400
LMTO [18]	8.5	6.67	4900
Pseudo [24]	7.4	6.80	4400
LCAO [25]	5.7	6.92	–

The agreement of the different calculations with each other and with experiment is remarkable. The work of Yin and Cohen [23] on the structural energies and selected phonon frequencies of silicon leads us to hope that accurate, truly ab-initio calculations of phonon spectra will be feasible within the next few years, even for more complicated systems like transition metals.

4. γ-α TRANSITION IN CERIUM

The most prominent feature of cerium and some of its compounds (CeP, $CeAl_2$) is an isostructural volume collapse transition with about 15 % loss in volume. For γ-cerium at room temperature this transition occurs at a pressure of 8 kbar. The first theoretical explanation of this phenomenon, en vogue now for thirty years, was the promotional model, suggested independently by Zachariasen and Pauling. The basic idea of this model is, that the single 4f-electron present in the atomic ground state of cerium as well as in the high volume γ-phase should move under pressure above the Fermi level, thereby going into the 5d-valence-band and increasing the valence from three to four. Therefore the transition was also termed a valence transition. Later theoretically refined versions of this model were consistent with the thermodynamics of the γ-α transition, but they were all based on ad

hoc assumptions about the pressure dependence of the 4f-level and involved substantial transfer of the localized 4f-electron to the 5d-band. This was never observed experimentally, and the promotional model was questioned by several authors on the basis of experimental evidence, including positron annihilation [26,27], cohesive properties [28,29], atomic promotion energies [29] and in the last year Compton scattering [30]. Instead of f→d promotion to explain the volume collapse transition Gustafsson et al. [27] suggested a Mott-transition, where the localized 4f-electron delocalizes into a 4f-band in the α-phase. This idea has been strongly supported by a detailed investigation of Johansson [29]. In the remainder of the paper we shall focus on which solution band theory can provide for the γ-α transition.
In dealing with f-electron systems self-consistency of potential and charge density is essential. All published self-consistent band structure calculations for fcc cerium [31-34] yield, depending on lattice constant, f-band widths of the order of 1 eV and occupation number close to unity in the 4f-band, almost independent of volume. These f-bands are much wider than has been assumed in the various promotional models, and, in fact, the partially filled 4f-band can contribute to cohesion. This bonding contribution can be estimated using a rectangular density of states as we have done for the pressure in section 2. The result is

$$E_{bond}^{f} = -\frac{\Delta_f}{2} n_f (1-n_f/14) \qquad (12)$$

With our band widths of 90 and 60 mRy for α- and γ-cerium, respectively, and $n_f = 1$ in both cases we find the corresponding f-bond energies of -0.6 and -0.4 eV. If the Mott-transition picture of Johansson [29] is correct, this bond energy has to be weighed against the energy to destroy the local f-moment in the γ-phase. The latter has been estimated from spin-polarized atomic calculations to be 0.3 eV [35], and hence we see that α-cerium is on the delocalized side. A similar estimate has been successfully applied to the localization of the 5f-electrons in the actinide series [36].
In zero temperature band theory such localizing transitions can occur through ferromagnetic spin polarization. The first transition of this kind was found by Andersen et al. [10] in a study of fcc γ-iron, which like γ-cerium is a high temperature phase, where unfortunately no ferromagnetism can be observed. Nevertheless this zero temperature magnetic transition for γ-iron seems to be related to martensitic transformations in steels. Guided by these results, we studied also the possibility of ferromagnetism in cerium [33]. Starting with α-cerium and expanding the lattice towards the γ-phase, we found a ferromagnetic instability in the 4f-band close to the observed lattice constant of γ-Ce and at a negative pressure of -90 kbar (see Fig. 2 in Ref. [33]). At this point the occupied part of the 4f-band is completely polarized, and, according to Equ. (11), this yields a magnetic

pressure of about 10 kbar, which is far too small in order to compensate for the large negative paramagnetic pressure. Consequently we have found a "Mott-transition" but at an unrealistically large negative pressure. This negative pressure is almost entirely due to the f-bond contribution (Equ. (10)). Even though the f-bond pressure is needed to understand the compressibility in the high pressure α'-phase [33], where the f-band width is several eV, it seems to be

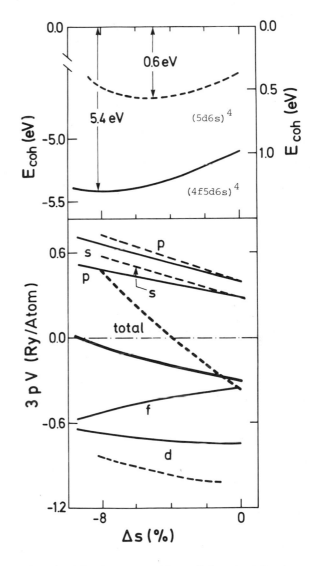

Fig. 5: Cohesive properties of fcc cerium as function of atomic radius. Δs = 0 % corresponds to γ-Ce and Δs = -6 % to α-Ce. Upper panel: cohesive energy in eV; lower panel: partial pressures in Ry/atom. The solid (dashed) lines correspond to the f-band (no f-band) calculation. See text for further explanation.

overestimated in the very narrow band regime. Skriver in another contribution at this conference [37] found in his investigation of the neighbouring element praseodymium with a $4f^2$-configuration a similar transition, but about 200 kbar too low as compared with experiment. This systematic overestimate of the bonding contribution from very narrow energy bands hints at possible deficiencies of the local density approximation. The related case of 5f-localization in americium is dealt with in the paper of Johansson at this conference [38].

In order to decide between the $(5d6s)^4$-configuration suggested by the promotional model and the $4f^1(5d6s)^3$-configuration given by self-consistent band theory, we recently performed total energy calculations for α-Ce. The promotional state was realized by neglecting the f-partial wave in the LMTO wave function expansion. The results for both configurations are displayed in Fig. 5. The f-band calculation yields a binding energy of 5.4 eV (with respect to the atomic ground state $4f^15d^16s^2$) in good agreement with the experimental value of 4.3 eV. The no-f-band cohesive energy curve (dashed line in Fig. 5) is far too shallow, and the minimum of -0.6 eV is at variance with the experimental value. Furthermore the pV-relation for this configuration is much steeper than for the f-band case (see lower part of Fig. 5). The corresponding bulk modulus of 1200 kbar is typical for a quadrivalent transition metal but about seven times larger than the observed value for α-cerium (160 kbar), whereas the f-band calculation yields 270 kbar in reasonable agreement. Also the low temperature electronic properties like specific heat and magnetic susceptibility are consistent with the f-band picture. A detailed account of our work on cerium including an interpretation of the Compton scattering experiments [30] will be given in a forthcoming publication [39].

5. SUMMARY

We have given a survey on recent electronic pressure calculations for a variety of systems, including simple and transition metals as well as semiconductors and the f-band metal cerium. The calculated cohesive properties: equilibrium lattice constants, cohesive energies and bulk moduli agree quite well with experiment. For cerium we found a ferromagnetic transition in the 4f-band which can be regarded as the low temperature analogue of the γ-α transition. These findings support the Mott-transition picture of Johansson [29], and similar transitions have been calculated for americium [40] and praseodymium [37]. The remainder of the rare earth metals remain on the localized side for presently attainable pressures.

As is obvious from Fig. 2 and the unrealistic transition pressure found for cerium, the bonding in very narrow 3d or 4f bands is less well described by local density theory than for broad band (4d or 5d) systems. This hints at a deficiency of the presently employed energy functio-

nals, and it seems, that more elaborate, possibly non-local functionals [41] are needed even for ground state properties. From the experimental side detailed measurements of X-ray and neutron form factors as well as Compton scattering cross sections for these narrow band materials would be very helpful in evaluating present day and future energy functionals.

REFERENCES

[1] Moruzzi, V.L., Janak, J.F. and Williams, A.R., Calculated properties of metals (Pergamon, N. Y. 1978).

[2] Glötzel, D. and Andersen, O.K., to be published.

[3] Skriver, H.L., Andersen, O.K. and Johansson, B., Phys. Rev. Lett. 41 (1978) 42.

[4] v. Barth, U. and Hedin, L., J. Phys. C5 (1972) 1629.

[5] Rudge, W.E., Phys. Rev. 181 (1969) 1033. Ross, M. and Johnson, K.W., Phys. Rev. B2 (1970) 4709. Averill, F.W., Phys. Rev. B4 (1971) 3315. Trickey, S.B., Green, F.R. and Averill, F.W., Phys. Rev. B8 (1973) 4822.

[6] Liberman, D.A., Phys. Rev. B3 (1971) 2081.

[7] Pettifor, D., Commun. Phys. 1 (1976) 1.

[8] Andersen, O.K., Phys. Rev. B15 (1975) 3060.

[9] Mackintosh, A.R. and Andersen, O.K., The electronic structure of transition metals, in Springford, M. (ed.), Electrons at the Fermi surface (Cambridge Univ. press 1980).

[10] Andersen, O.K., Madsen, J., Poulsen, U.K., Jepsen, O. and Kollár, J., Physica 86-88b (1977) 249.

[11] Andersen, O.K., Skriver, H.L., Nohl, H. and Johansson, B., Pure and Applied Chemistry 52 (1979) 93

[12] Andersen, O.K., Europhysics News 12 (1981) 4. Jepsen, O. and Andersen, O.K., to be published.

[13] Glötzel, D. and McMahan, A.K., Phys. Rev. B20 (1979) 3210.

[14] Poulsen, U.K., Kollár, J. and Andersen, O.K., J. Phys. F6 (1976) 241.

[15] Gunnarsson, O., Physica 91B (1977) 329.

[16] Wang, C.S. and Callaway, J., Phys. Rev. B15 (1977) 298.

[17] Hölscher, H. and Franse, J.J.M., Physics Lett. 75A (1980) 401.

[18] Glötzel, D., Segall, B. and Andersen, O.K., Solid State Commun. 36 (1980) 403.

[19] Hamann, D.R., Phys. Rev. Lett. 42 (1979) 662.

[20] Harmon, B. and Weber, W., to be published.

[21] Verges, J.A. and Tejedor, C.,
Phys. Rev. B20 (1979) 4251.

[22] Zunger, A., Phys. Rev. B21 (1980) 4785.

[23] Yin, M. and Cohen, M.L., Phys. Rev. Lett.
45 (1980) 1004.

[24] Yin, M. and Cohen, M.L., unpublished;
quoted by Bachelet, G.B., Greenside, H.
and Schlüter, M., Techn. Memor. Bell Labs.
(1981).

[25] Zunger, A. and Freeman, A.J.,
Phys. Rev. B15 (1977) 5049.

[26] Gustafson, D.R. and Mackintosh, A.R.,
J. Phys. Chem. Solids 25 (1964) 389.

[27] Gustafson, D.R., McNutt, J.D. and Roellig,
L.O., Phys. Rev. 183 (1969) 435.

[28] Wittig, J., Festkörperprobleme XIII
(1973) 375.

[29] Johansson, B., Phil. Mag. 30 (1974) 469.

[30] Kornstädt, U., Lässer, R. and Lengeler, B.,
Phys. Rev. B21 (1980) 1898

[31] Kmetko, E.A., in Benett, L.H. (ed.),
Electronic Density of States, Nat. Bur.
Stand. (US) Spec. Publ. 323 (1971) 67.

[32] Hill, H.H. and Kmetko, E.A.,
J.Phys. F5 (1975) 1119

[33] Glötzel, D. J.Phys. F8 (1978) L163

[34] Pickett, W.E., Freeman, A.J. and Koelling,
D.D., Phys. Rev. B23 (1981) 1266

[35] Glötzel, D., unpublished.

[36] Johansson, B., Skriver, H.L., Martensson,
N., Andersen, O.K. and Glötzel, D.,
Physica 102B (1980) 12.

[37] Skriver, H.L. , this conference.

[38] Johansson, B., Skriver, H.L. and
Andersen, O.K., this conference.

[39] Podloucky, R. and Glötzel, D., to be
published.

[40] Skriver, H.L., Andersen, O.K. and Johansson, B., Phys. Rev. Lett. 44 (1980) 1230.

[41] Gunnarsson, O., Jonson, M. and Lundqvist,
B.I., Phys. Rev. B20 (1979) 3136.

PHYSICS OF SOLIDS UNDER HIGH PRESSURE
J.S. Schilling, R.N. Shelton (editors)
© *North-Holland Publishing Company, 1981*

5f-DELOCALIZATION IN NEPTUNIUM LAVES PHASES UNDER PRESSURE

J. Moser[1], W. Potzel[1], B.D. Dunlap[2], G.M. Kalvius[1], J. Gal[3],
G. Wortmann[1*], D.J. Lam[2], J.C. Spirlet[4] and I. Nowik[5]

1 Physik Dept. Technische Univ. München, D-8046 Garching, Germany

2 Solid State Division, Argonne Nat. Lab., Argonne Ill. 60439, USA

3 Nuclear Research Center Negev and Ben Gurion Univ. Beer Sheva, Israel

4 Institute for Transuranium Elements, D-7500 Karlsruhe, Germany

5 Hebrew University, Jerusalem, Israel

The influence of pressure on electronic and magnetic properties of the
cubic Laves phases $NpOs_2$ and $NpAl_2$ has been studied by Mössbauer spectroscopy. A strong decrease of the isomer shift, the hyperfine field
and the magnetic ordering temperature was observed. This behaviour
points towards increasing 5f-electron delocalization due to the larger
5f-orbital overlap with smaller interatomic distance. Additional information was obtained from the line shape of the Mössbauer spectra,
which indicates the presence of relaxation phenomena. Our data are interpreted in terms of a dynamical model, which predicts pressure dependent fluctuations ($\sim 10^9$ Hz) between a highly localized (magnetic) and
an itinerant (nonmagnetic) electron state.

1. INTRODUCTION

One of the basic features of light actinide compounds is their strong volume
dependence of magnetic properties. Although the metallic elements are non-
magnetic, a large variety of their intermetallic compounds show magnetic
ordering. This behaviour has been attributed to their different inter-actinide
spacing (1).

While the 4f-orbitals of rare earth
systems are known to lie deep inside the
ion, which consequently results in highly localized magnetic moments, the
spatial distribution of the 5f-electrons
is much larger (2,3). Therefore the
lattice constant is one of the major
factors determining the degree of 5f-
electron overlap. if the actinide separation is small like in the elemental
metals, the overlap of the 5f-orbitals

produces a 5f-band too broad to support
an ordered magnetic moment. However, in
many intermetallic compounds the distance
between neighbouring atoms is much larger. Thus, the 5f-bandwidth becomes narrow and magnetic ordering occurs. Hence
the magnetic properties of light actinide
systems tend to change from strongly
itinerant (4) to more localized (5) behaviour as the volume increases.

Such a correlation between lattice constant and magnetic behaviour has been
observed previously in the cubic Neptunium Laves phase intermetallics NpX_2
(X = Al,Os,Ir,Mn,Ru) (6,7). Within these
compounds, the lattice constant changes
from 7.785 Å for $NpAl_2$ to 7.320 Å for
$NpMn_2$.

To investigate the influence of volume
change on the magnetic properties we
performed high pressure Mössbauer

experiments on two compounds of the
series, namely $NpOs_2$ and $NpAl_2$. They
have previously been shown to have the most
itinerant and localized magnetic characteristics, respectively.

The 60 keV Mössbauer resonance in ^{237}Np
provides an excellent tool for investigating the electronic and magnetic properties of such materials by means of
the isomer shift S (electron density
inside the nucleus) and the magnetic
hyperfine field B_{eff} (spin density inside the nucleus).

In addition to the hyperfine parameters
S and B_{eff}, the temperature dependence
of the hyperfine field has been used
to determine the volume dependence of
the magnetic ordering temperature. Experimental details have been described
previously (8,9).

2. RESULTS

Mössbauer velocity spectra for $NpOs_2$
and $NpAl_2$ at various pressures are presented in figs. 1 and 2. One can notice
a decrease of the hyperfine field and
the onset of relaxation line broadenings
with pressure.

Fig. 3 shows the pressure dependence of
the Curie temperature T_c, the effective
magnetic hyperfine field B_{eff} and the
isomer shift S as observed in $NpOs_2$ and
$NpAl_2$. The application of pressure results in a decrease of all three quantities, whereby the reduction of isomer
shift means an increase of electronic
charge density inside the nucleus.

In fig. 4 we have plotted B_{eff} vs. S at
different pressures. For both compounds
we obtain a linear correlation. It proofs
that the hyperfine field and with it the
magnetic moment at the Np-site decreases

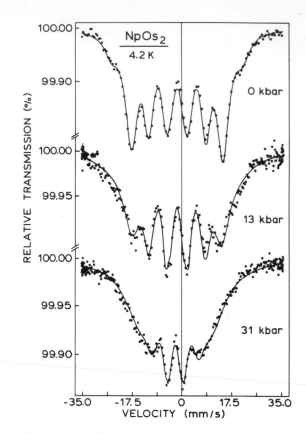

Figure 1: Mössbauer spectra of $NpOs_2$

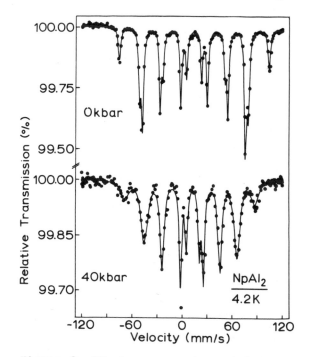

Figure 2: Mössbauer spectra of $NpAl_2$

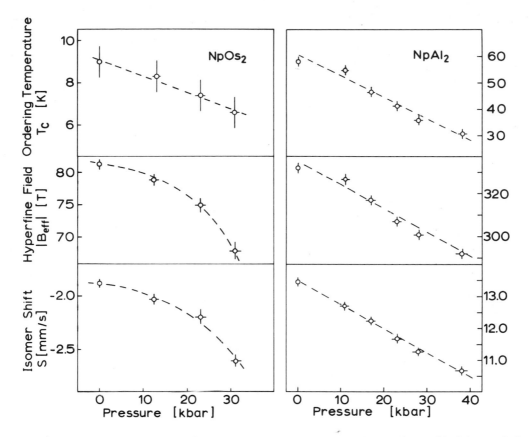

Figure 3: Pressure dependence of ordering temperature, hyperfine field, and isomer shift for $NpOs_2$ and $NpAl_2$. The hyperfine fields are extrapolated to T = 0 K.

as the electron density inside the nucleus becomes larger.

The hyperfine spectrum for the ^{237}Np Mössbauer transition in a magnetically ordered material consists of sixteen resonance lines, some of them being not resolved. They are positioned symmetrically with respect to the isomer shift (see figs. 1 and 2).

A high resolution scan of the inner four lines reveals that this basic symmetry is not maintained at high pressures. An assymmetric broadening of the lines at lower velocity is observed which becomes increasingly pronounced as the lattice constant is reduced (see fig. 5). Such line broadenings indicate fluctuations of the hyperfine field at a rate comparable with hyperfine frequencies.

However, ferro-(and para-)magnetic relaxation always result in the same broadening for two lines positioned symmetrically with respect to the center of the spectral pattern. The behaviour observed here is clearly different. Its basic features may bee understood with the aid of a two state dynamic model, which will be described in the following section.

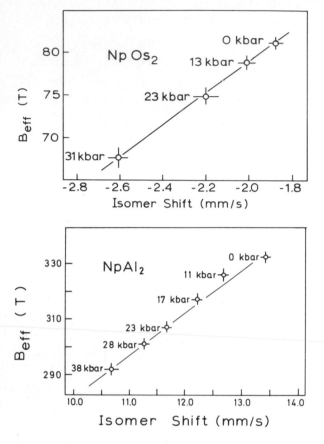

Figure 4: Hyperfine field (at T = 0) vs. isomer shift for NpOs$_2$ and NpAl$_2$ at various pressures

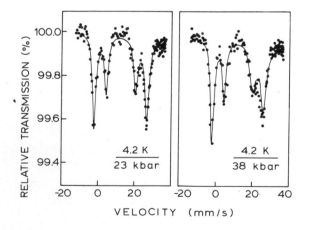

Figure 5: Central section of the Mössbauer spectra of NpAl$_2$ at various pressures

3. DISCUSSION

The systematic variation of magnetic properties in the cubic Laves phases NpX$_2$ as it was found in previous investigations (6) has been attributed to the different Np separation in these compounds. The correlation between the lattice parameter and the magnitude of hyperfine field and magnetic moment as well as magnetic ordering temperature was explained by increasing overlap and delocalization of 5f-electrons with reduced inter-actinide spacing.

The results of our high pressure experiments strongly support this picture. A quantitative discussion of the volume dependence of the 5f-overlap and the resulting effects is not possible at the present stage of this work, because compressibility data are lacking for the compounds studied.

However, from the linear correlation between the variations of hyperfine field and isomer shift with pressure in NpOs$_2$ and NpAl$_2$ (see fig. 4) we can extract important information on the changes in electronic structure with decreasing volume:

(i) The decrease of the isomer shift may be ascribed to an increase of itinerant character of the 5f-electrons. The assumption that it is this change in character which mainly produces the observed difference in electronic charge density inside the nucleus, rather than a simple compression of the valence electron shell, is supported by the fact that the curvature of the isomer shift of NpOs$_2$ under pressure exhibits the opposite sign than the one which is expected from the presence of a non-linear compressibility.

(ii) The increase of electron density

at the nucleus is acompanied by a reduction of the hyperfine field. In first
approximation (6,7) this also means
that the magnitude of the magnetic moment at the Np site is reduced with decreasing volume. This behaviour again
can be explained by a progressive delocalization of 5f-electrons due to an
increasing width and hybridization of
the 5f-band.

The response of these materials to
pressure follows the trend established
for 3d-transition metal systems, where
compression leads to d-electron delocalization and loss of magnetic moment
(10). It contrasts with the results for
rare earth Laves phase intermetallics
such as DyX_2. There no significant variation of the hyperfine field over a
wide range of values for the lattice
parameter is observed (11). In fact,
the hyperfine field and with it the
magnetic moment of the Dy^{3+} ion in these
compounds remains near the free ion
value within a few percent. Such a behaviour indicates highly localized
magnetism. Pressure experiments on
typical rare earth intermetallics are
not available. However, we have carried
out recently such measurements on another intermetallic Neptunium compound,
$NpCo_2Si_2$ (12). Neutron diffraction data
indicate that in this material the magnetic moment is well localized at the
Np site. We find the hyperfine field in
this compound to be indeed independent
of applied pressure.

(iii) The magnetic ordering temperature
in both $NpOs_2$ and $NpAl_2$ decreases
markedly with pressure. Again this can
be understood as a consequence of magnetic delocalization. In the localized
compound $NpCo_2Si_2$ we observe a weak increase of Neél temperature with pressure.

This effect can be explained qualitatively within the Rudermann-Kittel-
Arott-Model (13).

The observed volume dependences of isomer shift, hyperfine field and transition temperature in $NpOs_2$ and $NpAl_2$
points towards an increasing delocalization of 5f-electrons in these compounds. It cannot provide any information about the microscopic mechanism
connected to these phenomena. However,
from the analysis of line shape of the
Mössbauer spectra we are able to obtain
additional information in this direction.
As stated earlier, the assymmetric line
broadening in the inner section of our
spectra cannot be caused primarily by
ferromagnetic relaxation processes which
have previously been proposed in these
compounds (14). It may rather be understood on the basis of a dynamic two
state model. Within this model the electronic structure of the Np ion is
assumed to fluctuate rapidly between a
fully localized magnetic and a totally
itinerant (nonmagnetic) state. The hyperfine parameters are then given by

$$B_{eff} = B_{loc} \quad \text{and} \quad S = S_{loc}$$

for the localized configuration and

$$B_{eff} = 0 \quad \text{and} \quad S = S_{it}$$

for the itinerant configuration. The
two states have the populations P_{loc}
and $P_{it} = 1 - P_{loc}$.

If the fluctuation rate ω_r between the
two configurations is fast in comparison
with hyperfine frequencies ω_{hf}, the observed spectral parameters will be the
average, weighted with the respective
populations:

$$\overline{B}_{eff} = P_{loc} \cdot B_{loc} \quad \text{and}$$
$$\overline{S} = P_{loc} \cdot S_{loc} + (1 - P_{loc}) S_{it} .$$

In the lower limit ($\omega_r \ll \omega_{hf}$) the Möss-bauer spectrum would consist of a super-position of a nonmagnetic single line and a magnetically split hyperfine pattern.

If the fluctuation rate is of the same order as the hyperfine frequencies ($\sim 10^9$ Hz), the spectral lines start to broaden. Since the fluctuation occurs between states having different isomer shifts, lines which are positioned sym-metrically to the center of the pure magnetic pattern will broaden different-ly. This is what we have observed. The solid lines in fig. 5 represent least square fits to the line shapes within this two state model. From the correlat-ion between B_{eff} and S within the inter-metallic series of NpX_2 (see fig. 6) the isomer shift S_{it} for the fully itinerant configuration (vanishing hyperfine field) has been estimated to be -21 mm/s (with respect to $NpAl_2$).

From fits of the type shown in fig. 5 we find for the localized configuration the parameters S_{loc} = +6 mm/s and B_{loc} = 430 T. The population numbers of the localized state in $NpAl_2$ and $NpOs_2$ at the extreme pressures measured are given in the following Table:

	NpAl$_2$		NpOs$_2$	
Pressure (kbar)	0	38	0	31
P_{loc}	0.77	0.67	0.20	0.16

Besides it illustrates, that the elec-tronic structure of $NpAl_2$ at ambient pressure is not as far localized as it was presumed previously (7).

The correlation of fig. 6 clearly

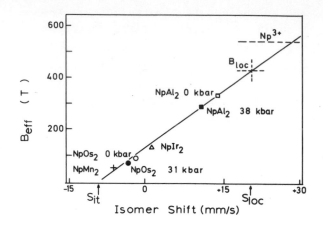

Figure 6: Hyperfine field vs. isomer shift for cubic Laves phase intermetallics and $NpAl_2$ and $NpOs_2$ at the extreme pressures measured

demonstrates that the variation of the hyperfine field and the isomer shift within the chemical series of NpX_2 as well as under pressure can be explained in terms of our two state fluctuation model, although it is certainly over-simplified. More accuracy is expected if fluctuations between more than two states are taken into account.

4. CONCLUSIONS

It has been demonstrated that high pressure Mössbauer experiments on the intermetallic Laves phases NpX_2 provide microscopic information about the volume dependence of electronic structure and magnetic behaviour of such systems. In particular we got the following results:

1. A more de-localized magnetic charac-ter is observed with smaller lattice constants.

2. The pressure dependence of hyperfine parameters indicate that increased overlap of 5f-electrons with reduced

volume of unit cell is the driving force for more itinerant magnetic behaviour.

3. Line shape dependence on pressure shows the presence of rapid fluctuation phenomena, which are assumed to arise from relaxation between a fully localized and a strongly delocalized (itinerant) state of the Np ion.

4. The populations of the two states and the relaxation rate are both influenced by the 5f-orbital overlap.

5. The Laves phase with the most local moment magnetism known ($NpAl_2$) is still not as fully localized as its rare earth counter parts.

6. ACKNOWLEDGEMENT

We would like to thank Dr. L. Asch and Dr. G. K. Shenoy for stimulating discussions and assistence in this work. Furthermore, we gratefully acknowledge financial support by the BMFT, Fed. Rep. Germany, and the U.S. Department of Energy.

* Present address: Institut für Atom- und Festkörperphysik, FU Berlin, D-1000 Berlin

REFERENCES:

(1) H.H. Hill, in: Plutonium 1970 and Other Actinides, W.N. Miner, ed. (Met. Soc. AIME, New York, 1970)p.2.

(2) A.J. Freeman and D.D. Koelling, in: The Actinides, Electronic Structure and Related properties, A.J. Freeman and J.B. Darby,Jr., eds. (Academic Press, New York, 1970)p.51.

(3) A.J. Freeman, Physica 102B (1980) 3.

(4) M.B. Brodsky and R.J. Trainor, Physica 91b (1977) 271.

(5) B.D. Dunlap and G.H. Lander, Phys. Rev. Lett. 33 (1974) 1046.

(6) A.T. Aldred, B.D. Dunlap, D.J. Lam, and I. Nowik, Phys. Rev. B10 (1974) 1011.

(7) A.T. Aldred, B.D. Dunlap and G.H. Lander, Phys Rev. B14 (1976) 1276.

(8) J. Gal, J. Moser, G. Wortmann, C. de Novion, J.C. Spirlet and G.M. Kalvius, J. de Phys. C1, Tome 41 (1980) 129.

(9) J. Moser, J. Gal, W. Potzel, G. Wortmann, G.M. Kalvius, B.D. Dunlap, D.J. Lam and J.C. Spirlet, Physica 102B (1980) 199.

(10) J.S. Schilling, Adv. Phys. 28 (1979) 657.

(11) S. Ofer, I. Nowik, and S.G. Cohen in: Chemical Application of Mössbauer Spectroscopy, V.I. Goldanskji and R.H. Herber, eds. (Academic Press, New York, 1968) p.426.

(12) W. Potzel, J. Moser, G.M. Kalvius, C. de Novion, J.C. Spirlet and J. Gal, to be published in Phys. Rev.

(13) A. Arott, in: Magnetism, Vol. IIB, G.T. Rado and H. Suhl, eds. (Academic Press, New York, 1966) p. 296.

(14) J. Gal, Z. Hadari, U. Atzmony, E.R. Bauminger, I. Nowik and S. Ofer, Phys. Rev. B8 (1973) 1901.

PHYSICS OF SOLIDS UNDER HIGH PRESSURE
J.S. Schilling, R.N. Shelton (editors)
© *North-Holland Publishing Company, 1981*

ELECTRONIC TRANSITIONS IN PRASEODYMIUM UNDER PRESSURE

H.L. Skriver

Risø National Laboratory
DK-4000 Roskilde, Denmark

The pressure-volume relation for praseodymium has been obtained from self-consistent, spin-polarised band calculations. In the relative volume range covered, i.e. from 1.0 to 0.45, we identify two distinct features: A delocalization of the 4f electrons and a depletion of the 6s band. We compare our results with similar calculations on other rare earths, and with high pressure experiments.

Several phase transitions take place in the rare earth metal praseodymium as a function of pressure. At room temperature and low pressures the metal forms in the double-hexagonal close packed structure which at a pressure around 40 kbar is transformed into the face centered cubic structure [1]. Around 75 kbar it exhibits a phase transformation which may involve a distortion of the fcc lattice [2]. Finally, near 190 kbar one finds a transition which appears to be the analogue of the isostructural γ-α transition found in cerium [2]. In addition to the findings from static pressure measurements shock wave experiments [3-5] indicate that praseodymium similar to several other metals exhibits an anormally which takes the form of a stiffening of the so-called Hugoniot. Usually the position of the anormally is deduced from the interception of two straight lines fitted to the conventional shock velocity, u_S, versus particle velocity, u_p, plots of the shock data. For praseodymium one finds the anormally at a pressure around 280 kbar and at a relative volume, V/V_O, of around 0.6. The experimental results are summarised in Fig. 1.

In the following we shall view these experimental results in the light of self-consistent energy band calculations performed as functions of atomic volume. The band structures are obtained by the linear Muffin-Tin Orbitals (LMTO) method [7] in analogy with those for the actinides [8,9], the xenon core is frozen, relativistic effects except spin-orbit coupling are included, and the effect of spin-polarization is allowed for. Some results of the band calculations are shown in Fig. 2. From the lower panel we observe that the bottom of the s-band, B_s, moves towards the Fermi level under compression and hence that s-electrons are transferred into the d-band. This observation is

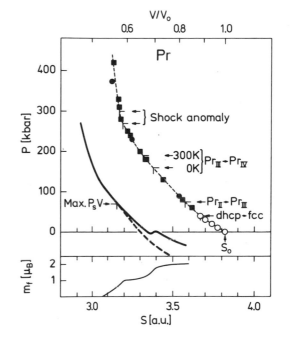

Figure 1: Measured and calculated pressure volume relation for proseodymium. The open circles are the static measurements of Ref. [6] while the filled circles and squares are the shock data [4,5] which constitutes the Hugoniot. The solid line is the calculated pressure volume relation including the effect of spin-polarization. The broken line is the result of a non-polarized calculation. The spin-polarization of the 4f states is shown in the lower panel. S_O is the equilibrium atomic radius equal to 3.8184 a.u. and $V_O = (4\pi/3) S_O^3$ is the corresponding atomic volume.

substanciated by the variation of the occupation numbers, upper panel. We note in passing that when the s-d transition is running out near $V/V_o = 0.5$ d-electrons start to transfer into the f-band. From the partial ℓ-pressures, middle panel, we observe that when s-electrons are being lost at the greatest rate near $V/V_o = 0.6$ the s-pressure developes a maximum. This constitutes a dramatic softening of the s-pressure. It is the subsequent stiffening which occurs near the volume where the s-band is completely depleted which was found to mark the range of the shock anomaly in lanthanum [10]. From the partial f-pressure we note the effect of spin-polarization and the decrease in the pressure occurring when the spin-polarization suddenly drops from 2 to 1 μ_B near $V/V_o = 0.7$, cf. Fig. 1. It is this decrease in f-pressure which was found to be responsible for the volume collapse in the γ-α transition in cerium [11] and americium [9], and for the lattice spacing anomaly occurring in the actinides as a function of atomic number [8].

The calculated pressure-volume and spin-polarization-volume curves are shown in Fig. 1. The comparison with experiments indicates that the calculated pressures are 100 to 200 kbar too low in the volume range $V/V_o = 0.8$-0.6. However, if the 4f-states are included in the core the pressures are found to be 100 to 200 kbar too high in the same volume range. Hence, the effect of treating the 4f-electrons on the same footing as the 6s- and 5d-electrons has the correct sign, but the contribution to the binding is overestimated. A similar deficiency is found in calculations for the neighbouring metal cerium [11] but not for the actinides [8]. It may be traced to the use of the local spin density approximation for exchange and correlation [12].

We shall now discuss the dhcp to fcc transition, the Pr_{III} to Pr_{IV} transition and the shock anomaly in terms of the energy band calculations. Duthie and Pettifor [13] found by means of canonical band theory a correlation between the sequence of crystal structures observed in the rare earths as a function of atomic number and the d-band occupation, n_d. In the present fully hybridized theory an n_d of 1.95 would signify the change from the dhcp to the fcc structure. By this criteria we find in agreement with experiments that La, β-Ce, Pr, and Nd should have the dhcp structure, $n_d < 1.95$, while α-Ce should form in the fcc structure, $n_d > 1.95$. If we use the same criteria to signify structural changes under pressure we find that La, Pr and Nd transform to the fcc structure at relative volumes of 0.93,

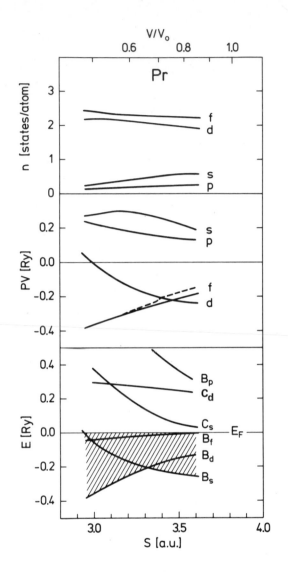

Figure 2: Results of band calculations: Partial occupation numbers, upper panel; partial pressures, middle panel; Fermi level, E_F, bottom of bands, B_ℓ, and centres of bands, C_ℓ, lower panel. The dashed curve, middle panel, is the 4f-pressure including the effect of spin-polarization.

0.81, and 0.67, respectively. Similarly, we find that α-Ce should transform to the dhcp structure at $V/V_o = 1.20$. Although these trends are correct the actual effect is overestimated because the 4f-states in the local density approximation tend to have slightly too high occupation numbers at the expense of d-electrons.

The Pr_{III} to Pr_{IV} transition at a relative volume around 0.7 is found experimentally to be similar to the $\gamma-\alpha$ transition in cerium [2]. From band theory
we find, cf. Fig. 1, that praseodymium
metal should go from a high volume,
highly spin-polarized state to a low volume, slightly polarized state. The situation is theoretically found to be analogous to the delocalization observed in
cerium [11], and the calculation strongly
support the viewpoint that the $Pr_{III} \rightarrow$
Pr_{IV} and $\gamma-\alpha$ transitions in Ce are two of
a kind, and that the 4f-electrons become
itinerant in the process. The self-consistent spin-polarized results may be
interpreted in a Stoner picture which in
its simplest form implies that the spin-

polarization will vanish when the state
density at the Fermi level $N(E_F)$ falls
below $1/I$ where I is the effective
Stoner interaction parameter. In this
picture the delocalization is connected
to the variation of the bandwidth $W \sim$
$1/N(E_F)$ with volume, and comparison with
experiments show that the local density
theory predicts the correct relative
volume for the onset of the delocalization in Ce and Pr, and in the actinide
series.

In the analysis of the shock data for
lanthanum [10] it was found that the
shock anomaly should be considered not
as a single interception of two straight
lines fitted to the data but as an anomalous behaviour over the range of rapid
stiffening caused by the termination of
the s-d transition. It was also found
that the anomaly was preceeded by a
softening of the s-pressure which as a
function of volume developed a maximum.
Judged from Fig. 2 praseodymium follows
the same pattern, and short of a complete calculation including the important core contributions we use the volume
at the maximum s-pressure to approximate
the onset of the anomaly. Hence the
anomalous behaviour should start at a
relative volume of 0.56 in fair agreement with experiments, cf. Fig. 1. We
have summarized the high pressure calculations of the first few rare earths
in Fig. 3, and conclude that local density theory correctly describes the
trends observed experimentally.

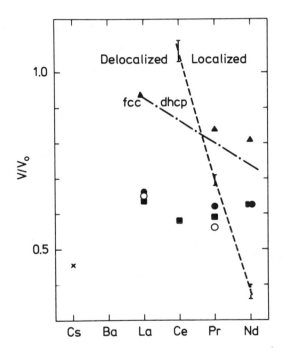

Figure 3: Summary. The dashed-dot line is
the calculated fcc-dhcp phase line, the
dashed line is the calculated line of 4f-
delocalization, the bars indicate the
range of significant change in the spin
polarization, and the open circles indicate the volume of maximum s-pressure.
The filled triangles are the results of
Ref. [1] and the filled circles and
squares are shock data from Refs. [4]
and [5], respectively. For comparison, we
have shown by a cross the volume at which
the X_3-level in cesium becomes occupied
roughly indicating the onset of the isostructural transformation caused by the
s-d transition in that metal [14].

REFERENCES:

[1] Piermarini, G.J. and Weir, C.E.,
    Allotropy in Some Rare-Earth Metals
    at High Pressure, Science 144 (1964)
    69-71.
[2] Wittig, J., Possible Observation of
    Localized 4f Shell Breakdown in Pra-
    seodymium Under Pressure, Z. Physik
    B-Condensed Matter 38 (1980) 11-20.
[3] Al'tshuler, L.V., Bakanova, A.A.,
    and Dudoladov, I.P., Effect of
    Electron Structure of the Compressi-
    bility of Metals at High Pressure,
    Soviet Physics JETP 26 (1968) 1115-
    1120.
[4] Gust, W.H. and Royce, E.B., New
    Electronic Interactions in Rare-
    Earth Metals at High Pressure, Phys.
    Rev. B8 (1973) 3595-3609.
[5] Carter, W.J., Fritz, J.N., Marsh,
    S.P. and McQueen, R.G., Hugoniot
    Equation of State of the Lanthani-
    des, J. Phys. Chem. Solids 36 (1975)
    741-752.
[6] Vaidya, S.N. and Kennedy, G.C., Com-
    pressibility of 22 Elemental Solids
    to 45 kbar, J. Phys. Chem. Solids
    33 (1972) 1377-1389.

[7]   Andersen, O.K., Linear Methods in
      Band Theory, Phys. Rev. $\underline{B12}$ (1975)
      3060-3083.
[8]   Skriver, H.L., Andersen, O.K., and
      Johansson, B., Calculated Bulk
      Properties of the Actinide Metals,
      Phys. Rev. Lett. $\underline{41}$ (1978) 42-45.
[9]   Skriver, H.L., Andersen, O.K. and
      Johansson, B., 5f-electron Deloca-
      lization in Ancericium, Phys. Rev.
      Lett. $\underline{44}$ (1980) 1230-1233.
[10] McMahan, A.K., Skriver, H.L., and
      Johansson, B., The s-d Transition
      in Compressed Lanthanum, Phys. Rev.
      $\underline{B23}$ (1981) 5016-50.
[11] Glötzel, D., Ground State Proper-
      ties of f-Band Metals: Lanthanum,
      Cerium and Thorium, J. Phys. $\underline{F8}$
      (1978) L163-L168.

[12] von Barth, U., and Hedin, L., A
      Local Exchange-Correlation Poten-
      tial for the Spin Polarized Case:
      I, J. Phys. $\underline{C5}$ (1972) 1629-1642.
[13] Duthie, J.C. and Pettifor, D.G.,
      Correlation between d-Band Occu-
      pancy and Crystal Structure in the
      Rare Earths, Phys. Rev. Lett. $\underline{38}$
      (1977) 564-567.
[14] Glötzel, D. and McMahan, A.K.,
      Relativistic Effects, Phonons and
      the Isostructural transition in
      Cesium, Phys. Rev. $\underline{B20}$ (1979) 3210-
      3216.

PHYSICS OF SOLIDS UNDER HIGH PRESSURE
J.S. Schilling, R.N. Shelton (editors)
© *North-Holland Publishing Company, 1981*

PRESSURE-INDUCED VALENCE INSTABILITY IN PRASEODYMIUM METAL AND DILUTE
LA PR AND Y PR ALLOYS: A COUNTERPART OF THE CERIUM CASE? *

Jörg Wittig

Institut für Festkörperforschung, KFA Jülich, D-517o Jülich, Germany

An intriguing phase transformation occurs in praseodymium metal under a
pressure of approximately 2ookbar. The transition is accompanied by
dramatic resistance anomalies similar to those at the γ/α transition in
cerium. The atomic volume decreases by 19%. This volume collapse is
therefore quite comparable to the one at the Ce γ/α transition. The
lattice symmetry changes from orthorhombic to hexagonal in (slight) contrast to the isostructural transition in Ce . The depression of the
superconducting $T_c(P)$ of lanthanum due to the presence of Pr impurities shows a resonance curve - like maximum in the same pressure range
in close analogy to Ce-doped systems at lower P . Recent superconductivity data for very dilute Y Pr alloys reveal a similar dramatic increase in pair-breaking power above 2oo kbar. This latter result is
particularly interesting since it proves that the phenomenon occurs in
single Pr impurity atoms even if they are imbedded in a non-f-level
host matrix (in contrast to the, in our opinion, rather strongly
4f-hybridized conduction band of the aforementioned La host metal). The
increased direct overlap of the $4f^2$ wave functions in pure Pr under
pressure may therefore not be the decisive cause for the electronic phase
transition.

These experiments provide convincing evidence for a pressure-induced
$4f^2$ shell instability in pure Pr as well as in dilutely dissolved Pr
atoms. This new 4f-shell instability has remarkably common features
with the well-known valence instability of Ce at lower P. One can therefore hope that the present findings will help to settle the still
puzzling Ce problem. Finally, our discovery opens the perspective that
several other light rare earth metals (or perhaps even all rare earth
metals) may undergo similar transitions if they are subjected to
pressures up into the megabar range. This will be a worthwhile enter-
prise particularly in view of possible systematics with the corresponding actinide metals.

1. INTRODUCTION

High pressure experiments have widened
our understanding in many areas of materials research. It has thus been discovered that in particular many metallic
4f (and also 5f) electron systems show
unusual and sometimes even dramatic
effects. The metal cerium is perhaps so
far the most prominent example. Fig. 1
shows the P-T phase diagram [1] being
based on the work of many authors. We
recall that, at zero pressure, the γ
phase is magnetic corresponding to the
$4f^1$ configuration. The "collapsed" α
phase, which possesses an approximately
17% higher density, is a strong Pauli-
type paramagnet. There is another interesting phase transition to the $\alpha'\alpha''$...
regime which, according to our data
[1], occurs at pressures as low as

37 kbar at room temperature. Since so
far all investigations have revealed
phase mixtures of two or three coexist-
ing crystallographic phases, the thermodynamically stable structure of Ce above
37 kbar is not known at present. It is
definitively known however that Ce is a
relatively "good" superconductor in the
$\alpha'\alpha''$...range, whereas it is a millikel-
vin-superconductor in the α phase, see
Fig. 2. Early in 1973, I published a
paper in which I postulated that the
pressure-induced metamorphosis of Ce
from a magnetic material to a superconductor must be linked to a delocalization of the $4f^1$ configuration from a
localized to an itinerant 4f electron
[2]. At that time almost nobody was in
favour of completely nonmagnetic
itinerant 4f electrons [3] although the
idea was not new. Those days were the

---

* Most of the work reported here was performed while the author was with the IBM T.J.
Watson Research Center, Yorktown Heights, as an IBM World Trade Visiting Scientist.

golden age of the "promotional" models
which were based upon the assumption
that pressure causes a change of the 4f
count from 1 to o ; the 4f electron was
thought of being literally "squeezed"
into the 6s 5d band. Conventionally,
this was called a valence change from a
trivalent to tetravalent state. Carrying
this picture to its logical conclusion,
$\alpha'\alpha''$ ...-Ce should have the electronic
structure $(6s\ 5d)^4$ which is the same as
for Hf. Fortunately, Jayaraman had measured the melting curve of Ce up into
the $\alpha'\alpha''$...range (Fig. 1). A comparison
of the melting points of $\alpha'\alpha''$...-Ce and
Hf revealed that the melting point of
$\alpha'\alpha''$...-Ce was at least 1ooo K too low
[2,4]. Because of this catastrophically
low melting point $\alpha'\alpha''$...-Ce could not
be a $(sd)^4$ band metal in my opinion as
requested by the promotional models
[2,4]. It ought to be a 6s 5d 4f band
metal with probably not a too narrow
4f bandwidth, since it is a "good"
superconductor [1,4]. We also postulated
that lanthanum is quite a similar 4f
band metal (with a smaller 4f share of
the band charge of course). This concept, which was a plain experimentalist's approach in order to understand a

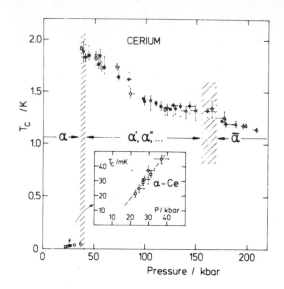

Figure 2 : Superconducting Tc vs. P for
Ce after Probst and Wittig [1]. Note the
abrupt change from a low-Tc to a rather
"good" superconductor at the α to
$\alpha'\alpha''$... phase boundary.

given physical situation still provides
a coherent picture as far as the origin
of the various physical anomalies of La
and Ce is concerned despite the criti-
cism which it originally met [4] and
still meets [8]. We have reiterated
this conception of nonmagnetic 4f band
states so extensively here because, from
the data presented below, it is now very
likely that even Pr's $4f^2$ shell can be
delocalized at higher P at a transition
which is probably the counterpart of the
Ce γ/α transition.

A band structure calculation by Glötzel
[5] marked the first progress on the
theoretical side. According to his results there is only a negligible change
of the number of 4f electrons at the
γ/α transition. This was confirmed in a
beautiful Compton scattering experiment
by Kornstädt et al. [6] demonstrating
that the Compton profile was essentially
the same for the α and γ phases, thereby
ruling out the promotional models.
Glötzel's results have recently been
confirmed by Pickett et al. [7]. The
present situation can be characterized
by saying that there is considerable
evidence now, at least from the experimental side, that the 4f electron itself
becomes itinerant in α-Ce. Apparently,
the 4f electron contributes to the metallic bonding in α-Ce and this is reflected in the volume collapse. Nevertheless it should be stressed that we

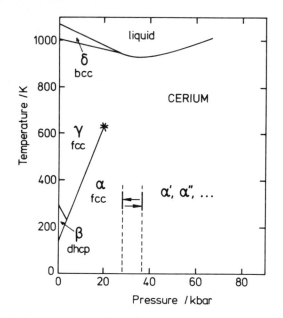

Figure 1 : P-T diagram of Ce. Note that
the phase boundary between the α phase
and the $\alpha'\alpha''$...regime lies between ~3o
and 37 kbar at 3ooK as indicated by the
hysteresis interval [1]. This low transformation pressure is at variance to
what has been elsewhere reported in the
literature.

are far from the point where all possible questions about the Ce problem have satisfactorily been answered yet. For instance, it is still a question, whether it is the increased 4f-4f overlap which is responsible for the delocalization. New data which bear on this particular point are presented below. Also, there is a fairly large number of Ce intermetallics which are believed to be mixed-valent materials with the atomic-like $4f^1$ and $4f^0$ configurations being degenerate. The relationship of these materials to pure Ce is at present not clear at all.

In the following three sections of the paper we report on three pieces of evidence which, taken together and by comparison with the in several respects analogous case of Ce, prove rather convincingly that Pr undergoes a valence transition under a pressure of approximately 2oo kbar.

2. RESISTANCE ANOMALIES IN PRASEODYMIUM

Figure 3 shows the room temperature resistance as a function of pressure [9]. A large discontinuous resistance decrease indicates a phase transition at 21o kbar. There is a striking similarity to the R-P dependence of Ce around its γ/α transition [1o]. In Figure 4, two different families of R-T isobars can be distinguished for the phases Pr III and Pr IV [9]. In Pr III, the resistivity behaves highly anomalously, so to speak α-plutonium-like. The temperature coefficient is extremely small between 3oo K and 1oo K. Below 5o K, the resistivity drops precipitously and obeys a strong T^2-law below $\simeq$5K [9]. There is a striking similarity to the R-T curves of β-Ce and, as far as the small temperature coefficient near room temperature is concerned, also with γ-Ce [11]. In marked contrast, Pr IV exhibits a rather "normal", i.e. an almost linear temperature characteristic as also observed for Ce in the "collapsed" α-phase, see Fig. 5. It was on the basis of this striking similarity between Ce and Pr that we suggested [9] that the Pr III-IV transition may be the direct counterpart of the Ce γ/α transition.
As will be further substantiated below, the Pr III-IV transition is in our view [9,12] a valence transition (of the non-promotional type) with the valence of Pr changing from 3 to 5 since the conduction band gains 2 itinerant 4f electrons. Supposing this interpretation is correct, one wonders, whether the imminent fundamental change of the elec-

tronic structure at the phase transition has precursor effects especially at finite temperatures. We think this is clearly revealed by the family of curves for Pr III in Fig. 4. It is seen that the pathological R(T) behaviour increases dramatically with P toward the phase transition. We have tentatively ascribed the anamalous R-T curves in Pr III to unusually soft and anharmonic phonons which further soften toward the transition [9,12].

Besides phonon scattering other mechanisms such as spin fluctuation scattering other mechanisms such as spin fluctuation scattering, as often discussed for Pu and Pu Al$_2$, can equally play a role [13]. In brief summary, there remains the strong and perhaps not unjustified hope that it is the same physical effect which causes the anamalous R-T characteristics in Pr III, β-Ce, Pu, Pu Al$_2$ and Ce Al$_2$. A thorough search for soft phonon modes in Ce Al$_2$ (especially at low temperatures) would hence be extremely interesting as a test for our proposal [9,12].

Figure 6 shows a hypothetical P-T diagram for Pr at higher pressure. The minimum in the melting curve is conjectural and has been drawn in connection with the above ideas of mode softening at a valence transition and in analogy to Ce (Fig. 1). Carter et al. have investigated Pr by shock wave techniques [14]. They found an anomaly at 27o kbar and 14oo K which they attributed to melting (Fig. 6, open circle). The published Hugoniot is also displayed in Figure 6 by the dotted curve. It intersects our proposed melting line. We must hence conclude that there is either no such dip in the melting curve or those authors have overlooked the solid-liquid phase change, presently expected at much lower pressure and temperature. We take it as a good omen for the existence of the predicted minimum that the melting line of Pr does not significantly increase with pressure up to ~7o kbar the highest pressure to which it has been investigated [15].

3. VOLUME COLLAPSE AT THE PR III-IV TRANSITION

In this section we describe the results of a collaborative effort together with Mao, Hazen and Bell [16] in order to determine the crystal structures and the lattice parameters as a function of pressure for the phases Pr III and Pr IV. Debye-Sherrer x-ray diffraction patterns where obtained in situ at high

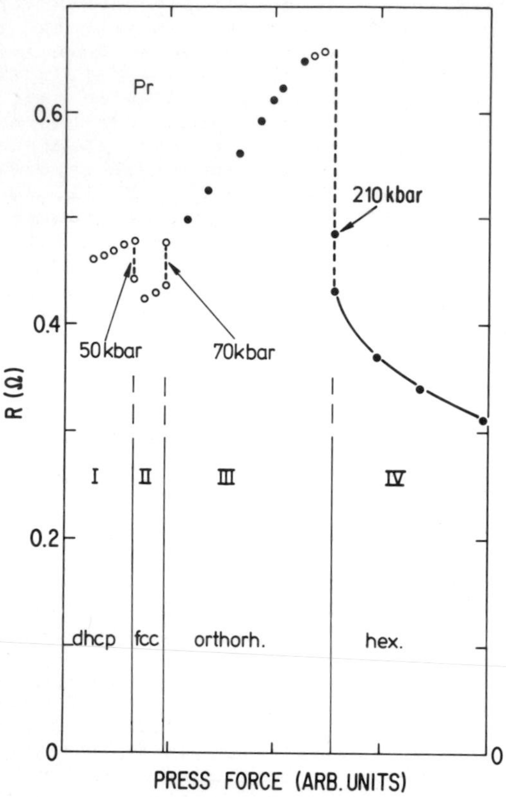

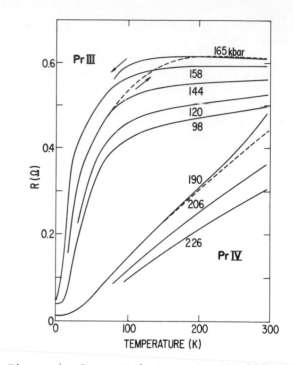

Figure 4: R vs. T isobars for Pr III and Pr IV. Pressures over 12o kbar are underestimated by ~1o% with respect to the ruby scale [9].

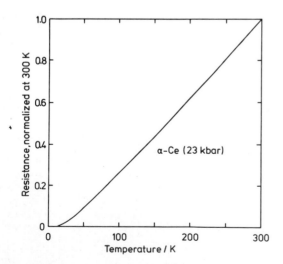

Figure 3: Room temperature resistance vs. pressure for Pr [9]. The Pr III-IV transition pressure has been corrected to 21o kbar. Crystal structures are listed.

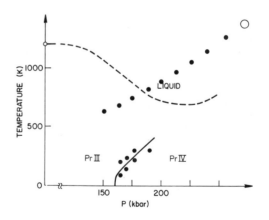

Figure 6: P-T diagram for Pr at higher P. The minimum in the melting curve is hypothetical.

Figure 5: R vs. T for pure α-Ce at 23 kbar (after Probst and Wittig).

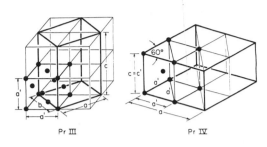

Figure 7: Relationship of orthorhombic
Pr III to ideal fcc and of hexagonal
Pr IV to ideal hcp.

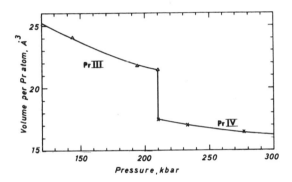

Figure 8: Volume vs. pressure for Pr.
The relative volume change amounts to
19% at the phase transition.

pressure in a diamond anvil press. The
x-ray powder diffraction patterns of
Pr III could be indexed for an orthorhombic phase being a distorted fcc
lattice. Figure 7 shows the unit cell
of Pr III and how it is related to the
fcc subcell: $a \simeq b \simeq \sqrt{2}$ a' and c $\simeq$ 2a'.
Actual atomic positions are however
distorted from the ideal positions.
Diffraction patterns of Pr IV at and
above $\simeq$21o kbar could be indexed for
a distorted hcp cell with axes a$\simeq$2a'
and c$\simeq$c' . Figure 7 shows the relationship of the unit cell to the primitive
unit cell of the ideal hcp structure.
Figure 8 shows the volume per Pr atom
as a function of pressure. The large
relative volume change at the phase
transition, calculated to be 19%, exceeds the one in Ce.
In contrast to the isostructural transition in Ce, the Pr transition is accompanied by a change of symmetry. We
think this is probably a minor point.
Since both Pr III and Pr IV are dis-

torted, however essentially twelve-fold
coordinated close-packed structures, the
observed volume collapse must be attributed to a collapse of the metallic radius, in close analogy to the situation
in Ce. The overall decrease of the metallic radius in Pr IV at 21o kbar
amounts to $\simeq$2o% as compared to normal
pressure. We had therefore until recently presumed [9,12] that it is the enormous increase of 4f-4f overlap which is
at the root of the phenomenon. Data to
be presented below, however, show that
this assumption may be in error.

We think these experiments provide
strong evidence for the Pr III - IV
transition being the $4f^2$ counterpart of
the $4f^1$ delocalization transition in Ce.
As pointed out by colleagues, it is also
possible that only one 4f electron of
Pr becomes delocalized by promotion
encountering an intermediate valence
situation in Pr IV with the $4f^2$ and $4f^1$
configuration being degenerate .
Although we strongly feel this will not
be the final answer, the possibility
that Pr undergoes a pressure-induced
$4f^2 \rightarrow 4f^1$ transition cannot be ruled out
at present.
It is worth pointing out that the
crystal structure of Pr IV is different
from the one of the corresponding 5f
band metal protactinium (Pa). It would
be extremely interesting to investigate
whether the Pa-crystal structure eventually occurs in Pr at higher pressure
as a result of corresponding electronic
states for the itinerant f states. Such
a phase, if it exists, will probably be
a superconductor like Pa [9].

4. ANOMALOUS PRESSURE DEPENDENCE OF T_c
   IN La̲ PR AND Y̲ PR ALLOYS

Cerium's pressure-induced metamorphosis
from a magnetic to a completely non-
magnetic material in the $\alpha'\alpha''$...regime
has interesting counterparts in several
dilute Ce alloys. In La̲ Ce for instance,
the depression of T_c (due to the Ce impurities), $\Delta T_c = T_{c,La\ host} - T_{c,LaCe\ alloy}$,
passes through a marked maximum as a
function of pressure near 15 kbar and
becomes relatively small and constant
at P>1oo kbar, cf. Figure 9 [17]. This
behaviour has been attributed to a
pressure-induced transition of the $4f^1$
shell from a magnetic to a nonmagnetic
virtual bound state the latter giving
rise to only a small depression of T_c
at high pressure [17]. In pursuing
the question further, whether Pr's $4f^2$
configuration can also be delocalized
by pressure, we have studied the
pressure dependence of T_c of La̲ Pr

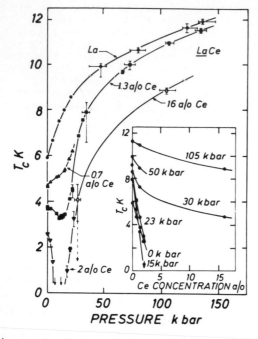

Figure 9: T_c vs. P for pure La and four La Ce alloys. Insert shows T_c vs. concentration x (after Maple et al. [17]).

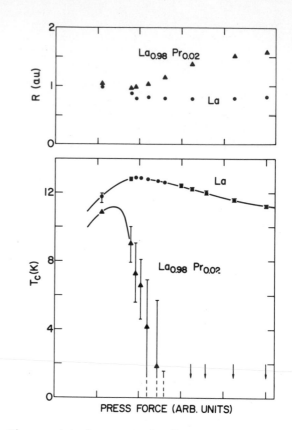

Figure 10: Lower part: T_c vs. press load for La and a 2 at/o La Pr alloy. The data cover the pressure range from ≃120 to ≃230 kbar.
Upper part: "Residual resistance" of the same sample pair. The Pr-doped sample shows a strong increase of the residual resistivity whereas it seems to be roughly independent of P for pure La (see text).

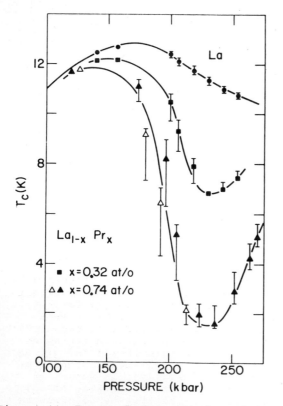

Figure 11: T_c vs. P for pure La and two dilute La Pr alloys.

alloys in order to search for a corresponding anomaly as in the <u>La</u> Ce system
[18]. The lower part of Figure 1o shows
T_c as a function of pressure between
$\simeq$12o kbar and $\simeq$23o kbar for a 2 at/o <u>La</u>
Pr alloy and for the pure host metal.
In the pressure range between 15o and
2oo kbar, the T_c of the Pr-doped sample
falls precipitously. At higher pressures
the sample remains normal down to a
lowest temperature of 1.2 K as indicated
by arrows. Figure 11 shows the pressure
dependence of T_c for two dilute <u>La</u> Pr
alloys and also for pure La to pressures
as high as 27o kbar. The two deep minima
at $\simeq$23o kbar correspond to pronounced
maxima of the depression ΔT_c as a function of P. There is no doubt that ΔT_c
will continue to decrease above
27o kbar.
In the upper part of Figure 1o we present another interesting piece of information. Here the normal-state low-temperature resistance (just above T_c or at
1.2 K, respectively) is plotted as a
function of pressure. Strictly speaking,
this resistance is not the residual resistivity contribution at the temperatures involved) but it comes practically very close to it. One can see that
the "residual resistivity" of the Pr-
doped sample increases drastically with
pressure whereas no noticeable change
occurs in the "residual resistivity" for
La. (Disregard some scatter for the data
points at low P). A conservative estimate reveals a value of at least 15$\mu\Omega$cm/
at/o Pr for the incremental resistivity
due to the Pr impurities at the highest
pressure of 23o kbar. The effect has
qualitatively also been observed for the
more dilute o.7% Pr sample. Apparently,
Pr gains strong resonant scattering
power in La at 23o kbar. This is consistent with our picture of a gradual
$4f^2$ shell delocalization, since a narrow,
virtual bound $4f^2$ state on Pr, imbedded
in a matrix of trivalent La atoms with
their only fractionally filled 4f reso-
nances, should give rise to strong resonant scattering.

We now turn to a more detailed discussion of the results in Figure 11 and how
we interpret them. Finally, we briefly
go over other interpretations. We believe that we are very probably dealing
with the analogous effect as in the
<u>La</u> Ce system (cf. Figure 9), namely,
a continuous pressure-induced demagnetization of the $4f^2$ shell in a one to one
correspondence to a demagnetization in
the pure metals (which, of course, still
remains to be shown to occur in the case
of Pr IV!). It should be noted, however,
that the specific depression $\Delta T_c/x$ at
the highest pressure of 27o kbar amounts

to $\simeq$7 K/at/o. Although this is an appreciable reduction from the value at
maximum pair breaking ($\Delta T_c/x \simeq$13 K/at/o
at 23o kbar) the Pr impurities certain-
ly cannot be classified as "nonmagnetic"
at that pressure. In the <u>La</u> Ce system
for comparison, the initial rate
$(dT_c/dx)_{x=o}$ drops to $\simeq$o.24 K/at/o at
the highest pressure so far studied
[17]. Figure 11 suggests that appreciably higher pressures, perhaps of the
order 35o kbar, will be necessary in
order to establish that the specific
depression eventually becomes relatively small as for the nonmagnetic Ce impurities in La. It remains further to
be shown that the shape of the T_c vs.
x curves exhibits an upward curvature
with concentration x at a fixed pressure above, say, 35o kbar. Such a curvature of $T_c(x)$ is considered to be
typical for nonmagnetic impurities
giving rise to pair-weakening [19].
Although the truly nonmagnetic state
of the Pr impurities lies, **unfortunate**-
ly, beyond our pressure limit, we think
the data of Figure 11 demonstrate
rather convincingly that the major portion of the magnetic-nonmagnetic transition has been surveyed in this
experiment.
Two explanations have been advanced for
the maximum of T_c in the <u>La</u> Ce system.
Müller-Hartmann and Zittartz have suggested that the Kondo temperature, T_K,
of the system may increase from
$T_K \ll T_{c,La}$ to $T_K \gg T_{c,La}$ with pressure.
Such a change would readily explain the
maximum of ΔT_c from their theory on the
influence of the Kondo effect in superconductors [2o]. A pressure-induced
magnetic-nonmagnetic transition seems
equally capable of accounting for the
effect [17]. The initial increase of
$\Delta T_c/x$ in the <u>La</u> Ce system with pressure
has been attributed to an increase of
the (negative) exchange parameter
$|J|=|J_o+J_1|$ with P [21]. Under certain
approximations one has

$$|J_1| \propto \frac{V_{kf}^2}{|E_f - E_F|}$$ with V_{kf}^2 determining the

strength of the mixing interaction between the conduction states and the localized $4f^1$ shell and $J_o > o$ for the
normal intraatomic exchange. Here
$|E_f - E_F|$ is the energy difference between the localized 4f level and the
Fermi energy. An increase of V_{kf}^2 or,
according to the original suggestion
[21], a decrease of $|E_f - E_F|$ with P
may thus lead to an increase of $|J|$
and, in turn, of $\Delta T_c/x$ [21]. Such a
pressure-dependent $|J|$ together with
the assumption of an eventually

smooth transition of the magnetic $4f^1$ shell to a delocalized nonmagnetic virtual bound state at high pressure can thus easily account for a maximum in ΔT_c. Both explanations rely upon a strong mixing interaction between the localized $4f$ electrons and the extended states leading to the Kondo effect which is observed in this system at low pressure [22].

If one of these explanations should essentially hold for the La Pr system, one would perhaps also expect a Kondo effect in La Pr particularly at those pressures where the specific depression $\Delta T_c/x$ strongly increases with pressure, i.e. between 150 and 230 kbar. Such a picture includes the assumption of a change of the sign of the effective exchange parameter J at a certain pressure (below 150 kbar) where $|J_1(P)|$ starts to override J_o. From our present data we are unable to answer the question whether a Kondo effect occurs. We have in fact observed $d\rho/dT < o$ at liquid He temperatures in two different samples (2 and 5.7 at/o Pr) at $\simeq 230$ kbar, the effect however being very small. The weak resistance minimum was absent at an unknown higher pressure in the 5.7% alloy. Below 230 kbar, $d\rho/dT$ was always positive. It is unknown whether this is an intrinsic effect due to the Pr impurities, or, must be, quite trivially, attributed to the onset of superconductivity in the sample which would mask, of course, any possible Kondo scattering. Finally, Pr may possess a nonmagnetic crystal-field ground state [23] in which case a Kondo effect should be expected to occur at higher temperatures only when magnetic crystal-field levels of the J=4 ground multiplet are thermally populated [24]. A precision study of the R-T dependence above liquid He temperatures may thus be a worthwhile experiment in order to clarify whether a high-temperature Kondo effect occurs in a certain pressure range or not.

In this connection, we would like to draw attention to an interesting discovery in the system $(\underline{Zr}\ Pr)B_{12}$. Fisk and Matthias observed a Kondo effect and reported an anomalously large specific depression $\Delta T_c/x \simeq 13$ K/at/o which is just as large as in La Pr at maximum pair breaking at $\simeq 230$ kbar [25]. The authors suggested that the large Pr atom in substituting the small Zr atom experiences high "lattice pressure" which they estimated to be roughly 200 kbar in surprisingly good agreement with the findings of this paper. As a critical test of the concept of "lattice pressure" one would like to know whether the Pr impurities can be forced into a nonmagnetic state by additional static pressure. It would also be quite interesting to investigate, whether Pr impurities will acquire strong pair-breaking power if they are under "lattice pressure" in simple superconducting matrices consisting out of atoms with smaller metallic radii. Possible candidates as host metals are perhaps in the fourth column, e.g. zirconium or hafnium and among the pentavalent transition metals (Nb, Ta). Ion implantation at low temperatures may be a means of introducing the Pr impurities by brute force in order to achieve the goal.

Could crystal-field effects be responsible for the resonance curve - like depression of T_c in Figure 11? The influence of crystal-field split rare earth impurity ions on superconductivity is rather well understood [24]. The major effect is, that $\Delta T_c/x$ is smaller than it would be without a crystal field, i.e. a situation in which we simply have a (2J+1)-fold degenerate state. According to the theory, a decrease of the splitting energy between the two lowest levels could in fact cause an increase of ΔT_c [24]. There are two arguments, however, which rule out that crystal-field changes play a significant role. 1. In order to explain the data, one must primarily look for a mechanism which is capable of explaining the anomalously large specific depression of $\simeq 13$ K/at/o at maximum pair-breaking. This value is more than a factor of 2 larger than what has been observed for all other La-based rare earth alloys. Assuming de Gennes-factor scaling [26], the specific depression $\Delta T_c/x$ should be, on the contrary, appreciably smaller for Pr than for Gd impurities. In the latter case a value of 5.3 K/at/o has been reported [26]. The influence of the crystal field can therefore be excluded because its possible changes under pressure can in no way account for the magnitude of the resonance-like depression of T_c. 2. As reported below, a similarly strong depression of the superconducting T_c is observed in the system $\underline{Y}$ Pr above $\simeq 200$ kbar. The crystal structure of Y is not yet known under these conditions [27]. Nevertheless we have little doubt [27] that it will be different from the anticipated [23] orthorhombic Pr III-type structure of La. If this turns out to be correct, the $\underline{Y}$ Pr data would prove the irrelevance of crystal-field effects, since

it appears rather unlikely that the same
strong decrease of the level splitting
could occur in two different crystal
structures at the same P. The Y Pr experiment points to the fact that we are
dealing with an instability of the Pr
impurity $4f^2$ shell and not with a
change of the properties of the matrices.

The study of the Y Pr-system was begun,
since here the Pr atoms are substituted
in a non-f-level host in contrast to
La metal. It was our hope to find a
clue, whether the (fractionally occupied) La 4f shells play an important
role for the valence transition of the
Pr atoms or not [12]. Fig. 12 shows a
few data points, T_c vs. press force, for
the high-P phase Y II and a 1 at/o Y Pr
alloy. The pressure varies from $\simeq 15o$
to 3oo kbar. It is seen that the T_c of
pure Y increases strongly to over 8 K.
This is a remarkable result and an
interesting problem in its own right
[28]. The Pr-doped sample showed a 1o%
resistance decrease at the lowest T of
1.2 K due to the onset of superconductivity. No sign of superconductivity was
observed (arrows) in the two runs at
the highest pressures.

Fig. 13 shows our "best guess" for the
T_c-P dependence of Y, based on at present still incomplete data. The T_c of
YI rises steeply from o.o65 K at
$\simeq 47$ kbar to about 2.5 K at $\simeq 14o$ kbar
as known from previous work [1].
At the Pr I-II phase transition [27] T_c
seems to drop discontinuously by several
tenths of a degree (the exact value is
difficult to ascertain). T_c continues
its strong monotonous increase in phase
II. Also shown are data for a dilute
o.35 at/o Y Pr alloy. The sample had
transformed to the Y II phase. It is
seen that T_c passes a maximum near
2oo kbar. Although the data, admittedly,
are incomplete, we think it is already
obvious that Fig. 13 is the direct
counterpart of Fig. 11. It should be
noted that the maximum of ΔT_c has not
quite been attained at the highest
pressure of 25o kbar, since the T_c of
the matrix rises steeply. The specific
depression $\Delta T_c/x$ is estimated to be of

the order of 12K/at/o at 25o kbar. This
is very close to what has been found
for the La Pr system. A negative $d\rho/dT$,
indicative for a Kondo effect, has not
been observed in any run in the liquid
He temperature range.
This experiment demonstrates in a clean
way that natural $4f^x$-$4f^2$ overlap (as in
the La Pr system) is _unimportant_ for
the valence transition in the first
approximation [29]. It is the single,

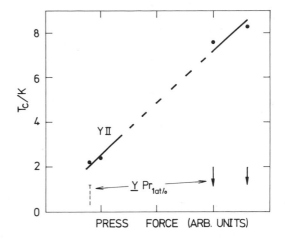

Figure 12: T_c vs. press force for the
high pressure phase Y II [27]. The data
cover the pressure range between
$\simeq 15o$ and $\simeq 3oo$ kbar. A 1 at/o Y Pr alloy
showed an onset of a superconducting
transition at the lowest P and remains
normal down to 1.2 K at the highest
pressures as indicated by arrows.

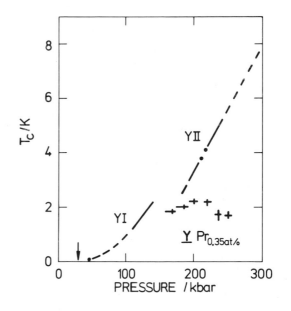

Figure 13: Preliminary data for the
T_c-P dependence of Y and a very dilute
Y Pr alloy. The phase transformation at
$\simeq 15o$ kbar is accompanied by a discontinuity in $T_c(P)$.

isolated $4f^2$ orbit, which delocalizes (if the picture, that we advocate, is correct) into a nonmagnetic $4f^2$ virtual bound state under compression. Our result lends considerable support to Fisk and Matthias' interpretation of the above-mentioned (Zr Pr) B$_{12}$ data [25] in terms of "lattice pressure".

In conclusion we would like to stress that these recent experiments provide considerable evidence that dilutely dissolved Pr atoms in a La or Y matrix undergo a valence transition at pressures over 2oo kbar. The exact nature of the valence transition deserves further investigation. Nevertheless, we have made it clear that we have a strong bias toward the (nonpromotional!) $Pr^{3+} \rightarrow Pr^{5+}$ possibility.

It has been pointed out to me by several colleagues that the anomalous pressure dependences of T_c (Figures 11 and 13) could also be caused by the onset of a (promotional) valence transition $Pr^{3+} \rightarrow Pr^{4+}$, or equivalently to a $4f^2 \rightarrow 4f^1$ transition, say, for $P \simeq$ 2oo kbar, because the Pr^{4+} configuration becomes energetically more favourable than Pr^{3+}. If the valence transition should in fact be of this kind and proceed with increasing pressure all the way through an intermediate valence regime to the stable $4f^1$ configuration for Pr, one should expect an appreciable depression of T_c at those pressures due to the magnetic $J = 5/2$ configuration. T_c-P studies to higher pressures than in the present work may therefore be useful to decide between this suggestion and our proposal of an, eventually, $4f^2$ nonmagnetic resonant state at pressures of the order of, say, 35o kbar.

5. SUMMARY AND OUTLOOK

The reported experiments have shown that the "electronic transition" of pure Pr metal has striking counterparts in dilute Pr alloys in the very same pressure range (or, perhaps more appropriately, under comparable compression?). The Pr single-impurity "electronic transition" is not triggered by the increased mutual $4f$-$4f$ overlap, since the effect occurs likewise for La Pr and Y Pr alloys. It is therefore quite likely that the $4f$-$4f$ overlap plays a secondary role in the electronic Pr III→IV transition of pure Pr metal also. In this case, we probably have to revise our previous position in which we placed emphasis on the increased $4f$-$4f$ overlap under P [9,12] The same conclusion may also hold for the analogous case of pure cerium [29].

It must therefore be the increase of the hybridization (V_{kf}^2) of the localized $4f^n (n=1,2)$ shells with the Bloch states which leads to a sufficiently broadened and hence nonmagnetic virtual bound $4f^n$ state, or, in the case of the pure metals, to a considerable effective $4f$ bandwidth under high P. This has not been known before with any degree of certainty. Generally speaking, this discovery is quite a striking example for the impact which high pressure experiments can have in materials research.

Based on the present knowledge it seems quite likely that even further $4f^n$ configurations $(n=3,4...)$ can also be delocalized (presumably by the same hybridization mechanism) under sufficiently high pressure. The investigation of other light rare earth metals under pressure up into the megabar range will therefore be an important project particularly in view of possible systematics with the corresponding 5f band metals. It is the author's urgent hope that the exciting subject of pressure-induced itinerant $4f$ electrons will have considerably advanced, before the next symposium, being devoted to the physics of solids under high pressure, takes place.

REFERENCES:

[1] Probst, C. and Wittig, J., Superconductivity, in Gschneidner, K.A. and Eyring, L. (eds.) Handbook on the Physics and Chemistry of Rare Earths Vol. I, p. 749 (North-Holland, Amsterdam, 1978).

[2] Wittig, J., The pressure variable in solid state physics: What about $4f$-band superconductors? in Queisser, H.J. (ed.) Festkörperprobleme XIII, p. 375 (Vieweg, Braunschweig, 1973).

[3] See for instance the articles by T.F. Smith and B. Coqblin et al. in Douglass, D.H. (ed.) Superconductivity in d- and f-band metals, AIP Conference Proceedings No. 4 (American Institute of Physics, New York, 1972). Freeman, A.J., Energy band structure, in Elliott, R.J. (ed.) Magnetic Properties of Rare Earth Metals, p. 245 (Plenum, London, 1972).

[4] Wittig, J., Comments on Solid State Physics 6 (1974) 13.

[5] Glötzel, D., J. Phys. F 8 (1978) L 163.

[6] Kornstädt, U., Lässer, R. and Lengeler, B., Phys. Rev. B21 (1980) 1898.

[7] Pickett, W.E., Freeman, A.J. and Koelling, D.D., Phys. Rev. B23 (1981) 1266.

[8] Wittig, J., to be published

[9] Wittig, J., Z. Physik B38 (1980) 11.

[10] Wittig, J., Phys. Rev. Lett. 21 (1968) 1250.

[11] Koskenmaki, D.C. and Gschneidner, K.A., Cerium, in Gschneidner, K.A. and Eyring, L. (eds.) Handbook on the Physics and Chemistry of Rare Earths Vol. 1, p. 337 (North-Holland, Amsterdam, 1978).

[12] Wittig, J., in Falicov, L.M. Hanke, W. and Maple, M.B., (eds.) Valence Fluctuations in Solids, p. 43 (North-Holland, Amsterdam, 1981).

[13] See for instance: Doniach, S., Many-electron effects in the actinides, in Freeman, A.J. and Darby, J.B. (eds.) The Actinides: Electronic Structure and Related Properties Vol. II, p. 51 (Academic, New York, 1974).

[14] Carter, W.J. Fritz, J.N., Marsh, S.P. and McQueen, R.G., J. Phys. Chem. Solids 36 (1975) 741.

[15] Cannon, J.F., J. Phys. Chem. Ref. Data 3 (1974) 781.

[16] Mao, H.K., Hazen, R.M., Bell, P.M. and Wittig, J., to be published.

[17] Maple, M.B., Wittig, J. and Kim, K.S., Phys. Rev. Lett. 23 (1969) 1375.

[18] Wittig, J., Phys. Rev. Lett. 46 (1981) 1431.

[19] Maple, M.B., Appl. Phys. 9 (1976) 179.

[20] Müller-Hartmann, E. and Zittartz, J., Z. Physik 234 (1970) 58.

[21] Coqblin, B. and Ratto, C.F., Phys. Rev. Lett. 21 (1968) 1065.

[22] Maple, M.B., DeLong, L.E. and Sales B.C., The Kondo Effect, in Gschneidner, K.A. and Eyring, L. (eds.) Handbook on the Physics and Chemistry of Rare Earths, Vol. 1, p. 797 (North-Holland, Amsterdam, 1978).

[23] There is now strong evidence (Heinrichs, M. and Wittig, J., unpublished) that the La host lattice symmetry is orthorhombic, namely Pr III-type, above ≃80 kbar. This is at variance with published Debye-Scherrer diffraction data (Syassen, K. and Holzapfel, B.W., Solid State Comm. 16 (1975) 533). The crystal-field splitting of the Pr ion is unknown under these conditions.

[24] Fulde, P., Crystal Fields, in Gschneidner, K.A. and Eyring, L. (eds.) Handbook on the Physics and Chemistry of Rare Earths, Vol. 2, p. 295 (North-Holland, Amsterdam, 1979) and references therein.

[25] Fisk, Z. and Matthias, B.T., Science 165 (1969) 279.

[26] We refer to the article by Maple et al. [22] for details.

[27] Y transforms to a high pressure phase Y II at a pressure of ≃150 kbar at room temperature as seen from a discontinuous resistance decrease (J. Wittig, unpublished data). Since Y is a pure d-band metal in contrast to the 4f-band metal La, we strongly expect a different sequence of crystallographic phases under P.

[28] This is probably linked to the pressure-induced s→d transfer.

[29] The first proof that an isolated 4f shell in a d-band host can be delocalized under P has been given for the system Y Ce by Maple, M.B. and Wittig, J., Solid State Comm. 9 (1971) 1611. $4f^x$ denotes the 4f occupation in La. Existing band structure calculations, [5] and Pickett, W.E. et al., Phys. Rev. B22 (1980) 2695, disagree considerably as to the value of x.

PHYSICS OF SOLIDS UNDER HIGH PRESSURE
J.S. Schilling, R.N. Shelton (editors)
© *North-Holland Publishing Company, 1981*

PRESSURE-INDUCED VALENCE AND STRUCTURE CHANGE IN SOME ANTI-Th_3P_4
STRUCTURE RARE EARTH COMPOUNDS

A. Werner*, H. D. Hochheimer*, A. Jayaraman[+], and E. Bucher[#]

*Max-Planck Institut für FKF, Stuttgart, Germany
[+]Bell Laboratories, Murray Hill, NJ 07974 USA
[#]Universität Konstanz, Konstanz, Germany

The anti-Th_3P_4 structure compounds Yb_4Bi_3 and Yb_4Sb_3 have been investigated to
350 kbar by high pressure X-ray diffraction, using the diamond anvil cell. From the
P-V data it is found that Yb_4Bi_3 and Yb_4Sb_3 are much more compressible, compared to
Sm_4Bi_3 before the valence transition. This suggests that a continuous change in the
valence state of Yb takes place with pressure in the two compounds and that they may
be in the mixed valent state already at ambient pressure. The "collapsed" anti-Th_3P_4
structure becomes unstable in Yb_4Bi_3 and Yb_4Sb_3 and new lines appear at high pressure,
that fit the NaCl structure. The latter structure change seems to occur also in the
electronically collapsed Sm_4Bi_3. The results will be presented and discussed.

1. INTRODUCTION

Many rare-earth germanides, antimonides and
bismuthides of the formula R_4B_3, where B is Ge,
Sb or Bi, are known to crystallize in the anti-
Th_3P_4 structure (cubic) (1-3), with the R
occupying the phosphorus sites (this has been
referred to as the Gd_4Bi_3 structure-type in
Reference 2). Among these, the bismuthides of
Sm, Eu and Yb are reported to have anomalously
large lattice parameters due to their divalent
character (1-4). It is believed that in
Sm_4Bi_3, Yb_4Bi_3 and Yb_4Sb_3 three of the rare
earth ions are divalent (2,4). The lattice
constant of Eu_4Bi_3 indicates that all the four
Eu ions are in the divalent state. Figure 1
shows the lattice parameter of the rare earth

Anti-Th_3P_4 structure compounds involving Sb and
Bi. The anomalously large lattice parameters
for Sm_4Bi_3, Eu_4Bi_3, Yb_4Bi_3 and Yb_4Sb_3 are
indicative of the presence of divalent rare
earth ions in these compounds. Hence a pressure-
induced valence change may be expected in them.
Indeed such a transition has been reported in
Sm_4Bi_3 at about 27 kbar (5). The pressure-
induced transition is first-order and is isostructural; the anti-Th_3P_4 structure is
retained in the high pressure phase, but the
lattice parameter decreases strikingly at the
transition pressure, from 9.70 to 9.4 Å.

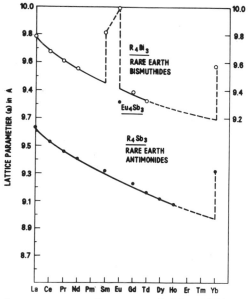

Figure 1: The lattice parameter of rare earth bismuthides and antimonides of the
formula R_4Bi_3 and R_4Sb_3. They crystallize in the anti-Th_3P_4 structure.

In the present study we have investigated the effect of pressure on Yb_4Sb_3 and Yb_4Bi_3, using high pressure X-ray diffraction techniques. From the pressure-volume relationship we find that both the Yb compounds exhibit a high compressibility, before transforming to a structure which we believe is defect NaCl type. We attribute the high compressibility to a gradual valence change of the divalent Yb. The structural instability is caused when the change in the radius ratio due to the valence transition crosses the limit of stability for the anti-Th_3P_4 lattice. These results will be presented and discussed.

2. EXPERIMENTS

The method of preparation of the samples has been described elsewhere (2,4). Since Yb_4Bi_3 and Yb_4Sb_3 are known to melt congruently, direct fusion of stoichiometric amounts of the rare earth metal and Bi or Sb in a sealed Ta container appears to work satisfactorily. The samples were polycrystalline, but quite hard and coherent. X-ray powder diffraction patterns showed that the material is single phase, with the anti-Th_3P_4 structure and lattice parameter in agreement with previously published values. The material has a silvery gray appearance.

Pressure-volume data were obtained using a gasketed diamond anvil cell (6) and the energy dispersion X-ray technique. Different pressure media were used as indicated in Fig. 2 and Fig. 3. The pressure generated was measured by the well-known ruby fluorescence technique (7).

With our setup it was possible to obtain a good powder pattern consisting of (310), (321), (420), (422), (510), (611) and (541) reflections of the cubic anti-Th_3P_4 structure. The cubic lattice parameter $\underline{a}$ was calculated at each pressure from the above reflections and the average value taken for computing the volume.

3. RESULTS AND DISCUSSION

The pressure-volume data are shown in Figs. 2 and 3. Calculated fit to the data in the range $0 \leq P \leq 100$ kbar using the Murnagan equation of state yields for Yb_4Bi_3 $B_0 \approx 320\pm25$ kbar and $B' \approx 7.0$ and for Yb_4Sb_3 $B \approx 430\pm25$ kbar and $B' \approx 7.0$. In Fig. 4 the pressure-volume data for Sm_4Bi_3 are shown. For pressures less than the first-order valence transition pressure, a similar fit to the data yields $B_0 = 523\pm50$ kbar. It has been shown by Anderson and Nefe (8) and Jayaraman et al (9) that bulk modulus scales with the volume, for related systems. If such a behavior is assumed for the anti-Th_3P_4 compounds, a bulk modulus of 580 and 630 kbar would be expected for Yb_4Bi_3 and Yb_4Sb_3. The presently obtained bulk moduli of 320 and 430 kbar respectively for the two compounds are thus strikingly small. We believe that the abnormally low value of the bulk modulus and B' of ~ 7 for Yb_4Sb_3 and Yb_4Bi_3 are due to a continuous valence collapse with pressure, and that the system is already in the mixed valence state at ambient pressure. An abnormal compression and hence low bulk modulus and high B' value are characteristic of mixed valence systems (10). For instance TmSe which is a

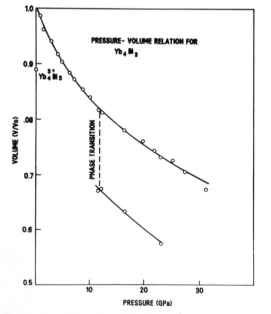

Figure 2: Pressure-volume data for Yb_4Bi_3. The decrease in volume near 120 kbar is due to a first-order phase transition (see text for the structure of the high pressure phase) triggered by unfavorable radius ratio.

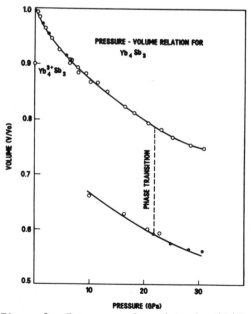

Figure 3: Pressure-volume data for Yb_4Sb_3. The abrupt change in volume near 220 kbar is due to first-order phase transition (see text for explanation) triggered by unfavorable radius ratio.

typical mixed valence system (11) has a bulk modulus of 375 kbar and B' ≈ 13 and undergoes a continuous compression (12). That the Yb compounds may be in a homogeneously mixed valent state at ambient pressure is also supported by the temperature dependence of the magnetic susceptibility (4).

At pressures higher than 120 kbar in the case of Yb_4Bi_3 and 220 kbar in the case of Yb_4Sb_3, two new peaks appear in the Debye-Scherrer pattern. These new peaks grow in intensity with pressure at the expense of the anti-Th_3P_4 peaks, but the latter never completely disappear even at 360 kbar, the highest pressure reached in the present study. The d-values of the observed reflections from the anti-Th_3P_4 structure at ambient pressure, ∿ 170 kbar and ∿ 310 kbar are given in Table I. The d-values of the two new peaks as measured at about 310 kbar are also given at the bottom of the table. Above ∿ 120 kbar for Yb_4Bi_3 and above 220 kbar for Yb_4Sb_3 the anti-Th_3P_4 structure becomes unstable. A similar transition is also seen in the case of Sm_4Bi_3 near 100 kbar (see Fig. 4). In fact this is to be expected in the event of a valence change in these compounds (see discussion later) and hence this itself is proof for the occurrence of pressure-induced valence change. We tentatively regard the sluggish phase transition at high pressure in both Yb compounds to be due to a deficiency NaCl-type structure. Accordingly, the two new peaks are assigned to the (200) and (222) reflections of the NaCl-lattice. The (220) and

higher index reflections from the NaCl phase should be present, but they probably overlap with the anti-Th_3P_4 peaks which are present even at the highest pressure. On the assumption that the new structure is to a deficiency NaCl-type (at the phase transition pressures of ∿ 120 and ∿ 220 kbar) in the case of Yb_4Bi_3 and Yb_4Sb_3 the volume change at the structural transition is about 15% and about 19% respectively.

4. STABILITY OF THE ANTI-Th_3P_4 STRUCTURE

From Fig. 1 it is clear that only the lighter rare earths crystallize in the anti-Th_3P_4 structure. There is no corresponding 4:3 compound in the case of heavy rare earths. In this connection Gambino (2) has proposed that the radius ratio of the rare earth ion to the pnictide controls the limit of stability of the anti-Th_3P_4 structure and the latter can exist only for radius ratios between the limits $0.520 > R_R/R_P > 0.417$. The Yb compounds crystallize in the anti-Th_3P_4 structure, only because the Yb is in the divalent state. When Yb ions are fully trivalent, the radius ratio would be 0.410 in the case of Yb_4Bi_3 and 0.419 in the case of Yb_4Sb_3. This indicates that $Yb_4^{3+}Sb_3$ would be marginally stable in the anti-Th_3P_4 structure and for $Yb_4^{3+}Bi_3$ the radius ratio of 0.410 falls clearly outside the limits indicated above. From this it can be inferred that Yb_4Bi_3 would be driven to the pressure-induced instability sooner than Yb_4Sb_3. The

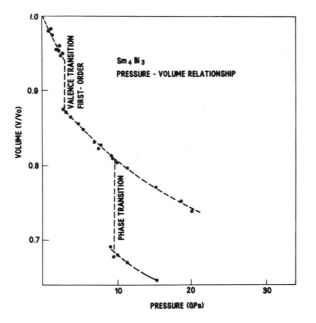

Figure 4: Pressure–volume data for Sm_4Bi_3. Isostructural first-order transition near 30 kbar is due to valence change. Transition near 100 kbar is due to the breakdown of the anti-Th_3P_4 structure due to unfavorable radius ratio.

TABLE I

Lattice parameter data at high pressure for the anti-Th_3P_4
phase and the high pressure phase of Yb_4Bi_3 and Yb_4Sb_3

Substance	Reflection	d-value (Å)		
		Atmospheric	170 kbar	312 kbar
Yb_4Bi_3	(310)	3.028	2.780	2.651
	(321)	2.549	2.354	2.245
	(420)	2.277	—	—
	(422)	1.961	1.802	1.721
	(510)	1.878	1.733	1.653
	(611)	1.551	1.428	1.354
	(541)	1.488	—	—
		Atmospheric	179 kbar	308 kbar
Yb_4Sb_3	(310)	2.947	2.747	—
	(321)	2.494	2.312	2.262
	(420)	—	1.946	1.884
	(422)	1.899	1.779	1.727
	(510)	1.827	1.710	1.654
	(611)	1.514	1.410	1.374
	(541)	1.437	1.340	

High pressure phase (NaCl?)   (200)   (222)

Yb_4Bi_3 - 2.437 and 1.411 at 312 kbar

Yb_4Sb_3 - 2.423 and 1.407 at 308 kbar

smaller is the radius ratio (within the indicated
limits) the lower would be the transition pressure. This is well borne out by the present
study. Yb_4Bi_3 undergoes the phase change near
120 kbar, whereas for Yb_4Sb_3 the phase transition sets in only near 220 kbar. Further, we
may also expect Yb_4Sb_3 to undergo the valence
transition fully before the structural
instability sets in, because the radius ratio
for $Yb_4^{3+}Sb_3$ lies still within the limits of
stability for the anti-Th_3P_4 structure. (The
radius ratios computed are for the ambient
pressure. At high pressure they would differ
by a small amount. But the arguments given
above are still valid in a relative sense.) It
is significant that the anti-Th_3P_4 structure
becomes unstable around $V/V_o \approx 0.8$ in all the
three cases (compare Figs. 2, 3, and 4). This
strongly suggests the role of radius ratio in
the phase change.

5.   ON THE NATURE OF THE HIGH PRESSURE PHASE

In the phase diagram of the rare earth pnictides
the NaCl-type phase occurs at the 1:1 composition, adjoining the field of stability of the
anti-Th_3P_4 structure. In systems involving
larger rare earth ions, the 1:1 compounds with
NaCl-type   structure melt incongruently whereas
the anti-Th_3P_4 structure compounds melt congruently. The reverse is true with smaller
rare earth ions (heavier rare earths); the
NaCl-type structure melts congruently and the
anti-Th_3P_4 structure incongruently. In some
sense this may be regarded as indicative of the
relative stability of the two structures with
regard to the size of the rare earth ion.
Accordingly the anti-Th_3P_4 structure would be
less stable with heavier rare earths. As
pointed out earlier, the Yb compounds crystallize in this structure only because of the
large size of the divalent Yb. When the radius
turns unfavorable because of the valence change,

the anti-Th$_3$P$_4$ structure should become unstable and from the phase diagram, the favored structure could very well be the NaCl-type. However it can be argued that the 4:3 composition is farther away from the 1:1 required for the NaCl-type structure. In this connection it is to be noted that some rare earth monosulfides exist in the NaCl-type phase as a deficiency structure, up to a limiting composition of $R_{1.0}S_{0.75}$, which would be R_4S_3 (13). Therefore it is not impossible that both Yb$_4$Sb$_3$ and Yb$_4$Bi$_3$ under pressure assume a deficiency NaCl structure. The volume of the NaCl-type phase has been calculated from the two new peaks assigned to (200) and (222) reflections and plotted in Figs. 2 and 3. The volume change is rather large and the density changes quite substantially in the right direction.

It is puzzling that the anti-Th$_3$P$_4$ phase appears to coexist over a large pressure range with the new phase; even at 350 kbar, our highest pressure reached in the present study. A rationalization for this behavior is, since the transition from anti-Th$_3$P$_4$ to the deficiency NaCl-type structure involves a drastic structural rearrangement it may be expected to be very very sluggish. It cannot be completely ruled out that we are dealing with a new high pressure phase different from the one we have postulated.

6.   COMMENTS ON CONTINUOUS VALENCE CHANGE WITH PRESSURE

While it is not uncommon for divalent Sm compounds to undergo a pressure-induced first-order valence transition (Sm$_4$Bi$_3$ itself undergoes a first-order valence change at 27 kbar pressure), so far no divalent Yb compound is known to undergo a first-order valence change (14). All the Yb chalcogenides including the recently investigated YbO (15) exhibit only a continuous valence change at high pressure. Thus the gradual valence change in the anti-Th$_3$P$_4$, Yb$_4$Bi$_3$ and Yb$_4$Sb$_3$ is consistent with the behavior of other divalent Yb compounds. The reason behind this, we believe, lies in the Hund's rule stability criterion for the rare earth valence. At half filling (4f^7) and for fully filled 4f^{14} configuration the Hund's rule coupling energy is quite strong, making the divalent state of both Eu and Yb very stable. Because of this any promotion of the 4f electron under the influence of pressure or other constraints would be strongly counteracted by an appropriate adjustment of the electronic levels preventing a catastrophic phase transition from taking place. For instance the 4f level may move away from the

5d band making further promotion difficult, without a further increase in compression energy. Hence any valence transition in divalent Eu and Yb systems may be expected to take place only continuously.

In any valence transition with rare earths, the compressibility of the lattice could be the dominant factor in determining whether a transition is first-order or continuous. We believe that this is the case for Sm chalcogenides; SmS shows a first-order valence transition while SmSe and SmTe exhibit a continuous valence transition (12). On the other hand in Eu and Yb compounds, the Hund's rule coupling energy is probably the dominant factor in controlling the order of the valence transition.

REFERENCES

[1]   Hohnke, D. and Parthe, E., Acta Cryst. 21 (1966) 435.
[2]   Gambino, R. J., J. Less Common Metals 12 (1967) 52.
[3]   Yoshihara, K., Taylor, J. B., Calvert, L. D. and Despault, J. G., J. Less Common Metals 41 (1975) 329-37.
[4]   Bucher, E., Cooper, A. S., Jaccard, D., and Sierro, J., in Parks, R. D. (ed.), Valence Instabilities and Related Narrow-Band Phenomena (Plenum, New York, 1977).
[5]   Jayaraman, A., Maines, R. G., and Bucher, E., Solid State Commun. 27 (1978) 709-11.
[6]   Weinstein, B. A. and Piermarini, G. J., Phys. Rev. B 12 (1975) 1172.
[7]   Barnett, J. D., Block, S., and Piermarini, G. J., Rev. Sci. Instrum. 44 (1973) 1.
[8]   Anderson, O. L. and Nefe, J. L., J. Geophys. Res. 70 (1965) 3951.
[9]   Jayaraman, A., Singh, A. K., Chatterjee, A. and Usha Devi, S., Phys. Rev. B 9 (1974) 2513.
[10] Penny, T. and Melcher, R. L., J. Physique Colloq. 37 (1976) C4-275.
[11] Batlogg, B., Ott, H. R., Kaldis, E., Thoni, W., and Wachter, P., Phys. Rev. B 19 (1979) 247.
[12] Werner, A. and Debray, D (to be published).
[13] Flahaut, J., in Gschneidner, E., Jr. and Eyring, L. (eds.), Handbook on the Physics and Chemistry of Rare Earths (North-Holland Publishing Co., Amsterdam and New York, 1979), Vol. 4, p. 17.
[14] Jayaraman, A., Dernier, P. D., and Longinotti, L. D., High Temps. High Press. 7 (1975) 1-28.
[15] Werner, A., Hochheimer, H. D., Jayaraman, A., and Leger, J. M., Solid State Commun. 38 (1981) 325.

PHYSICS OF SOLIDS UNDER HIGH PRESSURE
J.S. Schilling, R.N. Shelton (editors)
© North-Holland Publishing Company, 1981

PRESSURE INDUCED SEMICONDUCTOR-METAL TRANSITIONS IN Tm$_{1-x}$Eu$_x$Se

H. Boppart and P. Wachter

Laboratorium für Festkörperphysik, ETH Zürich
CH-8093 Zürich, Switzerland

In the Tm$_{1-x}$Eu$_x$Se pseudobinary system a transition from semiconductor to metal (SMT) is observed in function of composition. Here, we investigate the pressure induced SMTs for various Tm$_{1-x}$Eu$_x$Se compounds in a continuous measurement of the electrical resistivity and the specific volume. The SMTs are all continuous, but for one compound almost discontinuous.

1. INTRODUCTION

In the TmSe-EuSe system the rare earth ions are exchanged against each other. The substitution of the Eu ions in EuSe by smaller Tm ions leads to a reduction of the lattice constant as can be seen in Fig. 1. For x>0.25 the Tm$_{1-x}$Eu$_x$Se crystals are semiconductors and at x≈0.2 a compositionally induced semiconductor to metal transition (SMT) is observed (1-3).

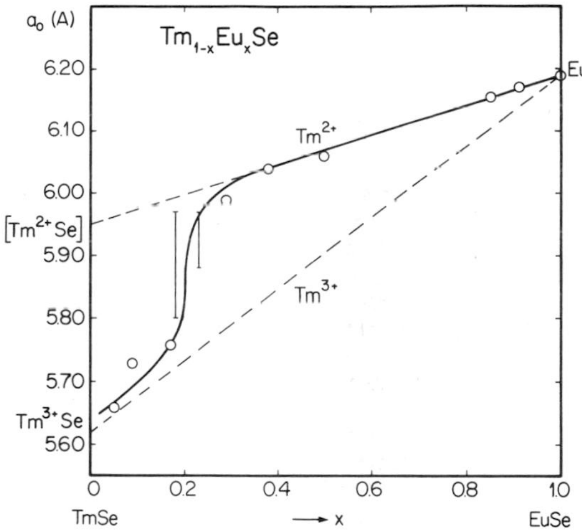

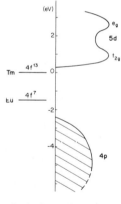

Fig.2: Energy-level scheme for semiconducting Tm$_{1-x}$Eu$_x$Se.

The gross features of the electronic-level scheme are explained in the following model: in the crystalline arrangement the 5d and 6s orbitals of the Tm and Eu ions overlap and form a common conduction band of some eV width. The rather localized 4f shells are either occupied by 7 (Eu) or 13 (Tm) electrons and their difference in energy (∿1.3 eV) is closely related to the difference of the third ionization energies for each rare earth ion (5).

Fig.1: Lattice constants (NaCl-structure) versus composition for Tm$_{1-x}$Eu$_x$Se. Isovalent Vegard-law lines indicate the change of Tm-valence.

EuSe is a semiconductor with an energy gap of 1.8 eV, the Eu ions are in a divalent state. This gap is determined by the separation in energy of the 4f^7 and the 4f^6(5d6s)^1 configurations. The substitution of Eu ions by Tm ions introduces Tm 4f^{13} states lying only 0.4 eV or less below the 5d6s conduction band, as determined by optical reflectivity measurements (4). In Fig.2 an energy-level scheme is proposed for semiconducting Tm$_{1-x}$Eu$_x$Se.

The Tm and the Eu ions in Tm$_{1-x}$Eu$_x$Se (x>0.3) are in a divalent state. This can be concluded from first, the results of the magnetic susceptibility measurements and second, the lattice constants (6). In Fig.1 the isovalent Vegard-law lines are drawn and it can be seen that for x>0.3 all compounds lie on the Tm^{2+}Se-Eu^{2+}Se line.

The reduction of the lattice constant due to the substitution of Eu ions by Tm ions increases the strength of the ligand field thus the crystal field split 5dt$_{2g}$ and 5de$_g$ parts of the conduction band move apart in energy keeping the 4f^{13}-5d center of gravity essentially the same.

This behavior can also be observed in the optical reflectivity spectra of $Tm_{0.5}Eu_{0.5}Se$ and $Tm_{0.15}Eu_{0.85}Se$ (7).

In the semiconducting compositions (x>0.25) a SMT can also be induced by modest external pressure. This SMT is driven by an increase of the ligand field splitting of the 5d conduction band due to a pressure induced reduction of the lattice constant. Consequently, the $4f^{13}$-5d energy gap is reduced and this leads to a valence change of only the Tm ions. The valence of Eu, however, would remain the same unless additional pressure (∿20 GPa) is applied to close the 1.8 eV gap.

We investigated the pressure induced SMT for some semiconducting $Tm_{1-x}Eu_xSe$ compositions. The two crucial quantities, characterizing such a SMT due to a valence change of the rare earth ion, have been measured:

1) The electrical resistivity, determined essentially by the number of its itinerant electrons.
2) The specific volume which is sensitive to the 4f configuration.

2.   EXPERIMENTAL

Hydrostatic pressure has been generated in a conventional piston-cylinder device with ether as pressure transmitting medium. Resistivity measurements were carried out using a conventional four probe method. The volume-pressure dependence was determined by means of a strain gauge technique. Two strain gauges were glued onto opposite faces of both the crystals to be measured and a reference of known compressibility (e.g. Cu). The four strain gauges are then connected to a resistance bridge. To eliminate the uncontrolled pressure effects on the strain gauge material (constantan) it is necessary to use a stabilized current for the bridge supply. Then the difference in the change of length of both crystals yields the volume-pressure dependence of the sample because the materials are cubic.

In Fig.3a the behavior of the resistivity under pressure for three different $Tm_{1-x}Eu_xSe$ compounds (x=0.5,0.38,0.29) is shown. From the pressure dependence of the resistivity ρ, a value for the activation energy for $4f^{13}$→5d excitation can be obtained. Assuming simple statistics for the number of carriers in the conduction band, ρ is given by $\rho(\Delta E,T)=\rho_o \cdot exp(\Delta E/kT)$, where ρ_o denotes the experimental value of ρ if the gap ΔE=0 and $\rho(\Delta E,T)$ the resistivity at ambient pressure. These experimental values and the numerical results are specified in Table 1. The exponential resistivity-pressure relation implies that the gap is closing linearly under pressure, thus,

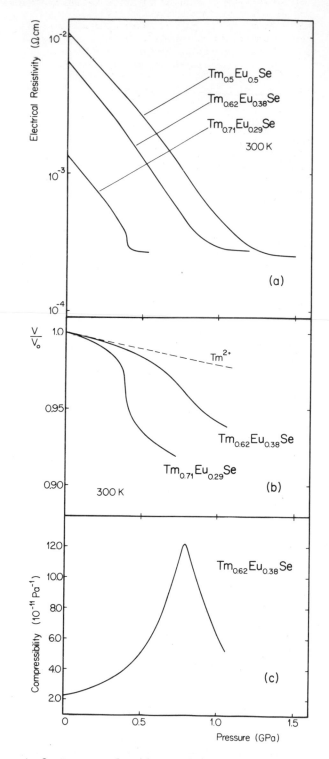

Fig.3: Pressure dependence of (a) the resistivity (b) the specific volume and (c) the compressibility of $Tm_{1-x}Eu_xSe$ at 300K.

$\Delta E(p)=\Delta E_o + (d\Delta E/dp)\cdot p$. This statement put in the above formula and the slope of the experimental curve yield the closing rate $d\Delta E/dp$ of the energy gap. Nearly the same closing rate in the investigated compounds is a striking feature of our data. A similar behavior was observed in Eu- and Yb- substituted SmS compounds (8).

The results of the volume-pressure measurement for two compounds (x=0.38,0.29) is shown in Fig.3b. The derivative of the volume-pressure curve yields the compressibility-pressure relation. In Fig.3c this relation is shown for Tm$_{0.62}$Eu$_{0.38}$Se. The initial compressibility κ for this compound is $2.3\cdot10^{-11}$Pa^{-1}. Already at low pressure the crystal gets markedly weaker and at about 0.8 GPa κ reaches the maximum value of $12\cdot10^{-11}$Pa^{-1}. Above this pressure, the crystal stiffens and κ decreases again. Pure EuSe has an initial compressibility of about $2\cdot10^{-11}$ Pa^{-1} (9). With increasing Tm fraction the initial compressibility is weakly enhanced as can be seen in Table 1.

A very interesting behavior is displayed by Tm$_{0.71}$Eu$_{0.29}$Se, a compound closest to the compositionally induced SMT. Both the volume-pressure and the resistivity-pressure relation indicate a tendency to a first order transition at about 0.4 GPa. We think only the stabilisation of the lattice due to divalent Eu ions prevents the discontinuous valence transition of the Tm ions.

3. DISCUSSION

The general feature of the TmSe-EuSe system lies in the introduction of small Tm ions into a stable EuSe lattice. This produces 4f levels lying little below the 5d conduction band. The combination of narrow energy gaps and relatively high compressibilities makes it possible to close the gaps with only modest pressure. While the gap is closing under pressure, we get a larger and larger admixture of 5d wavefunctions to the 4f^{13} states. This manifests itself in a strong reduction of the specific volume (see Fig.3b) and a softening of the crystal (see Fig.3c). The smaller the gap already is, the larger both effects are. Finally, we observe the highest compressibilities just when the resistivity indicates a nearly closed gap.

The dashed line in Fig. 3b shows the expected behavior of Tm$_{0.62}$Eu$_{0.38}$Se as variation of pressure with purely divalent Tm. The curve was calculated by the Birch equation of state with values of B_o=43 GPa and B_o'=4 (10). The main conclusion can be drawn if $\rho(p)$ and $V(p)$ are compared. This is possible since both quantities have been measured continuously on the same single crystal. (1) The valence of Tm is appreciably mixed already on the semiconducting side. (2) the valence varies continuously through the SMT and (3) stays intermediate in the metallic phase. Pressures of about 4-5 GPa are necessary to drive Tm completely trivalent. However, these pressures are much

Table 1.   Semiconducting Tm$_{1-x}$Eu$_x$Se (x>0.25)

x =	0.5	0.38	0.29
Lattice constant (Å)	6.06	6.04	5.99
Resistivity at ambient pressure (Ωcm)	$1.1\cdot10^{-2}$	$6.7\cdot10^{-3}$	$1.3\cdot10^{-3}$
Resistivity at high pressure (Ωcm)	$2.5\cdot10^{-4}$	$2.7\cdot10^{-4}$	$2.7\cdot10^{-4}$
Transition pressure (GPa)	1.3	1.0	0.4
($d\Delta E/dp$) for Tm 4f$^{13}\to$5d (meV·GPa^{-1})	-77	-81	-80
4f^{13}-5d energy gap for Tm from pressure data (meV)	100	80	40
Initial compressibility κ (10^{-11}Pa^{-1})		2.3	2.4
Maximum compressibility κ (10^{-11}Pa^{-1})		12	120
at pressure (GPa)		0.8	0.4

lower than the required pressure to change the
Eu valence ($\sim$20 GPa) and thus as long as the Tm
valence changes the Eu ions remain divalent.

Due to the increasing hybridization of the 4f
and 5d wavefunctions the Tm valence is already
mixed when the gap is not yet closed. We propose
that these hybridization effects in the still
semiconducting compounds cause significant deviations from the usual model as shown e.g. in
Fig.2. A strong mixing of 4f and 5d wavefunctions
will lead to a dispersive 4f-5d band, with mostly
4f character, still separated by a tiny gap from
a 5d-4f band, with mostly 5d character.

ACKNOWLEDGEMENT

The authors are very grateful to Dr. E. Kaldis
for growing and chemically characterizing the
single crystals.

REFERENCES

(1) B. Batlogg, E. Kaldis and P. Wachter
     J. de Phys. 40-C5, 370 (1979)
(2) B. Batlogg and P. Wachter, J. de Phys.
     41-C5, 59 (1980)
(3) E. Kaldis and B. Fritzler, J. de Phys.
     41-C5, 135 (1980)
(4) B. Batlogg, Phys. Rev. B 23, 650 (1981)
(5) L.R. Moss, J. Phys. Chem. 75, 393 (1971)
(6) H. Boppart and P. Wachter, J. Appl. Phys.
     52, 2161 (1981)
(7) H. Boppart, L. Frey and P. Wachter (un-
     published)
(8) A. Jayaraman and R.G. Maines, Phys. Rev.
     B 19, 4154 (1979)
(9) A. Jayaraman, A.K. Singh, A. Chatterjee
     and S. Usha Devi, Phys. Rev. B 9, 2513
     (1974)
(10) F. Birch, J. Geophys. Res. 57, 227 (1952)

PHYSICS OF SOLIDS UNDER HIGH PRESSURE
J.S. Schilling, R.N. Shelton (editors)
© *North-Holland Publishing Company, 1981*

HIGH PRESSURE DIFFRACTION STUDIES OF YbH_2 UP TO 28 GPa

J. Staun Olsen*, B. Buras*[†], L. Gerward**, B. Johansson[+], B. Lebech[†], H. Skriver[+++] and
S. Steenstrup*

*) Physics Laboratory II, University of Copenhagen, [†] Risø National Laboratory,
**) Laboratory for Applied Physics III, Technical University of Denmark, [+] Physics
Department, University of Aarhus, [++] Nordita, Copenhagen, Denmark.

It is expected that Yb in orthorhombic YbH_2 should undergo a phase transition, at
moderate pressure, from the divalent to a trivalent state, and that this transition
might be followed by a change to the fluorite structure characteristic of the trivalent
rare-earth hydrides.

For these reasons high pressure neutron and X-ray diffraction studies of YbH_2 have been
undertaken.

Neutron diffraction measurements up to 4 GPa and preliminary X-ray results up to 11 GPa
have shown no structural changes. Very recent X-ray diffraction experiments up to 28
GPa using synchrotron radiation have shown a clear structural change at about 14 GPa
from the orthorhombic phase to a hexagonal close packed with c/a = 1.34. The transition
is accompanied by a 5.4% decrease in volume. A preliminary interpretation of the transition is given.

1. INTRODUCTION

In several rare-earth compounds and intermetallics the valence state of the rare-earth ion
may be changed between the $4f^n(5d6s)^2$ and
$4f^{n-1}(5d6s)^3$ configurations by chemical mani-
pulations or by application of high pressure
[1]. By now quite a number of such unstable
valence systems have been studied, although the
main emphasis has been on chemically manipulated systems rather than on compressed systems.
Furthermore, no attention has been directed towards those cases where the valence change may
be accompanied by a change in crystals structure.

It can be expected that the rare-earth dihydride, YbH_2, is such a system. At zero pressure it shares the orthorhombic structure with
EuH_2, and in both compounds the ground state
configuration is $4f^n(5d6s)^2$. In contrast, the
ground state configuration of the other rare-
earth dihydrides is $4f^{n-1}(5d6s)^3$, and they all
form in the fcc fluorite structure. Hence, when
the valence state of Yb is changed to $4f^{n-1}$
$(5d6s)^3$ under pressure one would expect YbH_2 to
transform into an fcc structure.

Following these considerations we have undertaken high pressure experiments with YbH_2.

The preliminary measurements were performed at
Risø National Laboratory (Denmark) using the
neutron powder diffraction technique and the
high pressure cell described in ref. [2]. At
pressures up to 4 GPa no phase transformation
was observed. At the same time an attempt to
find the expected phase transformation was made
by means of a diamond anvil cell and the X-ray
white beam energy-dispersive technique [3].
Measurements up to 11 GPa performed at Physics
Laboratory II, University of Copenhagen, using
an X-ray tube did not show any phase transformation either. Only very recently experiments

at pressures up to 28 GPa have clearly demonstrated a phase transformation at about 14 GPa.
A preliminary account on these experiments and
the obtained results, followed by a discussion is
the subject of this paper.

2. EXPERIMENT

A diamond anvil high pressure cell was used and
an Ytterbium dihydride powder sample 200 μm in
diameter and 80 μm thick was enclosed in an Inconel gasket. A 4:1 methanol-ethanol mixture and
a ruby were added to the powder to allow for
hydrostatic pressure conditions and a proper
calibration of pressure, respectively.

The X-ray diffraction studies were performed at
the electron storage ring DORIS at DESY-HASYLAB
in Hamburg. The electron energy was 4 GeV (critical wavelength 1.06 Å) and the time averaged
electron current produced by 20 bunches was
usually between 20 and 40 mA. The white beam
energy-dispersive diffractometer [4] and the
triple axis spectrometer [5] working in the
energy dispersive mode were used. By applying
very fine slits (100 μm × 100 μm) in the incident
beam we avoided the diffraction by the gasket.
The very good collimation of the incident beam,
the small size of the illuminated part of the
sample, the relative large distance (35 cm)
between the sample and the detector, and a 200
μm slit in front of the detector resulted in a
small geometrical contribution to the total
resolution. The pure Germanium solid state detector had an energy resolution of 150 eV (FWHM)
at 5.9 keV. With a scattering angle of 2θ =
14.51°, used in this experiment, the FWHM of a
diffraction peak at 20 keV was 300 eV, corresponding to 10 channels.

The energy analysis of the diffracted X-rays was
carried out by means of a 2048-channel analyser
covering the range from 10 keV to 70 keV. The

exposure time was usually 500 s. The diffraction patterns were recorded on an x-y plotter and on paper tape or a floppy disk. The centroids of the diffraction maxima were determined by the Jupiter multichannel analyser system or by computer fitting.

The scattering angle was determined by fitting the spectra at zero pressure to the known structure of the sample. For this preliminary report this fitting and the ones reported below were done in a simplified way described in the next section. A more careful fitting will be done later, but it cannot change the basic results presented in this paper. However, the absolute values of the lattice constants may differ slightly from those presented here.

The pressure was determined by measurements of the frequencies of the two ruby lines, which could be clearly seen up to highest pressures. The accuracy of frequency determination corresponds to an uncertainty in the pressure less than 0.2 GPa. However, the uncertainty in the pressure of the sample might be larger due to pressure gradients in the cell.

3. RESULTS

Before presenting our results we recall the literature data concerning the structure of YbH_2 at atmospheric pressure. It is generally accepted that the structure is orthorhombic of space group Pnma (4 units YbH_2). Korst and Warf [6] on the basis of an X-ray study on $YbD_{1.98}$ found that with the origin at $\bar{1}$ the Yb atoms are in positions 4(c) with x = 0.240 and z = 0.110. The lattice constants measured by various authors are shown in Table I. The positions of the

Table I

Lattice constants of orthorhombic
Ytterbium dihydride at atmospheric pressure

Sample	Reference	a(Å)	b(Å)	c(Å)
YbD_2	6	5.871	3.561	6.763
YbH_2	7*	5.905	3.570	6.792
$YbH_{1.78}$	8*	5.895	3.574	6.801
YbH_2	this work	5.898	3.576	6.765

*) For reasons of comparison the lattice constants were interchanged.

hydrogen atoms were not determined but it was suggested [9] that their positions might be similar to those in CaH_2 [10].

The basic results of our high pressure measurements are presented in Table II and in Fig. 1 showing characteristic examples of diffraction spectra at atmospheric pressure, 14.3, 17.0, and 28.2 GPa. As can be seen, at pressures up to 11.5 GPa all diffraction peaks (except one discussed below) can be indexed in an orthorhombic unit cell. At pressures from 19.2 GPa all peaks (except two discussed below) can be indexed in a hexagonal close packed cell. At 14.3 GPa most of the peaks belong to the orthorhombic structure, however some already indicate

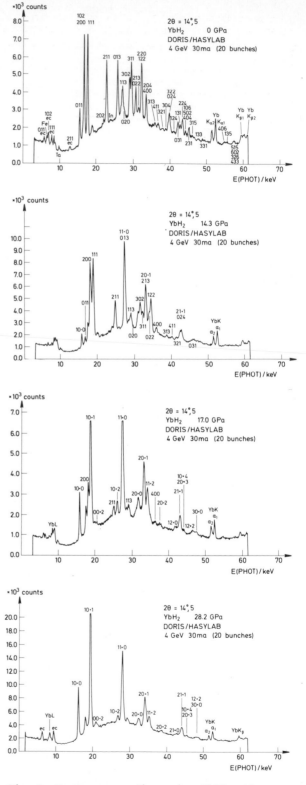

Fig. 1. X-ray energy-dispersive diffraction patterns of YbH_2 at several pressures.

Table II

Lattice constants of YbH$_2$ as function of pressure. $\bar{\Delta}$ is the mean difference (in channel numbers) between the observed and calculated peak positions for the 7 strongest lines in each spectrum.

Pressure	Orthorhombic phase					h.c.p. phase		
GPa	a	b	c	$\bar{\Delta}$		a	c	$\bar{\Delta}$
0.0	5.898	3.576	6.765	0.9				
9.3	5.521	3.410	6.482	0.9				
11.53	5.465	3.385	6.448	1.4				
14.3	5.335	3.405	6.382	1.4		3.614	4.848	1.6
17.0	5.183	3.450	6.317	1.5		3.595	4.802	0.6
19.2						3.574	4.753	0.6
21.9						3.560	4.718	0.9
24.5						3.542	4.674	0.7
25.8						3.533	4.652	0.4
27.5						3.521	4.622	0.7
28.0						3.517	4.614	1.0
28.2						3.516	4.610	0.6
17.95						3.584	4.787	1.1
15.4						3.606	4.830	1.3
4.9	5.710	3.470	6.607	2.7				
1.2	5.830	3.550	6.696	0.7				

the presence of the h.c.p. structure. At 17.0 GPa the situation is reversed. By decreasing pressure the h.c.p. structure turns back to the initial orthorhombic and thus the transition is reversible.

One small peak at about 19 keV in the orthorhombic phase cannot be indexed (it might be explained partly as an escape peak), and two small peaks at about 19 keV and 29 keV in the hexagonal phase cannot be indexed either. They have most probably another origin than YbH$_2$.

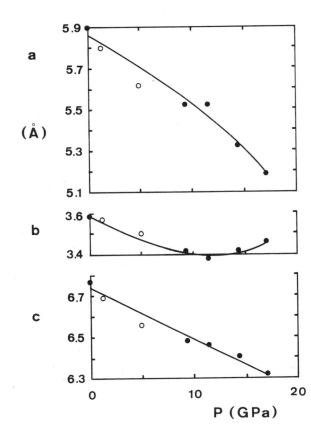

Fig. 2. Pressure dependence of the YbH$_2$ orthorhombic unit cell dimensions.

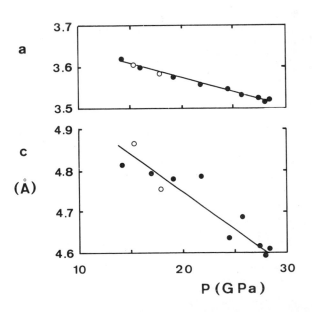

Fig. 3. Pressure dependence of the YbH$_2$ h.c.p. unit cell dimensions.

Fig. 2 presents the pressure dependence of the unit cell dimensions of the orthorhombic phase, and Fig. 3 of the h.c.p. phase. The indicated points (full circles for increasing pressure and open circles for decreasing pressure) represent a first approximation based on the position of three lines in the spectra. A subsequent refinement using the seven strongest lines results in lattice parameters indicated by the curve drawn through the points. These refined parameters are given in Table II.

Fig. 4 shows the pressure dependence of the unit cell volume divided by the number of Yb atoms in the respective unit cell (4 in the orthorhombic and 6 in the large h.c.p. unit cell).

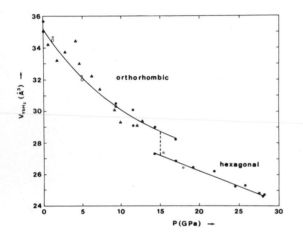

Fig. 4. Pressure dependence of the unit cell volume of YbH_2 divided by the number of Yb atoms in the respective unit cell (4 in the orthorhombic and 6 in the large h.c.p. unit cell).

On the basis of the above results we conclude that YbH_2 undergoes at about 14 GPa a pressure induces first order transition. The transition is accompanied by a 5.4% decrease in volume and a change in structure from an orthorhombic to a close packed hexagonal.

4. DISCUSSION

The high pressure phase of YbH_2 is thus not an f.c.c. as one could expect a priori on the basis of the analogy presented in the introduction. Knowing, however, that the high pressure phase is a h.c.p. we may try a posteriori to find some good reasons for it. Such an attempt is discussed below.

We translate the origin of the orthorhombic unit cell to the position of one Yb atom and renamed the basic vectors of the unit cell. At 14.3 GPa we have a_o = 3.405 Å, b_o = 6.382 Å and c_o = 5.335 Å (the index "o" stands for orthorhombic),

and the Yb atoms positions are: (0;0;0), (0.50; 0.78; 0.52) (0.50; 0.50; 0.02) and (0; 0.28; 0.50), if we use the values for x and z quoted earlier from ref. [7]. Fig. 5 shows the

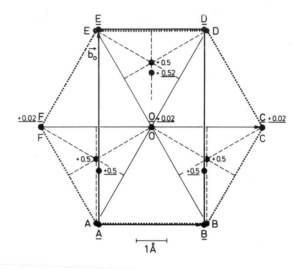

Fig. 5. The projections of the Yb atoms in the orthorhombic unit cell on the $(\vec{a}_o, \vec{b}_o)$ plane (underlined marking) and the Yb atoms in the h.c.p. large unit cell on the basal plane (not underlined marking). The drawing is in scale for the 14.3 GPa lattice constants. For details see text.

positions of the Yb atoms in the $(\vec{a}_o, \vec{b}_o)$ plane of the orthorhombic unit cell and the projections on it of the other atoms present in this unit cell (underlined marking). In the same figure the basal plane of the h.c.p. large unit cell of the high pressure phase of YbH_2 is drawn with the Yb atoms in this plane and the projection on it of the other atoms in this unit cell (not underlined marking). The drawing is made in scale for the lattice constants at 14.3 GPa where both phases are present:

$\underline{AB} = a_o = 3.405$ Å $\underline{AE} = b_o = 6.382$ Å

$AB = a_h = 3.614$ Å $AE = a_h \sqrt{3} = 6.260$

It is striking to notice that all Yb atoms in the basal plane of the hexagonal structure are very close to atoms lying in the $(\vec{a}_o, \vec{b}_o)$ plane or a bit above it (all marked by letters). Not much movement is thus needed for atoms lying in the $(\vec{a}_o, \vec{b}_o)$ plane or close to it in order to bring them in the hexagonal positions. As concerns the 3 Yb atoms with relative coordinates 0.50 and 0.52 in the $\vec{c}_o$ direction a similar statement could be made if c_o = 5.335 Å would be equal to c_h = 4.848 Å. It follows, however, from this discussion that the arrangement of the Yb atoms in the orthorhombic phase at 14.3 GPa is a slightly distorted hexagonal packing with c/a ≈ 5.335 Å: 3.405 Å ≈ 1.57. This statement is similar to that made in ref. [10] concerning the structure of CaH$_2$.

In view of that the discussed phase transformation in YbH$_2$ can be interpreted as a transition from a distorted hexagonal lattice with c/a ≈ 1.57 to a collapsed hexagonal lattice with c/a = 4.848 Å: 3.614 Å = 1.34. The collapse is most probably due to the change of the valence state from $4f^{14}(5d6s)^2$ to $4f^{13}(5d6s)^3$.

As already mentioned the results and the discussion presented here are of a preliminary nature. A comprehensive paper will be published later.

ACKNOWLEDGEMENT

It is a pleasure to thank Professor W.B. Holzapfel for valuable discussions concerning the diamond anvil high pressure cell and Professor B. Stalinski for making available the sample. We should like also to express our sincere thanks to HASYLAB-DESY for making possible the use of synchrotron radiation.

The financial help of the Danish Natural Sciences Research Council is gratefully acknowledged.

REFERENCES

[1]   See for instance A. Jayaraman: "Valence Changes in Compounds" in "Handbook of the Physics and Chemistry of Rare Earths", vol. 2, ed. by Karl A. Gschneidner Jr. and Le Roy Eyring (North Holland, Amsterdam 1979).

[2]   B. Buras, W. Kofoed, B. Lebech and G. Bäckström, Risø National Laboratory Report No. 357/1977.

[3]   See e.g. B. Buras, L. Gerward, A.M. Glazer, M. Hidaka and J. Staun Olsen, J. Appl. Cryst. 12, 531 (1979) and references therein.

[4]   J. Staun Olsen, B. Buras, L. Gerward and S. Steenstrup, Scientific Instruments 1981 (in press).

[5]   J. Als-Nielsen and B. Buras, Risø National Laboratory Report R-393, p. 43.

[6]   W.L. Korst and J.C. Warf, Acta Cryst. 9, 452 (1956).

[7]   J.C. Warf and K.I. Hardcasle, Inorg. Chem. 70, 1041 (1966).

[8]   C.E. Messer and P.C. Gianoukos, J. Less, Comm. Met. 15, 377 (1968).

[9]   R. Hazell (private communication).

[10] J. Bergsma and B.O. Loopstra, Acta Cryst. 15, 92 (1962).

PHYSICS OF SOLIDS UNDER HIGH PRESSURE
J.S. Schilling, R.N. Shelton (editors)
© *North-Holland Publishing Company, 1981*

LOW-TEMPERATURE ELECTRICAL RESISTIVITY OF UNSTABLE-VALENT $CeIn_3$, $CePd_3$ AND YbCuAl
UP TO 225 kbar

J.M. Mignot[+] and J. Wittig

Institut für Festkörperforschung der Kernforschungsanlage Jülich
Postfach 1913, D-5170 Jülich 1, W.-Germany

The high-pressure electrical resistance of several abnormal Ce and Yb compounds was investigated up to more than 200 kbar. The strong Kondo-type anomaly in the resistance-temperature characteristic of $CeIn_3$ disappears gradually above 100 kbar. The striking similarity with $CeAl_2$ is emphasized and the occurrence of a valence transition is inferred. The mixed-valent compound $CePd_3$ exhibits a similar, although less pronounced trend towards non-magnetism at high pressure. In YbCuAl, the results indicate a rapid decrease of the characteristic fluctuation temperature up to 80 kbar. At higher pressures additional anomalies appear below 30 K. These features are ascribed to the increasing stability of the magnetic trivalent state of ytterbium, and the possibility of a magnetically-ordered ground state is discussed.

1. INTRODUCTION

Whereas most metallic rare-earth materials are characterized by the high stability of one particular $4f^n$ configuration, the simplest description of "mixed-valence compounds" (MVC) implies two states with different f counts being nearly degenerate. Many examples of this situation have been found experimentally in cerium ($4f^0$-$4f^1$) and ytterbium ($4f^{13}$-$4f^{14}$) compounds [1]. The complete solution of the mixed-valence problem is a formidable theoretical challenge, and only simplifying models have been worked out so far [1]. From the experimentalist's point of view, one of the most useful concepts is the so-called Anderson-lattice model [2]. In this description, the magnetism of the open 4f shell results from the competition between the Coulomb-repulsion and f-d hybridization terms. In rare-earth materials, the Coulomb correlation energy U is very large (typically 6 eV) compared to the other energies of the problem. In this limit, the relevant parameter for the magnetic instability is the ratio E_f/Δ, where E_f is the 4f binding energy with respect to the Fermi level, and $\Delta = \pi V_{fk}^2 \rho(\varepsilon_F)$ the virtual-bound-state half width. For values of E_f/Δ of the order of unity, Kondo-type spin fluctuations are expected to set in, with a characteristic temperature T_K depending exponentially on the parameter E_f/Δ [3]:

$$T_K \simeq \frac{1}{2\pi} \sqrt{2\Delta U/\pi} \, \exp[\pi E_f/2\Delta]$$

Upon a further decrease of E_f, the true MV regime is achieved. This simple picture obviously contains serious shortcomings: both local screening effects and electron-phonon coupling are neglected. Furthermore, the translational invariance of the rare-earth-ion lattice is not considered. However, its simplicity makes it particularly suitable for a direct comparison with experiments and we will use it as a guideline in the following.

There is now a wealth of experimental results on MVC, covering almost the entire range of possible magnetic behavior. Unfortunately this variety of situations cannot be easily correlated with the variation of one particular physical quantity like the average valence. It is therefore an urgent task to look for experimental situations where this parameter can be varied continuously under controlled conditions. In a recent study, Mihalisin et al [4] succeeded in spanning the entire MV range by alloying the MVC $CePd_3$ with Rh or Ag. However, this method can produce undesired additional features due to deviations from stoichiometry and local environment effects.

In many respects, the application of an external -ideally hydrostatic- pressure would appear preferable. The promotion of a 4f electron into the conduction band reduces the shielding of the valence electrons from the nuclear potential and a decrease of the ionic volume results. This effect is observed, for instance, in SmS where a discontinuous transition occurs from the Sm^{2+} state to a MV state at a pressure of 6 kbar. Here, the pressure serves to initiate the valence instability. On the other hand, the pressure variable has not extensively been used so far for studying the MV state itself.

Recently, Eiling and Schilling [5] reported a detailed study of Anderson-lattice effects in CeAg up to 40 kbar. Apparently, only the entering stage of the mixed-valence regime could be achieved within the accessible pressure range. Previously, Probst and Wittig [6] had studied the electrical resistivity of $CeAl_2$ up to 130 kbar and detected strong changes of the R-T dependence pointing to an electronic transition. The transition is associated with a volume decrease of about 4 % [7] and bears striking resemblance to the classical $\gamma \rightarrow \alpha$ transition of pure cerium metal.

The experiments show the potential of high-pressure techniques in connexion with the

mixed-valence problem. They also touch on many
unsolved questions like the applicability of the
Anderson/Kondo picture, the degree of itineracy
of the cerium 4f-states in the high pressure limit and the description of the temperature-pressure phase diagrams. This situation calls for
further investigations in the widest possible
range of pressure involving all unstable-valent
rare-earth ions. In this paper, we present elec-
trical-transport measurements on the three compounds $CeIn_3$, $CePd_3$ and YbCuAl up to pressures
of $\simeq 225$ kbar.

2. EXPERIMENTS

The experimental set-up [8] consisted of a low-
temperature Cu-Be press with sintered-diamond
Bridgman anvils. Steatite, surrounded by a pyrophyllite gasket was used as pressure-transmitting
medium. Due to the brittleness of the substances
investigated, it was not possible to prepare bulk
samples of the desired dimension ($1 \times 0.1 \times 0.01$ mm^3)
from the available material. We therefore used
powder samples prepared by crushing small frag-
ments between WC anvils after carefully removing
the corroded surface layer. The load was increased in steps at room temperature, then the cell
was cooled nearly isobarically to liquid helium
temperature. During each low-temperature run,
the pressure was determined from the superconducting transition of a strip of lead foil connected in series with the sample. The detailed
procedure was described in a previous paper [9].
The experimental data were taken on the pressure-
increasing cycle, where the pressure inhomogeneity, $\Delta P/P$, read from the lead monometer did not
exceed 10 %. The only exception was YbCuAl below
100 kbar where $\Delta P/P$ reached 13 %.

3. RESULTS

3.1 $CeIn_3$

The compound $CeIn_3$ crystallizes in the cubic
$AuCu_3$ structure. The lattice constant [10] and
the Curie constant [11,12] indicate that cerium
is essentially trivalent. However, the large
paramagnetic Curie temperature ($\Theta \simeq 60$ K) [11,
12], the large linear heat capacity at low tem-
perature ($\gamma \simeq 130$ mJ/K$^2 \times$mole [13]) and the quite
anomalous dependence of the electrical resistiv-
ity [14] have been ascribed to spin fluctuations
[12] with a characteristic temperature of about
50-100 K. Below 10.2 K, the system orders magnetically in a ($\pi\pi\pi$) antiferromagnetic structure [15]. These properties are very similar to
those of $CeAl_2$, and both systems are actually
often cited as examples of Kondo lattices with a
magnetic ground state. This analogy was recently
substantiated by a resonant-photoemission study
of Allen et al. [16] which indicated that the
binding energy, the half-width of the 4f level,
and especially the ratio of these two quantities
are very similar in both compounds (E_f=2.8 and
2.1 eV, Δ=0.65 and 0.5 eV, E_f/Δ=4.3 and 4.2 for
$CeAl_2$ and $CeIn_3$, respectively), and consistent
with a Kondo-type model.

At the lowest pressure of 3 kbar, the resistance
of our sample (Fig. 1) exhibits a temperature-
dependence which is qualitatively very similar
to that reported in previous studies at ambient
pressure [14,17]. It first decreases slowly on
cooling, then goes through a shallow minimum
near 150 K, followed by a rounded maximum at
about 50 K. These features may be due to combined Kondo and crystal-field effects [14,18].
The onset of the magnetic order is indicated by
a kink at 9 K. The rather poor resistivity ratio
in the present measurements (R(295)/R(1.5)<4) is
probably due to the use of a powder specimen. As
in previous studies, the resistance dropped below 3 K due to the superconducting transition of
a small amount of free indium [14,18]. In the
normal state, however, the presence of this second phase does not seriously affect the results,
as indicated by the reasonable agreement with
the results of other groups at ambient pressure.

The resistance at 295 K is displayed in Figure 2
as a function of the press load. The maximum
load corresponds to a pressure of 226 kbar. The
striking qualitative feature of these data is
the existence of a resistance maximum at an estimated pressure of about 85 kbar. This effect
was observed both on a virgin sample and, as for
the present curve, after a high-pressure cycle
up to 180 kbar. Recalling the existence of such
a behavior in $CeAl_2$, where the existence of an
electronic transition is now well established,
we take this resistance maximum as an indication
that a similar phenomenon occurs in $CeIn_3$. A
crude comparison between these two sets of data
suggests that the transition takes place in
about the same pressure range, typically
100 kbar.

The low-temperature data are shown in Fig. 1 for
five different pressures (3, 59, 124, 182, and
226 kbar). In order to facilitate comparison between curves obtained in different high-pressure
cycles, each of them was normalized to the room-
temperature value of Fig. 2 for the same applied
load. Within the accuracy of the measurements
the curves were found to be reversible during
cooling and warming. The essential feature of
the data is the gradual transition from the
anomalous R-T characteristic of $CeIn_3$ to a completely normal behavior near 200 kbar, comparable, for instance, to the resistivity of the
non-magnetic isostructural compound $LaIn_3$ [17].
There is a striking similarity indeed between
these results and those obtained previously for
$CeAl_2$. Moreover, as it was discussed in ref.[6],
the low- and high-pressure curves closely resemble the R-T characteristics of pure cerium
metal in its (magnetic) β and (non-magnetic) α
phases, respectively. Therefore, the question
arises whether the pressure-induced transitions
in these different systems may have a common
origin. One widely accepted view [19] of the
$\gamma \rightarrow \alpha$ transition in pure cerium implies a partial
promotion of the initially localized 4f elec-
tron into the conduction band, leaving the ion
in a mixed-valent state. In this model, the

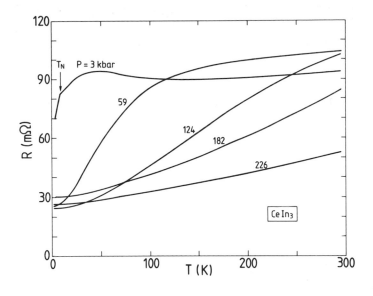

Figure 1. Temperature dependence of the resistance
of CeIn₃ at various pressures

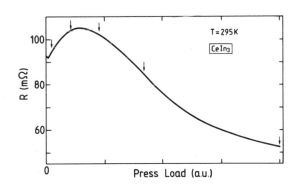

Figure 2. Room temperature resistance of CeIn₃
vs. press load (data taken during an increase of
the pressure). The highest pressure is 226 kbar.
The pressure at which the maximum occurs is about
85 kbar. The arrows indicate the values at which
the curves of Fig. 1 have been normalized

anomalous resistivity curve at low pressure is
ascribed to a Kondo effect superimposed on
the normal phonon resistivity. The same interpretation was invoked by Croft and Jayaraman [7]
in their work on CeAl₂. An alternative mechanism
was proposed by Hill [20]: This author argued
that if the distance between neighboring cerium
ions is smaller than a critical value d_{cr}, the
4f band becomes too broad to support "localized-
moment" magnetism. This interpretation was apparently supported by the empirical observation
that most cerium compounds can be grouped into
regions of "magnetic" (d>3.4 Å) or "non-magnetic"
(d<3.4 Å) behavior according to the average spacing between the ions (the so-called "Hill
plots"). Interestingly enough, the Ce-Ce spacing
is only slightly larger than the critical value
for both Ce metal and CeAl₂. This made it plausible that a transition from the magnetic to the
non-magnetic region could be induced by application of pressure.

In CeIn₃, the Ce-Ce spacing at ambient pressure
is very large compared to the preceding systems:
d=4.68 Å [10], and this will be even true at
pressures of the order of 100 kbar for any reasonable estimate of the compressibility. The anticipated high-pressure transition in CeIn₃
therefore clearly violates the Hill picture and
the original idea that direct 4f overlap is responsible for the magnetic-non-magnetic transition of cerium cannot apply to this system. We
believe that the transition is primarily due to
the mixing of the 4f states with the conduction-

band states. Whether the same holds for $CeAl_2$ (and α-Ce ?) is a major question but the similarity of the resistive behaviors makes it very plausible. A more detailed experimental comparison between the high-pressure properties of these two systems is clearly needed. In the case of $CeIn_3$, the most urgent task is to look for a volume anomaly in the 100 kbar range. A careful study of the magnetic ordering under pressure would also be important for testing the validity of the Anderson-lattice picture. Previous experience on CeAl [21] suggests that hydrostatic pressure may be important for this purpose.

3.2 $CePd_3$

The compound $CePd_3$ crystallizes in the same structure as $CeIn_3$ but in contrast to the latter it is believed to be in a mixed-valent state already at ambient pressure. The lattice constant (a_o=4.12 Å) indicates an intermediate valence of about 3.5 [22]. Recent resonant-photoemission experiments [16] indicate that, unlike $CeIn_3$ or $CeAl_2$, $CePd_3$ exhibits a non-lorentzian emission line impinging on the Fermi level, thus supporting the mixed-valence idea. The low-temperature properties are also characteristic of a MVC with a nearly temperature-independent susceptibility varying from 1.1 to 1.5 10^{-3} emu/mole between 300 and 20 K [22]. The electrical resistivity is quite unusual: it first rises from about 120 $\mu\Omega$cm at room temperature to a maximum value of 165$\mu\Omega$cm

near $T_M \simeq 130$ K and then drops to a residual value of a few 10 $\mu\Omega$cm [23,24]. Resistivity measurements [25] on the quasi-binary systems $Ce(Pd,Ag)_3$, where the substitution of Pd by Ag is believed to shift the cerium valence towards the 3+ state [4], revealed a drastic shift of the maximum to lower temperatures along with a strong increase of $\rho(T_M)$. This experimental observation suggests that T_M is in some way related to the fluctuation temperature of the system.

The results of the present measurements are shown in Fig. 3. Whereas the initial decrease at R(295) (insert) is certainly due to the compacting of the powder, the gradual increase of higher pressure probably includes contributions from both a change of the geometrical factor and an increase of the resistivity [26]. No attempt was made to separate the various effects and the different R-T characteristics are simply normalized at 295 K.

Already at 2 kbar, the curve exhibits an anomalously large residual resistance leading to a resistance ratio which is indeed smaller than 1 (R(295)/R(1.2)=0.8). This is in contrast to the resistivity of the bulk sample measured at ambient pressure in a conventional set-up where the resistance ratio was found equal to 6 [27]. A similar effect has been previously reported for non-stoichiometric ($CePd_{3+x}$) or La-substituted ($La_xCe_{1-x}Pd_3$) compounds [23] and, more

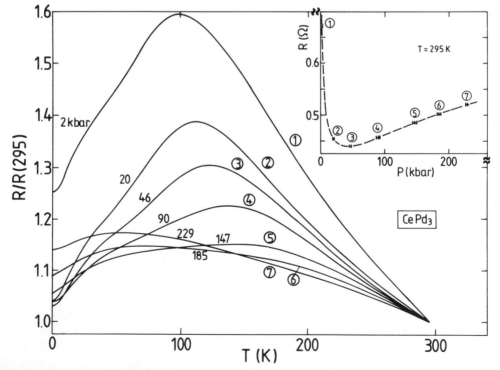

Figure 3. Temperature dependence of the resistance of $CePd_3$. Insert: Room-temperature resistance vs. pressure. The steep initial decrease is due to the compression of the powder

recently, for specimen subjected to a "wrong" annealing procedure [24]. In all cases, the introduction even of a small amount of disorder (x~1 %) in the pure compound produces a drastic reduction of the resistivity drop below T_M, while the high-temperature side of the curve remains essentially unaffected. It is therefore not surprising that the lattice defects produced by powdering or pressurizing the sample suffice to appreciably increase the resistivity at T=0. It ought to be noted, however, that in contrast to the cation-substituted compounds of ref. [24], our sample did not exhibit any upturn of the resistivity at low temperature.

Upon increasing the pressure, the main changes are (i) a rapid decrease of the total amplitude of the resistance anomaly , and (ii) a progressive shift of T_M to higher temperatures, up to 50 % at 100 kbar. Above 150 kbar, the latter effect is obscured by the appearance of a broad anomaly eventually leading at the highest pressure to another maximum near 50 K. We attribute this effect to lattice defects which may be generated by cold working. The increase of T_M, as well as the rapid decrease of the negative temperature coefficient, clearly indicate that like CeIn$_3$, CePd$_3$ tends to a non-magnetic state at high pressure. This conclusion is supported by previous susceptibility measurements at room temperature showing a linear decrease of χ up to 15 kbar ($\frac{1}{\chi}\frac{d\chi}{dP}$ = -0.43 % per kbar) [28]. The fact that the R-T characteristic of CePd$_3$ remains anomalous at 229 kbar even though this compound was already MV at ambient pressure is in contrast to the rapid transition towards a normal behavior as for CeIn$_3$ and CeAl$_2$. This difference may be partly explained by the unusually large bulk modulus of the latter compound [29] (C_B = 10.25 10^{11}erg/cm^3). From the present study, it appears that the intrinsic high-pressure properties of CePd$_3$ at low temperature can be ascertained only under improved experimental conditions, i.e. by using bulk specimens and reducing the pressure inhomogeneity.

3.3 YbCuAl

The results presented in the preceding sections, along with those of other studies [26,28], support the idea of a general trend towards nonmagnetism in anomalous cerium compounds at high pressure. If the conclusions drawn for CeIn$_3$ can be carried over to the other materials, this phenomenon primarily reflects the increasing admixture of the Ce 4f states with the conduction band states. Interesting questions arise if one now considers the case of ytterbium: Ytterbium is located symmetrically in the rare earth series with respect to cerium. Like cerium, ytterbium can occur in two different valence states, one magnetic ($4f^{13}$) with one hole in the 4f shell, and the other non-magnetic ($4f^{14}$) with a closed 4f shell. In many compounds, the valence of ytterbium has also be shown to be intermediate [30]. However, there are some fundamental differences between these two elements.

First, the localized character of the 4f orbitals is known to increase across the lanthanide series, so that the 4f electrons cannot make any significant contribution to the bonding in ytterbium systems as they might do, for instance, in cerium metal. Another intriguing feature already noted by Klaasse and coworkers [30] is the absence of magnetic order down to helium temperature in most metallic trivalent ytterbium compounds [31], whereas cerium, despite its even smaller de Gennes factor, forms many magnetically-ordering compounds. Finally, the magnetic (open-shell) configuration has the larger ionic radius in cerium but the smaller one in ytterbium.

Recently, Zell et al [32] systematically subjected MV ytterbium compounds to pressures up to 12 kbar. In all cases, a linear increase of the magnetic susceptibility vs. pressure was obtained at room temperature, which is exactly the opposite effect to what was observed in the case of cerium [28]. Furthermore, in the two compounds that were studied at low temperature (YbInAu$_2$ and YbInPd), pressure was found to decrease the temperature of the characteristic susceptibility maximum. The authors concluded that the magnetic state of ytterbium can be enforced under pressure if one squeezes out one of the 14 electrons of the 4f shell. However, much higher pressures are clearly needed to complete the transition.

In the present work, we studied the compound YbCuAl because its valence is very close to 3 already at ambient pressure [33], although its low-temperature properties, especially the magnetic susceptibility [34] and the electrical resistivity [35], exhibit anomalous features that are characteristic of MVC. In preliminary susceptibility measurements up to 4.5 kbar, Mattens and coworkers [35,36] observed an increase of the low-temperature susceptibility (T<15 K) of 3 % per kbar, along with a shift of the susceptibility maximum dT_{max}/dP=-1K/kbar. Considering the relatively low value of T_{max} at P=0 (T_{max}(0)=28 K) , a complete transition to the 3+ state appears possible within our experimental pressure range.

Figure 4 shows the temperature dependence of the resistance of YbCuAl under pressures up to 200 kbar. The R-P curve at 295 K (insert) is very similar to that for CePd$_3$ and all R-T characteristics were again normalized at this temperature. In contrast to the latter system, only a moderate reduction of the resistance ratio (from 2.1 to 1.6) was observed with respect to the bulk sample value [35]. The shape of the low-pressure R-T curve is identical to that reported previously [35], with a nearly temperature-independent part between 300 and 100 K followed by a steady decrease upon cooling down to the lowest experimental temperature of 1.2 K.

At higher pressure, a faint minimum appears at 150 K and the low-temperature resistance decrease is now preceded by a rounded maximum. The temperature of this maximum decreases rapidly

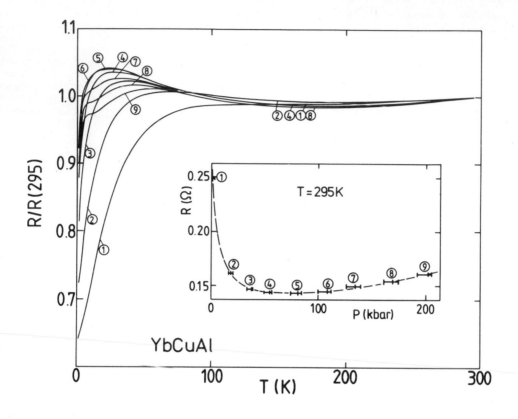

Figure 4.   Temperature dependence of the resistance of YbCuAl at various pressures
(from   1   to   9): 1.5, 19, 38, 55, 81, 108, 134, 169, and 202 kbar. Insert: Room-
temperature resistance vs pressure (see Fig. 3)

with increasing pressure and reaches 22 K at 81
kbar. At this pressure, only about 30 % of the
expected decrease is achieved above 1.2 K. Above
100 kbar, a qualitative change occurs and new
features appear in the R-T curve (some of them
are better seen in Fig. 5 where the low temperature region is displayed on an expanded T-
scale): first, the maximum shifts back to higher
temperatures (~50 K at 202 kbar) and $R(T_M)$ decreases accordingly. Furthermore, a shoulder appears on the low-temperature side of this maximum and becomes more and more pronounced as the
pressure is increase. At the maximum pressure,
the slope dR/dT is almost zero between 9 and
10 K. Finally, the resistance drops very steeply
below 5 K down to our limiting temperature.

In experimental studies on MVC, the temperature
of the susceptibility maximum is commonly taken
to represent the characteristic energy of the
magnetic fluctuations. At our lowest pressure of
1.5 kbar, the temperature, as deduced from the
measurements of Mattens, is 26 K. At this temperature, the resistance has already dropped by
13 %. In order to get a feeling for the pressure

dependence of the magnetic fluctuations, we can
empirically define a characteristic temperature
T^* by $R(T^*) = (1-0.13) \times R(hi-T)$. At 19 kbar, we
find $T^* \simeq 11$ K, hence $dT^*/dP = -0.9$ K/kbar. The
reasonable agreement with the variation measured
by Mattens et al [36] ($dT_{max}/dP = -1$ K/kbar up to
4.5 kbar) lends some support to the present
analysis. At higher pressures, T^* further decreases and becomes smaller than 1 K at 81 kbar.

This dramatic decrease of T^* suggests that the
system is close to a magnetic instability, as
also indicated by the 50 % increase of the linear specific heat coefficient γ between 0 and
10 kbar reported by Bleckwedel and Eichler [37].
It is therefore very tempting to ascribe the
regime change observed above 100 kbar to the appearance of a magnetic ground state of ytterbium associated with the trivalent configuration.
This view is illustrated by the striking resemblance of the highest-pressure curve with the
ambient-pressure R-T characteristic of magnetic
$CeIn_3$ (Fig. 6). As in the latter case, the faint
minimum and the maximum near 50 K could result
from combined Kondo-lattice and crystal-field

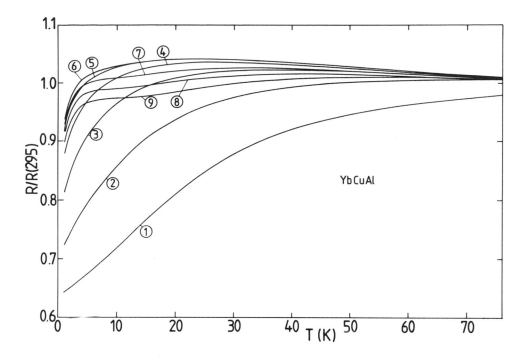

Figure 5.   Low-temperature part of Fig. 4 (expanded T-scale)

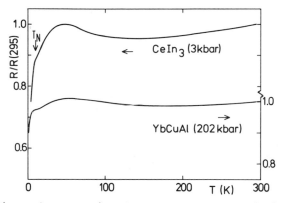

Figure 6.   Comparison between R-T characteristics for CeIn$_3$ (3 kbar) and YbCuAl (202 kbar)

effects. The shoulder at 5 K is very suggestive of a magnetic ordering. It is interesting to note that from the variation of the ordering temperature with the de Gennes factor at the end of the RCuAl series [33], a value of a few Kelvin is extrapolated for an hypothetical Yb^{3+}CuAl [38].

This interpretation remains of course speculative and needs to be confirmed by other types of measurements. Considering the rapid initial decrease of T*, crystal-field effects are not unlikely to appear within the experimental pressure range accessible for inelastic neutron scattering (P=20-30 kbar). In order to confirm the occurrence of a magnetic phase transition near 5 K, sensitive susceptibility measurements have to be performed above 100 kbar. Finally, resistance measurements at even higher pressure would be necessary to see if a normal trivalent behavior like in YCuAl is ultimately recovered.

4.   CONCLUSION

We have discussed some preliminary results on the effects of pressure in prototype unstable-valent rare-earth compounds. The outstanding issues of this study are the observation of a continuous γ→α-type transition in the electrical resistance of CeIn$_3$, and the possible appearance of a magnetic ground state in YbCuAl above 100 kbar. These observations are consistent with the widespread notion that pressure favors the higher-valence state, which is non-magnetic in cerium but magnetic in ytterbium. However, many fundamental questions remain unsolved and a quantitative treatment of the transport coefficients of MVC at finite temperature is well beyond the scope of existing theories. Even in the restricted range where the valence is very close to 3, the validity of analyses based on the additivity of a phonon term and a Kondo-type anomaly is certainly doubtful. These reservations are substantiated by the present YbCuAl results where the resistance above 100 K remains essentially constant on going from a mixed-valent state ($T^* \simeq 28$ K) to what could be regarded as a concentrated Kondo regime ($T^* \lesssim 1$ K).

ACKNOWLEDGEMENTS

We are indebted to J. Palleau, H. Schneider, and W.C.M. Mattens for providing samples of $CeIn_3$, $CePd_3$, and YbCuAl, respectively. Informative discussions with Prof. D.K. Wohlleben and Prof. P. de Châtel are appreciated. One of us (J.M.M) would like to thank the Deutsche Forschungs-gemeinschaft for supporting his stay at the Kernforschungsanlage Jülich and the Institut für Festkörperforschung for the hospitality shown him during the time this work was performed.

This work was funded in part by SFB 125, Deutsche Forschungsgemeinschaft.

REFERENCES

+ On leave from Centre de Recherches sur les Très Basses Températures, CNRS, Grenoble, France

[1] Many experimental and theoretical references can be found in the recent review article by Lawrence, J.M., Riseborough, P.S. and Parks, R.D., Reports on Progress in Physics 44 (1981) 1.

[2] Varma, C.M. and Yafet, Y., Phys. Rev. B 13 (1976) 2950.

[3] Haldane, F.D.M., J. Phys. C 11 (1978) 5015.

[4] Mihalisin, T., Scoboria, P. and Ward, J.A., Phys. Rev. Lett. 46 (1981) 862.

[5] Eiling, A. and Schilling, J.S., Phys. Rev. Lett. 46 (1981) 364.

[6] Probst, C. and Wittig, J., J. Magn. Magn. Mat. 9 (1978) 62.

[7] Croft, M. and Jayaraman, A., Solid State Commun. 29 (1979) 9.

[8] Wittig, J. and Probst, C., in Chu, C.W. and Woollam, J.A. (eds.), High Pressure and Low Temperature Physics (Plenum, New York, 1978), p. 433.

[9] Wittig, J., Z. Phys. B 38 (1980) 11.

[10] Harris, I.R. and Raynor, G.V., J. Less-Common Metals 9 (1965) 7.

[11] Tsuchida, T. and Wallace, W.E., J. Chem. Phys. 43 (1965) 3811.

[12] Lawrence, J.M., Phys. Rev. B 20 (1979) 3770.

[13] Van Diepen, A.M., Craig, R.S. and Wallace, W.E., J. Phys. Chem. Solids 32 (1971) 1867.

[14] Van Daal, H.J. and Buschow, K.H.J., Phys. Stat. Sol. (a) 3 (1970) 853.

[15] Lawrence, J.M. and Shapiro, S.M., Phys. Rev. B 22 (1980) 4379; Benoit A., Boucherle, J.X., Convert, P., Flouquet, J., Palleau, J. and Schweizer, J., Solid State Commun. 34 (1980) 293.

[16] Allen, J.W., Oh, S.-J., Lindau, I., Lawrence, J.M., Johansson, L.I. and Hagström, S.B., Phys. Rev. Lett. 46 (1981) 1100.

[17] Elenbaas, R.A., Schinkel, C.J. and van Deudekom, C.J.M., J. Magn. Magn. Mat. 15-18 (1980) 979.

[18] Elenbaas, R.A., Thesis, Amsterdam Univ. (February 1980).

[19] For a detailed discussion of the properties of elemental cerium, see, Koskenmaki, D.C. and Gschneider, K.A.,Jr., in Gschneider, K.A.,Jr. and LeRoy Eyring (eds.) Handbook on Physics and Chemistry of Rare Earths, Vol. 1 (North Holland, Amsterdam, 1978).

[20] Hill, H.H., in Miner, W.N. (eds.), Plutonium 1970 and other Actinides (The Metallurgical Society of the AIME, New York, 1970), p. 2.

[21] Peyrard, J., Doctorat d'Etat Thesis, Grenoble Univ. (June 1980).

[22] Gardner, W.E., Penfold, J., Smith, T.F. and Harris, I.R., J. Phys. F 2 (1972) 133.

[23] Scoboria, P., Crow, J.E. and Mihalisin, T., J. Appl. Phys. 50 (1979) 1895.

[24] Schneider, H. and Wohlleben, D., Electrical and thermal conductivity of $CePd_3$, YPd_3, $GdPd_3$ and some dilute alloys of $CePd_3$ with Y and Gd (to be published).

[25] Ward, J., Crow, J.E. and Mihalisin, T., in Crow, J.E., Guertin, R.P. and Mihalisin, T.W. (eds.), Crystalline Electric Fields and Structural Effects in f-Electron Systems (Plenum, New York, 1980), p. 333.

[26] Croft, M., Lewine, H.H., Neifeld, R., Jayaraman, A. and Mains, R., this conf.

[27] Schneider, H., private communication.

[28] Zell, W., Boksch, W., Roden, B. and Wohlleben, D., Spring Meeting of the German Physical Society (Münster, 1981), unpublished.

[29] Takke, R., Assmus, W., Lüthi, B., Goto, T. and Andres, K., in ref.[25], p. 321.

[30] Klaasse, J.C.P., de Boer, F.R. and de Châtel, P.F., Physica 106 B (1981) 178.

[31] One exception is $YbBe_{13}$ which orders anti-ferromagnetically at 1.28 K [Kappler, J.P., Doctorat d'Etat Thesis, Strasbourg Univ. (December 1980)].

[32] Zell, W., Pott, R., Roden, B. and Wohlleben, D., Pressure and Temperature Dependence of the Magnetic Susceptibility of some Ytterbium Compounds with Interme-diate Valence (to be published).

[33] Oesterreicher, H., J. Less-Common Metals 30 (1973) 225.

[34] Mattens, W.C.M.,, Elenbaas, R.A. and de Boer, F.R., Commun. Phys. 2 (1977) 47.

[35] Mattens, W.C.M., Thesis, Amsterdam Univ. (February 1980).

[36] Mattens W.C.M., Hölscher, H., Tuin, G.J.M., Moleman, A.C. and de Boer, F.R., J. Magn. Magn. Mat. 15-18 (1980) 982.

[37] Bleckwedel, A. and Eichler, A., this conf.

[38] Mattens, W.C.M., private communication.

PHYSICS OF SOLIDS UNDER HIGH PRESSURE
J.S. Schilling, R.N. Shelton (editors)
© *North-Holland Publishing Company, 1981*

L_{III} EDGE PROVES PRESSURE-INDUCED VALENCE TRANSITION IN Yb METAL

K. Syassen[+], G. Wortmann, J. Feldhaus, K.H. Frank, and G. Kaindl

Institut für Atom- und Festkörperphysik, Freie Universität Berlin, Germany

[+]Physikalisches Institut III, Universität Düsseldorf, Germany

The L_{III}-adsorption edge of Yb metal was investigated at pressures up to 330 kbar using a diamond-anvil high-pressure cell and a bent-crystal X-ray monochromator on a conventional 2-kW source. At 330 kbar a 6.5-eV shift of the L_{III}-edge to higher binding energy as compared to ambient pressure is observed. This shows that a valence transition occurs in Yb metal from divalent at ambient pressure to trivalent (or almost trivalent) at 330 kbar. In the intermediate pressure range structure and position of the L_{III}-edge reflect a mixed-valent state.

Ytterbium is divalent in its metallic state under ambient conditions |1|. Some time ago Johansson & Rosengren predicted for Yb metal an electronic transition to the trivalent state at an external pressure of about 135 kbar on the basis of pseudopotential calculations |2|. While earlier equation-of-state data obtained from shock-wave experiments |3| did not indicate such a transition, more recent volume compressibility measurements by the energy-dispersive X-ray diffraction method in combination with a diamond-anvil cell seemed to confirm such a valence transition |4|. These latter results indicate a continuous valence transition, while theoretically both continuous and discontinuous transitions have been discussed. It is quite clear that more data relevant to the valence state of Yb metal at high pressure are needed, especially those which are more directly related to the valence of the Yb ion, before the above question can be settled.

As was shown already in 1965 by Vainshtein et al. |5|, information on the mean valence of rare-earth ions in solids may be obtained in a rather direct way from position and structure of the L edges in X-ray absorption spectra. Due to an increase in the nuclear potential the peaked structure at the L edges (corresponding to transitions from the L shell to mostly d-like unoccupied states slightly above the Fermi level) shifts by approximately 7 eV to higher energies, when the rare-earth ions increase their valence from 2+ to 3+. With the recent availability of intense continuous X-ray sources in the form of synchrotron radiation such X-ray absorption measurements have increasingly been applied to the study of mean bulk valences of mixed-valent rare-earth solids at ambient pressures |6-9|. The present paper reports on an X-ray absorption study of Yb metal at pressures up to 330 kbar using a diamond-anvil high-pressure cell and a conventional X-ray source. The L_{III} edge absorption spectra obtained clearly show that Yb metal has changed its valence to 3+ or almost 3+ at a pressure of 330 kbar. The experiments reported represent the first high-pressure L-edge study, and demonstrate that such measurements are feasible with conventional laboratory X-ray sources.

The experiments were performed with a conventional X-ray absorption spectrometer based on a Rowland-type bent-crystal X-ray monochromator in combination with a 2-kW molybdenum X-ray source. An intense tunable beam of monochromatic X-rays ($\Delta E = 4.5$ eV at 8.9 keV) was produced by a bent Ge(220) crystal of the Johansson type. Details of the spectrometer are presented elsewhere |10|. External pressures up to 330 kbar at the Yb metal foil (99.99% purity; 5μm thickness) were generated in a diamond-anvil cell of the Syassen-Holzapfel type |11|, using an inconel gasket and silicon oil as the pressure transmitting medium. The ruby-fluorescence method served for an in-situ determination of the applied pressure with an accuracy of about 3%. At the highest pressure achieved the pressure variation over the sample area was less than 10% of the mean pressure. Due to the small area of the sample (0.2 mm diameter) and the large photoabsorption in the diamond-anvils (in the geometry used the anvils are transmitted by the X-rays over a length of $\simeq 5$ mm in axial direction), the photon flux was reduced to about 10 counts per second at the NaI(Tl) detector. For comparison, the counting rate achieved with the present spectrometer in normal transmission experiments at ambient pressure is usually of the order of 10^5 counts per second.

L_{III}-absorption spectra of Yb metal and Yb_2O_3, both at ambient pressure and room temperature, are shown in Fig. 1 as examples for divalent and trivalent Yb systems, respectively. The spectra exhibit a peaked structure at the absorption edge ("white line"), which is typical for L_{II} and L_{III} edges of 4f systems. In the trivalent Yb_2O_3 the position of the peaked structure at the absorption edge is shifted by 7 ± 1 eV to higher energies as compared to divalent Yb metal. This characteristic difference in the energy position of the L edges forms the basis for applications of L-edge absorption

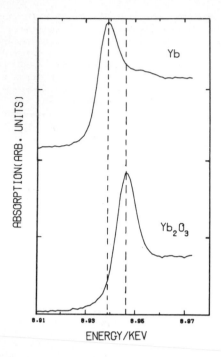

Figure 1: L_{III}-absorption spectra of Yb metal and Yb_2O_3 at ambient pressure.

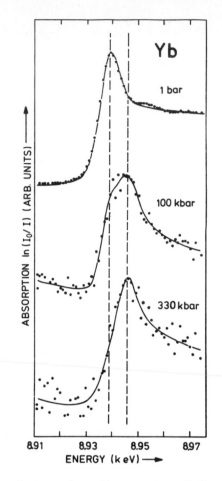

Figure 2: L_{III}-absorption spectra of Yb metal at the indicated pressures. The solid lines represent the result of least-squares fits of the experimental spectra (see text).

spectroscopy to a determination of mean valences in mixed-valent materials. The relative intensity of this white line and the additional features in the absorption spectra up to about 20 eV above the edge are characteristic for the respective system and reflect its unoccupied-band structure. In 4f systems with integral valence the width of the white line is mainly determined by lifetime effects and/or the energy resolution of the spectrometer, while band-structure effects play only a minor role.

The effects of high external pressure on the L_{III}-absorption spectra of Yb metal are clearly visible in Fig. 2. With increasing pressure the center of gravity of the L_{III}-edge peak shifts to higher energy indicating a pressure-induced change of the Yb valence. The spectrum obtained at a pressure of 100 kbar exhibits features characteristic for a mixed-valent system. In this case the peaked structure at the adsorption edge can be described by a superposition of two peaked edges displaced by about 7 eV and caused by divalent and trivalent Yb ions, respectively. Their relative intensities are indicative for the mean valence of the mixed-valent system |5-9|. Such a superposition of L-absorption edges is expected and indeed observed for mixed-valent materials |5-9|, since the characteristic probing time of X-ray absorption experiments is of the order of 10^{-16} s, quite similar to that of photoemission measurements. Therefore, X-ray absorption spectra represent a momentary situation in a valence-fluctuating solid. They principally do not allow a distinc-

tion between homogeneous and inhomogeneous mixed-valent solids |12|. The L-absorption edge in the 330-kbar spetrum is shifted to higher energies by 6.5 eV relative to that of the ambient-pressure spectrum. It clearly shows that Yb metal at 330 kbar has changed its valence almost to 3^+, with a small divalent component of about 10% still present in the L-edge spectrum.

The solid lines in Fig. 2 represent the results of least-squares fits of a superposition of two subspectra - corresponding to divalent and trivalent Yb - to the experimental data points. Each subspectrum is described by a step function folded with a Lorentzian of variable width and peak height (corresponding to the white line). The results of the analysis of the spectra taken at five different pressure points are summarized in Table 1. From the Yb^{2+}/Yb^{3+} intensity ratios given in column 2 the mean Yb valences presented in the last column were derived. To achieve this it was assumed that the

pressure (kbar)	Yb^{2+}/Yb^{3+} ratio	mean valence
0	98/2	2.0
30	90/10	2.1
100	50/50	2.5
220	35/65	2.7
330	10/90	2.9

Table 1: Pressure dependence of the Yb^{2+}/Yb^{3+} intensity ratio (second column) and of the derived valence (third column) of Yb-metal obtained from a least-squares fit analysis of the L_{III}-absorption spectra. The error bar for the mean valence at high pressure is estimated as 0.1.

$2^+/3^+$ intensity ratio of an L-edge absorption spectrum represents the mixed-valent situation in the initial state of the rare-earth solid. In this way final-state screening effects due to the positive core hole, which are known to play an important role in 3d and 4d X-ray photoemission spectra of rare-earth solids [13,14], are neglected. It may be expected, however, that such final-state effects are much less visible in L-edge absorption spectra as compared to X-ray photoemission spectra due to the localized nature of the final state in the electronic transition accompanying the L-absorption process. A further justification of the applied procedure is given by the agreement (within 0.1) between mean valences derived from L-absorption spectra and those obtained from other initial-state methods, as e.g. Mössbauer isomer-shift and lattice-constant measurements, for several other mixed-valent systems [6-9,12].

The values for the mean Yb valence listed in column 3 clearly show the occurrence of a pressure-induced valence transition in Yb metal from divalent at ambient pressure to trivalent at 330 kbar. Although, the error bars of the present results - due to limited statistical accuracy of the data - are still quite large, the valence transition seems to cover a rather broad pressure range. Even though the analysis of our 30-kbar spectrum results in a mean valence of $v=2.1\pm0.1$, we do not assume that Yb is already mixed valent in its fcc phase up to the structural phase transition (fcc→bcc) at 40 kbar [15,16]. Similarly, we do not claim that a mixed-valent state is still existing at 330 kbar, where the molar volume of Yb has already reached a value typical for trivalent rare-earth ions [4]. At 100 kbar and 220 kbar, however, the present data unambiguously reveal the presence of a mixed-valent state.

In conclusion, we have shown that L-edge ab-

sorption spectroscopy in combination with conventional high pressure techniques is a new tool to study pressure-induced valence transitions in rare-earth solids [17]. Such experiments are feasible with conventional X-ray sources, even though the use of synchrotron radiation should allow a considerable improvement of the statistical accuracy of the data. An important aspect of the method, which should be studied with high accuracy, is the question of final-state screening effects on the L-edge absorption spectra.

ACKNOWLEDGMENT

The authors acknowledge fruitful discussions with W.-D. Schneider. This work was supported in part by the Bundesministerium für Forschung und Technologie (grant No. 05 115 KA).

REFERENCES

[1] A. Jayaraman, in: Handbook on the Physics and Chemistry of Rare Earths, ed. by K.A. Gschneidner, Jr. and L. Eyring (North-Holland Publishing Company, Amsterdam, 1978), p. 707.

[2] B. Johansson and A. Rosengren, Phys. Rev. B11, 2836 (1975) and Phys. Rev. B14, 361 (1976); A. Rosengren and B. Johansson, Phys. Rev. B13, 1468 (1976).

[3] W.H. Gust and E.B. Royce, Phys. Rev. B8, 3595 (1973).

[4] K. Syassen and W.B. Holzapfel, in: High-Pressure Science and Technology, ed. by K.D. Timmerhaus and M.S. Barber (Plenum Publ. Corp., New York, 1979), Vol. 1, p. 223.

[5] E.E. Vainshtein, S.M. Blokhin, and Y.B. Paderno, Sov. Phys. Solid State 6, 2318 (1965).

[6] H. Launois, M. Rawiso, E. Holland-Moritz, R. Pott, and D. Wohlleben, Phys. Rev. Lett. 44, 1271 (1980).

[7] R.M. Martin, J.B. Boyce, J.W. Allen, and F. Holtzberg, Phys. Rev. Lett. 44, 1275 (1980).

[8] R. Nagarajan, E.V. Sampathkumaran, L.C. Gupta, R. Vijayaraghavan, Bhaktdarshan, B.D. Padalia, Phys. Lett. 81A, 397 (1981).

[9] K.R. Bauchspiess, W. Baksch, E. Holland-Moritz, H. Launois, R. Pott, and D. Wohlleben, in: Valence Fluctuations in Solids, ed. by L.M. Falicov, N. Hanke, M.P. Maple (North-Holland Publishing Comp., Amsterdam, 1981), p. 417.

[10] J. Feldhaus, PhD-Thesis, Freie Universität Berlin (1981), unpublished.

[11] G. Huber, K. Syassen, and W.B. Holzapfel, Phys. Rev. B15, 5123 (1977).

[12] J.M. Lawrence, P.S. Riseborough, R.D. Parks, Rep. on Progr. in Physics 44, 1 (1981).

[13] J.C. Fuggle, M. Campagna, Z. Zolnierek, R. Lässer, and A. Platau, Phys. Rev. Lett. 45, 1597 (1980).

[14] W.D. Schneider, C. Laubschat, I. Nowik, and G. Kaindl, Phys. Rev. B, in print.

|15| D.B. McWhan, T.M. Rice, and P.H. Schmidt,
     Phys. Rev. <u>177</u>, 1063 (1969).

|16| H. Katzmann and J.A. Mydosh, Z. Physik <u>256</u>,
     380 (1972).

|17| Further high-pressure L-edge absorption
     studies of a rare-earth system (SmS) are
     reported at this conference: K.H. Frank
     et al., contribution to this conference
     and R. Ingalls et al., contribution to
     this conference.

PHYSICS OF SOLIDS UNDER HIGH PRESSURE
J.S. Schilling, R.N. Shelton (editors)
© *North-Holland Publishing Company, 1981*

SPECIFIC HEAT MEASUREMENTS ON INTERMEDIATE-VALENT YbCuAl UNDER HIGH PRESSURE

Axel Bleckwedel and Andreas Eichler

Institut für Technische Physik
Technische Universität
D-3300 Braunschweig

A recently developed method for the measurement of low-temperature specific heat of metals under high pressure, which has already been applied successfully to simple, comparably soft metals, is utilized in an investigation on the intermediate-valent compound YbCuAl. The experimental implications resulting from the mechanical and thermal properties of YbCuAl are briefly described. Under pressures up to 10 kbar a dramatic increase of the specific heat by about 50 % is found. This result is supposed to be a consequence of valence change with pressure.

1. HIGH-PRESSURE CALORIMETRY

High pressures of the order of 10^4 bar have become a widely used experimental means to influence the physical properties of solid materials, in order to check our present theoretical models of the solid state. Nearly all experimental methods of investigation, which are in use at normal pressure, are also applied at high pressures. Measurements of the specific heat, especially at low temperatures, however, are rarely performed under pressure despite their valuable informations on the type and distribution of excitations in the solid material as a function of the lattice parameters. [1] The reason for this deficiency of calorimetric investigations are technical difficulties arising from the high pressure conditions. On selecting the proper calorimetric method one has to make allowance for the fact that the sample cannot be thermally decoupled from its environment because of the pressure transmitting medium. Even under these circumstances the sample heat capacity by itself may be determined by ac- or pulse techniques; for precise measurements, however, a way of quantitatively correcting for the addenda contribution of the pressure medium has to be found, which is always present to some extent. Apart from this serious problem there is another one arising from the necessity that in ac- or pulse methods the measuring elements heater and thermometer have to be in intimate thermal contact to the sample, which means that they have to be mounted inside the high pressure cell. They have to combine, therefore, mechanical strength with small heat capacity, i.e. small dimensions. Moreover, the thermometers must have high sensitivity to the lowest temperature even at high pressure in order to enable precise quantitative calorimetric investigations. In our institute a technique has been developed recently which presents a solution of these problems. [2] It is based on the ac-method [3], in which the sample is heated sinusoidally with frequency ω and the sample capacity C can be calculated from the amplitude $\tilde{T}$ of its temperature oscillation. One essential feature of our technique is the use of finely grained diamond powder as a pressure transmitting medium in a piston-cylinder cell. Because of its low thermal conductivity it nearly decouples the sample from the pressure apparatus. Secondly, its extremely small specific heat is most advantageous, since its contribution may be neglected at all. Both heater and thermometer are prepared as tiny slices cut from standard carbon resistors; the sensitivity of these thermometers at low temperatures surpasses that of thermocouples by far. As reported earlier [2], careful investigations on indium metal were performed to determine the accuracy of this method. By comparison at normal pressure it could be shown that our data agreed with data from the literature within ± 3 %. In a subsequent measurement on gallium, the pressure range was extended to 35 kbar, and the specific heat of the high-pressure phase Ga II was determined. [4] This clearly demonstrates the power of our technique. Indium and gallium, however, are exceptional materials, as their preparation for high-pressure calorimetric samples is particularly easy. Since indium is extremely soft and gallium melts near room temperature, the measuring elements can be entirely integrated into the sample with ease, and they experience nearly homogeneous stresses under pressure. It remains to prove, therefore, that the technique can also be applied to hard and high melting materials. This would open up a wide field of interesting physical problems to be treated by this method. The present investigation on YbCuAl represents a step towards this aim.

2. ADAPTATION OF THE TECHNIQUE TO YbCuAl

YbCuAl as sample material in high-pressure calorimetric measurements of the type described above introduces two new difficulties. Firstly, it is hard and brittle, so that the measuring elements cannot be mounted inside the sample. Therefore, gallium was chosen as embedding medium for both the elements and the YbCuAl sample in form of irregularly shaped lumps. (Fig. 1) Gallium metal is particularly suited as embedding substance, as it is easily melted and, moreover, is distinguished by a comparably high thermal diffusivity and small specific heat at

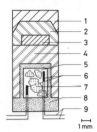

1mm

Fig. 1. Cross section of the high-pressure cell.
1 CuBe antiextrusion ring, 2 teflon spacer,
3 indium manometer, 4 teflon capsule, 5 gallium
embedding medium, 6 carbon heater and thermometer, 7 YbCuAl lumps, 8 diamond powder, 9 electric leads. The whole assembly is mounted in a
pressure cylinder.

low temperatures. As can be seen from the
successful measurements on gallium itself [4],
it evidently constitutes a quasihydrostatic environment for the measuring elements under pressure. The second problem arises from the low
thermal diffusivity n of YbCuAl. As a conse-
quence, the thermal wavelength $\lambda \propto \sqrt{(n/\omega)}$ in the
usual frequency range of our method is of the
same order of magnitude as the diameter of the
YbCuAl lumps. Therefore, the contribution of the
YbCuAl heat capacity C_Y to the measured ac-sig-
nal will be frequency dependent, and the frequency response of $\tilde{T}$ will be of an entirely different shape compared to the former investigations (Fig. 2). [2] Whereas normally always
$\tilde{T} \omega \leq P_O/C$ (P_O effective heating power), the
equality being valid in an intermediate fre-
quency range $1/\tau_1 < \omega < 1/\tau_2$, now values
$\tilde{T} \omega > P_O/C$ may also be found. As a consequence
the heat capacity calculated from the measured
$\tilde{T}$ may be smaller than the true sample capacity,
since above frequencies of about 20/s the YbCuAl
pieces are increasingly thermally decoupled
from the gallium and the measuring elements.

For that reason the optimum measuring frequency
is n o t at the maximum of $\tilde{T} \omega$ - as in our
former experiments [2] - but in the frequency
range of the plateau in $\tilde{T} \omega$ (around 15/s). This
choice of ω can be qualitatively justified by
means of an extended thermal model, which takes
into account the peculiarities of the sample
arrangement and the thermal properties of
YbCuAl (Fig. 3). It is hard to tell, however,
if the equality sign is exactly valid in the
frequency range of the plateau, or if there are
deviations caused by non negligible thermal
shunts via the electric leads of the measuring
elements (R_1 and R_2 in fig. 3). As some of the
parameters of the model can neither be experimentally determined nor unambiguously calculated
from a numerical fit, it is not possible to compute exact correction terms for those deviations.
Even more complications arise from the fact that
the shape of $\tilde{T} \omega$ has been found to be somewhat
pressure dependent, the plateau being worse defined at zero pressure than at the highest
pressures reached. We believe this to be a consequence of the pressure dependence of the
thermal boundary resistance between gallium and
the YbCuAl pieces. Presumably, these difficulties in finding the optimum frequency are the
reason for the comparably large scatter and uncertainty of our experimental points.
In order to get the specific heat of YbCuAl by
itself the addenda contributions (Ga, heater,
thermometer) have to be subtracted. Except for
the epoxy coatings of the measuring elements
this presents no problem. For instance, the Ga
capacity is well known, and even its pressure
dependence can be estimated with sufficient
accuracy from the Grüneisen constants. The overall addenda contribution is smaller than 6 %,
that of the measuring elements only about 0.5 %
of the total heat capacity. Only the cubic term
of the specific heat of YbCuAl is affected by
some uncertainty, since the addenda - mainly
the epoxy coatings - contribute about 20 % to
this term, the exact magnitude as well as the
pressure dependence of this proportion being
unknown.
The pressure is determined from the superconducting transition temperature of a separately mount-

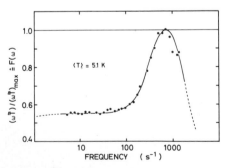

Fig. 2. Normalized frequency dependence of the
ac temperature amplitude $\tilde{T}$. The experimental
points have been fitted by a function derived
from the thermal model shown in fig. 3. Contrary
to our former experience, correct heat capacities are not determined at the frequency of the
maximum in $\omega\tilde{T}$, but on the plateau near 15/s
instead.

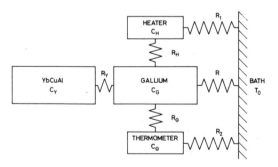

Fig. 3. Thermal model applicable to the sample
arrangement of fig. 1, showing the different
heat capacities (C) and thermal resistances (R)
involved. It should be borne in mind, that C_Y is
to be treated as frequency dependent.

ed indium sample (Fig. 1). Since it is separated from the calorimetric sample by layers of teflon and diamond powder, some degree of pressure difference cannot be excluded. This has been allowed for by the error bars in fig. 4, which are based on the empirical experiences from our investigation on the gallium high pressure phase. [4]

3. RESULTS AND DISCUSSION

There is strong experimental evidence that in YbCuAl the Yb ions are in a non-integer so-called intermediate valence state, since the two configurations $4f^{14}5d^0$ and $4f^{13}5d^1$ are nearly degenerate. [5,6] The anomalous (negative) thermal expansion of YbCuAl at low temperature indicates that the mean valency is temperature dependent, shifting closer to 3 with increasing temperature. The two involved integer-valence states, Yb^{2+} and Yb^{3+}, respectively, differ in their volumes appreciably, the latter being about 10 % smaller. Therefore, a strong influence of pressure on the mean valency and - in connection with it - on the physical properties of YbCuAl is to be expected. This is confirmed by the present results on the specific heat under pressure. A dramatic increase with pressure is found, which is mainly due to the rise of the already large electronic part $C_{el} = \gamma T$. Up to 10 kbar, γ increases by about 50 % (Fig. 4), leading to an averaged relative pressure dependence $d\ln\gamma/dp = 5 \times 10^{-5}$/bar. We do not know of any other example of such a large rise of γ within a single crystallographic modification. Our result is in satisfactory agreement with what is expected from other experimental findings on YbCuAl. From the anomalous low-temperature thermal expansion one arrives - via the Grüneisen relation - at $d\ln\gamma/dp = 8.7 \times 10^{-5}$/bar. [6] The pressure dependence of the magnetic susceptibility χ [7], which should be equal to that of γ from theoretical arguments [8], yields a somewhat smaller value of $d\ln\chi/dp = 3 \times 10^{-5}$/bar.
It is interesting to compare with measurements on alloys in the system $Sc_xYb_{1-x}CuAl$. [5] The substitution of some Yb ions by the smaller

Sc ions simulates the application of pressure on the matrix. It can be derived from measurements of the magnetic susceptibility that a Sc concentration of x = 0.1 approximately corresponds to an external pressure of 11 kbar, that is about the maximum pressure reached in our experiments. In an alloy of this concentration Mattens found $\gamma = 375$ mJ/mole K^2 [5], which is surprisingly close to our value at 10.6 kbar. The results of Mattens on YbCuAl and $Sc_{0.1}Yb_{0.9}CuAl$ are included in fig. 4 as triangles.
The low-temperature properties of YbCuAl can be described in a Fermi liquid model, which treats the f-states of Yb as a local impurity, constituting a scattering resonance for the conduction electrons of width Δ and distance (E_F-E_f) from the Fermi energy. [9] Since in YbCuAl the width Δ of the resulting structure in the density of states near E_F is only about 30 K, one will expect electronic contributions also to the cubic term in the low temperature specific heat. The occurrence of such contributions have been shown by comparing the specific heat data of YbCuAl and of integer-valent LuCuAl, by what means the electronic contribution to the cubic term in YbCuAl could be determined as about 67 %. [6] We find a rise of the total cubic term β under pressure, which must be explained as an increase of the electronic contribution β_{el}. In principle, from γ and β_{el} the parameters Δ and $(E_F - E_f)$ of the Fermi liquid model can be calculated. Because of the rather large uncertainties in our results on β, instead of giving precise numbers we only wish to mention that the resonance seems to get narrowed and shifted closer to E_F under pressure, in agreement with the phenomenological picture that pressure tends to stabilize the trivalent state.

We thank R. Pott, II. Phys. Inst., Univ. Köln, for kindly making the samples available to us. The suggestion of this experiment by D. Wohlleben and helpful discussions with W. Gey are gratefully acknowledged. This work has been made possible by financial support by the Deutsche Forschungsgemeinschaft.

REFERENCES

[1] Loriers-Susse, C., High Temp.- High Pressures 12 (1980) 119-129.
[2] Eichler, A. and Gey, W., Rev. Sci. Instrum. 50 (1979) 1445-1452.
[3] Sullivan, P. F. and Seidel, G., Phys. Rev. 173 (1968) 679-685.
[4] Eichler, A., Bohn, H. and Gey, W., Z. Physik B 38 (1980) 21-25.
[5] Mattens, W. C. M., Thesis, Univ. of Amsterdam (February 1980).
[6] Pott, R., Schefzyk, R., Wohlleben, D. and Junod, A., to be published.
[7] Mattens, W. C. M., Hölscher, H., Tuin, G.J. M., Moleman, A. C. and de Boer, F. R., J. Magn.Magn.Mat. 15-18 (1980) 982-984.
[8] Lustfeld, H. and Bringer, A., Solid State Commun. 28 (1978) 119-122.
[9] Newns, D. M. and Hewson, A. C., J. Phys. F: Metal Phys. 10 (1980) 2429-2445.

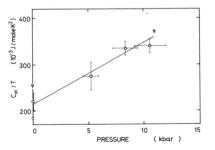

Fig. 4. Pressure dependence of the electronic specific heat coefficient of YbCuAl. Circles: this work; triangles: results of Mattens [5] on YbCuAl at P = 0 and $Sc_{0.1}Yb_{0.9}CuAl$, corresponding to a pressure of 11 kbar.

PHYSICS OF SOLIDS UNDER HIGH PRESSURE
J.S. Schilling, R.N. Shelton (editors)
© *North-Holland Publishing Company, 1981*

EFFECTS OF PRESSURE ON THE MIXED VALENT PROPERTIES OF Ce AND ITS INTERMETALLIC COMPOUNDS

S. H. Liu

Solid State Division, Oak Ridge National Laboratory*
Oak Ridge, Tennessee 37830, U.S.A.

The rare earth element Ce and a number of rare earth compounds exhibit valence transition under pressure. This has been one of the most active research topics in recent years. There also exist several intermetallic compounds of Ce that do not undergo abrupt valence changes. Instead, the valence varies continuously with pressure or temperature. A systematic high pressure study of these compounds will further elucidate the phenomenon of valence fluctuation. For these materials, the single parameter which is most sensitive to the pressure is the position of the Ce 4f level relative to the conduction bands. Therefore, it should be possible to monitor the valence with pressure and follow the evolvement of the system from a local moment state to a nonmagnetic state. The latter may be a spin compensated Kondo lattice state of a lattice of virtual bound states. We review the experimental evidence of these exotic states and suggest new experiments to investigate their physical properties.

1. INTRODUCTION

Cerium metal and many of its intermetallic compounds exhibit anomalous thermal, magnetic and transport properties which can be attributed to the fact that the 4f level of Ce is near the Fermi level. In the crystalline solid the 4f state is hybridized with the conduction electron states so that the 4f level assumes a finite width. Whenever the separation between the 4f level and the Fermi level is comparable to this width, the conduction electrons at the Fermi level will take on the 4f character, such as large effective mass and electronic specific heat. At the same time the occupation of the 4f level is nonintegral, somewhere between zero and one. This situation is often described as a quantum mechanical mixing of two different valent states of Ce, i.e., Ce^{4+} and Ce^{3+}. The term valence is not well defined when the 4f state is hybridized with the band states, and attempts to give a precise number to the valence of Ce often leads to confusion, especially when the 4f occupation is close to zero or one. Therefore, as an operational definition, we will refer to a system as mixed-valent whenever the 4f level is at such a position that its finite width affects the properties of the system.

In addition to the position and the width of the 4f level, the properties of mixed valent systems are further influenced by the Coulomb repulsion between 4f electrons in opposite spin states, the crystal field effects, and the magnetic

interaction between the 4f electrons on different sites. The interplay of these interactions gives rise to the great diversity of physical properties observed in mixed valent systems. On the other hand, the complexity of the interaction Hamiltonian makes the development of an all-encompassing theory extremely difficult. Even with drastic simplifications the model has been solved approximately only in limiting cases. Many manifestations of the mixed-valent phenomenon are understood qualitatively by interpolating or extrapolating known theoretical results outside their domains of reliability. There are also uncertainties in the model parameters so that one can not predict a priori which system will exhibit what behavior. One must cross correlate different experimental results in order to build a physical description for each material. In this paper we propose to show how high pressure techniques can be employed to gain crucial information which will help our understanding of these fascinating systems. We will use Ce and some of its intermetallic compounds as examples, although the principles apply to other mixed valent rare-earth and actinide systems as well.

There exist in literature a number of review articles on the phenomena and the theories of mixed valence.[1-7] The reviews by Koskenmaki and Gschneidner,[1] Jayaraman,[2] and Probst and Wittig[3] are especially relevant to the high pressure effects discussed in this paper.

---

*Operated by Union Carbide Corporation under contract W-7405-eng-26 with the U.S. Department of Energy.

2. BASIC INTERACTIONS AND THEIR EFFECTS

The interactions that are relevant to the phenomena of mixed valence can be summarized in the following Hamiltonian

$$H = H_d + H_f + H_{fd} + \text{others} . \qquad (1)$$

The first term H_d is the energy of the band electrons, mainly of d character,

$$H_d = \sum_{\underline{k}\sigma} \varepsilon_{\underline{k}} \, d_{\underline{k}\sigma}^{\dagger} \, d_{\underline{k}\sigma} \qquad (2)$$

where the annihilation operator for a band electron with wave vector $\underline{k}$ and spin σ is denoted by $d_{\underline{k}\sigma}$, and the energy of this state is $\varepsilon_{\underline{k}}$, which is independent of the spin. This band Hamiltonian is the vastly simplified version of the real situation which involves many bands of different characters. The second term H_f is the energy of the f levels

$$H_f = \varepsilon_f \sum_{i\sigma} f_{i\sigma}^{\dagger} f_{i\sigma} + U \sum_{i} f_{i\uparrow}^{\dagger} f_{i\uparrow} f_{i\downarrow}^{\dagger} f_{i\downarrow} , \qquad (3)$$

where we denote the annihilation operator of an f electron at site i with spin σ by $f_{i\sigma}$. The f electron wave functions on different sites do not overlap. The sum on i is over all the Ce sites, which corresponds to the entire lattice in the Ce metal and a sublattice in Ce compounds. The two particle interaction term is the Coulomb repulsion between a pair of f electrons on the same site with opposite spins. The interaction strength U is so large that at any time the f level can at most be singly occupied. Here again, the model is simplified by leaving out the degeneracy of the f orbital. The term H_{fd} represents the mixing of the d and f electrons

$$H_{fd} = \frac{1}{\sqrt{N}} \sum_{\underline{k}\sigma i} [V_{\underline{k}} d_{\underline{k}\sigma}^{\dagger} f_{i\sigma} e^{-i\underline{k}\cdot\underline{R}_i} + \text{h.c.}] , \qquad (4)$$

where N is the number of Ce sites and $\underline{R}_i$ is the position of the ith site. This interaction allows the f electrons to become itinerant, and as a result the f level acquires a finite width. The term labelled "others" in Eq. (1) includes the spin-orbit coupling and the crystal field interaction of the f electrons.

The complexity of the mixed valent problem becomes apparent when one studies the first three terms of the Hamiltonian. Consider the Ce sites as isolated impurities, then in the trivial case of U = 0 the f-d interaction gives rise to the f-level width(8)

$$\Gamma = \pi \, N_d(\mu) \, \langle V_{\underline{k}}^2 \rangle \qquad (5)$$

where $N_d(\mu)$ is the band density of states at the Fermi level μ, and $\langle V_{\underline{k}}^2 \rangle$ is the average f-d interaction strength over the Fermi surface. The f occupation is found to be

$$\langle n_\sigma \rangle = \frac{1}{\pi} \tan^{-1}[\Gamma/(\varepsilon_f - \mu)] . \qquad (6)$$

This shows that whenever $|\varepsilon_f - \mu| \stackrel{\sim}{=} \Gamma$ the occupation is fractional. The level is less than half-filled when it is above the Fermi level and is more than half-filled otherwise. If the full lattice of f levels is considered the Hamiltonian is again soluble for U = 0. The new energy bands are the f-d hybridized bands

$$\varepsilon_{\underline{k}}^{(\pm)} = \frac{1}{2} (\varepsilon_{\underline{k}} + \varepsilon_f) \pm [\frac{1}{4} (\varepsilon_{\underline{k}} - \varepsilon_f)^2 + |V_{\underline{k}}|^2]^{1/2}. \qquad (7)$$

These are two bands separated by $2|V_{\underline{k}}|$ at the point where the d band crosses the f level. The energy range in which the f and d levels mix is again measured by Γ, and the occupation of the f level is fractional whenever the f level is within the range Γ from the Fermi level.

In the presence of the U term the one impurity problem has no simple solution. The mean-field theory of Varma and Yafet(9) gives the following result for the f level width

$$\Gamma_\sigma = \pi(1 - \langle n_{-\sigma} \rangle) N_d(\mu) \langle V_{\underline{k}}^2 \rangle . \qquad (8)$$

The meaning of this result is that an electron of spin σ can occupy an f level provided that the level is not already occupied by an electron with the opposite spin. The occupation number is given by

$$\langle n_\sigma \rangle = (1 - \langle n_{-\sigma} \rangle) \frac{1}{\pi} \tan^{-1}(\frac{\Gamma_\sigma}{\varepsilon_f - \mu}) . \qquad (9)$$

The mean-field theory is probably valid only when the f orbitals are slightly occupied, i.e., when the f level is sufficiently far above the Fermi level. Nevertheless, it illustrates one feature of mixed valence which is probably valid in general, that is, the f level tends to narrow when it is partially occupied. In the band case the mean-field theory again predicts hybridized f-d bands similar to those in Eq. (7) except that the effective f-d hybridization interaction is $[1 - \langle n_{-\sigma} \rangle]^{1/2} V_{\underline{k}}$. In both the impurity case and the band case one finds equal population of spin-up and spin-down electrons, i.e. the system is nonmagnetic. In later discussions we will refer to this nonmagnetic state as the virtual bound state.

Whether the system should be described by the impurity model or the band model may depend on the temperature. For $kT \gg \Gamma$ the individual f sites are uncorrelated and the impurity model applies; but when $kT \ll \Gamma$ it is energetically more favorable for the f sites to become correlated and form band states. The thermodynamics of the problem is not well understood because of the large Coulomb interaction. For instance, it is not known whether the breakdown of the inter-site correlation with increasing temperature takes place continuously or by a phase transition. Regardless of the description, the f-d hybridization model predicts a high specific heat coefficient γ and a high, finite and temperature dependent magnetic susceptibility. These properties are seen in α-Ce, $CeSn_3$, $CeAl_3$, $CePd_3$ and $CeBe_{13}$.[10]

Stable 4f moments are formed on the Ce sites when the f level is sufficiently below the Fermi energy and is occupied by nearly one electron. This situation requires different considerations from the virtual bound state case. One finds by applying the Schrieffer-Wolff transformation to the first three terms of Eq. (1) that the local moment interacts with the spin of the conduction electron via an effective exchange interaction of the strength[11]

$$J = \langle V_k^2 \rangle / (\varepsilon_f - \mu) \; . \qquad (10)$$

Since $\varepsilon_f - \mu < 0$, the sign of J is negative (antiferromagnetic), and this is the condition for the Kondo scattering of the band electrons. As long as the temperature is much greater than the Kondo temperature T_K, which is related to J, the f electrons act independently. At low temperatures the system may go into a number of different states. The f-d exchange gives rise to the RKKY interaction between the f electrons on different sites. As a result the f moments can order magnetically, and the internal magnetic field thus produced quenches the Kondo scattering. This situation is seen in the β and γ phases of Ce,[1] $CeIn_3$, $CeAl_2$ and others.[12] Another possibility involves the crystalline field effect which makes the spin flip scattering inelastic and thus inhibits the Kondo effect at low temperatures. Since Ce has the total spin of 5/2, the Kramers degeneracy persists under crystal fields of all symmetries. Hence the low temperature state of this system is expected to be magnetic. A third and most intriguing possibility is that the Kondo effect may overcome the RKKY exchange so that the low temperature state becomes nonmagnetic. The nature of the nonmagnetic ground state of the Kondo lattice is not well understood. A system with any of these properties will be said to be in a local moment state.

Finally, the spin-orbit coupling of the f orbitals puts the unhybridized f electrons in eigenstates of the total angular momentum. This has the effect of changing the size as well as the spatial distribution of the magnetic moment of the f electron. The neutron form factor measurements on $CeAl_3$[13] and $CeSn_3$[14] confirm that the spin-orbit coupling is unaffected by mixed valence.

Having enumerated the various factors which influence the physical properties of the mixed valent system, we are in the position to discuss how these factors may be manipulated by applying pressure to the system.

3. EFFECTS OF PRESSURE

Under hydrostatic pressure the widths of the conduction bands tend to increase because of the increased overlap of the wave functions of neighboring atoms. Under normal circumstances the broadening of the bands brings about a decrease in the density of states at the Fermi energy, roughly by 1% per kbar of pressure for d band metals. For the pressure that transforms γ-Ce to α-Ce, the change in the band density of states is not significant. The increase of electron density under pressure increases the screening of the nuclear charge, and results in upward shifts of the d and f levels relative to the s level. However, the d level shift is compensated by the increase of d band width so that only the f level shift is apparent.[15] The size of the f level shift is of the order 1 meV/kbar. In mixed valent systems the f level is within about 20 meV of the Fermi energy so that the f level shift produced by just a few kbar of pressure is significant. The f level width also increases somewhat for two reasons: the f orbital on one Ce atom overlaps more strongly with the d orbitals of the nearest neighbor atoms, and the decreased f population improves the likelihood of d-f mixing as explained in the last Section, Eq. (8).

The best known high pressure effect on mixed valent systems is the abrupt change in the lattice parameter in Ce and SmS under pressure. The phase transition is accompanied by a sudden depletion of the f level occupation, which occurs when the f level moves up through the Fermi level. In many alloy and compound systems the physical properties change smoothly with pressure because the f level is initially either above the Fermi energy or too far below the Fermi energy. The two kinds of systems react to pressure initially in opposite ways. The Type I system, such as α-Ce, has the f level above the Fermi energy and exhibits weaker mixed valent properties under pressure. The Type II system, such as γ-Ce, has the f level below the Fermi energy and exhibits stronger mixed valent properties under pressure. In the following we will discuss in some detail what can be learned from probing these smooth and less dramatic pressure effects.

3.1 Magnetic Susceptibility

The magnetic susceptibilities of both α-Ce[17] and $CeSn_3$[18] are found to decrease under pressure. This is in accordance with the Type I behavior because at low temperatures the susceptibility reflects the density of states at the Fermi level, which decreases when the 4f level moves up under pressure, while at high temperatures the susceptibility reflects the 4f population, which also decreases under pressure. The broad susceptibility maximum of $CeSn_3$ at around 150 K is reduced at 6 kbar, making the temperature dependence more Pauli-like. However, the upturn in the susceptibility at low temperatures persists and is seemingly unaffected by pressure. This lends support to the suspicion that it is not an intrinsic property of the crystal.

The pressure effects on the magnetic properties of Type II systems $CeAl_2$, $CeIn_3$ and $CeAl_3$ have been studied.(19) The results are of strong theoretical interest because they provide clues to the mechanism which limits the susceptibility at low temperatures. If the susceptibility is limited by the quenching of the local moments due to the Kondo effect, the increased $|J|$ under pressure should cause the susceptibility to decrease. If, as suggested by some authors,(20) that the low temperature properties of $CeAl_3$ is explainable in terms of virtual-bound 4f states, then the low temperature susceptibility should increase under pressure because the Fermi level is closer to the peak of the virtual state. The measured values of the susceptibility under 6 kbar of pressure for both $CeAl_2$ and $CeAl_3$ are lower than the zero pressure values, but for $CeIn_3$ there appears to be no change. This probably indicates that both mechanisms are present, and one is more or less important than the other in different systems. It is also possible that all systems undergo a transition from the local moment behavior to the virtual bound state behavior as the pressure is varied. One should make these measurements with fine increments of pressure so that no detail is missed, especially near the phase boundary between Type I and Type II behaviors.

The Curie-Weiss part of the susceptibility is dominated by the 4f population, so it is expected to diminish under pressure.

3.2 Electrical Resistivity

The electrical resistivity of the Type II system $CeAl_2$ has been measured over a wide range of temperature and pressure.(21,22) Under ambient pressure the magnetic part of the electrical resistivity, which is obtained by subtracting the electrical resistivity of $LaAl_2$ from that of $CeAl_2$, shows a broad maximum around 50 K due to the combined effect of Kondo scattering and crystal field splitting, and a sharp maximum around 4 K due to magnetic ordering. In the pressure range 0 - 16 kbar the magnetic resistivity is found to increase, the temperature of the broad maximum to decrease, and the magnetic ordering temperature to decrease. The increased resistivity is consistent with the stronger Kondo scattering as the f level approaches the Fermi energy. The decreased temperature of the resistivity maximum indicates a reduction in the crystal field splitting. The decrease in the magnetic ordering temperature is also consistent

with what is observed in heavy rare earth systems.(23)

At about 65 kbar of pressure an abrupt change of the volume takes place and the Ce loses its local moment.(24) Under higher pressures the behavior of the electrical resistivity is totally different from the low pressure case, i.e. the resistivity decreases with increasing pressure and the maximum is not evident. This can be understood in terms of the virtual bound state model. At high temperatures the electrons that are trapped in the virtual-bound f states act like independent scattering centers of the current carriers. When the f level moves up from the Fermi energy, the number of scattering centers as measured by the f population goes down, hence the reduced resistivity. The temperature dependence reflects the change of the degree of coherence of the f electron wave functions on different Ce sites, as discussed in the previous Section.

The behavior in the intermediate pressure range is determined by two conflicting effects. The f level moves up close to the Fermi energy and loses some of its population, and this causes a reduction of the number of scattering centers. Simultaneously the effective exchange parameter $|J|$ increases, and this gives stronger scattering. The delicate interplay of these two effects can cause the resistivity to go either up or down with pressure. The resistivity decreases with pressure in the 28-65 kbar range, but a maximum in the magnetic part is still discernible. Thus, the system displays the local moment behavior and the virtual bound state behavior concurrently. The theoretical problem associated with this phenomenon is intriguing because the exchange parameter J must be renormalized by the f level width. A concerted theoretical and experimental investigation using fine pressure increments will yield important information about mixed valent systems.

The same experiments should be carried out on $CeIn_3$, which is also of Type II.

The behavior of the Type II system $CeAl_3$ offers new theoretical challenge.(25) The high temperature resistivity increases with pressure at a higher rate than $CeAl_2$. The resistivity below 10 K depends on pressure in the opposite manner, which can not be explained in terms of the crystal field effect. If we consider the susceptibility data under pressure, we may rationalize that the reduction in the low temperature resistivity is probably due to more effective quenching of the local moments with the increased $|J|$. If this is the case, one should also see this effect in $CeAl_2$ somewhere in the pressure range 16-65 kbar and in $CeIn_3$ in some appropriate pressure range.

The temperature dependence of the resistivity of $CeSn_3$ resembles that of $CeAl_2$ above 65 kbar.(18) Thus we expect $CeSn_3$ to follow the Type I behavior.

3.3 Heat Capacity

The electronic contribution to the heat capacity of Type I systems is expected to decrease with pressure, as has been observed in α-Ce,(26) for the reason that the pressure shifts the center of the virtual bound state away from the Fermi energy and thus reduces the density of electron states. Both α-Ce and $CeSn_3$(27) exhibit an upturn in the heat capacity at low temperatures, reminiscent of the low temperature upturn of their susceptibilities. Just like the latter, the upturn in the heat capacity of α-Ce shows no clear pressure dependence.

For those Type II systems which order magnetically the low temperature heat capacity has a magnetic part which can be subtracted out from the data. The remaining heat capacity has the linear temperature dependence at low temperatures, and the coefficient γ is determined jointly by the density of electronic states and the spin fluctuation. Above the magnetic ordering temperature the γ value is expected to be a measure of the density of electron states alone. If this is the correct picture, then the data on $CeAl_2$, $CeIn_3$ and $CeAl_3$ can be qualitatively understood.(19) It seems that the high temperature value of γ increases with pressure because the density of electron states increases when the f level approaches the Fermi energy from below, and the low temperature value of γ decreases under pressure because of the increase of $|J|$. This interpretation correlates well with the susceptibility and resistivity data for these systems.

3.4 Thermoelectric Power

The effect of high pressure on the thermoelectric power of Ce has been investigated by following through the γ to α transition at various temperatures.(28) At a constant temperature the thermoelectric power increases rapidly with pressure until the γ-α phase boundary is reached, and at this point the thermoelectric power drops abruptly. In the α phase the thermoelectric power is observed to decrease steadily with increasing pressure and to change sign under sufficient pressure. The authors attributed the observed phenomenon to the pressure shift of the virtual-bound f states. The electrons that are temporarily trapped in 4f states contribute to the thermoelectric power by scattering the conduction electrons, and this contribution increases when the f level approaches the Fermi energy from below and changes sign when the f level crosses the Fermi energy. When combined with the normal positive thermoelectric power, the total result changes sign at a higher pressure than the critical value for γ-α transition. The investigation was carried out at room temperature and above, so it missed the interesting range in which the long range coherence of the f electrons on different sites takes place. It is difficult to do the low temperature experiment on Ce for metallurgical reasons, but there should be no difficulty with Ce intermetallic compounds.

3.5 Neutron Scattering

The pressure induced γ-α phase transition of Ce at room temperature has been probed by inelastic neutron scattering.(29) The goal of the experiment was to observe the diffuse paramagnetic scattering of the 4f moments and deduce the d-f interaction from the linewidth. It was found that in the γ phase the linewidth increases with pressure, indicating a stronger exchange parameter J. The authors reported no trace of paramagnetic scattering in the α phase, but this result has been reinterpreted to mean that the linewidth is exceedingly large, perhaps an order of magnitude larger than that for the γ phase.(30) The spin relaxation widths have also been observed in $CeAl_3$,(13,31) $CePd_3$(32) and $CeSn_3$.(33) As a general rule, the Type II systems have small ($\sim$ 5 meV) and temperature sensitive linewidths, in contrast to the large ($\sim$ 50 meV) and temperature insensitive linewidths for Type I systems. Clearly in Type I systems the spin relaxation must take place by a different mechanism, presumably by electrons fluctuating in and out of the virtual f levels rather than by spin fluctuation as in Type II systems.(33) We anticipate a continuous transition from the local moment behavior to the virtual bound state behavior by putting a Type II system under pressure.

The neutron form factors of the Type II systems $CeAl_2$(34) and $CeAl_3$(13) have been reported. This quantity is the Fourier transform of the magnetization density, and for $CeAl_2$ and $CeAl_3$ the magnetic electron is clearly the 4f electron. In contrast the magnetic field induced moment of the Type I system $CeSn_3$ has the 4f form factor above 40 K but a mixture of 4f and 5d moments below this temperature.(14,35) The authors explained this result on the basis of the virtual bound state model. In particular the long range coherence of the 4f wave functions on different sites makes the band description more suitable at low temperatures, and the added d character in the neutron form factor is a direct manifestation of d-f hybridization. If this is the correct interpretation, one should see a transition from pure 4f form factor to a mixed d and f form factor by subjecting a Type II system to pressure.

Since the γ-α phase transition is associated with a large change in the lattice parameter, it has been anticipated that the phonon spectrum in mixed valent systems should be anomalous near the valence transition. So far the anomaly has only been seen in the semiconducting system $Sm_{.25}Y_{.75}S$,(36) and the corresponding phenomenon in metallic systems has not been established. It may be feasible to use pressure as another parameter to bring the system close to the transition point and render the phonon anomaly visible.

3.6 De Haas-van Alphen Effect

It is difficult to conduct the de Haas-van Alphen experiment because it requires high quality single crystal specimens. So far the only successful experiments on mixed valent materials have been carried out on the Type I material $CeSn_3$.(37-39) Compared with the non-mixed-valent system $LaSn_3$, $CeSn_3$ possesses many more orbital frequencies, and the electrons in these orbits have large masses, typically ten times the free electron mass. This result establishes beyond any doubt that $CeSn_3$ has energy bands of the f character at the Fermi energy, and that there is long range coherence of the f electron wave functions on different Ce sites.

The pressure dependence of the de Haas-van Alphen frequencies should be studied to assess how successful the present band calculation techniques can elucidate the nature of the 4f bands. The band calculation accounts for the observed orbital frequencies in $LaSn_3$(37) and goes a long way toward analyzing the orbital frequencies of $CeSn_3$.(40) However, the calculated density of states at the Fermi level corresponds to a specific heat coefficient γ that is much too low compared with the experiment, and the theoretical neutron form factor disagrees completely with the measured one. The latter has been the basis of the speculation that the local spin density approximation for the exchange-correlation potential in band calculations fails to include certain important correlation effects in mixed valent systems.(27) It is hoped that the pressure effect will give helpful clues to this puzzle.

It has been conjectured that the strong spin fluctuations could destroy the de Haas-van Alphen effect in Type II systems. One should regard this prediction as a challenge rather than a discouragement. There is no fundamental reason why high quality single crystal specimens of Type II systems can not be prepared. Due to the translational symmetry there must exist a Fermi surface in such specimens at sufficiently low temperatures. The observation of the Fermi surface by whatever means will be extremely exciting.

4. CONCLUDING REMARKS

We have left out in the foregoing discussion a number of experimental techniques, notably optical absorption, photoemission and resonant techniques such as electron spin resonance, nuclear magnetic resonance and Mössbauer effect. The optical experiments have contributed a great deal to the mixed valent problem, but they suffer from low resolution and thus are not suitable for probing the small changes in physical properties induced by high pressure. The spin relaxation rate in mixed valent systems is too fast, and consequently the resonant methods

have only been effective as alternative ways to measure the static magnetic susceptibility. Nevertheless, the Knight shift measurements on the non-mixed-valent components of Ce inter-metallic compounds can give information on how the mixed valence of Ce affects its neighbors. This information is valuable in improving the microscopic model for mixed valent compounds.(41)

REFERENCES:

[1] Koskenmaki, D.C. and Gschneidner, K.A., Jr., in Gschneidner, K.A., Jr. and Eyring, L. (eds.), Handbook on the Physics and Chemistry of Rare Earths (North-Holland, Amsterdam, 1978), Vol. 1, Chap.4.

[2] Jayaraman, A., ibid., Vol. 2, Chap. 20.

[3] Probst, C. and Wittig, J., ibid., Vol. 1, Chap. 10.

[4] Maple, M.D., Delong, L.E. and Sales, B.C., ibid., Vol. 1, Chap. 11.

[5] Parks, R.D., Valence Instabilities and Related Narrow Band Phenomena (Plenum, New York, 1977).

[6] Varma, C.M., Rev. Mod. Phys. 48 (1976) 219.

[7] Lawrence, J.M., Riseborough, P.S., and Parks, R.D., Valence Fluctuation Phenomena, preprint.

[8] Anderson, P.W., Phys. Rev. 124 (1961) 41.

[9] Varma, C.M. and Yafet, Y., Phys. Rev. B 13 (1976) 2950.

[10] See Refs. 1 and 7.

[11] Schrieffer, J.R. and Wolff, P.A., Phys. Rev. 149 (1966) 491.

[12] See Refs. 1 and 7.

[13] Edelstein, A.S., Child, H.R. and Tranchita, C., Phys. Rev. Lett. 36 (1976) 1332.

[14] Stassis, C., Loong, C.-K., Harmon, B.N., Liu, S.H., and Moon, R.M., J. Appl. Phys. 50 (1979) 7567.

[15] Pickett, W.E., Freeman, A.J. and Koelling, D.D., Phys. Rev. B 23 (1981) 1266.

[16] Ratto, C.F., Coqblin, B. and D'Agliano, E.G., Advan. in Phys. 18 (1969) 489.

[17] MacPherson, M.R., Everett, G.E., Wohlleben, D. and Maple, M.B., Phys. Rev. Lett. 26 (1971) 20.

[18] Sereni, J.G., J. Phys. F. 10 (1980) 2831.

[19] Berton, A., Chaussy, J., Chouteau, G., Cornut, B., Flouquet, J., Odin, J., Palleau, J., Peyrard, J. and Tournier, R., J. de Phys. 40 (1979) C5-326.

[20] Andres, K., Graebner, J.E. and Ott, H.R., Phys. Rev. Lett. 35 (1975) 1779.

[21] Nicolas-Francillon, M., Percheron, A., Achard, J.C., Gorochov, O., Cornut, B., Jerome, D. and Coqblin, B., Solid State Commun. 11 (1972) 845.

[22] Probst, C. and Wittig, J., J. Magn. Magn. Mat. 9 (1978) 62.

[23] Jayaraman, A., in Gschneidner, K.A., Jr. and Eyring, L. (eds.), Handbook on the Physics and Chemistry of Rare Earths (North-Holland, Amsterdam, 1978), Vol. 1, Chap. 9.

[24] Croft, M. and Jayaraman, A., Solid State Commun. 29 (1979) 9.

[25] Percheron, A., Achard, J.C., Gorochov, O., Cornut, B., Jerome, D., and Coqblin, B., Solid State Commun. 12 (1973) 1289.

[26] Phillips, N.E. and Gschneidner, K.A., Jr., private communciation.

[27] Liu, S.H., Stassis, C. and Gschneidner, K.A., in Proceedings of the International Conference on the Valence Fluctuation in Solids, Santa Barbara, CA (North-Holland, Amsterdam, 1981).

[28] Ramesh, T.G., Reshamwala, A.S. and Ramaseshan, S., Pramana (India) 2 (1974) 171; Solid State Commun. 15 (1974) 1851.

[29] Rainford, B.D., Buras, B. and Lebech, B., Physica 86-88B (1977) 41.

[30] Shapiro, S.M., Axe, J.D., Birgeneau, R.J., Lawrence, J.M. and Parks, R.D., Phys. Rev. B 16 (1977) 2225.

[31] Edelstein, A.S., Majewski, R., Sinha, S.K., Brun, T., Pelizzari, C.A. and Child, H.R., in Proceedings of the Conference on Neutron Scattering, Gatlinburg, TN, 1976.

[32] Holland-Moritz, E., Loewenhaupt, M., Schmatz, W., and Wohlleben, D., Physica 86-88B (1977) 239: Phys. Rev. Lett. 38 (1977) 983.

[33] Loewenhaupt, M. and Holland-Moritz, E., J. Magn. Magn. Mat. 14 (1979) 227.

[34] Barbara, B., Rossignol, M.F., Boucherle, J.X., Schweizer, J. and Buevoz, J.L., J. Appl. Phys. 50 (1979) 2300.

[35] Liu, S.H., Harmon, B.N., Stassis, C. and Symeonides, S., J. Magn. Magn. Mat. 15-18 (1980) 942.

[36] Mook, H.A., Nicklow, R.M., Penney, T., Holtzberg, F., and Shafer, M.W., Phys. Rev. B 18 (1978) 2925.

[37] Johanson, W.R., Crabtree, G.W., Koelling, D.D., Edelstein, A.S. and McMasters, O.D., J. Appl. Phys. 52 (1981) 2134.

[38] Crabtree, G.W., Johanson, W.R., Edelstein, E.S. and McMasters, O.D., in Proceedings of the International Conference on Valence Fluctuations in Solids, Santa Barbara, CA (North-Holland, 1981).

[39] Johanson, W.R., Crabtree, G.W., Edelstein, A.S. and McMasters, O.D., Phys. Rev. Lett. 46 (1981) 504.

[40] Koelling, D.D., private communication.

[41] Malik, S.K., Vijayaraghavan, R., Garg, S.K., and Ripmeester, R.J., Phys. Stat. Sol. (b) 68 (1975) 399.

PHYSICS OF SOLIDS UNDER HIGH PRESSURE
J.S. Schilling, R.N. Shelton (editors)
© North-Holland Publishing Company, 1981

ELECTRONIC INSTABILITIES IN RARE EARTH COMPOUNDS:
SPIN FLUCTUATIONS IN Ce COMPOUNDS

M. Croft, H.H. Levine, and R. Neifeld
Rutgers University
Serin Physics Laboratory
Piscataway, N.J. 08854

Resistivity results on a number of Ce compounds, all of which have been previously shown to undergo pressure or alloy induced electronic instabilities, are presented. The points emphasized in the discussion are that spin fluctuations (due to the Kondo effect) play an important role in these compounds and that a dramatic increase in the spin fluctuation energy scale heralds the approach of the electronic instability.

The energy difference between the $4f^n$ and the $4f^{n-1}$ configurations of Ce, Sm, Eu, Tm and Yb is small. This enables these elements to occur in a $4f^n$, a $4f^{n-1}$ or an intermediate $4f^{n-\delta}$ ($\delta<1$) configuration depending on the specific chemistry of the compound in which they are placed. (1,2, 3) (As discussed below, the situation for Ce is actually more complicated). Since the ionic volume of the $4f^n$ configuration is 25-30 % larger than the $4f^{n-1}$ configuration, pressure will, in general, induce a crossover from the $4f^n$, to the $4f^{n-\delta}$ and then to the $4f^{n-1}$ configuration. In some cases, like elemental Ce and SmS, (5) this crossover takes the form of a first order instability; in others, like SmSe (5) and $CeSn_3$ (6), it is quite continuous.

This paper and the appended "recent results" communication (7) address two physical processes which arise near a configurational instability. The first deals with spin fluctuations in Ce systems in the regime immediately precursive to the configuration change. The second deals with cooperative charge fluctuations in a Eu system at its configurational instability. See (7).

At ambient pressure (T=300K) elemental Ce is in its high-volume, local magnetic moment γ phase. Near 7 kbars of pressure it transforms to its low volume, Pauli paramagnetic α phase (4). Most theoretical schemes explain the volume collapse at this transition in terms of a decrease in the 4f charge density screening the Xe core. It is unclear at present whether this decrease in 4f-screening is accompanied by promotion of charge density into a conduction d-band or into delocalized f states. (3,8) Consequently, for Ce the pertinent electronic variable appears to be a modified f-level occupation number which represents the integrated 4f charge density within some screening radius of the core (rather than the total count of electrons of f-character).

In terms of their magnetic state and effective volume, most Ce compounds can be characterized as γ-Ce-like or α-Ce-like. (9, 10) As indicated in the first paragraph, one might expect to find an f shell instability analogous to the $\gamma \to \alpha$ transition in the high pressure phase diagrams of γ-like Ce systems. To date such transitions have been observed in CeP (near 100 kbars), in $CeAl_2$ (near 70 kbars) and CeS (near 120 kbars). (11, 12, 13, 14).

Interest in the physics of the γ-like to α-like crossover at low temperatures has lead to extensive studies of γ-like Ce systems suspected to be on the verge of transforming to an α-like state. (The temperature induced $\gamma \to \alpha$ transition in elemental Ce precludes such studies in pure Ce.(9)) In what follows emphasis is placed on the fundamental role that spin fluctuations (due to the Kondo effect) play in such γ-like Ce systems.

The Kondo effect involves the flipping of local 4f magnetic moments via an antiferromagnetic interaction with the conduction electrons.(15) The presence of a low temperature resistivity minimum followed by a region with a negative temperature coefficient of resistivity (TCR) has been the longest standing evidence for Kondo effect-related spin fluctuations in Ce compounds.(16)

In figure 1 we show the low temperature resistivity of the three γ-like systems CeP, $CeAl_2$, and CeS with known pressure induced instabilities into α-like phases.(11,12,13,14) These compounds all order magnetically with their ordering temperature on the low temperature side of the resistivity maximum. (17,18,19) Above their ordering temperatures the spin fluctuation related local resistivity minimum and region of negative TCR are clearly evident. The persistance of the negative TCR upon Ce dilution rules out critical scattering as a cause for these structures in all but a few Ce compounds (see reference 14). Also shown in comparison is the resistivity of the $CeAl_2$ based alloy $Ce_{0.6}Th_{0.4}Al_2$ in which spin fluctuations have previously been shown to be substantially quenched. (16,19)

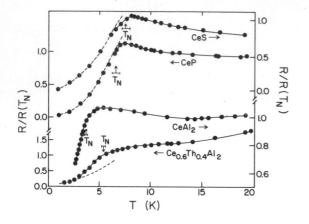

Figure 1: The resistivity versus temperature of several Ce compounds. The residual resistivity has been subtracted out and the curves normalized to $R(T_N)$, the resistivity at their ordering temperatures. The T_N values are taken from the literature. The solid lines are guides to the eye and the dashed lines result from fits of the lowest temperature data to a T^2-functional form.

Recently, it has been suggested that a positive T^2 behaviour in the resistivity of Ce compounds below their ordering temperatures was associated with incoherent spin fluctuations.[20] While this is only a hypothesis, it is interesting to note that the low temperature resistivity of CeS and CeP follow closely a T^2 law, as indicated by the dashed lines in figure 1. The lower ordering temperature of $CeAl_2$ precludes a similar fit over this temperature range, however, the coexistence of spin fluctuations with magnetic order has been long established for $CeAl_2$.[21] Again for comparison a T^2 fit to the lowest temperature resistivity in the spin fluctuation-quenched $Ce_{0.6}Th_{0.4}Al_2$ compound is seen to deviate strongly from the experimental data at higher temperatures (see figure 1).

Although similar spin fluctuation effects are observed at ambient pressures in CeP, CeS and $CeAl_2$, it is important to note two points. Firstly, the conduction band population per Ce atom is much less than one in CeP, equal to one in CeS, and much larger than one in $CeAl_2$.[14] In the limit of very strong spin fluctuations, the ground states in these three cases are, respectively, believed to be a magnetically ordered, a non-magnetic Kondo insulating, and a nonmagnetic metallic state [22,23]. Thus this very similar ambient pressure behaviour should change to a very divergent behaviour at high pressures where the spin fluctuation effects are stronger. Secondly, stoichiometric variations in CeS and especially in CeP will cause variations in conduction band population which could have strong effects on the magnetic properties.[14,17,18]

Under increasing pressure, the low temperature

properties of γ-like compounds are dominated by the rapid increase in the spin fluctuation energy scale (the spin fluctuation temperature, T_{SF}). A number of authors have dicussed the influence of this increasing T_{SF} on low temperature properties such as the magnetic ordering temperature, the specific heat and the average magnetic moment. [24,25,26] At high temperatures the increase in T_{SF} under pressure is also manifested in the pressure-dependent resistivity. Since the single-impurity spin fluctuation (Kondo) resistivity (R) is a monotonically decreasing function of the variable $x = \ln(T/T_{SF})$ (i.e. $dR/dx<0$) it is clear that an increasing T_{SF} (at constant T) leads to an increasing resistivity.[15] Although coherent band behaviour mandates that concentrated systems show a decreasing resistivity at low temperatures, well above T_{SF} such single-impurity type behaviour is observed in even concentrated Ce compounds.[27]

Referring to figure 2a, one sees the room temperature, pressure dependent resistivity of three materials (CeS, $CeIn_3$ and $Ce_{0.6}Y_{0.4}Al_2$) all of which have been driven through an f shell instability in high pressure or alloy studies. [13, 14,25,26,28,29] To 40 kbars the monotonic increase in resistivity of all these systems is caused by pressure-induced increase in T_{SF} mentioned above. Pressures greater than 40 kbars are required to induce the transformation to the α-like state for these compounds. In contrast, the compound $CeSn_3$ shows a continuous pressure-induced crossover into an α-like phase.[6] Accordingly, the resistivity of $CeSn_3$ (figure 2b), like that of elemental Ce and $CeAl_2$, once they have started their transformations to an α-like phase, is monotonically decreasing under pressure [4,13]

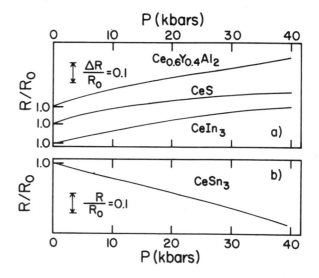

Figure 2a and 2b: The pressure-dependent resistivity of several Ce compounds at room temperature. The curves are normalized to R_o, their ambient pressure resistivity.

Two of the authors (30) have carried out an extensive study (discussed at length elsewhere) of a series of alloys $(Ce,R)Al_2$ with R=Sc,Y, and $La_{1-y}Y_y$. In these studies R is always 3+ (isoelectronic with Ce in $CeAl_2$). The choice and percentage of R is used to vary the effective volume of the material. In figures 3a and 3b the scaled resistivity of selected compounds from this study are shown. Two parameters, the total activity of the magnetic scattering ($\tilde{R}$) and the temperature scale factor T_{SF}, were adjusted to collapse the experimental data as shown in figures 3a and 3b.

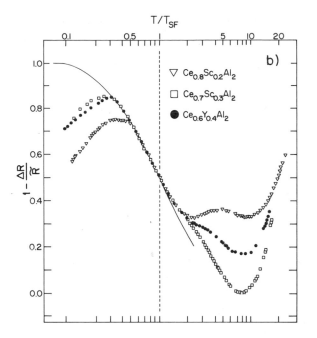

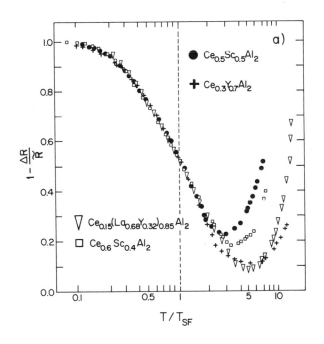

Figure 3a. The resistivity versus $\ell n\, T/T_{SF}$ of several $CeAl_2$-based alloys scaled and normalized to show a universal low temperature resistivity form. ΔR is the low temperature saturation resistivity minus the resistivity. $\tilde{R}$ represents the total activity due to the spin fluctuations and along with T_{SF} is adjusted so as to collapse the data. The high temperature upturns in figures 3a and 3b are due to photon scattering.

Figure 3b. The resistivity versus $\ell n(T/T_{SF})$ of several $CeAl_2$-based alloys scaled and normalized to follow the universal resistivity behaviour (from figure 3a) near T_{SF}. The deviation from the universal empirical curve (as indicated by the solid line) at $T/T_{SF}<0.5$ is due to a combination of magnetic interaction and intersite coherency effects. The structure near $T/T_{SF}<2.0$ is interpreted as crystalline electric field effects.

The more Ce-dilute compounds and those with higher T_{SF} values (shown in 3a) follow a monotonic, universal low-T behaviour similar to that observed previously for other Kondo systems (15,31). The high temperature upturns in all samples is due to phonon scattering. The more concentrated Ce compounds and those with lower T_{SF} values, shown in figure 3b, deviate at low temperatures from universal behaviour either due to coherent band or magnetic interaction effects (also as previously seen in concentrated Kondo systems (32)). The additional structure near $T/T_{SF}\sim5$ in figure 3b is caused by crystalline electric field effects. (9)

In figure 4, the spin fluctuation temperatures of these alloys, derived from the above-mentioned scaling procedure, are plotted versus the experimentally determined reduced volume of the specific alloy involved. Also included in figure 4, from the work of Steglich et.al. (32), are estimates of T_{SF} (based on a T^2-type analysis) in the $R=La_{1-y}Y_y$ alloy system. The approximate volume at which a collapse to an α-like state is observed in these alloys is indicated by the hashed region in figure 4. (12,29,30)

It is apparent from figure 4, that the spin fluctuation temperature of (Ce,R)Al$_2$ increases by
more than two orders of magnitude as the volume
is reduced towards an α-like state. Moreover,
this behaviour appears surprisingly insensitive
to Ce concentration and the chemical make up of
R. It should be noted that no real sign that
this dramatic increase in T$_{SF}$ is abating is visible in these results. Thus by virtue of this
large increase in energy scale, careful consideration must be given to the participation of
spin fluctuations in the low volume limit of
γ-like Ce compounds and also in the high volume
limit of α-like Ce compounds.

The systematic spin fluctuation behaviour emphasized above constitutes a basic (but often overlooked) part of the γ-like to α-Ce like crossover. Any truly successful theory for Ce compounds should quantitatively account for such
spin fluctuation effects.

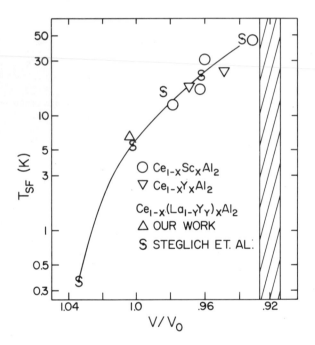

Figure 4. The spin fluctuation temperatures, T$_{SF}$,
of the compounds shown in figures 3a and 3b
versus the unit cell volume of the compound (normalized to the unit cell volume (v$_o$) of CeAl$_2$).
T$_{SF}$ was fixed by the temperature scale variable
required to collapse the data as shown in figures
3a and 3b. Also included (indicated by S) are
T$_{SF}$ values estimated in reference (32).

Acknowledgements

This research was funded in part by the Rutgers
Research Council and the Research Corporation.
The authors would like to thank Mr. Schildwachter
for help in sample preparation and characterization.

References

(1) See numerous articles in "Valence Instabili-
    ties and Related Narrow Band Phenomena",
    edited by R.D. Parks, (Plenum, New York,
    1977)

(2) See numerous articles in "Valence Fluctua-
    tions in Solids", edited by L. Falicov,
    W. Hanke, and M.P. Maple, North-Holland
    Publishing Co., New York, 1981)

(3) D. Wohlleben, ibid 2. p.1. and references
    therein.

(4) A. Jarayaman, Phys. Rev. 137, A179 (1965)
    (and references therein).

(5) A. Jarayaman, A. Singh, A. Chatterjee, and
    S. Usha Devi, Phys. Rev. B. 9, 2513 (1974).

(6) J. Beille, D. Block, J. Voiron, and G. Pari-
    sot, Physica 86-88B, 231 (1977).

(7) See Croft, Hodges, Kemly, Krishnan, Murgai,
    Gupta, Parks, in these proceedings.

(8) J. Wittig, ibid 2, p. 43.

(9) B. Coqblin, G. Gonzales-Jimenez, A. Bhatta-
    charjee, J. Iglesias-Sicardi, B. Cornut, and
    R. Jullien, J. Magn and Mag. Mater. 3. 67
    (1976) and references therein.

(10) M. Croft, J. Weaver, D. Peterman, and A.
     Franciosi, Phys. Rev. Lett. 46, 1104 (1981).

(11) A. Jarayaman, W. Lowe, L. Longinotti, and
     E. Bucher, Phys. Rev. Lett. 46, 1104 (1981).

(12) M. Croft and A. Jarayaman, Sol. St. Comm.
     29, 9 (1979).

(13) C. Probst and J. Wittig, Jour. Magn. and
     Mag. Mat. 9, 62, (1979).

(14) M. Croft and A. Jarayaman, Sol. St. Comm.
     35, 203 (1980).

(15) See the review article G. Gruner and Z.
     Zawadowski, Rep. Prog. Phys. 37, 1497 (1974).

(16) K. Bushow and H. van Daal, Phys. Rev. Lett.
     23, 408 (1969).

(17) F. Hulliger, B. Natterer, and H. Ott, Magnet.
     Magn. Mat. 8, 87 (1978).

(18) F. Hulliger and H. Ott, Zeit. Phys. B 29, 47
     (1977).

(19) M. Croft and H.H. Levine, Phys. Rev. B. 22,
     4366 (1980).

(20) N. Sato, T. Komatsubara, S. Kunii, T. Suzuki,
     and T. Kasuya, ibid. 2. p. 259.

(21) See F. Stelich, C. Bredl, M. Lowenhaupt, and K. Schotte, J. Physique C5-40, 301 (1979) and references therein.

(22) C. Lacroix and M. Cyrot, Phys. Rev. B20, 1969 (1979).

(23) R. Jullien, P. Pfeuty and B. Coqblin, ibid. 2, p. 169.

(24) M. Croft, R. Guertin, L. Kupferberg, and R.D. Parks Phys. Rev. B 20, 2073 (1979) and references therein.

(25) See A. Benoit, J. Boucherle, J. Flouquet, F. Holtzberg, J. Schweizer, and C. Vettier, ibid 2 p. 197 (and references cited therein).

(26) A. Eiling and J. Schilling, Phys. Rev. Lett., 46, 364 (1981) and references therein.

(27) F.D.M. Haldane, ibid 1, p. 191.

(28) J. Lawrence and D. Murphy, Phys. Rev. Lett. 40, 961 (1978).

(29) J. Aarts, F. deBoer, S. Horn, F. Steglich, and D. Mechede, ibid 2, p. 301.

(30) H.H. Levine and M. Croft, ibid 2, p. 279, M. Croft and H.H. Levine to be published.

(31) J. Schilling, Adv. in Phys. 28, No. 5, p. 657 and references therein.

(32) F. Steglich, W. Franz, and W. Seuken, Physica 86-88B, 503 (1977).

PHYSICS OF SOLIDS UNDER HIGH PRESSURE
J.S. Schilling, R.N. Shelton (editors)
© North-Holland Publishing Company, 1981

ELECTRONIC INSTABILITIES IN RARE EARTH COMPOUNDS:
"RECENT RESULTS" ON COOPERATIVE CHARGE FLUCTUATIONS
IN $EuPd_2Si_2$

M. Croft, J.A. Hodges,
E. Kemly and A. Krishnan
Rutgers University
Serin Physics Laboratory
Piscataway, N.J. 08854

V. Murgai, L. Gupta and R. Parks
Polytechnic Institute of New York
33 Jay Street
Brooklyn, New York 11201

The results of the temperature dependent ^{151}Eu Mössbauer spectroscopy on dilute Eu in $LaPd_2Si_2$ and $EuPd_2Si_2$ are reported. Comparison of these two studies emphasizes the role of cooperative charge fluctuations in the concentrated compound. An extension of existing phenomenology to include these cooperative effects is constructed to quantitatively describe the data.

The experimental situation for 4f shell instabilities in Eu compounds stands in sharp contrast to that outlined earlier for Ce compounds (see (1) in these proceedings). First the nature of the instability for Eu is accepted to be a promotion of a 4f electron to the conduction band. (2) Second, there is no experimental evidence for Kondo effect type spin fluctuations (which are so prevalent in Ce compounds (1)) in Eu^{2+} compounds. Indeed recent theoretical work seems to preclude such spin fluctuations in most Eu compounds due to the large angular momentum (J=7/2) of Eu^{2+}.(3) Finally the thermal variation of configuration in the Eu compounds, studied thus far, is well described by a phenomenology based on a non-cooperative Boltzmann distribution among ionic configurations.(2,4,5,6) In this "recent results" appendum, experimental results on $EuPd_2Si_2$ are presented and interpreted by extending previous phenomenology to include cooperative effects. Although these results are at ambient pressure they are directly pertinent to the character of pressure induced valence (configuration) transitions in Eu based compounds.

In the configurational crossover field it has proved useful to summarize experimental results using a phenomenology based on two ionoc configurations separated by an excitation energy E. (eg. for Eu these are $Eu^{2+}4f^7$ and $Eu^{3+}4f^6$). (3,4,5,6) The occupation probabilities for these configurations are given by the Boltzmann distribution

$$P_2 = \frac{8e^{-E/\tau}}{Z} \text{ and } P_3 = \frac{1+3e^{-480/\tau}}{Z} \quad (1)$$

with the partition function Z being defined by the sum of the Eu^{2+} and Eu^{3+} probabilities (P_2 and P_3 respectively). The measured ^{151}Eu Mössbauer isomer shift is an average over the two configurations as the time scale of the Mössbauer measurement is longer than the charge fluctuation time. If the two configurations have isomer shifts i_2 and i_3 the average value i is given by

$$i = i_2 P_2 + i_3 P_3. \quad (2)$$

Quantum mechanical tunneling between these two ionic configurations is simulated in this model by setting $\tau = \sqrt{T^2 + T_F^2}$. Here kT_F is supposed to represent the tunneling-induced configuration energy width.(2,4,5,6) (The reader is referred to the appendix for a matrix model justification of including interconfigurational tunneling in this way).

In figure 1 (solid boxes) we show the temperature depend ^{151}Eu isomer shift for 1 % Eu in the compound $LaPd_2Si_2$. (Details of the Mössbauer spectroscopy and sample preparation will be presented elsewhere (7)). This compound is a dilute analog of the newly discovered interconfigurational system $EuPd_2Si_2$.(8) The dashed line in figure 1 represents the predictions of the ionic model with E chosen to be 458 K, $i_2 = -10.6$ and $i_3 = -1.28$ (in mm/sec). The values of i_2 and i_3 are close to those found in isomorphous integral valent compounds(8). The value of $T_F = 50K$ is the same as previously used for dilute Eu in $ScAl_2$. (6) It is apparent that this dilute interconfigurational compound is adequately described by a single excitation energy over this temperature range.

The authors have also performed Mössbauer spectroscopy on $EuPd_2Si_2$. In addition to the main Mössbauer resonance reported previously (8) additional, much weaker, absorption intensity (representing sites or regions in the material which are more toward the 2+ state) has been observed. The following discussion deals only with the main line.

In figure 1 the temperature dependent isomer shift of the main line in $EuPd_2Si_2$ is shown. The temperature variation of this line is much more nonlinear than the dilute sample and can not be fitted by a single excitation energy. This dramatically increased nonlinearity (compared to the dilute compound) provides direct evidence for a cooperative Eu-Eu interaction operating in the concentrated system.

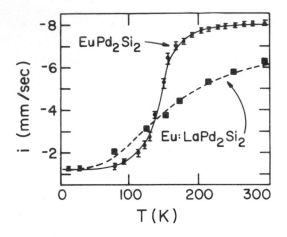

Figure 1. [151]Eu isomer shift versus temperature for 1 % Eu in LaPd$_2$Si$_2$ (solid squares) and for EuPd$_2$Si$_2$ (points with error bars). The dashed and solid lines are predictions of theoretical models discussed in the text.

Such a cooperative effect can be incorporated into the existing ionic state phenomenology by invoking a cooperative coupling of the excitation energy to either, or both, the average volume, (eg. a "compression shift" mechanism after reference (9)) and the average valence (eg. a Coulombic shift mechanism after reference (10)).

For interconfigurational Eu compounds the valence, volume and [151]Eu isomer shift are all directly coupled. To a first approximation this coupling is usually assumed to be linear.(2,4,5,6) Therefore the cooperative excitation energy coupling can be written in terms of the isomer shift as

$$E = E_o + \alpha (i - i_o) \qquad (3)$$

where E_o and i_o are the model excitation energy and isomer shift at T = 0.

Combining (3) with (1) and (2) one arrives at a transcendental equation-of-state for i(T) which must be solved numerically. This equation of state for i(T) is reminiscent of a meanfield treatment of ferromagnetism (in an external field) (11). It is not surprising therefore that this equation of state (like its magnetic analogue) yields continuous, first order, or second order valence transitions, depending on the magnitude of the interaction parameter α and on the excitation energy E_o.

Adjusting the parameters in this cooperative ionic model to fit the isomer shift data of EuPd$_2$Si$_2$ one arrives at the solid line figure 1. The ad-

justed values of E_o, i_o and α respectively are 589 K, -1.29 mm/sec, and 58.2 $\frac{k \cdot sec}{mm}$ The choices of i_2, i_3, i_o and T_F have been set at the as in the dilute case. The theoretical line provides a good fit to the data over the entire temperature range. The i(T) curve obtained is substantially sharpened by cooperativity however the valence transition is still continuous.

The proximity of the configurational change in EuPd$_2$Si$_2$ to a first order phase transition can be illustrated by considering variations in the parameters in the fitted isomer shift equation-of-state. Systematically decreasing Eo, while holding the cooperativity parameter α constant in this model depresses the transition temperature and sharpens the nonlinearity into a first order phase change. Alternatively increasing E_o (ie. stabilizing the 3+ state as would occur with increasing pressure) increases the transition temperature and decreases its nonlinearity. Thus the transition in EuPd$_2$Si$_2$ should move to higher temperatures and become more gradual at high pressure. On the other hand, the degree of cooperativity present in this system make it probable that a Eu system (perhaps an alloy of EuPd$_2$Si$_2$ with a lower E) may be found which has a first order valence transition line terminated by a critical point in its pressure temperature phase diagram.

Acknowledgements: The authors would like to thank Dr. N. Koller for support and T. Yang for technical assistance in this work. This work was supported in part by the National Science Foundation.

Appendix: Ionic Configuration Based Phenomenology

The existing interconfigurational model as used above contains an effective configurational width which has been introduced phenomenologically (2, 4,5,6) Some understanding of one contribution to this effective width can be obtained via a related matrix model described below.

Consider the case of the Hund's rule ground states of Eu^{2+} (J=7/2) and Eu^{3+} (J=0). This is a 9 dimension vector space. If m$_j$ labels the z component of angular momentum in the Eu^{2+} state we will label the first 8 states in this space by n=m$_j$ + 9/2 and the last state (n=9) will be associated with the Eu^{3+} configuration. In the absence of conduction electron mediated transitions the 9x9 Hamiltonian matrix (H) is diagonal with
$$H_{nn} = \begin{cases} E & n = 1,8 \\ O & n = 0 \end{cases}$$

where E is the excitation energy of the Eu^{2+} state relative to the Eu^{3+} state. In the process of configurational crossover the value of E goes from negative (Eu^{2+} ground state), to zero, to positive. (Eu^{3+} ground state).

In this simplified ionic model the 4f-conduction-electron-coupling can be approximated by off diagonal matrix elements connecting the Eu^{2+} and Eu^{3+} states. Since the ability of the conduction band to absorb or emit an electron of different angular momenta will in general be different, we must assume that the effective off diagonal matrix elements (Δ_n) will be different when connecting the Eu^{3+} state to Eu^{2+} states with differing $|m_j|$. That is

$$H_{9n} = H_{9n}{}^* = \Delta_n \qquad n = 1,8 \qquad \text{and}$$

$$\Delta_n = \Delta_{9-n} \qquad n = 1,8$$

We assume that all matrix elements connecting Eu^{2+} states vanish since these elements are pertinent to spin fluctuations rather than charge changes. ie.

$$H_{nm} \equiv 0 \quad n \neq m, \; n,m < 8$$

The energy eigenvalues and eigenfunctions for this matrix are a sevenfold degenerage set labeled

$$E_i = E, \; \psi_i{}^* = (a^i{}_1, \ldots, a^i{}_9) \qquad \text{with}$$

$$a^i{}_1 = 1/\sqrt{2} \text{ and } a_j{}^i = -\delta_{j,i+1}/\sqrt{2} \qquad j = 2,8.$$

There are also two split off states with

$$E_{\pm} = \frac{E}{2}\{1 \pm (1 + 4\Gamma^2/E^2)^{1/2}\},$$

$$\psi_{\pm}{}^* = (a_1{}^{\pm}, \ldots, a_9{}^{\pm}) \qquad \text{with}$$

$$a_9^{\pm} = (E-E_{\pm})/(\Gamma^2 + (E-E_{\pm})^2)^{1/2} \qquad \text{and}$$

$$a_j^{\pm} = \Delta_j \, a_9/(E-E_{\pm}) \qquad j=1,8.$$

Here $\Gamma^2 = \sum\limits_{n=1}^{8} \Delta_n{}^2$.

The schematic energy diagram for this model is shown in Figures A1a) for a Eu^{3+} ground state and A1b) for a Eu^{2+} ground state. The solid circles represent the states with no off diagonal elements and the open circles represent the states with interconfigurational coupling present. The states i=1,7 are mixtures of purely Eu^{+2} states with different m_j's and have the same energy, E, as the original (nonmixed) Eu^{2+} state. The two states E+ and E- are mixtures of Eu^{2+} and Eu^{3+} and are split off in energy from the original unmixed ionic states.

In this treatment the ground state, ψ_g (either ψ_+ or ψ_g for E<0 and E>0 respectively) (ψ_g) is always interconfigurational with a degenerate pure Eu^{2+} state and another interconfigurational state lying at higher energies. The singlet ψ_g

is also nonmagnetic to first order with all Eu^{2+} states with $+|m_j|$ and $-|m_j|$ mixed into it with equal amplitudes. (ie. there is no T=0 susceptibility divergence).

One could proceed with the thermodynamics of this model however the authors prefer to make contact with the existing phenomenological ionic model.

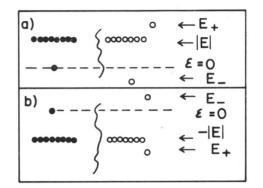

Figure A1a and A1b. Schematic energy level diagram for the matrix model (see text) for a Eu^{3+} (a) and Eu^{2+} state (b) lowest in energy.

The previously used ionic model (1,4,5,6) starts off by forming the partition function (Z) for two pure ionic configurations separated by an energy E. The replacement $T \to \tau = \sqrt{T^2 + T_F{}^2}$ in Z is then made to simulate the effect of finite widths of the energy levels. By taking the T=0 limit of equations (1) in the text one can deduce that the ground state of the previously used ionic model (ϕ_g) is

$$\phi_g{}^* = (b_1, \ldots, b_9) \qquad \text{with}$$

$$b_9 = (1 + 8e^{-E/kT_F})^{-1/2} \qquad \text{and}$$

$$b_i = b_9 \, e^{-E/2kT_F}$$

This ground state is equivalent in construction to the ψ_g found in matrix model and like that model is both <u>interconfigurational</u> and <u>nonmagnetic</u>. The role of the various parameters in the two models can be seen equating their T=0, Eu^{2+} and Eu^{3+} configurational probabilities. Although this matrix model avoids the ad hoc replacement $T \to \sqrt{T^2 + T_F{}^2}$ to achieve the proper ground state its spirit (ie. including the band degrees of freedom via a phenomenological parameter) is basically similar to that of the existing ionic model. Also, the simplicity of this matrix model is lost when Hunds rule excited states are included. Therefore in the text we use the existing model evoking the contact to the matrix model as a partial justification of the replacement $T \to \tau = \sqrt{T^2 + T_F{}^2}$.

References

(1) See M. Croft, H.H. Levine, R. Neifeld, and
    A. Jarayaman and R. Mains in these proceed-
    ings.

(2) See D. Wohlleben in "Valence Fluctuations
    in Solids", edited by L. Falicov, W. Hanke,
    and M.P. Maple (North-Holland Pub. Co.,
    N.Y., 1981) p1. and references therein.

(3) Ph. Nozieres and A. Blandin, J. Physique $\underline{41}$,
    193 (1980)

(4) B. Sales and D. Wohlleben, Phys. Rev. Lett.
    $\underline{35}$, 1240 (1975) B. Sales, J. Low. Temp. Phys.
    $\underline{28}$, 107 (1977).

(5) E. Bauminger, D. Froindlich, I. Nowik, S. Ofer,
    I. Felner, and I. Mayer, Phys. Rev. Lett. $\underline{30}$,
    1053 (1973).
    E. Bauminger, I. Felner, D. Levron, I. Nowik,
    and S. Ofer, Phys. Rev. Lett. $\underline{33}$, 890 (1974).

(6) W. Franz, and F. Steglich, and W. Zell and
    D. Wohlleben, and F. Pobell, Phys. Rev. Lett.
    $\underline{45}$, 64 (1980).

(7) M. Croft, J.A. Hodges, A. Krishnan, and
    E. Kemly, to be published.

(8) E. Sampathkamaran, R. Vijayaraghavan, K. Go-
    palakrishnan, R. Pilley, H. Devare, L. Gupta,
    B. Post, and R. Parks, ibid 2, p. 193.

(9) B. Coqblin and A. Blandin, Adv. in Phys. $\underline{17}$,
    281 (1968).

(10) L. Falicov and J. Kimball, Phys. Rev. Lett.
     $\underline{22}$, 997 (1969).

(11) See C. Kittel, Introduction to Solid State
     Physics, 5th edition (John Wiley and Sons,
     N.Y., 1976) Chapter 15.

PHYSICS OF SOLIDS UNDER HIGH PRESSURE
J.S. Schilling, R.N. Shelton (editors)
© *North-Holland Publishing Company, 1981*

SOME RECENT RESULTS IN MAGNETISM UNDER HIGH PRESSURE

James S. Schilling

Experimentalphysik IV
Universität Bochum
463 Bochum, W. Germany

High pressures have the potential not only to create or destroy magnetism, but also to cause a well-defined continuous variation of the magnetic properties of a single-phase system. The later capability is particularly useful in a critical test of theories of matter. Recent experimental work exemplifying both types of studies are presented. The evolution of a stable-moment system from magnetism to superconductivity is outlined. The topics discussed include: pressure dependence of the exchange interaction and magnetic ordering temperature in dilute and concentrated stable-moment systems, use of pressure as a spectroscopic tool, transition from stable to unstable moment (mixed-valent) behavior, and the effect of pressure on the magnetic susceptibility of weak-itinerant magnetic and strongly Stoner-enhanced systems like $ZrZn_2$, $TiBe_2$, and Pd. Suggestions are made for future work.

1. INTRODUCTION

In condensed-matter physics, high pressures can be used not only to create entirely new forms of matter out of old, but also to generate small but well-defined changes in the relevant physical parameters of a particular phase. It is this dual role which makes the pressure technique so valuable in both applied and basic studies in all of condensed-matter physics.

In the present paper I will attempt to illustrate how high pressures can be used to advantage in studies of the magnetic properties of matter. Before discussing in detail some recent results, perhaps it is well to first step back and ask whether there are any general statements one can make about how high pressures affect magnetism. In the case of the transport properties, for instance, it is well known that all substances will eventually become metallic if subjected to sufficiently high pressures, the last one probably being solid He at pressures ~100 Mbars (1000 GPas)!(1) How does magnetism fare? The situation here is considerably more complicated, or perhaps I should say, more interesting. To some extent one may liken the condensation of atoms out of the gaseous phase to applying high pressure because in both cases atoms are brought closer together. Whereas the vast majority of free neutral atoms possesses a magnetic moment, only in a few rare cases is the magnetism retained in the condensed phase. This is shown in Figure 1. It would, however, be incorrect to then conclude that compressing an arbitrary magnetic solid would inevitably purge it of its magnetism. A good counterexample is elemental Yb-metal which can only *become magnetic* if pressures sufficiently large to remove an electron from the filled $4f^{14}$-shell are applied. The existence of a non-filled inner shell is a prerequisite for local magnetism. The overriding effect of compressing a solid is to reduce

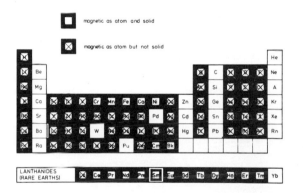

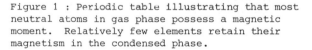

Figure 1 : Periodic table illustrating that most neutral atoms in gas phase possess a magnetic moment. Relatively few elements retain their magnetism in the condensed phase.

the number of electrons in the inner atomic shells, i.e. pressure progressively breaks up the atomic shell structure until ultimately, for $P>10^6$ GPa, only a Thomas-Fermi electron gas is left.(2) The local magnetism will mirror the momentary state of the atomic shell structure, rising and falling as new shells are broken open and depleted. Only after the last electron has been squeezed out of the innermost shell does the local magnetism finally disappear for good (of course, the extremely weak nuclear magnetism would still be present). The foregoing is illustrated schematically in Figure 2. Band magnetism can also react in very interesting ways to applied pressure. It would appear that the overriding effect is a *reduction* of band magnetism under pressure because band-broadening often leads to a decrease in the density of states at the Fermi surface, $N(E_f)$, which normally dominates over

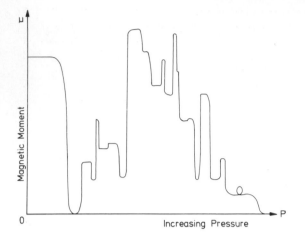

Figure 2 : Schematic representation of magnetic moment μ versus pressure for a given solid. Pressure gradually breaks up the atomic shell structure, which is reflected in $\mu(P)$. At far right $P \sim 10^6$ GPa, the atomic shell structure is destroyed and only Thomas-Fermi electron gas is left.

the exchange interaction increase.(3) It does, however, occur that $N(E_f)$ increases under pressure (4); indeed, in some systems pressure stabilizes the band magnetism, at least over a certain pressure range.(5)

The pressure-induced magnetic transitions discussed above and illustrated in Figure 2 can be grouped into three categories:

(a) local→local, eg. Gd 4f^7(5d^{1}6s^2)→4f^6(5d^{2}6s^2) (note that band electrons are given in parentheses);

(b) local→band, eg. Am 5f^6(6d^{1}7s^2)→(5f^{6}6d^{1}7s^2);

(c) band→band, eg. Fe (3d-band) or ZrZn$_2$(4d-band). For a further discussion of the above types of transition, I refer the reader to the excellent papers in the present volume by Johansson, Wohlfarth, and others. It suffices here to point out that the richness of the magnetic behavior of a *single* element over an arbitrarily large pressure range should easily rival that exhibited by all the elements in the periodic table at ambient pressure.

To put the discussion on a more concrete footing, let us focus our attention on the far-left peak in our magnetic phase diagram in Figure 2, which is shown on an expanded scale in Figure 3. This portion of the total μ versus P phase diagram corresponds to the special case where pressure totally *destroys* the magnetic moment. We could just as well have shown that part of the phase diagram corresponding to Yb where pressure *creates* magnetism, or to Tm where pressure changes μ from one nonzero value to another.

Starting at the far left in Figure 3, the magnetic moment of a very "stable moment" system is seen to be initially independent of pressure, but to eventually fall to zero within the "unstable moment" or "mixed-valent" regime at much higher pressures. The magnetic ordering temperature T_o, which in a local-moment system, is a function of the size of the moment μ and the strength of the exchange interaction $\mathcal{Y}$, would be expected to exhibit only a relatively weak pressure dependence in the "stable moment" regime where μ is a constant.(6) As a given system, however, is "pressured" close to the unstable region, a sensitive increase in one component of the exchange interaction leads to an anomalous rise and subsequent fall in $T_o(P)$.(7) Such Anderson-lattice behavior will be discussed in more detail later. At higher pressures, the magnetic moment itself becomes unstable and falls to zero; T_o will also vanish, probably at a more rapid rate than μ, as indicated in Figure 3.

The disappearance of the magnetism is favorable for the occurrence of superconductivity. However, between the magnetic and superconducting regions there is a wasteland containing systems, like Pd, which are almost magnetic or superconducting, but really neither. These systems can have a very large Stoner enhancement factor which makes them possible candidates for p-wave superconductivity.(8) Going further to the right to the superconducting systems, a possible initial rise of the superconducting transition temperature T_c with pressure as found, for example, for La(9) and La$_3$Se$_4$(10,11) due to say, an increase in $N(E_f)$, must be followed at some pressure by a decrease in T_c which is at least partly due to lattice stiffening. No quantity, including $N(E_f)$, can, after all, increase forever! Note that one should not infer that the superconducting systems in Figure 3 would become magnetic if subjected to negative pressure (going to the left in the diagram). Such studies of superconductivity under pressure have the potential to shed light on the important question of what determines the maximum value of T_c for a given system or class of systems. I will not discuss studies of superconductivity further here but refer the interested reader to the relevant papers by Segre and Braun, Eiling et al., Wittig, Razavi and Schilling, and Smith in the present volume.

The arrows in Figure 3 roughly locate the ambient-pressure position of selected systems on the phase diagram. Systems with initially very stable moments (eg. Gd, AgMn) can be estimated to require enormous pressures >100 GPa to reach instability, whereas relatively unstable systems like β-Ce, CeAl$_2$, or ZrZn$_2$ need only a few GPas to destroy the magnetism. In the following I will outline some recent results which I hope will exemplify what kinds of information can be obtained by pressure experiments in both stable and unstable moment regimes.

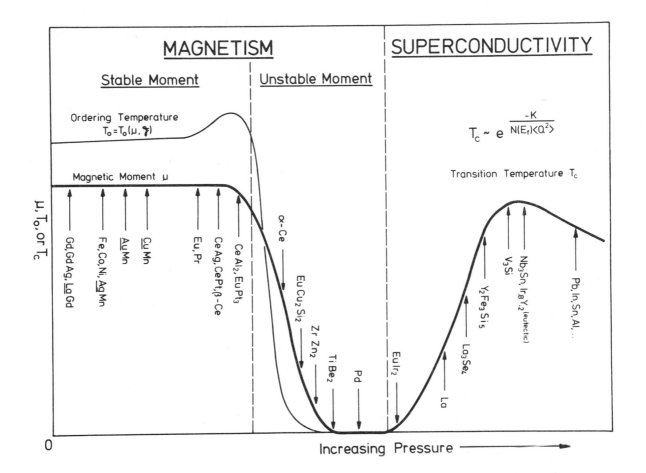

Figure 3 : Expanded version of far-left peak in Figure 2 showing initial destruction of magnetic moment μ under pressure followed by a rise of superconductivity. Also shown are possible pressure dependences of magnetic ordering temperature T_O and superconducting transition temperature T_C. Arrows mark approximate initial positions of selected systems on diagram at ambient pressure.

2. STABLE LOCAL MOMENTS

2.1 Exchange Interaction in Single Impurity Systems

Perhaps the "simplest" problem in all of metallic magnetism is that of a single stable magnetic impurity in a nonmagnetic host--eg. Gd in La or Mn in Ag. Such systems are initially located to the far left in the phase diagram in Figure 3. The magnetic moment μ_i of the impurity experiences an exchange interaction J with the magnetic moment μ_e of the host conduction electrons

The impurity magnetic moment μ_i, because of its extreme stability here, does not change when modest pressures are applied. What *does* change under compression is the magnitude of the exchange interaction J.(6) Such studies would thus allow a sensitive test of the few existing first-principle calculations of the exchange interaction(12); a valid theory must be able to not only predict correctly the magnitude of J, but also its pressure dependence $\partial J/\partial P$. The situation here is analogous to that for superconductivity where it is very difficult to calculate T_C or $\partial T_C/\partial P$ for a given system accurately, even though superconductivity itself is well understood.

The exchange interaction J, which in principle is a function of both the magnitude and direction of the electron wave-vectors $\underline{k}$ and $\underline{k}'$ before and after the exchange scattering, can be written in a certain approximation(6,13,14) in the form

$$J = J_oP_o(\cos\theta) + 3J_1P_1(\cos\theta) + 5J_2P_2(\cos\theta) + 7J_3P_3(\cos\theta) + \cdots \quad , \quad \{1\}$$

where $P_\ell(\cos\theta)$ is the ℓth Legendre polynomial and θ is the scattering angle. To determine the pressure dependence of each J_ℓ-component individually requires studying various properties, eg. NMR, resistivity, EPR, etc. under pressure.(6,13). Resistivity studies, for instance, can be used to estimate the pressure dependence of the so-called "Kondo" exchange component, $\partial J_k/\partial P$, where $J_k=J_2$ for 3d-impurities like Mn or Fe and $J_k=J_3$ for 4f-impurities like Gd or Ce.(6) In Figure 4 we display the volume-dependence of the J_k-component $\partial\ln|J_k|/\partial\ln V\equiv\partial\ln|J_k|/(K\partial P)$, where K is the compressibility, for several systems.(6) We see that, with the exception of LaGd, the magnitude of J_k increases about 2-4 times faster than the volume decreases, which is a rather sizeable dependence compared to what would be expected for a simple

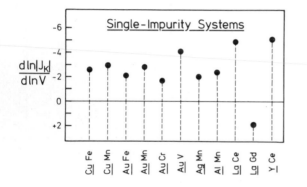

Figure 4 : Volume dependence of the "Kondo" J_k-component of the exchange interaction between magnetic impurity and host conduction electrons. For 4f-impurities, $J_k=J_3$ and for 3d-impurities, $J_k=J_2$ (see Equation 1). For LaGd, a non-Kondo system, J_k is an appropriate average of J_o, J_1, J_2, and J_3. See Reference 6 for details.

wavefunction-overlap exchange integral.(6) The reason for this could be that $|J_k|\propto 1/E_{exc}$, where E_{exc} is the energy which must be expended to destabilize the local moment, as shown in Figure 5 for AgMn. Since pressure pushes the stable-moment impurity ever closer to the instability, $|E_{exc}|$ decreases and $|J_k|$ rises rapidly. The increase of $|J_k|$ leads to a sharp increase in the Kondo temperature T_k, since we have(15)

$$kT_k \simeq E_f\exp(-1/N(E_f)|J_k|) \quad . \quad \{2\}$$

The magnitude of the T_k-increase with pressure can be determined from the shift of the large resistivity anomaly along the logarithmic temperature scale(see Figure 5), allowing the determination of the values of $\partial\ln|J_k|/\partial\ln V$ given in Figure 4. The increasing importance of the Kondo-like anomalies under pressure is thus simply a warning that the magnetic impurity is approaching

the instability region. See Reference 6 for more details, including a discussion of the anti-Kondo system LaGd.

2.2 High Pressure "Spectroscopy"

From the preceding discussion it follows that a knowledge of how rapidly $|E_{exc}|$ decreases under pressure is of prime importance when attempting to estimate in which pressure range a given system will enter the "unstable-moment" regime in Figure 3. Johannson and Rosengren(16) derived a simple approximate expression for the pressure dependence of E_{exc}

$$dE_{exc}/dP \simeq \Omega_n - \Omega_{n+1} \quad \{3\}$$

in terms of the difference between the equilibrium atomic volumes of the low(n) and high(n+1) valence states of the magnetic components. Since the atomic volume of the higher valence state (eg. Mn: $3d^4(4s^24p^1)$) is always less than that of the lower valence state (eg. Mn: $3d^5(4s^2)$), as shown in Figure 5, dE_{exc}/dP is always positive. This implies that if $E_{exc}<0$, pressure decreases $|E_{exc}|$ and if $E_{exc}>0$, pressure increases $|E_{exc}|$, i.e. pressure pushes the magnetic level through the Fermi energy, as shown in Figure 5. Inserting the appropriate atomic volumes in Equation 3, one obtains for Eu, $d|E_{exc}|/dP\simeq-100$ meV/GPa; a more exact estimate by Johannson and Rosengren(16) gives the value -125 meV/GPa which corresponds to the volume dependence $d|E_{exc}|/d\ln V\simeq+1.8\times10^4$K. This value is in excellent agreement with that($+1.6\times10^4$K) obtained by Abd-Elmeguid et al. on metallic $Eu_{1-x}Pt_xPt_2$ compounds(17) and in only fair agreement with the value $\sim+5\times10^4$K from optical studies by Wachter on semiconducting EuO, EuS, EuSe, and EuTe.(18) One obtains similarly from Equation 3 for γ-Ce or β-Ce, $d|E_{exc}|/dP\simeq-60$ meV/GPa or $d|E_{exc}|/d\ln V\simeq+1.6\times10^4$K. This value is in reasonable agreement with the value $+1.2\times10^4$K which can be estimated from magnetic susceptibility studies by Sereni et al.(19) on numerous Ce-compounds where E_{exc} is determined as a function of the Ce ionic radius.

It would be of interest to write Equation 3 in a somewhat more general form

$$dE_{exc}/dP \simeq \alpha(\Omega_n/\Omega_{i,n})(4\pi/3)(r_{i,n}^3 - r_{i,n+1}^3) \{4\}$$

where α is the ratio of the compressibility of the given compound or alloy to that of the pure magnetic component in the appropriate valence state. α thus is small for "soft" magnetic impurities substituted in a very stiff compound which absorbs most of the applied pressure, not transmitting much of it to the magnetic impurity. $r_{i,n}$ and $r_{i,n+1}$ are the ionic radii for the magnetic component in the n and n+1 ionic states, respectively, and the ionic volume $\Omega_{i,n}=4\pi r_{i,n}^3/3$. In Equation 4 it is also implicitly assumed that $\Omega_n/\Omega_{i,n}$ is independent of the value of n. If we are interested in only a rough estimate of dE_{exc}/dP, then we can set $\alpha=1$ and $\Omega_n/\Omega_{i,n} \simeq 7$,(20) to obtain

$$dE_{exc}/dP(meV/GPa) \simeq 200(r^3_{i,n}-r^3_{i,n+1}) , \quad \{5\}$$

where r is in Å. This simple expression thus allows us to get a quick rough estimate of dE_{exc}/dP using only tabulated ionic radii.

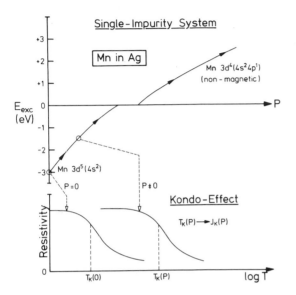

Figure 5 : Hypothetical pressure dependence of the excitation energy E_{exc} for Mn impurities in Ag. As E_{exc} decreases in magnitude, the Kondo resistivity anomaly shifts rapidly to higher temperatures, giving a warning that the moment is about to become unstable.

Viewing Figure 5, if it is assumed that dE_{exc}/dP is independent of the momentary value of E_{exc}, then, because J_k and T_k are functions of $1/E_{exc}$, T_k should increase ever more rapidly as pressure drives E_{exc} to zero. It would thus seem conceivable that the measured value of dT_k/dP could be used to estimate the momentary value of E_{exc} itself—high pressure spectroscopy! If we ignore any pressure dependence of E_f, $N(E_f)$, or the mixing matrix element V_m, where $J_k \sim |V_m|^2/E_{exc}$, it follows from Equation 2 that

$$dE_{exc}/dP =(E_{exc}/(lnT_k/T_f))(dlnT_k/dP). \{6\}$$

Combining Equations 5 and 6 I obtain

$$E_{exc}(meV) =200(r^3_{i,n}-r^3_{i,n+1})(lnT_k/T_f)(dlnT_k/dP)^{-1} \{7\}$$

Let's apply this formula to the classic dilute system CuFe where $T_k \simeq 24K(21)$ and $dlnT_k/dP \simeq +0.1$ $GPa^{-1}(6)$, $200(r^3_{1,2}-r^3_{1,3}) \simeq +30$ meV/GPa, and $T_f=8.2x10^4K$, to obtain $E_{exc} \simeq -2$ eV, certainly a reasonable value. For AgMn I obtain $E_{exc} \simeq -3$ eV which is in excellent agreement with photoemission and optical studies.(22) For LaCe I find $E_{exc} \simeq -0.2$ eV. However, recent photoelectron spectroscopy results indicate that for a number

of Ce-compounds, $E_{exc} \simeq -2$ eV,(23) an order of magnitude different from our rough estimate. It would appear that here, at least, one or more of the assumptions leading to Equation 7 is invalid. It is certainly possible that here, the mixing matrix element V_m, and not E_{exc}, depends strongly on pressure(23) as would be the case, for instance, if Ce undergoes a pressure-induced 4f local-to-4f band transition. Recent band structure calculations on Ce indicate that E_{exc} changes very little with pressure,(24) in agreement with lattice pressure experiments by Croft et al.(23) Further properties of Ce and Eu compounds, which belong in or near the "unstable moment" regime, will be discussed in a later section.

2.3 Concentrated Magnetic Systems

When the concentration of the magnetic component in a "single-impurity" system is increased, the strength of the interactions $\mathcal{J}$ between impurities also increases, opening up the possibility that magnetic ordering occurs in the temperature range of a given experiment. There are numerous distinct mechanisms by which magnetic impurities may interact, eg. direct exchange due to orbital overlap, dipole-dipole interaction, indirect RKKY-interaction, superexchange, etc. The magnitude of the interaction for each individual mechanism changes in its own characteristic way if pressure is applied(6) (for example, for the dipole-dipole interaction, $\mathcal{J}_{dd} \sim 1/R^3$, so that $\partial ln\mathcal{J}_{dd}/\partial lnV=-1$). It is thus possible, if we have preknowledge of these dependences, to identify the dominant interaction mechanism responsible for a given ordered magnetic state by measuring the pressure dependence of the ordering temperature—the "guilty" interaction mechanism will betray itself by its characteristic pressure dependence (recall that in the "stable-moment" regime, the impurity moment is independent of pressure). If one has determined which type of magnetic interaction is dominant in a given magnetic system, one is in a better position to set up a correct theoretical model of its magnetism. Furthermore, it is of interest in its own right to follow how the pressure dependence of a given system evolves as the concentration of the magnetic component is varied over a wide range.

I would like to illustrate these ideas by discussing briefly some recent results on spin-glasses under pressure. In Figure 6 are shown the first high pressure measurements of the magnetic susceptibility of a classical spin-glass, here AgMn. The peak in the susceptibility, which marks the freezing temperature T_0, is shifted slowly to higher temperature by pressure.(25) Such non-dramatic behavior is typical for stable-moment systems,(6) as indicated by the slow increase of T_0 with pressure in the phase diagram in Figure 3. In Figure 7 we display the volume dependences of T_0 for all classical spin-glasses measured.(25,26) We see that the magnetic Grüneisen parameter $\gamma_m=\partial lnT_0/\partial lnV$ can be a function not only of the host matrix, but also of the type of impurity.

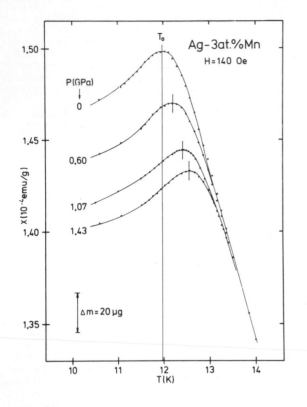

Figure 6 : Static magnetic susceptibility of Ag-3at.%Mn versus temperature at four pressures. The spin-glass freezing temperature T_O shifts reversibly to higher temperature with pressure at the rate $\partial T_O/\partial P = +420\pm40$ mK/GPa.

In no case does $\gamma_m \approx -1$, which shows clearly that the dipole-dipole interaction is not of dominant importance. Where they can be compared, the present direct pressure experiments are in good agreement with recent thermal expansion work,(27) summarized by Smith in the present volume. Although in the spin-glass phase γ_m appears to have little concentration dependence,(27) the present results on Au-3at.%Fe a spin-glass, Au-20at.%Fe an inhomogeneous ferromagnet, and α-Fe a band magnet,(26) show that γ_m *is* strongly dependent on the *type* of magnetic order. In fact, entirely different interaction mechanisms are probably dominant in each of the above three $Au_{1-x}Fe_x$ systems. In contrast to this situation is the behavior of the rare earth system $La_{1-x}Gd_x$ under compression where $\gamma_m \approx +2$ for both amorphous spin-glass $(La_{.7}Gd_{.3})_{.8}Au_{.2}$(28) or ferromagnetic Gd phases. Presumably here the RKKY-interaction is dominant over the entire concentration range, as would be expected from the strong localization of the $4f^7$-state of Gd.

We can learn more by comparing the present results on concentrated systems to those in Figure 4 on

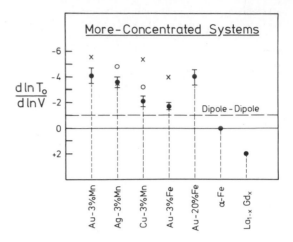

Figure 7 : Volume dependence of magnetic ordering temperature for various systems(concentration is in at.%). ($\bullet$) volume dependence of spin-glass freezing temperature T_O(for Au-20at.%Fe, α-Fe, and Gd, T_O is Curie temperature) by measuring static susceptibility, as in Figure 6. (o) results of thermal expansion studies by Simpson and Smith(Reference 27). (x) values of the dependence $\partial \ln N(E_f)J_2^2/\partial \ln V$ from Figure 4(see also Reference 6). The LaGd spin-glass system is amorphous $(La_{.7}Gd_{.3})_{.8}Au_{.2}$.(28) Horizontal dashed line marks volume dependence (-1) of the dipole-dipole interaction.

single-impurity alloys where $\partial \ln |J_2|/\partial \ln V$ was determined for several 3d-impurities. If the RKKY-interaction is responsible for the spin-glass freezing at T_O, then(6,13)

$$T_O \sim \mathcal{J}_{RKKY} \sim (J_o - 3J_1 + 5J_2 + \cdot \cdot)^2 N(E_f) . \quad \{8\}$$

In the RKKY-interaction several J_ℓ-components are, in principle, equally important for 3d-impurities, namely J_o, J_1, and J_2. Unfortunately, from the resistivity studies in the very dilute limit only the pressure dependence of the J_2-component can be determined. EPR and NMR studies under pressure would be useful here to extract the pressure dependence of the other components.(13) What we can do is determine whether the J_2-component is dominant in $\mathcal{J}_{RKKY}$ or not; if it dominates, then we have

$$\partial \ln T_O/\partial \ln V \approx \partial \ln(J_2^2 N(E_f))/\partial \ln V . \quad \{9\}$$

The value of $\partial \ln N(E_f)/\partial \ln V$ is here approximately equal to +1.(4) From Figure 7 we see that Equation 9 holds exactly for AgMn and nearly for AuMn, indicating the dominance of the J_2-component in these systems.(25,26) Davidov et al.(13) reached the same conclusion for AgMn analyzing resistivity, NMR, and EPR work which indicated that $(J_o-3J_1)<<5|J_2|$ in Equation 8. Apparently, for CuMn and AuFe, J_2 does not dominate. The relatively rapid pressure dependence of J_2 is "diluted" by the sluggish dependence of the non-Kondo exchange components J_0 and J_1, causing $\partial \ln T_0 / \partial \ln V$ to be less than $\partial \ln(J_2^2 N(E_f)) / \partial \ln V$.

I point out in conclusion that for the rare earth system $La_{1-x}Gd_x$, $\partial \ln(J^2 N(E_f)) / \partial \ln V \simeq \partial \ln T_0 / \partial \ln V \simeq +2$ for all values of x, where $\partial \ln J^2 N(E_f) / \partial \ln V$ was determined in the dilute limit $(x < .1)$ from the *positive* resistivity slope at low temperatures, (6,29) and/or from the superconducting pair-breaking.(30) Apparently, over the entire concentration range of this system, a single exchange component, probably J_2, is responsible for the positive resistivity slope, superconducting pair-breaking, and the RKKY-interactions between Gd-spins. This is certainly a simple and satisfying result.

3. UNSTABLE MOMENT SYSTEMS

We now move to the right in the general phase diagram in Figure 3 and consider systems within or near the unstable-moment or mixed-valent regime. In such systems E_{exc} may well be only a fraction of an eV; so, assuming $\partial |E_{exc}| / \partial P \simeq -100$ meV/GPa, as estimated earlier for Eu, it is clear that one can generate sizeable reductions in the magnetic moment μ within an easily accessible pressure range. In this regime, therefore, one expects drastic changes in μ, the magnetic interactions, and the magnetic ordering temperature, as is indicated in Figure 3. In the stable-moment regime, I pointed out that calculations of exchange interactions and their pressure dependences are in a rather preliminary state; this is certainly even more so the case in the unstable-moment regime where far more complicated phenomena are taking place. For this reason it would seem prudent to first attempt to understand the interesting transition region between stable and unstable moments before proceeding further.

In the Introduction I pointed out that, whereas Ce-compounds will *lose* their magnetic moments under pressure $(\partial \mu / \partial P < 0)$ because the single $4f^1$ electron will be squeezed out of the inner shell, for Yb-compounds pressure should *create* a magnetic moment $(\partial \mu / \partial P > 0)$ because a filled shell is broken into, i.e. $4f^{14} \to 4f^{13}$. Because at a given temperature, the static magnetic susceptibility is proportional to the square of the moment $\chi \sim \mu^2$, χ should increase with pressure for Yb-systems and decrease for Ce-systems. This behavior is, in fact, observed in recent experiments at room temperature by Zell et al.,(31) as shown in Figure 8. Particularly interesting is a plot of $\partial \ln \chi / \partial P$ versus the *valence* of various Yb-compounds, as shown in Figure 9. It is seen that the increase of χ with pressure is most rapid at valence 2.5 where Yb is half-way along to developing its full moment for the $4f^{13}$-configuration. A corresponding result would also be expected for Ce-systems, as indicated by the drop in $\mu(P)$ in Figure 3. A number of further new results on Yb-systems are discussed in the present volume.

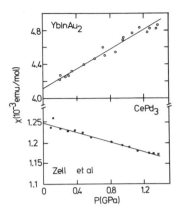

Figure 8 : Static magnetic susceptibility at 300K versus pressure for $YbInAu_2$ and $CePd_3$ from Zell et al(Reference 31).

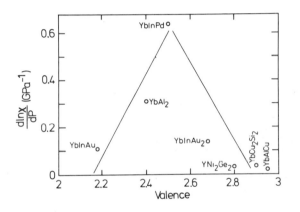

Figure 9 : Relative change of static magnetic susceptibility at 300K with pressure for various Yb-compounds a function of their valence, from Zell et al.(Reference 31).

In Figure 10 I show further anomalies which can occur when concentrated or dilute stable-moment systems are squeezed into the unstable-moment or mixed-valent regime. A volume-collapse or at least compressibility-anomalies often signals the entry of a concentrated system into the later regime,(32) as does the ability of individual unstable impurities to suppress superconductivity and generate large resistivity-anomalies. It is interesting to observe that, as seen in Figure 10 for Ce-systems, all three types of anomaly occur at approximately the same critical pressure ($\sim$1 GPa). This clearly suggests that mixed-valent behavior is principally a *local phenomena*; the first step in understanding mixed-valency in concentrated magnetic systems is to study the anomalies of single-impurity systems. Indeed, it has been shown that single Eu-ions in $ScAl_2$

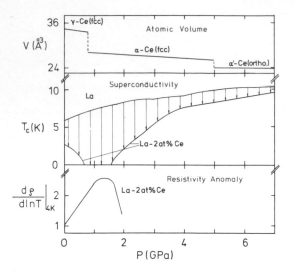

Figure 10 : Anomalies in the atomic volume V of Ce,(33) the superconducting transition temperature T_c,(34) and the negative resistivity slope $\partial\rho/\partial\ln T$ (35) of a La-2at.%Ce alloy as a function of pressure.

can show typical mixed-valent behavior.(36) Eu-impurities in Chevrel-compounds show similar anomalies under pressure to those found in Ce-systems, as discussed by Chu et al. in the present volume. Although it has not yet been possible to push Eu-metal far enough to the right in the phase diagram in Figure 3 for it to become superconducting above 2.2K,(37) this has been accomplished for Ce.(38) $EuIr_2$ is superconducting even at ambient pressure.(39)

Wittig has recently shown that at considerably higher pressures ($\sim$20 GPa) than for Ce, Pr-ions in both Pr-metal and La or Y host metals enter the unstable-moment regime and show closely analogous behavior to Ce.(38) It would seem reasonable to assume that both Ce and Pr are demagnetizing in a very similar way, be it by having 4f-electrons squeezed out of the shell into a 5d-band or by forming a 4f-band. If Pr can be shown to become superconducting at pressures somewhat above the demagnetizing pressure, this would give strong evidence that a 4f-band is formed, for the following reason--if one of Pr's two 4f-electrons were to be squeezed into the 5d-band, one electron would be left behind whose magnetism would presumably destroy any possibility for superconductivity, unless the Pr ordered antiferromagnetically.

From the above it is clear that for Ce, Pr, or Eu impurities to demagnetize, they require neither other Ce's, Pr's, or Eu's as neighbors, nor strong f-character in their host-metal band structure at the Fermi surface. At first sight

this observation would seem to support the promotional model 4f→5d. It is, however, certainly conceivable that the overriding effect of pressure is to squeeze the 4f-electron wave-functions out of their inner shell. If the wave-functions reach out to each other, as in concentrated systems like pure Ce, they join hands and form a 4f-band. But if they reach out as single ions in a nonmagnetic matrix, they experience, because of their extension, sizeable hybridization with the conduction electrons of the host, which causes further extension, etc. So it could well be that 4f-band formation or 4f→5d promotion are just two aspects of the same thing.

As I mentioned before, it is of considerable interest to examine the behavior of magnetic ordering phenomena in the transition region between stable and unstable moments. As pressure squeezes Ce-ions towards instability, E_{exc}→0, and the exchange component $J_3\sim1/E_{exc}$ gets very large in magnitude. According to Equation 8 this would be expected to lead to an *enhancement* of the ordering temperature T_O as the unstable-moment regime is approached, as is indicated in Figure 3. However, the Kondo model tells us that as $|J_3|$ gets large, the tendency for the conduction electron spins to compensate the impurity spin also increases, which would weaken the interaction between impurity spins. This is the essence of the Kondo-lattice model(7) which is the "hard-spin" version of the more general Anderson-lattice model where as $|J_3|$ increases with pressure, T_O should pass through a maximum and fall to zero, as indicated both in Figure 3 and in the insert in Figure 11. Also shown in Figure 11 is the pressure dependence of the Curie temperature of CeAg which rises rapidly at first and passes through a maximum at $\sim$0.7 GPa.(40) This behavior should be contrasted with that of GdAg, a very-stable-moment system, whose ordering temperature rises relatively slowly, showing no particular structure as a function of pressure. It should also be mentioned that Debray et al.(41) have observed the Neel temperature of TmSe to pass through a maximum at $\sim$1.7 GPa.

Another interesting feature of the CeAg data is that once T_O has passed through its maximum value, the normally sharp kink in the resistivity at T_O becomes progressively more rounded.(40) This could be evidence for a change from long-range ferromagnetic to a shorter-range type of order, perhaps even spin-glass like. If the Kondo spin-fluctuations are strong enough to lower the ordering temperature T_O, they could also be capable of changing the *nature* of the magnetic order. The temperature dependence of the electrical resistivity of CeAg is highly anomalous and shows dramatic changes under hydrostatic pressure.(40)

For a more comprehensive discussion of the mixed-valence problem, including examples of unstable rare earth and actinide systems, the reader is referred to numerous other papers in the present volume.

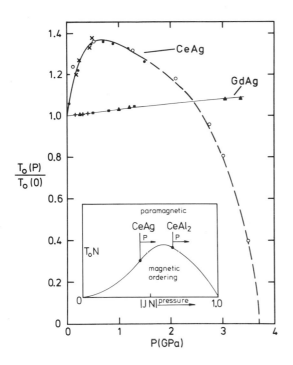

Figure 11 : Relative Curie temperature of CeAg ($T_o(0)=5.6K$) and Neel temperature of GdAg ($T_o(0)=140K$) versus hydrostatic pressure.(40) Solid lines are drawn for clarity; dashed line for CeAg marks pressure-region where exact T_o-value is uncertain due to broadening of transition. Insert shows schematic dependence of Neel temperature T_o of Kondo-lattice on $|JN|$, where $N\equiv N(E_f)$, according to Doniach.(7)

4. BAND MAGNETS--STRONG, WEAK, and ALMOST

Fe, Ni, and Co are generally accepted to belong to the class of strong or nearly strong band magnets. The stability of this magnetism is underscored by the extremely small decrease of the saturation magnetic moment under pressure, such that nominally several 100 GPas(Mbars) would be necessary to destroy the magnetism.(42) For this reason these systems are listed at the far left in Figure 3. α-Fe helps itself out by undergoing a magnetism-destroying structural phase transition from α(bcc) to ϵ(hcp) at about 14 GPa at helium temperatures.(43) ϵ-Fe remains paramagnetic down to at least 2K.(44) Wohlfarth has argued that ϵ-Fe may remain paramagnetic to 0K and even become a superconductor at temperatures well below 1K (45); this prediction has yet to be checked experimentally.

Local moment systems with unstable moments, like Eu, have their counterpart in band magnetism in the so-called "weak itinerant ferromagnets", like $ZrZn_2$ and various Pt-Ni alloys.(5) These systems

possess only a fraction of a μ_B per formula unit which quickly drops to zero under the application of pressure. In $ZrZn_2$, for instance, 0.8 GPa suffices to reduce the Curie temperature from 25K to 0K.(5,46) At pressures above 0.8 GPa $ZrZn_2$ remains paramagnetic at all temperatures but possesses a large Stoner enhancement factor which could trigger p-wave superconductivity. However, Cordes et al. have recently studied the magnetism of $ZrZn_2$ to 2 GPa and 25mK and found no evidence for superconductivity.(46) It could be that any p-wave pairing is quenched by defect scattering in the $ZrZn_2$ sample introduced by shear stresses arising within the pressure medium after freezing. He-gas techniques would be preferable here. Such pressure studies are of considerable importance because they can examine the delicate interplay between magnetism, spin fluctuations, and various possible forms of superconductivity, all on a single sample.

Pure Pd has a Stoner enhancement factor $S\simeq 10$. Brodsky and Freeman(24,47) have recently reported the observation of an enormous enhancement in the Stoner enhancement factor S for Pd in Au-Pd-Au sandwiches from 10 to values over 10,000! The magnetic susceptibility in such a system is given by

$$\chi = \mu_B^2 N(E_f) S = \frac{\mu_B^2 N(E_f)}{1-\bar{I}N(E_f)} \qquad , \qquad \{10\}$$

where $\bar{I}$ is the screened electron-electron interaction. In Pd, $\bar{I}N(E_f)\simeq 0.9$ giving $S\simeq 10$. Brodsky and Freeman argue that the Au-layers, with their 5% larger lattice parameter compared to Pd, *stretches* the Pd-lattice by $\sim$2.5%, resulting in the strong S-enhancement. Indeed, thermal expansion measurements(4) and band structure calculations(48) indicate that $N(E_f)$ should increase as the lattice parameter of Pd is isotropically increased; an increase of $\bar{I}N(E_f)$ by only 11.1% would give $S\simeq 10,000$!

One way to test these ideas is to *decrease* the lattice parameter by applying hydrostatic pressure. A large *increase* in S for negative pressures should be accompanied by a rapid *decrease* in S, and therefore χ, for positive pressures. In Figure 12 are shown the results of recent experiments by Gerhardt et al.(49) which show that the susceptibility of Pd decreases only *very little* when hydrostatic pressure is applied. At 4K, $\partial ln\chi/\partial lnV\simeq +2.8\pm 0.5$, which implies that a 2.5% increase in lattice parameter would only increase χ by about 20%, which is orders of magnitude less than the apparent increase reported for the Au-Pd-Au sandwiches. In the sandwiches, apparently, the tetragonal nature of the strain and/or changes in the electron structure at the Au-Pd interfaces are of prime importance; the effect of isotropic lattice expansion plays an insignificant role here. We have also searched for superconductivity in Pd to 3.7 GPa and 1.3K, with a negative result(49); at atmospheric pressure Pd shows no superconductivity to 1.7 mK.(50)

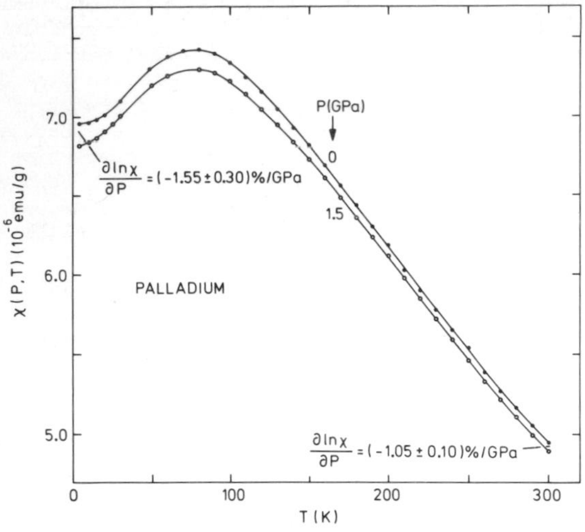

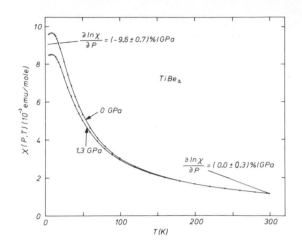

Figure 12 : Temperature dependence of static magnetic susceptibility of pure Pd from 4-300K at
hydrostatic pressures O and 1.5 GPa, from Gerhardt et al.(49) A magnetic field of 5.6 Tesla
was applied.

Figure 13 : Temperature dependence of static magnetic susceptibility of $TiBe_2$ from 3-300K at hydrostatic pressures O and 1.3 GPa, from Gerhardt
et al.(55) A magnetic field of 2.1 Tesla was
applied.

Using the value $\partial \ln N(E_f)/\partial \ln V \simeq +2.2$ for Pd from
thermal expansion work,(4) which is about twice
that estimated from a band structure calcula-
tion,(48) and $\partial \ln \chi/\partial \ln V = +2.8 \pm 0.5$ from our measurement, it follows from Equation 10 that
$\partial \ln \overline{I}/\partial \ln V = -2.1$ and $\partial \ln S/\partial \ln V = +0.6 \pm 0.5$. The exchange interaction $\overline{I}$ thus increases with pressure
at about the same rate that the density of states
$N(E_f)$ decreases, leading to the very slight change in S observed. It is interesting to note that
Franse et al. have reported a similar tendency
for the Uranium compounds UCo_2 and UAl_2.(51)
Spin-dependent band structure calculations would
be interesting here to establish to what extend
this near cancellation in the pressure dependence
of $N(E_f)$ and $\overline{I}$ is a general result. Such investigations are of obvious importance for magnetism
in general.

$TiBe_2$ was reported a year ago by Matthias et al.
to be the "first itinerant antiferromagnetic" on
the basis of a peak in the magnetic susceptibility at ∿9K.(52) Since then the label "antiferromagnetic" has been contested.(53) Very recent
NMR results(54) find no evidence for antiferromagnetism to 1.4K but indicate that $TiBe_2$, like
Pd, is an "almost-ferromagnet" with a very large
Stoner enhancement factor S≃100. In view of this
there is considerable interest in comparing the
pressure dependence of the magnetic susceptibility of $TiBe_2$ to that of Pd. In Figure 13 are
shown very recent results of Gerhardt et al.(55)
The most striking feature of the data is the
large *temperature dependence* of $\partial \chi/\partial P$! Whereas
at 300K, $\partial \ln \chi/\partial P \simeq (O \pm 0.3)\%/GPa$, at 4K it is found

$\partial \ln \chi/\partial P = (-9.6 \pm 0.7)\%/GPa$ which is, respectively,
much less and much greater than the volume depen-
dences found for Pd(see Figure 12). Such a dependence would be expected if E_f lay near a very
sharp peak in the density of states which becomes
thermally-smeared out at 300K. The field dependence of the magnetization at low temperatures
is also anomalous.(53) To bring out the anomaly,
we plot in Figure 14 the derivative $\Delta M/\Delta H$ versus
H which shows a peak at ∿5 Tesla (50 kOe). Under
pressure, this peak is seen to decrease in height,
broaden, and shift to somewhat higher fields.
Such measurements of the pressure, temperature,
and field dependence of the magnetization of
$TiBe_2$ should provide a stringent test of future
calculations of the electron and magnetic properties of this fascinating system.

If one takes systems like $ZrZn_2$, $TiBe_2$, or Pd
and squeezes the spin-fluctuations out of them,
one would expect that at some point regular s-
wave superconductivity would appear, as indicated
in the phase diagram in Figure 3. $T_c(P)$ would
then presumably rise initially with pressure
before eventually passing through a maximum and
falling off as the lattice stiffening effect
becomes dominant. I refer the reader to other
papers in this volume for a discussion of superconductivity under pressure.

It would be beautiful indeed to take a stable-
moment system, like $Gd(4f^7)$ and squeeze it all
the way across the phase diagram in Figure 3,
until finally the superconductivity disappears
and the magnetism rises again after the second
4f-electron has been squeezed out, leaving
$Gd(4f^5)$. Numerous 100 GPas (Mbars) would be
needed to accomplish this task; however, such
pressures are not yet and may never be available

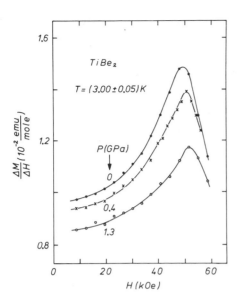

Figure 14 : Field derivative of magnetization of TiBe$_2$ versus field H at 3.0K for hydrostatic pressures 0, 0.4, and 1.3 GPa, from Gerhardt et al.(55)

in a static pressure experiment. Perhaps the behavior of Ce under pressure is about the closest we will ever come to the above goal. Of course in Ce, as in most concentrated magnetic systems, structural phase changes "disturb" the smooth evolution of the magnetization or ordering temperature with pressure idealized in Figure 3; for Ce numerous jumps in the parameters occur, as seen in Figure 10. For this reason it is especially attractive to choose for such studies close-packed metallic systems which are not prone to phase transition. Diluting the magnetic component will also help reduce the chance that a magnetic phase transition will lead to a structural one.

I hope that in this paper I have succeeded in illustrating how pressure is a powerful tool for bringing out the many magnetic states of matter using only a single system. From a careful study of the evolution of a given system from one magnetic phase to the next one can learn much towards answering the general questions of why magnetism exists where it does and how to create magnetism where it doesn't.

5. ACKNOWLEDGMENTS

It is a pleasure to thank the Deutsche Forschungsgemeinschaft for providing partial support of this work. The author would also like to acknowledge stimulating discussions with innumerable colleagues, in particular A. Eiling, W. Gerhardt, U. Hardebusch, S. Methfessel, and H. Micklitz. Special thanks are due D. Wohlleben and W. Zell both for helpful discussions and for permission to show some of their most recent data prior to publication.

REFERENCES

(1)   Ross, M. and McMahan, A.K. (see paper in present volume).

(2)   Teller, E., Photonic Compression, Plenary Lecture, 6th AIRAPT International High Pressure Conference, Boulder, Colorado, USA (July 1977).

(3)   Kortekaas, T.F.M. and Franse, J.J.M., J. Phys. F 6 (1976) 1161.

(4)   Barron, T.H.K., Collins, J.G. and White, G.K., Adv. Phys. 29 (1980) 609.

(5)   Wohlfarth, E.P.(see paper in present volume).

(6)   See discussion and references in: Schilling, J.S., Adv. Phys. 28 (1979) 657.

(7)   Doniach, S., in Valence Instabilities and Related Narrow-Band Phenomena, in Parks, R.D. (ed.) (Plenum, N.Y., 1977)p. 169.

(8)   Fay, D. and Appel, J., Phys. Rev. B 22 (1980) 3173.

(9)   Balster, H. and Wittig, J., J. Low Temp. Phys. 21 (1967) 377; Wittig, J., Phys. Rev. Lett. 46 (1981) 1431.

(10) Shelton, R.N., Moodenbaugh, A.R., Dernier, P.D. and Matthias, B.T., Mat. Res. Bull. 10 (1975) 1111.

(11) Eiling, A., Schilling, J.S. and Bach, H. (see paper in present volume).

(12) See, for example, Podloucky, R., Zeller, R. and Dederichs, P.H., Phys. Rev. B 22 (1980) 5777.

(13) Davidov, D., Maki, K., Orbach, R., Rettori, C. and Chock, E.P., Solid State Commun. 12 (1973) 621; Davidov, D., Rettori, C., Orbach, R., Dixon, A. and Chock, E.P., Phys. Rev. B 11 (1975) 3546.

(14) Watson, R.E., Freeman, A.J. and Koide, S., Phys. Rev. 186 (1969) 625.

(15) Kondo, J., Prog. theor. Phys. 32 (1964) 37.

(16) Johansson, B. and Rosengren, A., Phys. Rev. B 14 (1976) 361.

(17) Abd-Elmeguid, M., Micklitz, H. and Buschow, K.H.J., Solid State Commun. 36 (1980) 69.

(18) Wachter, P., Handbook on the Physics and Chemistry of Rare Earths, in Gschneider, K.A., Jr. and Eyring, L. (eds.) (North-Holland, N.Y., 1978)Vol II, p. 507.

(19) Sereni, J.G., J. Phys. F 10 (1980) 2831.

(20) The ratio of the atomic to ionic volume appears to be approximately 7±2 for a number of elements.

[21] Loram, J.W., Whall, T.E. and Ford, P.J., Phys. Rev. B 2 (1970) 857.

[22] Meyers, H.P., Wallden, L. and Karllson, A., Phil. Mag. 18 (1968) 725; Norris, C. and Wallden, L., Solid State Commun. 7 (1969) 99.

[23] Croft, M., Weaver, J.H., Peterman, D.J. and Franciosi, A., Phys. Rev. Lett. 46 (1981) 1104.

[24] Freeman, A.J. (see paper in present volume).

[25] Hardebusch, U. Gerhardt, W. and Schilling, J.S., Phys. Rev. Lett. 44 (1980) 352; Hardebusch, U., Diplom-Thesis (University of Bochum, 1981).

[26] Gerhardt, W., Diplom-Thesis (University of Bochum, 1981); Gerhardt, W. and Schilling, J.S. (unpublished).

[27] Simpson, M.A. and Smith, T.F., J. Phys. F 11 (1981) 397; see also paper by Smith, T.F. in present volume.

[28] Razavi, F. and Schilling, J.S. (unpublished).

[29] Larsen, U., Peukert, H. and Schilling, J. S., Z. Physik B 37 (1980) 115.

[30] Smith, T.F., Phys. Rev. Lett. 17 (1966) 386.

[31] Zell, W., Dissertation (University of Cologne, 1981); Zell, W., Boksch, W., Roden, B. and Wohlleben, D., Verhandl. DPG 16 (1981) 314; for Yb-compounds, see also, Zell, W., Pott, R., Roden, B. and Wohlleben, D., Solid State Commun. (to be publ.)

[32] Jayaraman, A., Handbook on the Physics and Chemistry of Rare Earths, in Gschneider, K.A., Jr. and Eyring, L. (eds.) (North-Holland, N.Y., 1978) Vol. I, p. 707 and Vol. II, p. 575.

[33] Bridgman, P.W., Proc. Amer. Acad. Arts Sci. 79 (1951) 149.

[34] Maple, M.B., Wittig, J. and Kim, K.S., Phys. Rev. Lett. 23 (1969) 1375.

[35] Kim, K.S. and Maple, M.B., Phys. Rev. B 2 (1970) 4696.

[36] Franz, W., Steglich, F., Zell, W., Wohlleben, D. and Pobell, F., Phys. Rev. Lett. 45 (1980) 64.

[37] Bundy, F.P. and Dunn, K.J. (see paper in present volume).

[38] Wittig, J. (see paper in present volume).

[39] Matthias, B.T., Fisk, Z. and Smith, J.L., Phys. Lett. 72A (1979) 257.

[40] Eiling, A. and Schilling, J.S., Phys. Rev. Lett. 46 (1981) 364.

[41] Debray, D., Kahn, R., Decker, D.L., Werner, A., Loewenhaupt, M., Holland-Moritz, E. and Ray, D.K., Valence Fluctuations in Solids, Falicov, L.M., Hanke, W. and Maple, M.B. (eds.) (North-Holland, N.Y., 1981).

[42] Tatsumoto, E., Fujiwara, H., Tange, H. and Kato, Y., Phys. Rev. 128 (1962) 2179; Kouvel, J.S. and Hartelius, C.C., J. Appl. Phys. 35 (1960) 940.

[43] Williamson, E.L., Bukshtan, S. and Ingalls, R., Phys. Rev. B 6 (1972) 4194.

[44] König, K., Wortmann, G. and Kalvius, G.M., Proc. of the Inter. Conf. on Mößbauer Spectroscopy, Hrynkiewicz, A.Z. Sawicki, J.A. (eds.) (Univ. Publ. House, Cracau, 1975).

[45] Wohlfarth, E.P., Phys. Lett. 75 A (1979) 141.

[46] Cordes, H.G., Fischer, K. and Pobell, F., Proceedings of the 16th International Conf. on Physics at Low Temperatures (North-Holland, N.Y., 1981).

[47] Brodsky, M.B. and Freeman, A.J., Phys. Rev. Lett. 45 (1980) 133; Brodsky, M.B., J. Appl. Phys. 52 (1981) 1665.

[48] Das, S.G., Koelling, D.D. and Mueller, F. M., Solid State Commun. 12 (1973) 89.

[49] Gerhardt, W., Razavi, F., Schilling, J.S., Hüser, D. and Mydosh, J.A. (to be published in Phys. Rev. B, Rapid Communications).

[50] Webb, R.A., Ketterson, J.B., Halperin, W. P., Vuillemin, J.J. and Sandesara, N.B., J. Low Temp. Phys. 32 (1978) 659.

[51] Franse, J.J.M., Frings, P.H., de Boer, F. R. and Menovsky, A. (see paper in this volume).

[52] Matthias, B.T., Giorgi, A.L., Struebing, V.O. and Smith, J.L., Phys. Lett. 69A (1978) 221.

[53] Monod, P., Felner, I., Chouteau, G. and Shaltiel, D., J. Phys. 41 (1980) L-511.

[54] Takagi, S., Yasuoka, H., Huang, C.Y. and Smith, J.L., J. Phys. Soc. Japan 50 (1981) 2137.

[55] Gerhardt, W., Schilling, J.S. and Smith, J.L. (unpublished).

PHYSICS OF SOLIDS UNDER HIGH PRESSURE
J.S. Schilling, R.N. Shelton (editors)
© *North-Holland Publishing Company, 1981*

UNUSUAL OBSERVATION IN SOME UNUSUAL RARE-EARTH CHEVREL TERNARY COMPOUNDS

C. W. Chu, M. K. Wu, R. L. Meng, T. H. Lin, and V. Diatschenko

Department of Physics and Energy Laboratory
University of Houston
Houston, TX 77004 U S A

S. Z. Huang
IBM Research Laboratory, San Jose, CA 95193, U S A

The unusual rare-earth Chevrel ternaries Ce-Mo-S, Eu-Mo-(S-Se) and (Eu-Sn)-Mo-Se have been studied under pressure up to 120 kbar. Unusual magnetic and superconducting characteristics are observed under pressure. The competition between an antiferromagnetic, a Kondo and a superconducting state in Ce-Mo-S and related compounds will be discussed. Various possibilities for the pressure-induced superconducting in Eu-Mo-S are considered.

I. INTRODUCTION

The rare-earth Chevrel Ternaries, $RE_yMo_6X_8$ (or RE-Mo-X) (with y~1, X=S or Se) have been found to be superconducting[1], in spite of the presence of a relatively high concentration of magnetic RE-ions. Their crystal structure consist of clusters of Mo-atoms, arranging in a rhombohedral lattice with open channels to accomodate the RE-ions. The superconducting transition temperature (T_c) of these compounds varies[2] systematically with the RE-position in the periodic table and thus correlates well with the de Gennes factor. However, the deduced exchange interaction between the RE-ions is rather small[2], consistent with the suggestion that the superconductivity in this class of compounds is associated mainly with the Mo-clusters[3] and has very little to do with the magnetic RE-ions. Recent band structure calculations[4] which show little contribution from the RE-ions to the overall density of states lends further support to such a suggestion. Unfortunately, a close examination of the T_c-variation[2] reveals the existence of several unusual members of the ternary family. They are Ce-Mo-X, Eu-Mo-X and Yb-Mo-S. The first two are not superconducting in contrast to all expectations and the last one is superconducting but with a high T_c out of line with other members of the family. In the present report, preliminary results of our studies are presented for the first two cases.

The Ce-atoms[5] can be found in a magnetic, a Kondo or a non-magnetic state, depending on the Ce-Ce interaction strength. It has also been demonstrated[6] that Ce-Mo-S orders antiferromagnetically at ~2K and can be treated as a Kondo system in its dilute form. We therefore decided to investigate the competition between an antiferromagnetic (AF) state and a Kondo (K) state in a concentrated magnetic system by subjecting the Ce-Mo-S and related Chevrel compounds to high pressures. As pressure (P) increases, the AF-state has been observed to be first enhanced and then suppressed, and the resistance (R) to become logarithmically dependent on temperature (T) over a wide temperature range. Similar observations have also been made in La-doped and Ce-rich Ce-Mo-S samples except at different pressures. In addition, an R-drop signalling the onset of a superconducting transition was detected in Ce-rich Ce-Mo-S sample while R exhibited a strong ℓnT dependence. The results will be discussed and compared with predictions of the model on a concentrated magnetic system or Kondo lattice[7].

Recently, non-bulk superconductivity with an onset above 11K and a potentially high critical magnetic field was induced abruptly in Eu-Mo-S above ~7kbar but not in the isoelectronic compound Eu-Mo-Se down to 1.2K up to 80kbar[8]. In an attempt to understand such an observation, we have studied the Eu-Mo-(S-Se) and (Eu-Sn)-Mo-Se mixed compounds. We found a reduction of the critical pressure (P_c) to induce non-bulk superconductivity in Eu-Mo-(S-Se) and an unpredictable pressure effect on the T_c of (Eu-Sn)-Mo-Se. The results will be discussed along with various possibilities for the occurrence of superconductivity in Eu-Mo-S under P.

II. EXPERIMENTALS

We have measured the ac magnetic susceptibility (χ) and R of $Ce_{1.2}Mo_6S_8$, $(Ce_{1-x}La_x)_{1.2}Mo_6S_8$, $Ce_{1.2+x}Mo_6S_8$, $Eu_{1.2}Mo_6(S_{1-x}Se_x)_8$ and $(Eu_xSn_{1-x})_{1.2}Mo_6Se_8$ compounds as a function of T under different P's. All samples investigated were prepared by the sintering technique previously reported[1]. The x-ray diffraction patterns exhibit only a single Chevrel phase with a resolution ~5%. No intensity analysis of the diffraction pattern has been made. The hexagonal lattice parameters are, respectively, a=9.12 and c=11.47A for $Ce_{1.2}Mo_6S_8$, 9.16 and 11.52A for $Eu_{1.2}Mo_6S_8$, and 9.48 and 12.01A for $Eu_{1.2}Mo_6Se_8$, in very good agreement with published results(1). The quoted compositions agree with data of energy dispersive analysis of x-ray, with a rather poor resolution of ~15% due to the sintered nature of the samples. Bar-samples of dimensions $3 \times 1 \times 1mm^3$ were cut from the sintered compacted compounds for hydrostatic measurements up to ~20kbar. Powdered samples were used for quasi-hydrostatic pressure measurements up to ~120kbar. The hydrostatic environment was provided by the modified self-clamp technique, using the fluid mixture of 1:1 n-pentane and isoamyl alcohol as the pressure medium. The quasi-hydrostatic pressure was generated with a Bridgman anvil arrangement, employing the solid steatite as the pressure medium. R was determined by a standard four-lead technique and χ by an ac inductance-bridge method, both operating at ~23Hz. The hydrostatic P was directly measured by a superconducting Pb-manometer situated next to the sample. For the quasi-hydrostatic runs, P was determined indirectly from the load which was calibrated against a superconducting Pb-manometer in a separate run. Due to the variations in packing from run to run, the cited pressure could be underestimated by as much as 15% at our highest P. T was measured with a chromel-alumel thermocouple above 20K and a Ge-thermometer below.

III. RESULTS AND DISCUSSIONS

A. (Ce-La)-Mo-S

Some of the results discussed here have been briefly reported previously(9). The T-dependence of R previously observed in Ce-Mo-S(2) was reproduced in our samples. As shown in Fig. 1, on cooling, R first decreases smoothly, then rises slowly below ~15K, but finally drops sharply at ~2K. The R-minimum at ~15K

has been attributed to the Kondo-scattering and the R-peak at ~2K to an AF ordering which was later confirmed by specific heat measurements(6). Under P, the room temperature R decreases rapidly and the low temperature R-rise grows, with only a slight change in the minimum R as displayed in Figs. 1 and 2. The low temperature R-rise evolves into a distinct ℓnT dependent R over an ever increasing T-range as P increases. At the same time, the slope of $R-\ell nT$ curves increases continuously. For example, at 120kbar, R increases linearly with decreasing ℓnT from 200 to 4K, corresponding to an ~18 fold increase of R. The low temperature χ was also monitored in the hydrostatic pressure region and depicted in Fig. 3 as a function of T.

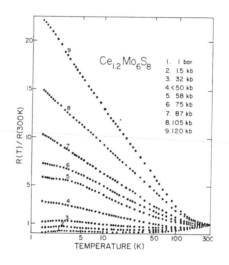

Figure 2

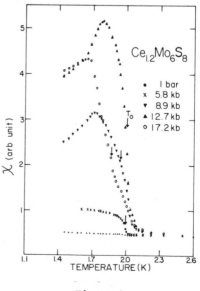

Figure 1

Figure 3

The small χ-rise, signalling the onset of an
AF-ordering increases in magnitude and shifts
toward higher T under P < 13kbar. At higher P,
the reverse is true. No R-anomaly indicative
of a P-induced phase transition at 300K was
detected as evident from Fig. 4. The R-minimum
at ~65kbar in this figure represents the minimum in the R-T curve in Fig. 1 being pushed to
above 300K by P. Similar behavior of R(T, P)
and χ(T, P) was also observed in (Ce-La)-Mo-S
and Ce-rich Ce-Mo-S provided proper scaling
was made on P. For example, the enhancement and suppression of the AF-state occurs at higher P in (Ce-La)-Mo-S than in

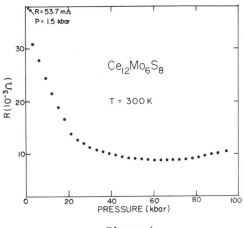

Figure 4

Ce-Mo-S, but at lower P in Ce-rich Ce-Mo-S.
However, an R-drop starts to appear above
~79kbar at ~3.5K, as shown in Fig. 5. An increase of the detecting current removes the

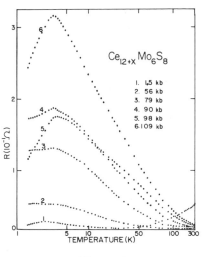

Figure 5

R-drop, suggesting that the R-drop is associated with a superconducting transition. The high
T_c and its behavior under P indicate that free
Ce can not be the cause. All of the above

mentioned observations are reversible on the
removal of P. For instance, curve 4 in Fig. 2
was taken after the high pressure run.
Unfortunately, the exact P could not be determined during the P-reduction cycle.

As described earlier, the R-drop and the χ-rise
in Ce-Mo-S represent a transition to an AF-
state. The temperature T_p of the R-peak and T_o
of the χ-rise as defined respectively in Figs.
1 and 3 are therefore a direct measure of the
Néel temperature T_N of Ce-Mo-S. As shown in
Fig. 6,

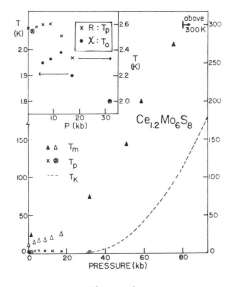

Figure 6

both T_p and T_o are initially enhanced and then
suppressed by P, forming the phase boundary between a paramagnetic and an antiferromagnetic
state(6). The R-minimum has been ascribed to
the Kondo-scattering(6). The temperature T_m of
the R-minimum defined in Fig. 2 can therefore
be considered as a qualitative measure of the
Kondo temperature T_K, which is often lower
than T_m. We therefore schematically represent
the T_K-P relation by a dashed curve in Fig. 6.
Both T_K and T_m are increasing function of P.
Since T_K is ∝ exp (1/JN) with J being the negative exchange interaction and N the density of
states at the Fermi level, P enhances $|JN|$,
i.e. ∂$|JN|$/∂P > 0 up to 120kbar. Similar con-
clusions can be drawn from the ever increasing
slope of the R-ℓnT plot with P, shown in Figs.
1 and 2. This is because the contribution to R
due to Kondo scattering is ∝JNℓnT(10). The pos-
itive P-effect on $|JN|$ can be understood by
considering the J which is ∝$|V_{kf}|^2$/$|E_f-E_F|$,

with V_{kf} measuring the mixing between the conduction- and 4f-electrons, E_f the 4f-level and E_F the Fermi level. A positive $\partial|JN|/\partial P$ can be realized either through the increasing mixing between the conduction- and 4f-electrons due to 4f-band broadening, or through the upward lift of E_f toward E_F by P. Such a positive $\partial|JN|/\partial P$ up to ~120kbar would also imply a positive $\partial T_N/\partial P$ up to ~120kbar, since T_N is $\propto J^2N$ for an RKKY antiferromagnet like Ce-Mo-S. It is in disagreement with our observation. This suggests that there must exist a competition between various magnetic interactions.

To examine the competition between different magnetic interactions, let us consider the model proposed by Doniach[7], for a concentrated magnetic system called Kondo lattice. According to this model, a transition from an AF to a K-state will result when J exceeds a critical value. Calculations[7] made on a one dimension analogue with one localized spin per cell at 0° K support the proposition. A qualitative argument for the proposition was also made by comparing the binding energies of the AF- and K-states. Since they are $\propto \exp(1/JN)$ and J^2N for the K- and AF- state, respectively, when $|JN|$ is small, the AF-state dominates, while large, the K-state prevails. This is in agreement with the phase diagram shown in Fig. 6. The suppression of the AF-state by P is therefore attributed to the interference between the AF- and K-states, instead of to the 4f-band broadening. Such a band broadening and the ensuing delocalization of the magnetic momenta would have prevented us from observing the continuous increase in Kondo scattering after the complete suppression of the AF-state. Results on (Ce-La)-Mo-S and Ce-rich Ce-Mo-S can be understood in terms of a change in J. Replacement of Ce by La (a few % only) reduces $|J|$ and therefore corresponds to reducing the effective P applied to the sample. Consequently, observations similar to Ce-Mo-S can only be made at high pressures. Addition of Ce to Ce-Mo-S increases $|J|$, as evident from the accompanying increase in T_p, although the atomic site for the extra Ce-atoms remains unknown. The general behavior of Ce-Mo-S under P again has been observed in Ce-rich sample but at a lower P. All these are consistent with the phase diagram shown in Fig. 6.

The observation of superconductivity above ~79 kbar in the Ce-rich Ce-Mo-S is highly unusual. Since both the superconducting signal and $|JN|$ (or $\partial R/\partial \ell nT$) of this compound has gone through a minimum as a function of P, it implies that the sample will loose its superconductivity and become a Kondo-system with higher T_K at P > 105 kbar. The appearance of superconductivity can not be explained in terms of a complete demagnetization of the 4f-electrons in Ce either due to 4f-band formation or 4f-level promotion to above E_F. By varying T, a superconducting state subtended by two normal states has been observed in a re-entrant superconductor[11]. A normal state sandwiched between two superconducting phases in (Eu-Sn)-Mo-S by varying H has also

been observed[12]. Since P, T, and H can be treated on an equal footing thermodynamically, our detection of superconductivity in Ce-rich Ce-Mo-S only in a limited P-range may not be surprising. However, the continuous increase of $|JN|$ with P on both sides of the superconducting state makes the Ce-rich Ce-Mo-S different from a re-entrant superconductor where superconductivity is known to be destroyed by the ferromagnetic ordering at low T. The underlying reason for our observation is unknown.

In the above discussion, the ℓnT dependence of R has been associated with the K-scattering. However the large T-range over which such a ℓnT dependence is valid and the large R-increase with decreasing T are in strong contrast to ordinary K-effect. In addition, the model on Kondo lattice[7] is yet to be proved theoretically for a three dimensional lattice. Therefore other possibilities for the ℓnT term can not be ruled out at the present time. These include a yet unknown type of magnetic excitation and a gap-opening of the type proposed recently by Jullien et al.[13]. The role of defects like Anderson localization[14], perhaps, should also be considered.

B. (Eu-Sn)-Mo-(S-Se)
The temperature dependence of R normalized to its value at 300K for $Eu_{1.2}Mo_6(S_{1-x}Se_x)_8$ is displayed in Fig. 7. The overall behavior of

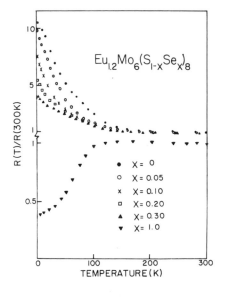

Figure 7

R(T) varies continuously from that characteristics of Eu-Mo-S[15] to that characteristics of Eu-Mo-Se[15] as x increases. In other words, $\partial R/\partial T$ at low T changes from negative to positive with increasing x. Consequently, the residual R at 1.2K decreases smoothly with x. No sign of superconductivity was detected down to 1.2K in any of these compounds at atmospheric

pressure. Under P, the residual R decreases
continuously and $\partial R/\partial T$ changes from negative to
positive for compounds with $x \lesssim 0.3$. A sudden
drop in R for these compounds, indicative of a
superconducting transition, occures at a critical pressure P_c lower than that for $x = 0$, when
$\partial R/\partial T$ is still negative prior to the R-drop.
Typical results are exemplified by those for
$x = 0.05$ in Fig. 8.

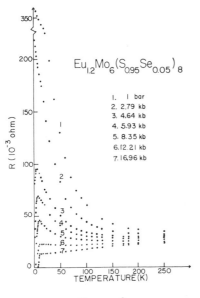

Figure 8

The resistive superconducting signal increases
and the transition width decreases with increasing P. The P-effect on the magnetic superconducting signal is similar to the resistive one,
except that the former has a smaller volume
fraction and appears only at a higher P_c and a

lower T_c. For example, a superconducting signal at 1.2K of ~1.5% of a perfect diamagnet
was not detected until P was enhanced to 11.3
kbar for a powdered sample with $x = 0.05$. In
contrast, the R-measurements showed a superconducting onset at ~4.6kbar and a ~50% drop at
~10kbar and 1.2K for the same sample. The maximum T_c obtained under P for Eu-Mo-(S-Se) decreases rapidly with x. The T_c of $(Eu_xSn_{1-x})_{1.2}$
Mo_6Se_8 has been found to vary smoothly with x.
At low x, the rate of T_c-suppression is small
but larger than the corresponding mixed sulfide
ternaries[1]. The P-dependence of T_c for different x's is depicted in Fig. 9. It is evi-
dent that $\partial T_c/\partial P$ is highly unusual, varying
from linear to nonlinear, and from negative to
positive, depending on x in a completely unpredictable way.

Eu-Mo-S and Eu-Mo-Se are different in their
volume and $\partial R/\partial T$ below ~120K[6]. The former
has a smaller unit cell volume and a negative

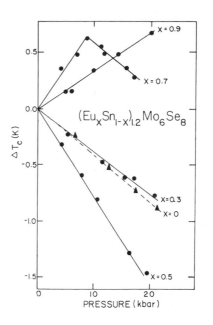

Figure 9

$\partial R/\partial T$ in contrast to the latter. Based on
these differences, several possibilities have
been proposed to account for the observed superconductivity under P in Eu-Mo-S but not in
Eu-Mo-Se, irrespective of the usually higher T_c
of the selenides than the sulfides[1]. The possibilities include: (1) the attainment of an
optimal volume, (2) the inducement of a magnetic Eu^{+2} to nonmagnetic Eu^{+3} valence transition, and (3) the generation of a semiconductor
to metal transition below 120K, by P.

The first possibility was made, assuming that
there might exist an optimal volume most favorable for superconductivity in RE-Mo-X. If the
optimal volume is smaller than those of Eu-Mo-S
and Eu-Mo-Se, it would be easier to attain such
a volume for Eu-Mo-S under P, because of its
smaller unit cell volume. Unfortunately, an
increase in the unit cell volume of Eu-Mo-S by
a slight substitution of S by Se shows a reduction in P_c. This is in contradiction with
the prediction.

The second possibility was made based on the
suggestion that the absence of superconductivity
in Eu-Mo-X was a result of the magnetic nature
of the Eu-ions[1], although the effect of the
magnetic RE-ions on the superconductivity of
this calss of compounds has been demonstrated
to be negligibly small. According to this suggested possibility, a P-induced change of the
Eu-ions from their magnetic divalent state to
their non-magnetic trivalent state and, consequently, P-induced superconductivity in Eu-Mo-X
are expected, since Eu^{+2} has a larger volume
than Eu^{3+}. $\partial T_c/\partial P$ of (Eu_x-Sn_{1-x})-Mo-Se should

consist of two parts, namely, $(\partial T_c/\partial P)_o(1-x) +$ $(\partial T_c/\partial P)_m x$, with $(\partial T_c/\partial P)_o$ negative and characteristics of Sn-Mo-Se(16) and $(\partial T_c/\partial P)_m$ positive resulting from the suppression of the magnetism of Eu-Mo-Se by P. This predicts a monotonically increasing $(\partial T_c/\partial P)$ from negative to positive as x increases. Unfortunately, such a prediction has not been borne out by our observations. The observed irregular variation of $\partial T_c/\partial P$ with x implies a complicated band structure for (Eu-Sn)-Mo-X.

The third possibility was made by emphasizing the difference in $\partial R/\partial T$ below ~120K between Eu-Mo-S and Eu-Mo-Se. The negative $\partial R/\partial T$ in Eu-Mo-S has been suggested to be associated with the possible presence of a semiconducting gap(8,15). Pressure will then suppress the gap, generate a semiconductor to metal transition and, consequently, induce superconductivity. It is known that a negative $\partial R/\partial T$ is necessary for the existence of a semiconducting gap but far from being sufficient. For instance, defect scattering and electron-tunneling through a thin insulating coating separating the metallic parts can contribute a negative $\partial R/\partial T$. However, the slight substitution of S by Se and the ensuing increase in defects was found to reduce the R at low T, instead. In addition, the R-rise in Eu-Mo-S at low T is too large to be accounted for by defect-scattering. The observation of a sign change from positive to negative under P in Ce-Mo-S (which was prepared similarly to Eu-Mo-X) and in one of the (Eu-Sn)-Mo-S strongly suggests that the negative $\partial R/\partial T$ in Eu-Mo-S can not be caused by electron-tunneling through an insulating coating but may be associated with a semiconducting gap at low T. Although band structure calculations predict a metallic behavior for Eu-Mo-S at 300K, the possibility of a lattice distortion and thus a semiconducting gap at low T can not be ruled out. Recent results on the Seebeck coefficient(17) of Eu-Mo-S indeed support such a possibility by exhibiting drastic variation in magnitude near 120K. Since the stoichiometry of the Chevrel phase is important to the band structure as evident from calculations(4), it should also be important to the above mentioned possible low T lattice distortion.

The absence of superconductivity in the metallic Eu-Mo-Se suggests the removal of a semiconducting gap is not sufficient for superconducting to occur and especially to occur at a relatively high T_c. The always non-bulk nature of the superconductivity observed in Eu-Mo-(S-Se) under P demonstrated that the samples investigated can not be homogenous. The appearance of this non-bulk superconductivity when $\partial R/\partial T$ is still negative leads us to the proposition that interfaces between the metallic and semiconducting parts, arising from concentration fluctuation, may have an important role in our observation. Interface mechanism has been disucssed(18) as one of the avenues leading to high T_c and high critical field.

Systematic studies on single crystal Chevrel compounds will help assess the role of interface in the superconductivity of Eu-Mo-S. Equally important, in this respect, is to examine the volume fraction of superconductivity and the transport properties of Eu-Mo-(S-Se) as a function of P. A maximum in the superconducting volume fraction as P varies can be considered as a strong evidence for the presence of interface superconductivity. A peculiar R-T behavior characteristic of excitations with interfaces may also be expected. Studies of this kind are being carried out.

Finally, the fact that a negative $\partial R/\partial T$ has been promoted and superconductivity induced by P in magnetic Ce-Mo-S suggests yet another possibility, i.e. a possible new type of magnetic excitations favorable for superconductivity. The reappearance of superconductivity recently observed in (Eu-Sn)-Mo-S(12) following its disappearance as the magnetic field increases suggests the possible existence of such type of magnetic excitations. Investigations are currently under way on Eu-Mo-S at high P.

Note Added to Proof-In this Conference, McCallum et al. reported results of their study on four Eu-Mo-S samples with different Mo-concentrations under P. An upper limit of < 5% of Mo was set for three samples. In two of these three samples, only fractional ac superconductivity signal was detected. None of the four exhibited any sign of superconductivity (> 1%) in dc magnetization measurements. They therefore concluded that the observed superconductivity was non-bulk in agreement with our observations. However, they further suggested that such non-bulk superconductivity was caused by Mo-filaments which could be connected under P to form close loops. We are in strong disagreement with such a suggestion for the following reasons: (1) The T_c (~11-13K) in compressed Eu-Mo-S is too high to be caused by Mo-filaments whose highest known T_c is only 6.5K, (2) the maximum T_c for Eu-Mo-(S-Se) under P decreases rapidly with Se-content instead of a constant characteristic of the suggested Mo-filaments, (3) P_c is the same for Eu-Mo-S prepared in different batches where distribution of Mo-filaments is expected to be different, and (4) the critical field is too large for weak-links which connect the various Mo-filaments to form a close electric current loop. In addition, preliminary measurements indicate that the critical field shows a maximum near ~11kbar in Eu-Mo-S with a P_c ~7kbar.

IV.　ACKNOWLEDGEMENT

We would like to thank D. L. Zhang, P. H. Schmidt, R. N. Shelton and B. Maple for discussion, and P. H. Hor for technical help. The work is supported in part by NSF Grant No. DMR 79-08486 and DMR 81-10504 and the Energy Lab of the University of Houston.

REFERENCES

1. For a Review, see φ. Fischer, Appl. Phys.
16 (1978).

2. M. B. Maple, L. E. DeLong, W. A. Fertig,
D. C. Johnston, R. W. McCallum and R. N.
Shelton, Valence Instabilities and Related
Narrow Band Phenomena, ed. by R. D. Parks
(Plenum, New York, 1977) 17.

·3. φ. Fischer, A. Treyvand, R. Chevrel and
M. Sergent, S. S. Comm. 17, (1975) 721.

4. T. Jarlborg and A. J. Freeman, Phys. Rev.
Lett. 44 (1980) 178.

5. S. H. Liu, This Conference.

6. R. W. McCallum, Ph.D. Thesis, University
of California at San Diego (1977).

7. S. Doniach, Physica 91B (1977) 231.

8. C. W. Chu, S. Z. Huang, J. H. Lin, R. L.
Meng, M. K. Wu and P. H. Schmidt, Phys. Rev.
Lett. 46 (1981) 276.

9. M. K. Wu, V. Diatschenko, P. H. Hor, S. Z.
Huang, T. H. Lin, R. L. Meng, D. L. Zhang and
C. W. Chu, preprint.

10. J. Kondo, Solid State Phys. 23 (1969) 183.

11. W. A. Fertig, D. C. Johnston, L. E. DeLong,
R. W. McCallum, M. B. Maple and B. T. Matthias,
Phys. Rev. Lett. 38 (1977) 987.

12. M. Isino, N. Kobayashi and Y. Muto, Ternary
Superconductors, ed. by G. K. Shenoy, B. D.
Dunlap and F. Y. Fradin (North Holland, New
York, 1981) 95.

13. R. Jullien and P. Pfeuty, J. Phys. F11
(1981) 353.

14. N. F. Mott, M. Pepper, S. Pollert, R. H.
Wallis and C. J. Adkins, Proc. Roy. Soc.
(London) A 345 (1975) 169.

15. C. W. Chu, S. Z. Huang, J. H. Lin, R. L.
Meng, M. K. Wu and P. H. Schmidt, Ternary
Superconductors, ed. by G. K. Shenoy, B. D.
Dunlap and F. Y. Fradin (North Holland, New
York, 1981) 103.

16. R. N. Shelton, Superconductivity in d- and
f-band Metals, ed. by D. H. Douglass (Plenum,
New York, 1976) 137.

17. S. Z. Huang et al., private communication.

18. See for example, T. H. Geballe and C. W.
Chu, Comments Solid State Phys. 9 (1979) 115.

PHYSICS OF SOLIDS UNDER HIGH PRESSURE
J.S. Schilling, R.N. Shelton (editors)
© *North-Holland Publishing Company, 1981*

HIGH PRESSURE STUDY ON A LINEAR CHAIN COMPOUND TaS_3

D.L.Zhang[*], M.K.Wu, T.H.Lin, P.H.Hor and C.W.Chu

Department of Physics and Energy Laboratory
University of Houston
Houston, Texas   77004 USA

A.H.Thompson

EXXON Research and Engineering Company
Linden, New Jersey 07036 USA

The resistance and thermopower of TaS_3 have been measured up to 120kbar down to 1.2K. The Peierls transition is first enhanced and then suppressed by pressure. Superconductivity is induced by a pressure above ~68 kbar. The results are discussed in terms of existing models.

I. INTRODUCTION

The transition-metal trichalcogenides MX_3 (M=transition-metal element; X=S or Se) form an interesting family of compounds, ranging from a semiconductor[1] to a superconductor[2]. They have highly anisotropic structures with parallel chains of trigonal prisms MX_6 which define the easy direction of electric current flow. Peierls transitions characteristic of a quasi-one-dimensional(1-D) system and accompanied by the formation of charge density-wave(CDW) have been detected in $NbSe_3$[3] and TaS_3[4]. The unusual observations[5-8] of the field- and frequency-dependent conductivity, the broadband and narrowband noise in the non-Ohmic region, and the large dielectric constant in these two compounds have attracted great attention and provided important insight to the understanding of the Physics of CDW. In spite of the gross similarity, subtle differences do exist between $NbSe_3$ and TaS_3. For instance, $NbSe_3$ has two incommensurate CDW modes [3] wheras TaS_3 has one commensurate CDW mode[4] below the Peierls transition temperature T_O. The small threshold field reported for non-Ohmic effect to occur in TaS_3[8] is therefore difficult to reconcile with the existence of a commensurate CDW mode[4] in this compound. On the other hand, unlike $NbSe_3$, TaS_3 displays a semiconducting behavior[4] below T_O. It implies a complete removal of the Fermi-surface and the condensation of all electrons in the CDW mode in $TaS3$. This makes TaS_3 a simpler system than $NbSe_3$. In an attempt to unravel further the nature of CDW and the interplay between various electronic instabilities, we therefore have examined the transport properties of TaS_3 under high pressure. We observed an initial enhancement and

final suppression of T_O by pressure, different temperature dependences of resistivity and thermopower near T_O, and the appearance of superconductivity above 68kbar. The results will be discussed in terms of existing models.

II. EXPERIMENTALS

We have measured the electrical resistance R of the single crystal TaS_3 up to ~120k bar down to 1.2K. The Seebeck coefficient S was also determined but only under hydrostatic pressure up to 19kbar. The samples investigated were from two different batches used for previous studies [8]. X-ray analysis indicated a single rhombohedral structure. The dimensions of the samples used were estimated to be $~0.6 \times ~0.08 \times ~0.008$ mm^3. R was determined with a standard four-lead ac technique operated at 71Hz (the results were identical to the dc ones). Electrical lead contacts were established with silver paste for the hydrostatic runs or by pressure for the quasi-hydrostatic runs. Low current $\lesssim$ 1µA was used to ensure the measurements free from non-linear or heating effect. S was measured with a technique similar to that for the linear chain organic compounds [9]. The pressure P was generated by a modified clamp technique [10]. With a piston-cylinder arrangement using 1:1 fluid mixture of n-pentane and isoamyl alcohol as medium, a hydrostatic P was achieved. On the other hand, using the Bridgman anvil arrangement with the solid steatite as medium, a quasi-hydrostatic P was obtained . P was determined for the hydrostatic case with a superconducting Pb-manometer and for the quasi-hydrostatic case on the basis of the load which was calibrated against a Pb-manometer but in a separate experimental run. Because of the

different packings from run to run, our
highest pressure could be underestimated
as much as ~15%. The temperature T was
measured with an alumel-chromel thermocouple above 20K and a Ge-thermometer
below.

III. RESULTS AND DISCUSSIONS

More than ten samples have been examined.
While behave similarly under P, they can
be different in details. For instance,
the resistivity and S at 300K and 1bar
can differ by as much as ~40% from sample
to sample and T_O by ~20K°. Figure 1

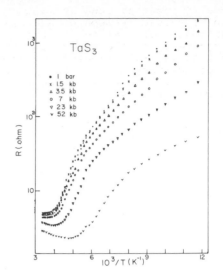

Figure 2

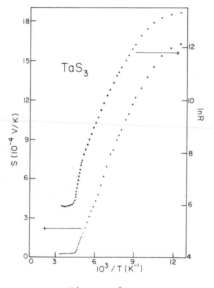

Figure 1

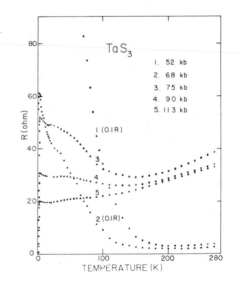

Figure 3

shows the typical T-dependences of R and
S at 1bar for one sample. As previously
reported [4,8], R decreases on cooling
through a minimum at T_m and then
increases slowly prior to a rapid rise
with a maximum $\partial R/\partial T$ at T_O. Such a rapid
R-rise had been associated with the
Peierls transition from X-ray study [4].
R exhibits at least two activation ener-
gies Δ_1 (near T_O) and Δ_2 (below ~200K)
as reflected in the two distinct slopes
in the ℓnR-$1/T$ plots in different T-
regions shown in Figs. 1 and 2. R
continues to increase below 77K although
at a slower rate. At 4.2K, R is estimated to be more than 200MΩ, too large
to be accurately measured with our
experimental arrangement. The P-effect
on R is shown in Figs. 2 and 3 for two
different samples. All curves in Fig. 2
and 1 and 2 in Fig. 3 are for one sample
and curves 3-5 in Fig. 3 for another.
The overall R is first enhanced by P but
then suppressed at higher P. The

activation energy Δ_1 near T_O responds to
P in a similar fashion as R, while Δ_2 at
low T continues to decrease with increasing P. At high P, an R-shoulder develops
at low T as evident from Fig. 3. At ~68k
bar, a drastic drop in R appears at ~3K,
following a large R-rise as displayed in
the same figure. Above ~68kbar, the R-
drop increases in magnitude and is
shifted toward higher T. A constant 95%
R-drop was obtained at 90 and 113kbar.
At the same time, the low T rise of R is
suppressed but leaving some distinct
structure in the R-T curves. An external
magnetic field shifts the R-drop at 113k

bar toward lower T, suggesting that the
R-drop is caused by superconducting
transition. The P-effect on the onset
temperature T_C of the superconducting
transition is ~+10^{-5}K/bar. The observed
superconductivity disappears upon the
removal of P. The P-effect on T_O and T_m
are shown in Fig. 4.

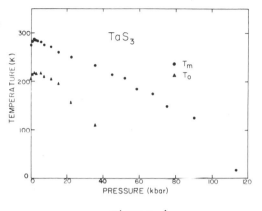

Figure 4

The values of S at room temperature range
from −30 to −60μV/K at ambient pressure.
It reaches a peak of ~1mV/K at 115K. In
spite of the appearance of an R-minimum
above T_O, S displays only a slow increase
with decreasing T before a sharp increase
at T_O as shown in Fig. 1. Therefore the
sharp S-increase coincides with the rapid
R-rise. Under P, S is monotonically
suppressed near room temperature, as
depicted in Fig. 5. The temperature at
which the sharp S-rise occurs is first
shifted up and then down, similar to T_O
and T_m shown in Fig. 4.

Peierls transition is known to result
from the nearly 1-D nature of a system
[11]. TaS₃ crystallizes in a quasi-1-D
fibrous form with the Ta conducting
chains along the c-axis. The metal-
semiconducting transition at T_O in TaS₃
demonstrates its weaker interchain
coupling than NbSe₃. The application of
P will enhance the interchain coupling at
a faster rate than the intrachain
coupling and thus reduce the 1-D character. A monotonically decreasing T_O is
hence expected as pressure increases.
The observed non-monotonic variation of
T_O suggests the possible existence of an
optimal interchain coupling for the
occurrence of Peierls transition and the
ensuing CDW formation. It should note
that a monotonically decreasing T_O with
P up to 17kbar was earlier reported [12].
The reason for the discrepancy remains
unknown.

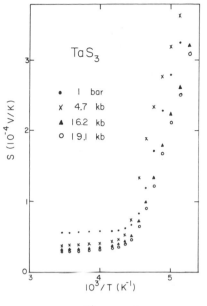

Figure 5

The slow increase of S with decreasing T
above T_O seems to suggest that TaS₃ is a
semimetal above T_O. The absence of a
minimum of S and its continuous decrease
with P, in contrast to R, may be attributed to the "zero-current" nature of S
and the ensuing insensitivity to the
dynamics of electrons. It is therefore
tempting to ascribe the R-increase
between T_m and T_O to fluctuations caused
by CDW formation. Detailed studies are
carries out to examine such a suggestion.

The S for TaS₃ observed above T_O seems
to be too large for ordinary metal.
TaS₃ can be considered as a linear chain
conductor because of its quasi-1D character. A narrow-band Hubbard chain model
has been developed [13] successfully for
S of the highly conducting 1:2 TCNQ
linear chain organic conductors. According to such a model,
$S = -(k_B/e)\ln[2(1-\rho)/\rho]$, where k_B is the
Boltzmann constant, e the electron charge,
and ρ the electron density per site.
The superstructure of TaS₃ below T_O shows
a four-fold increase in periodicity [4].
This implies a quarterly filled band for
TaS₃ and therefore a ρ of 0.5e per site.
Using this ρ, a value of −60μV/K is obtained for S, in fair agreement of our
observations.

The energy gap E_g resulting from the CDW
formation can be obtained from the slope
of the $\ln$R-1/T plot. E_g is twice the
activation energy in a semiconducting
state. As mentioned earlier, there
exists at least two activation energies

in TaS$_3$ in different temperature regions. We found that Δ_1 near T_O is almost sample independent but Δ_2 is, i.e. while Δ_1 varies only from 2750 to 3000K, Δ_2 between 550 and 950K. Therefore, one may conclude that extrinsic excitations is important below 200K and $2\Delta_1$ should be a better measure of E_g. However, in most previous studies [4,12], $2\Delta_2$ was taken as the E_g of CDW. By neglecting the hole-contribution, S can be written [14] as $-(k_B/e)(\Delta_S/k_B T + A_C)$, where Δ_S is the activation energy and A_C a constant dependent on the nature of the scattering centers. From Fig. 1, the slope below T_O gives $\Delta_S \sim 3000K$, in agreement with Δ_1 obtained from R-measurements. The results on R and S suggest that E_g is constant over a rather wide temperature range near T_O. This is in agreement with the constant threshold field [8] detected between T_O and 130K at 1bar, since the threshold field has been shown to be proportional to E_g for weak impurity pinning of CDW [15]. Unfortunately, E_g so obtained is too large for a $T_O \sim 220K$, even if fluctuations are taken into consideration [16].

The reversibility of the transition on pressure cycling suggests that the superconductivity observed in TaS$_3$ can not be caused by a P-induced decomposition of the compound. As suggested by the different T-behavior of R and S near and above T_O, fluctuations may play an non-negligible role in TaS$_3$. According to a model [17] of weakly coupled superconducting filaments or chains in the presence of 1-D fluctuations, $1/T_C \sim 1/T_{CO} + 0.1N(E_F)/CT_{CO}^2$, where T_{CO} is the transition temperature in the strong coupling limit and C is a measure of the coupling between filaments. P is expected to increase C and hence to enhance T_C as observed. The simultaneous detection of superconductivity and an activation energy in TaS$_3$ under P is indeed unusual. The possibility of a second metallic phase responsible for the superconductivity can not be ruled out at the present time, since the transition remains incomplete even under our highest P down to 1.2K. However, equally possible is a transition from a CDW-semiconducting state to a superconducting state. In fact, such a transition has been predicted [18,19] provided that the superconducting order-parameter is spatially dependent [19]. Systematic studies are current under way to elucidate some of the questions posed in this preliminary report.

ACKNOWLEDGEMENT

The work at Houston is supported in part by NSF Grant Nos. DMR 79-08486 and DMR 81-10504 and the Energy Laboratory of the University of Houston.

REFERENCES

*On leave from the Physics Institute, Chinese Academy of Sciences, Beijing, China.

[1].See for example, G.Perluzzo, A.A. Lakhani and S.Jandl, S.S.Comm. 35 301 (1980)

[2].T.Sambongi, M.Yamamoto, K.Tsutsumi, Y.Shiozaki, K.Yamaya and Y.Abe, J. Phys. Soc. Japan 42, 1421 (1977)

[3].R.M.Fleming, D.E.Moncton and D.B. McWhan, Phys. Rev. B18, 5560 (1978) and references therein

[4].T.Sambongi, K.Tsutsumi, Y.Shiozaki, M.Yamamoto, K.Yamaya and Y.Abe, S. S.Comm. 22, 279 (1977)

[5].P.Monceau, N.P.Ong, A.M.Portis, A. Meerschaut and J.Rouxel, Phys. Rev. Lett. 37, 602 (1976)

[6].G.Grüner, L.C.Tippie, J.Sanny, W.G. Clark and N.P.Ong, Phys. Rev. Lett. 45, 935 (1980)

[7].G.Grüner, A.Zawadowski and P.M. Chaikin, Phys. Rev. Lett. 46, 511 (1981)

[8].A.H.Thompson, A.Zettel and G.Grüner, Phys. Rev. Lett. 47, 64 (1981)

[9].P.M.Chaikin and J.F.Kwak, Rev. Sci. Instrum. 46, 218 (1975)

[10].C.W.Chu, A.P.Rusakov, S.Huang, S. Early, T.H.Geballe and C.Y.Huang, Phys. Rev. B18, 2116 (1978)

[11].R.E.Peierls, Quantum Theory of Solids, (Clarendon, Oxford, 1975) p.108

[12].M.Ido, K.Tsutsumi, T.Sambongi and N.Mori, S.S.Comm. 29, 399 (1979)

[13].P.M.Chaikin, Thermoelectricity in Metallic Compounds, ed. by F.J. Blatt and P.A.Schroeder (Plenum, New York, 1977) p.359

[14].H.Fritzsche, S.S.Comm. 9, 1813 (1971)

[15].P.A.Lee and T.M.Rice, Phys. Rev. B19, 3970 (1979)

[16].P.A.Lee, T.M.Rice and P.W.Anderson, Phys. Rev. Lett. 31, 462 (1973)

[17].G.Deutscher, Y.Imry and L.Gunther, Phys. Rev. B10, 4598 (1971)

[18].C.A.Balseiro and L.M.Falicov, Phys. Rev. B20, 4457 (1979)

[19].K.Machida, T.Koyama,and T.Matsubara Phys. Rev. B23, 99 (1981)

PHYSICS OF SOLIDS UNDER HIGH PRESSURE
J.S. Schilling, R.N. Shelton (editors)
© *North-Holland Publishing Company, 1981*

THE APPLICATION OF THERMODYNAMIC GRÜNEISEN PARAMETERS TO VOLUME DEPENDENCE STUDIES

T.F. Smith

Department of Physics, Monash University,
Clayton, Victoria 3168, Australia

The volume dependence of a characteristic interaction energy, E, may be conveniently expressed in terms of a Grüneisen parameter, $\gamma_E = - d \ln E/d \ln V$. Values for γ_E may be derived from direct measurements of the volume dependence of E, or from the thermodynamic relationship, $\gamma_E = V\beta_E B_S/C_P^E$ involving the contributions associated with the interaction to the specific heat capacity, C_P^E and the volume thermal expansion, β_E. B_S is the adiabatic bulk modulus.
The present paper describes some recent developments in the use of thermodynamic Grüneisen parameters for the study of the volume dependences of the magnetic interactions in weak magnetic systems and the elastic properties of A15 structure superconductors.

1. INTRODUCTION

The influence of change of volume on the energy of a physical system may be conveniently expressed in terms of a Grüneisen parameter,

$$\gamma_i = -d \ln E_i/d \ln V \qquad (1)$$

where E_i represents a characteristic energy of the system. In the case of phonons $E_i = \hbar\omega_i$, and for electrons $E_i = E_F$, the Fermi energy. In systems which involve the ordering of magnetic spins or electric dipoles E_i may be related to the ordering energy, which in turn may be related to the ordering temperature.

In a situation where the individual energies, E_i, associated with the components of the system may be considered to be independent (the Born-Oppenheimer adiabatic approximation) it can be readily shown that

$$\gamma_i = \beta_i VB_T/C_V^i = \beta_i VB_S/C_P^i \qquad (2)$$

where B_T, B_S are isothermal and adiabatic bulk moduli and β_i, C_V^i and C_P^i are the contributions to the volume expansion coefficient and molar heat capacities at constant volume and constant pressure respectively. Strictly speaking, although the heat capacity may be decomposed into its individual components, this is not true for the thermal expansion coefficient as this is a ratio of derivatives of the free energy [1]. However, the product of the thermal expansion and the bulk modulus is separable and in practice, since the bulk modulus is generally dictated by the static properties of the lattice and is therefore relatively temperature independent, particularly at low temperature, it is a reasonable approximation to treat it as an unseparated constant [2].

The application of (2) requires the identification of the individual components of the heat capacity and the thermal expansion. This may be achieved either by a knowledge of the appropriate temperature dependence or by the choice of the temperature range over which the required component is the dominant contribution. Clearly, this approach suffers considerable uncertainty if the contribution from the interaction of interest cannot be separated unambiguously from those of other interactions which may be present. In spite of its limitations the thermodynamic approach can often provide information which is not readily available from direct pressure measurements, i.e. the volume dependence of the density of electron states as represented by the electronic Grüneisen parameter.*

A comprehensive review of the low temperature thermal expansion of solids and thermodynamically derived Grüneisen parameters has recently been published by Barron, Collins and White [2]. From this review it is evident that the lattice and electronic Grüneisen parameters represent the most widely determined thermodynamic Grüneisen parameters. Apart from the magnetic Grüneisen parameters associated with the 'invar problem' there are relatively few systems for which thermodynamic Grüneisen parameters have been determined.

In principle, the thermodynamically derived value for γ_i should be equal to that derived from direct pressure dependence measurements in the limit of zero pressure. However, it is only for the lattice Grüneisen parameter, γ_ℓ, that there has been any extensive comparisons between thermodynamic values in the zero temperature limit and those derived from measurements of the pressure dependence of the elastic constants. In general the agreement between the two values

---

*A value for γ_e for a superconductor may be indirectly derived from measurements of the pressure dependence of the thermodynamic critical field.

is satisfactory.   A qualitative correlation between the magnetic Grüneisen parameter, γ_m, and the pressure dependence of the Curie temperature for a number of iron based alloys has been noted [2], but the numerical agreement is poor due to the uncertainty in separating the electronic and magnetic contributions to the thermodynamic parameters.

The present paper describes some recent developments in the use of thermodynamic measurements for the study of the volume dependences of the magnetic behaviour of some weak magnetic systems and the elastic properties in the A15 superconductors V_3Si and V_3Ge.

2.   WEAK MAGNETIC SYSTEMS

2.1   Kondo and Spin Glass Systems

In the limit of a single magnetic impurity the only magnetic contribution to the thermodynamic properties is due to the impurity-conduction-electron interaction; the Kondo effect. Since the thermodynamic functions associated with the Kondo effect are functions only of T/T_K [3], where T_K is the Kondo temperature, it follows that the magnetic Grüneisen parameter γ_K associated with the Kondo effect is given by

$$\gamma_K = -d \ln T_K/d \ln V \qquad (3).$$

As the magnetic impurity concentration increases the indirect (RKKY) interaction between the impurities via the conduction electrons begins to dominate the magnetic behaviour and eventually leads to the spin-glass regime which may be characterised by a temperature T_g, the glass temperature at which the orientations of the magnetic moments become randomly frozen [4]. The spin-glass state Grüneisen parameter is then given by

$$\gamma_g = -d \ln T_g/d \ln V \qquad (4).$$

The RKKY Hamiltonian

$$H = V_o \sum_{i,j} \cos(2k_F r_{ij})/(2k_F r_{ij})^3 \underline{S}_i \cdot \underline{S}_j \quad (5)$$

which expresses the magnetic energy associated with the interaction of the magnetic spin S_i with the surrounding moments S_j at the distance r_{ij}, leads to a further expression for γ_g [5,6],

$$\gamma_g = - d \ln V_o/d \ln V \qquad (6).$$

Here, V_o represents the mean interaction energy which is related to the RKKY exchange parameter, J_{RKKY} and the density of electron states at the Fermi energy, $N(E_F)$ by [7]

$$V_o \propto N(E_F)J_{RKKY}^2 \qquad (7)$$

from which it follows that

$$\gamma_g = -d \ln N(E_F)/d \ln V - 2d \ln J_{RKKY}/d \ln V$$

$$(8)$$

It may be noted that both γ_K and γ_g are expected to be independent of composition and temperature.

Dilute magnetic systems are particularly favourable for thermodynamic studies as the magnetic contribution tends to be dominant at low temperatures and the corrections for the lattice and electronic contributions of the host matrix are relatively straightforward.   While there have been relatively comprehensive studies of the heat capacity due to magnetic impurities there has been practically no measurements of the corresponding thermal expansion.

White first reported large low temperature contributions to the thermal expansion which resembled those found in the heat capacity for copper alloys containing 0.56 and 0.23 at% Mn [6] and 0.2 at% Fe [8].   From these data he obtained $\gamma_m \sim 3$.   A more detailed study of the expansion behaviour for CuMn and AgMn was made by Khan and Griffiths [9] from which they concluded that γ_m for CuMn decreased significantly with decreasing Mn content, in contradiction to the prediction based upon the RKKY Hamiltonian.   Both of these early investigations suffer from the drawback of having to combine their expansion data with heat capacity data taken in other, independent programmes for different samples.   In view of the sensitivity of the magnetic contribution to the thermodynamic properties and the uncertainties in composition, which are difficult to avoid, combining data from different samples is not a satisfactory procedure.

The author and co-workers have recently reported [10,11] the determination of thermodynamic magnetic Grüneisen parameters for several dilute magnetic impurity systems where all the required measurements were made on the same sample.   Three of these systems, MoFe, CuCr and CuFe were specifically chosen as being representative of different alloy systems in which the Kondo effect is likely to be significant (see table 1).   The reader is referred to the original publications for experimental details.

The measured glass temperatures, T_g, and the literature values for the Kondo temperature, T_K, for the alloys investigated are summarised in table 1.   The T_g values represent the mean of two values determined for specimens cut from regions close to each end of the alloy ingot.

<u>Table 1</u> T_g and T_K for spin-glass alloys

Alloy	$T_K(K)$	$T_g(K)$
Cu 1 at% Fe	24	5.7
Cu 0.2 at% Cr	2	<2.2
Mo 1 at% Fe	0.2	<1.7
Cu 0.2 at% Mn	0.012	<2.2
Cu 0.5 at% Mn	0.012	5.5
Cu 1.0 at% Mn	0.012	10.0
(Cu + 10 at% Al)0.2 at% Mn		<2.2
(Cu + 10 at% Al)0.5 at% Mn		3.9
(Cu + 10 at% Al)1.0 at% Mn		6.3
Ag 1.9 at% Mn	10^{-4}-10^{-16}	8.0

2.1.1 Magnetic contributions to heat capacity and thermal expansion

The form of the magnetic contribution to the thermal expansion and the specific heat capacity for the <u>Cu</u>Mn alloys is shown in figure 1.

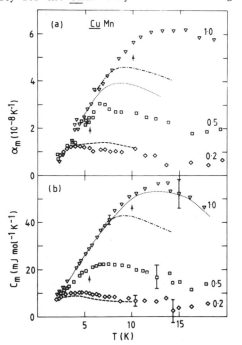

Figure 1:   (a) Magnetic contribution α_m to thermal expansion of <u>Cu</u>Mn alloys [11] Arrows indicate T_g. Cu 0.54 at% Mn, -----[9]; Cu 0.23 at% Mn,— — —, Cu 0.56 at% Mn, -·-·-   [6]

(b) Magnetic contribution C_m to the heat capacity of <u>Cu</u>Mn alloys. Arrows indicate $\overline{T_g}$. Cu 0.56 at% Mn, -·-·-; Cu 0.15 at% Mn, — — — [12].

The broad Schottky-like nature of both contributions is evident.   The height of the peak and the temperature at which the maximum value is attained, T_{max}, are both approximately proportional to the Mn concentration and where T_g could be determined it is approximately

$3/4$ T_{max}.   Similar curves were also obtained for the (Cu + 10 at% Al)Mn alloys.

It is common practice when presenting thermodynamic data for spin glasses to express these as universal functions of T/c, where c is the concentration of magnetic impurity [13].   Plots of α_m/c and C_m/c as functions of T/c for the <u>Cu</u>Mn alloys are shown in figure 2.   These plots

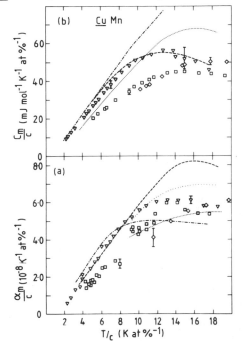

Figure 2:   (a) α_m/c as a function of T/c for (Cu)Mn.   0.23 at% Mn, ------; 0.56 at% Mn, — — — [6]; 0.54 at% Mn, ...; 1.26 at% Mn -·-·- [9]; 0.2 at% Mn ◇; 0.5 at% Mn,□; 1.0 at% Mn, ▽ [11].

(b) C_m/c as a function of T/c for (Cu)Mn.   0.56 at% Mn, -----; 1.15 at% Mn, -·-·- [12]; 1.2 at% Mn, — — — [14]; 0.2 at% Mn,◇; 0.5 at% Mn,□;   1.0 at% Mn, ▽ [11].

also incorporate different sets of published data and demonstrate a reasonable adherence to the universal scaling law.   However, the scatter in the data, which is taken to be an indication of the uncertainties in the homogeneity and composition of the samples, illustrates the limitation on making quantitative calculations based upon thermodynamic data taken on different specimens.

Figure 3 shows the magnetic contributions to the thermal expansion and the specific heat capacity for the Ag 1.9 at% Mn sample.   The plot for C_m illustrates the uncertainty which may arise in extracting the magnetic contribution when the effect of the impurity on the thermodynamic properties of the host matrix

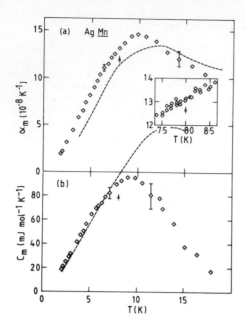

Figure 3: (a) Magnetic contribution α_m to
thermal expansion of Ag 1.9 at% Mn.
Arrow shows T_g for the specimen.
---- scaled Ag 1.05 at% Mn [9].
Insert, α_m in the vicinity of T_g.

(b) Magnetic contribution C_m to the
heat capacity of Ag 1.9 at% Mn.
Arrow shows T_g for the specimen.
---- scaled Ag 1.01 at% Mn [15].

is unknown. While the lattice spacing of
silver is almost unchanged with the addition of
1.9 at% Mn, the considerable mass difference
between the host and the impurity atoms can be
expected to result in some modification of the
lattice heat capacity. Du Chatenier and
de Nobel [15] have reported indications of a
decrease in the Debye temperature with the
addition of Mn to Ag from measurements of the
heat capacity.

The data shown in figure 3 were derived by
subtracting a pure silver background and the
relatively sharp drop in C_m above its maximum
value compared with the variation for α_m would
suggest that the background heat capacity has
been underestimated. This suspicion is
strengthened when the data are compared with
the curves representing values for α_m and C_m
which have been scaled from the data for
Ag 1.05 at% Mn [9] and Ag 1.01 at% Mn [15]
respectively, assuming α_m/c and C_m/c to be
universal functions of T/c. While there is
quite reasonable agreement between the scaled
and measured values for α_m, there is a marked
discrepancy for C_m with the measured values
falling well below the scaled values above
7.5 K.

The absence of any detectable feature in both
α_m and C_m at T_g is common to all of the alloys.
The insert to figure 3 provides a detailed plot
of α_m in the vicinity of T_g for the AgMn alloy.

Turning now to the systems in which single
partical effects can be expected to influence
the spin-glass ordering the data for α_m and
C_m for the MoFe, CuCr and CuFe are shown in
figures 4 and 5 [10]. C_m data for Cu 0.24 at%
Fe [17], Mo 0.84 at% Fe and Mo 0.51 at% Fe [16]
and α_m data for Cu 0.2 at% Fe [8] are also
included.

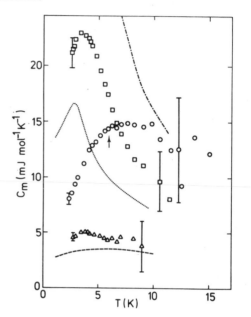

Figure 4: Magnetic contribution C_m to heat
capacity of Cu 1 at% Fe, ○; Cu 0.2
at% Cr, △; Mo 1 at% Fe, □ [10];
Mo 0.84 at% Fe, -·-·-; Mo 0.51 at%
Fe, ... [16]; Cu 0.24 at% Fe, ---
[17]. Arrow shows T_g for Cu 1 at% Fe.

The magnetic heat capacity for the Cu 1 at% Fe
and Cu 0.2 at% Cr shows the broad peak, falling
off slowly at high temperature characteristic
of the spin-glass systems considered above.
The magnetic contribution to the thermal expansion for the CuCr also resembles that for a
spin glass, but that for the CuFe exhibits the
opposite curvature, being concave downwards,
and has fallen to zero by about 3 K. The
magnetic contributions for the Mo 1 at% Fe
exhibit quite marked differences from those
for the spin-glass systems described here.
C_m has a much sharper peak and α_m undergoes a
change in sign at about 6.6 K.

It may be noted that the C_m data for the Mo
1 at% Fe alloy lie between those reported by
Amamou et al. [16] for Mo 0.84 at% Fe and Mo
0.51 at% Fe suggesting that there may have been
some loss of iron from the sample during preparation. Attention is also drawn to the α_m

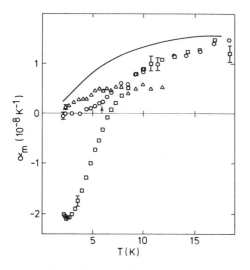

Figure 5: Magnetic contribution α_m to thermal
expansion of Cu 1 at% Fe,$\bigcirc$;
Cu 0.2 at% Cr,$\triangle$; Mo 1 at% Fe,$\square$ [10];
Cu 0.2 at% Fe,‗‗‗ [8].   Arrow shows
T_g for Cu 1 at% Fe.

data for Cu 0.2 at% Fe obtained by White [8],
which lies well above that for the Cu 1 at% Fe
below 20 K.   This difference is believed to
be due to competition between the spin-glass
ordering and the Kondo effect rather than a
gross failure in sample preparation.

2.1.2 Magnetic Grüneisen parameters

The magnetic Grüneisen parameters calculated
from α_m and C_m for the CuMn, (Cu + 10 at% Al)Mn
and AgMn are summarised in figures 6 and 7.

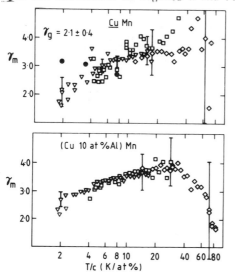

Figure 6: γ_m as a function of T/c for CuMn and
(Cu 10 at% Al)Mn alloys. 0.2 at%
Mn$\diamond$; 0.5 at% Mn,$\square$; 1.0 at% Mn,$\triangledown$ [11].

The solid symbols are the data from Khan and
Griffiths [9] and the bar from White [6].
These data, which represent average γ_m values
over a range of temperature, have been arbitarily plotted for T = 4.2 K.

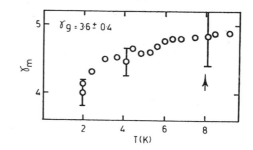

Figure 7: The magnetic Grüneisen parameter,
γ_m, as a function of temperature for
Ag 1.9 at% Mn [11].   The arrow
indicates T_g.

In the case of the Cu based alloys the data for
the different compositions have been combined
by plotting γ_m as a function of T/c.   The plot
for the CuMn alloys also includes the values
obtained by White [6] and Khan and Griffiths
[9], arbitarily plotted for a temperature of
4.2 K as the quoted values represent mean low
temperature values.

Contrary to the claim of Khan and Griffiths
there is little evidence to suggest any strong
composition dependence for γ_m, although it is
evident that γ_m does decrease with increased
Mn concentration.   It may also be noted that
γ_m shows a decrease with decreasing temperature
below 5 K for both the Cu based and Ag alloys.

The extent to which the apparent decrease of
γ_m above 7 K for the (Cu + 10 at% Al) 0.2 at%
Mn is real is debatable in view of the relatively large uncertainty which is attached to the
values.   It may also be noted that the uncertainty in the values of C_m restricts the
values for γ_m for the Ag 1.9 at% Mn to below
9 K.

Values of γ_g for a limited number of spin-glass
systems have been determined by Schilling and
co-workers from direct measurements of the
glass temperature, T_g, as a function of pressure.   In particular they obtain values of
3.6 ± 0.3 for Ag 3 at% Mn [18] and 2.1 ± 0.4
for Cu 3 at% Mn [19] which are in reasonable
agreement with the thermodynamically derived
values for γ_m particularly in view of the
higher concentrations and the higher temperatures corresponding to the glass-temperatures
of 12 and 20 K respectively.   It may also be
noted that the value of γ_g for the AgMn is in
good agreement with values derived less directly from pressure measurements of the resistivity, but that for the CuMn is significantly
lower than the earlier estimates of 7.0 ± 0.5
[20].

As discussed by Schilling [20] the value for γ_g (and γ_m) provides a good guide for distinguishing between possible mechanisms for spin-glass freezing.   The thermodynamic and pressure values for γ_m and γ_g both favour the RKKY interaction model, for which γ_g is expected from (8) to be of order 4, rather than the dipolar anisotropy model for which $\gamma_g = 1$.   However, the temperature dependence for γ_m does raise the possibility that the anisotropy interaction begins to dominate as the temperature is lowered.

In a situation where there are two independent interactions contributing to the thermodynamic properties such that the heat capacity may be expressed as $C = C_a + C_b$, then the total Grüneisen parameter γ_T associated with the combination of the two contributions is given as

$$\gamma_T = (C_a\gamma_a + C_b\gamma_b)/(C_a + C_b) \qquad (9)$$

where γ_a and γ_b are respectively the Grüneisen parameters associated with the contributions indexed by a and b.   Thus, although γ_a and γ_b may be independent of temperature any temperature dependence of C_a/C_b will result in a temperature dependence of γ_T.

Since anisotropy interactions are relatively weak, the interaction energy between near neighbours is only expected to influence the thermodynamic behaviour at temperatures below about 5 K [21].   Thus, the decrease in γ_m is consistent with a transition from a regime at a temperature at which the excitation energy is dominated by the RKKY interaction to one in which dipolar anisotropy energies become dominant.

As the strength of the dipolar interaction falls off as r^{-3} the same argument which leads to the scaling law for the RKKY interaction [13] predicts that γ_m should be a universal function of T/c.   As shown in figure 6 this appears to be the case.

The magnetic Grüneisen parameters for the CuFe, CuCr and MoFe are shown in figure 8.   The strong temperature dependence of γ_m, relative to the spin-glass systems described above is immediately evident.   Since $T_K \sim T_g$ for these systems, the possibility arises that single-particle interactions are responsible for the temperature dependence.   At temperatures well above either the glass or the Kondo temperature the spin-glass contribution to the heat capacity is predicted to decrease much more rapidly than that due to the single-impurity.   Thus, the magnetic Grüneisen parameter would be expected to approach the single-impurity value, γ_K, at high temperature.   As the temperature is lowered towards T_g the situation becomes complex, but γ_m could be expected to tend towards γ_g.

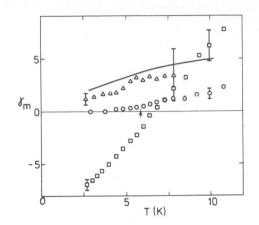

Figure 8: Magnetic Grüneisen parameter γ_m for Cu 1 at% Fe, O ; Cu 0.2 at% Cr, △ ; Mo 1 at% Fe, □ ; Cu 0.2 at% Fe, —— [10]. Arrow shows T_g for Cu 1 at% Fe.

Schilling and co-workers [20] have determined values for the volume dependence of V_0 and T_K for a wide variety of magnetic impurity systems and find that γ_K is significantly larger than γ_g.   Thus it would appear qualitatively that a systematic shift in dominance from the spin-glass to the Kondo regime would offer an explanation for the increase in γ_m with increasing temperature.

Only the CuFe system offers the opportunity for a quantitative comparison between the values of γ_g and γ_K derived from pressure measurements and thermodynamic γ_m values.   In this case the pressure measurements give values of 15 and 3.3 for γ_K and γ_g respectively which are considerably larger than the values for γ_m.   This considerable discrepancy highlights the fundamental weakness in a model which attempts to characterise an inherently complex state of two mutually interacting systems by two independent parameters. Attention is drawn to the larger values of γ_m for the Cu 0.2 at% Fe [8] for which the impurity-impurity interactions are expected to be less important and therefore γ_m should be closer to γ_K.

While no direct numerical comparison has been possible for the CuCr system, the similarity between the temperature dependence for this and the Cu 0.2 at% Fe would suggest that the behaviour for the two systems is very similar.

Since T_K for the MoFe system is much lower than for either the CuFe or the CuCr it could be expected that γ_m should have the weaker temperature dependence characteristic of a spin-glass, whereas it has the strongest.   The change in sign presents a further complication as it is impossible to account for opposite signs for γ_K and γ_g within the framework of the simple RKKY model.   It may be noted that Ford and Schilling [22] have observed an unusually large

pressure dependence for the impurity spin
scattering resistivity of a Mo 1 at% Fe alloy
indicating the presence of some strongly volume
dependent interaction.

2.2. Critical concentration alloys: Ni-Cr

The alloy systems with compositions close to
the critical composition for the formation of
the ferromagnetic state represent a class of
weak magnetic systems which have been extensively studied [23].   It is a common feature of
such solid solutions that the low temperature
heat capacity may be represented by

$$C_p = A + BT + CT^3 \qquad (10)$$

The anomalous 'constant' contribution to the
heat capacity has been variously attributed to
many-body spin density fluctuations (para-
magnons) or superparamagnetic clusters [23].
While the latter model is now the most widely
accepted, the actual nature of the superparamagnetic contribution is still a matter for
debate.   Basically two theoretical approaches
are commonly used to explain the superparamagnetic contribution to the heat capacity.
In one, the randomly distributed magnetic
regions are treated as independent superparamagnetic particles the orientation of which is
dictated by an anisotropy energy which leads to
the heat capacity contribution [24].   In the
other approach the heat capacity contribution
is attributed to the indirect exchange between
the superparamagnetic clusters giving rise to a
spin-glass-like behaviour [25].

Although the low temperature heat capacity has
been measured for a number of alloy systems
close to the critical composition for ferromagnetism there is only one published report of
a measurement of the thermal expansion [26].
While the magnetic contribution in these measurements for Ni_3Al and Ni-Pt alloys is not
specifically identified, there is nevertheless
clear evidence of a change in the sign of the
low temperature expansion coefficient on passing
from the ferromagnetic to paramagnetic regime.

An as yet unpublished study comprising measurements of the thermal expansion, heat capacity
and pressure dependence of the ferromagnetic
ordering temperature has been made for Ni-Cr
alloys in the vicinity of the critical composition ($\sim$12 at% Cr) [27].   A brief survey of
these measurements is given below.

2.2.1 Heat capacity and thermal expansion measurements

Low temperature specific heat capacity data for
9, 11 and 12 at% Cr are shown in figure 9.
Attention is first drawn to the absence of any
detectable anomaly at the ferromagnetic ordering
temperature, T_f, as defined by the maximum in
the low field a.c. susceptibility (see the
insert to figure 9), for the 11 at% alloy.

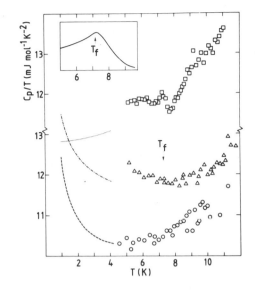

Figure 9:   Low temperature specific heat capacity for Ni-Cr alloys: 9 at% Cr,□;
11 at% Cr,△; 12 at% Cr,O.   The
curves represent the data of Moody
et al.[28].   9 at% Cr,...; 11 at%
Cr, ·-·-; 12 at% Cr, — —. Note
the displaced vertical scale for the
9 at% Cr data.   Insert: Low field
a.c. susceptibility for Ni 11 at% Cr.

Nor was any evidence of any anomaly found at the
ordering temperature (105 K) for the 9 at% alloy.
Estimates of the discontinuity in C_p/T based
upon a simple model of magnetic ordering of
superparamagnetic spins indicate that this should
be large enough to be resolved in the 11 at%
data, though it might be concealed in the ±1.5%
scatter for the 9 at% data.

Thermal expansion data for the 11 at% Cr sample
are given in figure 10.   Concentrating on the
temperature region in the vicinity of the
susceptibility maximum there is evidence of a
slight 'kink'.   Assigning a value of
$\sim 1 \times 10^{-9}$ K^{-2} to the discontinuity in α/T and
taking $dT_f/dP = -7.2 \times 10^{-10}$ K Pa^{-1} (see below)
the Ehrenfest equation gives $\Delta C_p/T \approx 0.2$ mJ
$mole^{-1}$ K^{-1} which would be too small to be
visible in the heat capacity measurements.
(see figure 10).

Unfortunately, the specific heat capacity data
were only available from a cryostat designed
to operate above 4K.   Consequently, the data
do not extend down into the temperature range
for which the magnetic contribution to the
specific heat capacity has been clearly resolved
[28].   However, when these data are compared
with values calculated from (10) using the
coefficients quoted by Moody et al. [28] (shown
in figure 9) the agreement is quite reasonable.

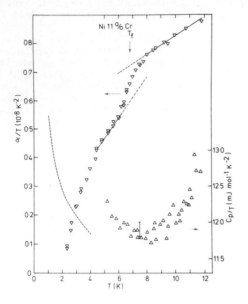

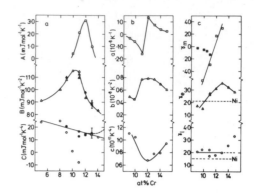

Figure 11 : (a) Coefficients for the specific
               heat capacity (open symbols
               from Moody et al. [28]),

           (b) coefficients for the thermal
               expansion,

           (c) Magnetic, electronic and
               lattice Grüneisen parameters,
               as a function of Cr concen-
               tration (open symbols),
               -d ln T_f/d ln V (closed symbols).

Figure 10: Low temperature specific heat
            capacity and thermal expansion data
            for Ni 11 at% Cr.   The broken
            curve represents the data of
            Moody et al. [28].   The bar shown
            in the specific heat capacity data
            represents the magnitude of the
            discontinuity calculated from the
            discontinuity in α/T using the
            Ehrenfest equation.

In view of the relatively high temperature
range of the data, fits of C_p/T to T can only
be expected to give relatively crude estimates
of the coefficients B and C (A being negligible
above 4 K).   Nevertheless, these coefficients,
which are shown in figure 11(a), agree quite
well with those obtained by Moody et al.

The thermal expansion data for the Ni-Cr alloys
are presented in figure 12.   A striking
difference in the expansion behaviour for the
paramagnetic and ferromagnetic alloys is
immediately evident.   Whereas the paramagnetic
alloys exhibit a turn-up in α/T, which is
characteristic of the heat capacity data for
both paramagnetic and ferromagnetic alloys
about the critical composition, the ferromagnetic alloys have the opposite behaviour.
Thus, it is immediately evident that the sign
of the magnetic contribution to the thermal
expansion changes sign on passing from the
paramagnetic to the ferromagnetic state.
Furthermore, the magnetic contribution to the
thermal expansion is largest in the immediate
vicinity of the critical composition.

In an effort to distinguish the various contributions to the thermal expansion all but the
11 at% data have been least squares fitted

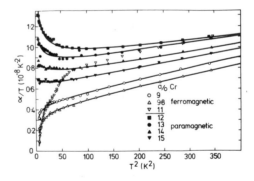

Figure 12: α/T as a function of T^2 for Ni-Cr
            alloys.

for T < 20 K to an expansion of the form (10)
adopted for the heat capacity data.   The
resulting fits are shown in figure 12 and the
coefficients (denoted by a,b,c) are plotted in
figure 11(b).   In the case of the 11 at% alloy,
because of the discontinuity in the temperature
dependence of the expansion coefficient at the
magnetic ordering temperature, only the data
up to 5 K were least squares fitted to the
expression $(\alpha-cT^3)/T = a/T + b$ where c was

taken to be $0.74 \times 10^{-11} \text{ K}^{-4}$ (see figure 11(b)).

While reasonable fits to the data are obtained, it is impossible to judge to what extent any physical significance can be attached to the compositional variation of the coefficients b and c, which are nominally associated with the electronic and lattice contributions respectively. It is suspected that the magnetic contribution to the specific heat capacity may make a significant contribution to the linear T term and such a deviation from the approximation of a constant term would also influence the T^3 coefficient [23]. This is illustrated by the negative and near zero values for the T^3 coefficient obtained from the least squares fitting of their heat capacity data by Moody et al. (figure 11(a)). Similar effects can be expected for the fitting of the expansion data. Nevertheless, such uncertainties do not detract from the change in sign of the magnetic contribution on passing from the ferromagnetic to the paramagnetic state.

2.2.2 Grüneisen parameters

The individual Grüneisen parameters calculated from the coefficients plotted in figures 11(a) and 11(b) using expression (2) are shown in figure 11(c). The extent to which the apparent composition dependence of the Grüneisen parameters identified with the electronic and lattice excitations is real is debatable since neither the lattice dynamics nor the one-electron excitations responsible for the electronic contribution to the heat capacity are likely to be very sensitive to composition. It is probable that the variations stem from the inadequate separation of the three contributions and it may be noted that only a relatively small adjustment to the compositional dependence of B, as indicated by the broken line in figure 11(a), would be sufficient to keep γ_ℓ constant at a value of about 2.

The electronic and lattice Grüneisen parameters for Ni are also shown in figure 11(c).

The most striking behaviour is seen in the magnitude and compositional dependence of γ_m. Also shown in figure 11(c) are values of $-d \ln T_f/d \ln V$ derived from measurements of the pressure dependence of T_f. Clearly, there is a marked discrepancy between the two sets of parameters which suggests that they are related to different magnetic energy terms.

2.2.3 Discussion

Whilst the detailed analysis and interpretation of the data described in the above sections is still in progress there are three very significant points evident;

(i)   the absence of any feature in the heat capacity at T_f,

(ii) the difference in the magnitude of γ_m and $- d \ln T_f/d \ln V$,

(iii) the change in sign of γ_m in going from the ferromagnetic to the paramagnetic regimes.

As already noted above, (ii) leads us to conclude that there are at least two magnetic energy terms having quite different volume dependences. For convenience we shall identify these as inter-cluster and intra-cluster, where the former is associated with the establishment of the magnetic ordering. These energy terms themselves may have complex compositions with direct and indirect exchange contributions to the inter-cluster energy and exchange and anisotropy contributions to the intra-cluster energy [29].

Observation (i) would suggest that the magnetic ordering at T_f either involves only a small number of the magnetic regions or that the number of degrees of freedom of each individual magnetic region is not significantly reduced by the onset of the ferromagnetism. It then follows that it is the intra-cluster energy which is primarily responsible for the low temperature magnetic contributions to the heat capacity and the thermal expansion. The magnitude and sensitivity of γ_m to the transition to the magnetically ordered state suggest that the intra-cluster energy represents a delicate balance between terms having comparable but opposite volume dependences.

3.   A15 SUPERCONDUCTORS

The application of thermal expansion measurements to the study of the lattice anharmonicity and its possible consequences for the lattice stability and superconductivity of the A15 compounds has been previously discussed in some detail [30]. Central to this discussion were the negative low temperature expansion coefficients for V_3Si and V_3Ge from which estimates for the total Grüneisen parameter of up to -100 and -40 respectively were obtained. The values in the zero temperature limit, were -25 and -11 respectively. In the case of V_3Si, this value was taken to represent the lattice Grüneisen parameter as the expansion data were taken in the superconducting state well below T_c where the electronic contribution is expected to be negligible. This assumption could not be made for the V_3Ge and no specific assignment of the origin of the Grüneisen parameter could be made. The above values for the total Grüneisen parameter may be compared with values for the lattice Grüneisen parameter derived from measurements of the elastic constants as a function of pressure. Carcia and Barsch have determined zero temperature values for γ_ℓ of 5 and -15 for transforming and non-transforming V_3Si [31] respectively and -0.052 for V_3Ge [32].

3.1   Thermal Expansion V_3Si and V_3Ge

Subsequent [33] to the above reports thermal expansion measurements on single crystal samples of V_3Si have shown that anisotropy in the expansion can exist to temperatures up to ~ 80 K.

This suggests that a departure from cubic symmetry may occur well above the ∿20 K generally
assigned to the transition temperature from
X-ray studies.   These observations raised the
possibility that the negative expansion
behaviour found in the polycrystalline samples
was associated with the cubic to tetragonal
transformation and was due to there being a
preference to c-axis formation in the direction
of measurement (c/a > 1), which was along the
axis of the cast-ingot.   To test this possibility, measurements of the thermal expansion
along each of three mutually orthogonal directions through a V 25.75 at% Si polycrystalline
specimen have been undertaken [34].

To make these measurements, the original ingot
was spark cut and lapped to form three right
prisms having faces parallel and perpendicular
to the ingot axis, which was parallel to the
hearth in which it was cast (see figure 13).
The data for each direction through the ingot
were taken by appropriately stacking the three
prisms to form a composite sample for the expansion cell.   The resulting expansion data as a
function of temperature are shown in figure 13.

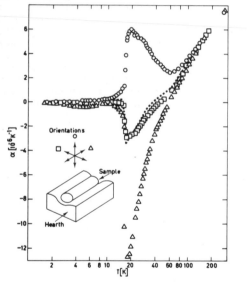

Figure 13: Linear thermal expansion coefficients
            for a polycrystalline sample of
            V 25.75 at% Si measured along three
            mutually orthogonal directions.   The
            broken curve is a plot of one-third
            of the coefficient of volume expan-
            stion [33].

At temperatures above 100 K the expansion coefficient is essentially the same for all three
directions, but at lower temperatures a marked
anisotropy develops.   In the most extreme case,
α for the direction perpendicular to the axis
and the hearth has the opposite sign to the
other two directions.   All three directions
have extremal values for α immediately above the
superconducting transition temperature.   Sharp
reductions in the magnitude of α occur at T_c,

but the expansion remains and isotropic down
to the low temperature limit of the measurements (figure 14).

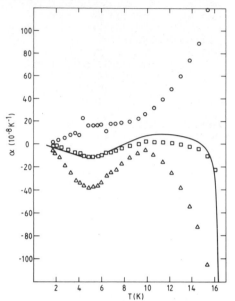

Figure 14: Expansion data for V 25.75 at% Si in
            the superconducting state.

The onset of the anisotropic expansion indicates
that the departure from cubic symmetry is
occurring at temperatures well above those
normally associated with the structural transition.   The appearance of the data suggests a
strong preferential alignment within the sample
such that c-axis behaviour is predominant along
the direction for which the positive peak is
observed.

Optical microscopy on the polished faces of
each of the prisms revealed a preferred
orientation in the grain growth which could lead
to a preference for a and c axis formation. This
structure, which is illustrated in figure 15,
consists of columnar-like grains fanning out
from the side of the ingot which was in contact
with the water-cooled furnace hearth during the
final melt.   Thus, the long axes of the grains
were preferentially aligned along the direction
for which the positive expansion peak was
observed.

From the summation of the expansion coefficients
for the three orthogonal directions the volume
coefficient, β, may be obtained.   This is
shown plotted as β/3 in figures 13 and 14.   The
fractional decrease in volume between the
appearance of the anisotropic behaviour and the
onset of the negative expansion (∿36 K) is
estimated to be 31×10^{-5}, while the fractional
increase between 36 K and 18 K is 8×10^{-5}.
This latter value is well within the upper
limit of one or two parts in 10^4 placed upon
the volume change accompanying the transition
by Batterman and Barrett [35].

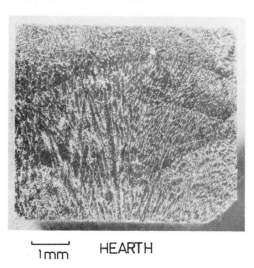

1mm   HEARTH

Figure 15: Optical micrograph for V 25.75 at% Si expansion specimen showing a section perpendicular to both the hearth and the axis of the ingot.

The establishment of a connection between the negative expansion behaviour for V_3Si and a departure from cubic symmetry raises the question of whether a similar effect arises for V_3Ge, the only other A15 structure compound for which a negative low temperature expansion behaviour has been observed [30]. To date no static distortion away from cubic symmetry has been reported for V_3Ge.

Expansion data taken [36] on prisms cut from a cast ingot with the same orientations as the V-Si prisms, are shown in figure 16. Again there

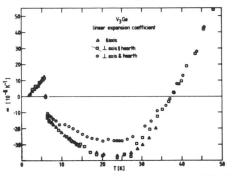

Figure 16: Linear thermal expansion coefficients for a polycrystalline sample of V_3Ge measured along three mutually orthogonal directions [35].

is clear evidence of anisotropy, though much smaller than in the case of the V-Si. While the expansion coefficients below approximately 38 K are negative in all three directions, there is a definite displacement towards more positive values for the measurements along the

direction perpendicular to both the axis of the ingot and the hearth.

3.2 Grüneisen Parameters

The observation of departures from cubic symmetry for the V_3Si and V_3Ge raises fundamental questions on the interpretation of measurements of their physical properties. The pressure dependence of the elastic constants for V_3Si, specifically $c' = (c_{11}-c_{12})/2$, in particular, could depend quite sensitively upon the direction of measurement even for nominally non-transforming crystals. This may explain the difference in the sign of γ_ℓ for transforming and non-transforming V_3Si reported by Carcia and Barsch [32]. It may be noted that dc'/dP changes from positive to negative for both the transforming and non-transforming crystals below 100 K, but becomes positive again for the nominally non-transforming crystal below about 50 K.

The expansion data for the V-Si below 5 K (figure 14), being well below T_c, can be expected to have a negligible electronic component and indicate a highly anisotropic lattice Grüneisen parameter. The thermodynamic Gruneisen parameters calculated assuming the Debye temperature of 300 K derived from the elastic constants have values of 69, -284 and -41 for the three directions giving an average value of -256 which is considerably larger than that derived from the elastic constant measurements. This is also the situation for the V_3Ge where the average $\gamma_G \sim -40$ just above T_c.

3.3 Discussion

The observation of a departure from isotropic expansion behaviour for the V-Si and V Ge raises fundamental questions in relationship to the interpretation of their physical properties, particularly where these have been determined for polycrystalline material. It also raises the question of whether anisotropy is an inherent property of the material, or whether is due to the physical characteristics of the sample.

In view of the grain structure for the V-Si sample it has been suggested that the anisotropic expansion behaviour could result from a residual stress field within the sample associated with defects or arising from the large thermal gradients occurring in the casting process [34]. Such a stress field σ coupled with a strong temperature dependence of the appropriate elastic compliance constant, s, would lead to a contribution to the thermal strain of the form

$$\varepsilon = \frac{\partial s}{\partial T} \sigma \Delta T \qquad (11).$$

Summed over a polycrystal material of cubic symmetry and no preferred orientation, the nett strain would be zero. However, where the sample possesses preferred orientation an anisotropic strain can be expected.

Using temperature derivatives of the elastic complicance constants calculated from published elastic moduli and assuming that the internal stress is constant it is estimated to be approximately 2 MPa for the V-Si and 5 MPa for the V_3Ge. Such values are well within the stress levels which are found in cast metals [37].

Acknowledgement

The financial support of the Australian Research Grants Committee is gratefully acknowledged.

REFERENCES

[1]   Fletcher, G.C., Physica 93B (1978) 149-64
[2]   Barron, T.H.K., Collins, J.G. and White, G.K., Ad. in Phys. 29 (1980) 609-730
[3]   Grüner, G. and Zawodowski, A., Rep. Prog. Phys. 37 (1974) 1497-1583
[4]   Mydosh, J.A., J. Magn. Magn. Mater. 7 (1978) 237-48
[5]   Liu, S.H., Phys. Rev. 127 (1962) 1889-91
[6]   White, G.K., J. Phys. Chem. Solids 23 (1962) 169-71
[7]   Yosida, K., Phys. Rev. 106 (1957) 893-8
[8]   White, G.K., Am. Inst. Phys. Conf. Proc. No. 3 (1971) 59-64
[9]   Khan, J.A. and Griffiths, D.J., Phys. F: Metal Phys. 8 (1978) 763-73
[10] Simpson, M.A. and Smith, T.F., J. Phys. F: Metal Phys. 11 (1981) 397-404
[11] Simpson, M.A., Smith, T.F. and Gmelin, E., J. Phys. F: Metal Phys. (1981) to be published
[12] Zimmerman, J.E. and Hoare, F.E., J. Phys. Chem. Solids 17 (1960) 52-6
[13] Souletie, J. and Tournier, R., J. Low Temp. Phys. 1 (1969) 95-108
[14] Wenger, L.E. and Keesom, P.H., Phys. Rev. B 13 (1976) 4053-9
[15] Du Chatenier, F.J. and de Nobel, J., Physica 32 (1966) 1097-109
[16] Amamou, A., Caudron, R., Costa, P., Gautier, F., Friedt, J.L. and Loegel, B., J. Phys. F: Metal Phys. 6 (1976) 2371-88
[17] Franck, J.P., Manchester, F.D., and Martin, D.L., Proc. R. Soc. A 263 (1961) 494-507
[18] Hardebusch, U., Gerhardt, W. and Schilling, J.S., Phys. Rev. Lett. 44 (1980) 352-5
[19] Schilling, J.S., private communication
[20] Schilling, J.S., Adv. Phys. 28 (1979) 657-715
[21] Window, B.J., Magn. Magn. Mater. 1 (1975) 167-79
[22] Ford, P.J. and Schilling, J.S., J. Phys. F: Metal Phys. 6 (1976) L285
[23] Hahn, A. and Wohlfarth, E.P., Helv. Phys. Acta. 41 (1968) 857-68
[24] Schröder, K., J. Appl. Phys. 32 (1961) 880-2
[25] Ododo, J.C. and Coles, B.R., J. Phys. F: Metal Phys. 7 (1977) 2393-400
[26] Kortekaas, T.F.M. and Franse, J.J.M., phys. stat. sol. (a) 40 (1977) 479-85
[27] Simpson, M.A., Thesis, Monash University, (1980) unpublished
[28] Moody, D.E., Staveley, M.G. and Kuentzler, R., Phys. Lett. 33A (1970) 244-5
[29] Soukoulis, C.M. and Levin, K., Phys. Rev. Lett. 39 (1977) 581-4
[30] Smith, T.F. and Finlayson, T.R. in High-Pressure and Low-Temperature Physics, ed. C.W. Chu and J.A. Woollam (Plenum, 1978) 315-336
[31] Carcia, P.F. and Barsch, G.R., phys. stat. sol. (b) 59 (1973) 595-606
[32] Carcia, P.F. and Barsch, G.R., Phys. Rev. B 8 (1973) 2505-15
[33] Fukase, T., Kobayashi, T., Isino, M., Toyoto, N. and Muto, Y., J. de Physique C6 (1978) 406-7; Milewits, M. and Williamson, S.J. ibid. 408-10
[34] Gibbs, E.E., Finlayson, T.R. and Smith, T.F., Sol. St. Commun. 37 (1981) 33-35
[35] Batterman, B.W. and Barrett, C.S., Phys. Rev. 145 (1966) 296-301
[36] Finlayson, T.R., Gibbs, E.E. and Smith, T.F., Proceedings LT 16 (1981) to be published
[37] Angus, H.T., Cast Iron: Physical and Engineering Properties (Butterworths, 1976) 392-412

PHYSICS OF SOLIDS UNDER HIGH PRESSURE
J.S. Schilling, R.N. Shelton (editors)
© *North-Holland Publishing Company, 1981*

NONLINEAR PRESSURE EFFECTS IN SUPERCONDUCTING RARE EARTH - IRON - SILICIDES

C.U. Segre and H.F. Braun [*]

Institute for Pure and Applied Physical Sciences [$]
University of California at San Diego
La Jolla, California 92093

The superconducting transition temperatures (T_c) of $RE_2Fe_3Si_5$ (RE = Sc,Y,Lu) vary in a nonlinear fashion under hydrostatic pressure up to 20 kbar. For the Sc and Lu compounds T_c decreases at an average rate of -7×10^{-5}K bar^{-1}, while T_c of $Y_2Fe_3Si_5$ increases rapidly (33×10^{-5}K bar^{-1}) and passes through a maximum whose value is twice that at ambient pressure. The results are compared with the effects of volume contraction across the series $(Y_{1-x}Lu_x)_2Fe_3Si_5$ and $(Y_{1-x}Dy_x)_2Fe_3Si_5$.

1. INTRODUCTION

The compounds of the series $RE_2Fe_3Si_5$ are superconducting or magnetic, depending on whether the element RE (Sc,Y,rare earth) has a magnetic moment or not [1,2], the members with Lu and Sc exhibiting the highest superconducting transition temperatures yet observed for compounds with an ordered iron sublattice.
These compounds crystallize in the primitive tetragonal $Sc_2Fe_3Si_5$-type structure [3] in which the iron occupies two sets of point positions. The iron atoms form clusters, in one set chains along the tetragonal c-direction and in the other squares parallel to the basal plane [4]. The iron atoms carry no magnetic moment [5,6].

Recently, the solid solutions between $RE_2Fe_3Si_5$ and the superconducting $Lu_2Fe_3Si_5$ were studied [7]. For magnetic rare earth metals RE, the degradation of T_c with RE concentration did not follow Abrikosov-Gorkov theory which previously has successfully explained the T_c-variation in similar systems of solid solutions between ternary rhodium borides $(Lu_{1-x}RE_x)Rh_4B_4$ [8]. It was found [7] that in the case of the iron silicides, magnetic and nonmagnetic RE atoms of similar atomic size led to similar T_c-degradation, indicating that the volume effect of alloying is significant.

A particularly clean way of examining volume effects on the superconducting critical temperature is the application of hydrostatic pressure. We present the results of a pressure study of the three superconductors in the iron silicide series, $Sc_2Fe_3Si_5$, $Y_2Fe_3Si_5$, and $Lu_2Fe_3Si_5$. The pressure results are compared with those obtained from solid solutions of $Y_2Fe_3Si_5$ with $Dy_2Fe_3Si_5$ and $Lu_2Fe_3Si_5$.

2. EXPERIMENTAL

The samples were prepared by arc melting stoichiometric amounts of the high purity elements (RE m3N, Fe m5N, Si m7N) in an ultra high purity argon atmosphere. The pseudoternary solid solutions $(Y_{1-x}RE_x)_2Fe_3Si_5$ were prepared in a two step process, first arc melting large master alloys $Y_2Fe_3Si_5$ and $(Y_{0.9}RE_{0.1})_2Fe_3Si_5$, homogenizing them at 1150°C for 4 days and combining them to synthesize the pseudoternary series by arc melting. All of the samples were homogenized at 1150°C for 4 days and subsequently annealed at 800°C for 8 days. The presence of the $Sc_2Fe_3Si_5$-type structure was confirmed by X-ray powder diffraction using Cu and Cr radiation. Photomicrographic examination served to establish sample homogeneity. The superconducting or magnetic transition temperatures were determined by low frequency (23 Hz) ac susceptibility measurements. Hydrostatic pressure was applied at room temperature using a one-to-one mixture of n-pentane and iso-amyl alcohol as the pressure transmitting medium. The desired pressure was sustained by a self-clamp technique. A superconducting Sn manometer was used to determine the pressure at low temperature [9]. Superconducting transition temperatures at zero pressure were found to be reproducible after pressure cycling.

3. RESULTS AND DISCUSSION

The superconducting critical temperatures of both $Lu_2Fe_3Si_5$ and $Sc_2Fe_3Si_5$ are rapidly depressed with increasing pressure (Fig. 1). In both cases, the T_c- degradation is non-linear. For the Lu compound, the pressure coefficient $|dT_c/dP|$ of the critical temperature increases with increasing pressure and becomes constant

[*] Present address: Département de Physique de la Matière Condensée, Université de Genève, 1211 Genève, Switzerland.

[$] Research supported by the National Science Foundation under contract NSF/DMR77-08469.

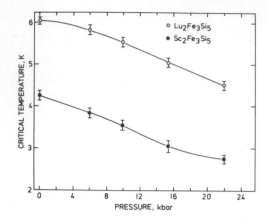

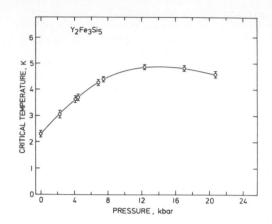

Fig. 1. The variation of T_C with pressure for
        $Lu_2Fe_3Si_5$ and $Sc_2Fe_3Si_5$. Error bars
        indicate transition widths.

Fig. 2. The variation of T_C with pressure for
        $Y_2Fe_3Si_5$.

above 8 kbar, while for the Sc compound, the
pressure coefficient is approximately constant
up to 12 kbar and decreases at higher pressures.
To extract values of the pressure coefficients
for a comparison with other compounds, we
estimate the slopes of the approximately linear
portions of the pressure dependence. Table 1
summarizes the compounds with the largest $|dT_C/dP|$
for selected binary and ternary systems.
The rates of depression for both iron silicides
are very large, equalled or exceeded only by
$AgSnSe_2$ [17] and $LaMo_6Se_8$ [12] . The behaviour of

$Y_2Fe_3Si_5$ under pressure is completely different
(Fig. 2). The critical temperature increases
smoothly from 2.3 K, reaches a maximum of almost
5 K at 15 kbar and decreases at higher pressures
reaching 4.6 K at 20 kbar. The initial rate of
increase of T_C is very high, 33×10^{-5} K bar^{-1},
while the rate of decrease above 18 kbar,
-6×10^{-5} K bar^{-1}, is comparable to the values
found for the Sc and Lu compounds.

The sensitivity of the superconducting $RE_2Fe_3Si_5$
-compounds to pressure is exceptionally high.
Values of $|dT_C/dP|$ exceeding 5×10^{-5}K bar^{-1} are
rare. For example, in the ternary rare earth
rhodium borides, the pressure coefficients are
below 2×10^{-5}K bar^{-1} [10]. The pressure coeffi-
ents of T_C for the iron silicides are similar in
magnitude to those of the binary La-chalcogenides [11] and Chevrel-phase compounds [12].
Some of the ternary Mo-chalcogenides (sulfides
with Cu,Zn and Cd, and selenides with Cu and Ag)
show a nonlinear variation of T_C with pressure
whose overall shape is similar to that observed
for $Y_2Fe_3Si_5$, with similar magnitude of the
initial dT_C/dP [12]. The non-linear pressure
dependence of T_C in these Mo-chalcogenides and
other high-T_C superconductors has been attributed to a crystallographic instability, which
may be temperature- or pressure-induced [10].
A pressure-induced lattice transformation
appears unlikely in the case of $Y_2Fe_3Si_5$, given
the smooth and continuous variation of T_C. The
variation of resistivity with temperature for
this compound shows no evidence for any crystallographic transformation [13], however, a
detailed low-temperature X-ray study has not
been conducted.

TABLE 1.

T_C at P=0 and dT_C/dP for selected compounds.

	T_C	dT_C/dP	Ref.
	K	10^{-5}K bar^{-1}	
Lu_2Fe_3Si	6.0	-10.2[a]	this work
$Sc_2Fe_3Si_5$	4.3	-7.7[a]	this work
$ThRh_4B_4$	4.5	-0.66	10
$LaMo_6S_8$	7.0	-14.2	12
$AgSnSe_2$	4.7	-7.3	17
La_3Se_4	7.6	$+30.0$[b]	11[c]
$Cu_{1.4}Mo_6Se_8$	5.9	$+50.0$[b]	12[c]
$Y_2Fe_3Si_5$	2.3	$+33.0$[b]	this work

[a] linear portion

[b] slope of initial increase

[c] data estimate from figure

The application of hydrostatic pressure leads
to changes in the crystalline properties, and,

in particular, to a reduction of the volume. There is no microscopic theory relating T_c to the effect of pressure, however, in view of the strong and non-linear variation in the iron silicides it appears reasonable to assume that electronic reasons are dominant, e.g. a variation of the density of states at the Fermi level under pressure.

The reduction of the unit cell volume can also be achieved by substituting one atomic species with an isoelectronic one of smaller atomic volume. The $Sc_2Fe_3Si_5$-type structure is, besides with Y, also found with the rare earth elements Sm and Gd through Lu [1]. We chose to substitute Y by Lu and Dy. With both substituents, complete series of solid solutions are formed. The pure effect of alloying will be seen with Lu while the magnetic exchange between the localized Dy moments and the conduction electrons should provide an additional reduction in T_c for this series of solid solutions. The unit cell volumes of $Y_2Fe_3Si_5$ and $Lu_2Fe_3Si_5$ differ by less than 4%, those of the Y and Dy compounds by less than 1% (Table 2). For both systems of solid solutions there is an initial increase in T_c which rises to maximum values of 3.0 K and 2.8 K for the Lu and Dy series, respectively, and then quickly falls off to temperatures below the measurement limits (Fig. 3). Suprising, at first sight, is the fact that the T_c is depressed more rapidly with the substitution of Lu than with the substitution of the magnetic Dy. Apparently, the alloying effect, which depends on the size differences between solvent (Y) and solute (Lu, Dy) atoms, has a much stronger influence on the T_c than has the magnetic exchange with the RE moment. This becomes evident from Fig. 4, where T_c is plotted vs the relative volume contraction. For the conversion of the T_c vs concentration dependence, we use "chemical compressibilities"

TABLE 2.

Lattice parameters of $RE_2Fe_3Si_5$-compounds. Estimated errors in parentheses.

RE	a $\overset{\circ}{A}$	c $\overset{\circ}{A}$	c/a	V $\overset{\circ}{A}^3$
Y	10.435(8)	5.475(8)	0.525	596
Lu	10.342(3)	5.385(3)	0.521	575
Dy	10.423(8)	5.465(8)	0.524	594

$(1/V)$ (dV/dx) of -4.1×10^{-3} and -33.9×10^{-3} for the Dy and Lu series, respectively, calculated from the measured lattice parameters. For the same volume change, the T_c's are lower and the T_c degradation is more rapid for the Dy series than for the Lu series. The difference between the two curves is attributed to the magnetism of the Dy ions. In the same diagram, the pressure data of Fig. 2 is plotted, assuming a compressibility of 0.5×10^{-6} bar^{-1} [14]. The initial rise of T_c is equal for both of the alloy series and within an order of magnitude compatible with the hydrostatic pressure result, suggesting that it is simply a manifestation of the volume contraction. It is clear, however, that for the alloy systems, effects other than volume change must contribute to the observed T_c-degradation. Impurity scattering may be the reason, smearing the density of states and contributing to a reduction of T_c. In addition, changes of the band structure with alloying appear possible. These effects begin to dominate that of the volume contraction at moderate values of volume change and prevent the maximum T_c in the solid solutions becoming as high as the peak seen in hydrostatic pressure experiments.

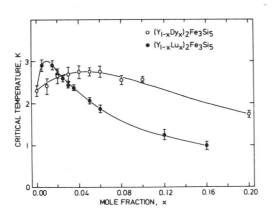

Fig. 3. T_c in the solid solutions $(Y_{1-x}Lu_x)_2$ Fe_3Si_5 and $(Y_{1-x}Dy_x)_2Fe_3Si_5$.

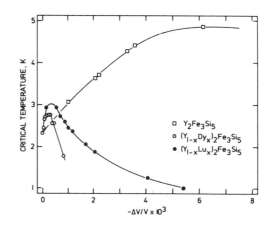

Fig. 4. The variation of T_c with relative volume change for hydrostatic pressure (■) and alloy systems (○ ●).

Why are the alloying effects in the iron silicides so much more dramatic than in the ternary
rhodium borides? We may attempt to understand
the differences in terms of the crystal structure. The RE atoms in the $Sc_2Fe_3Si_5$-type
structure form squares which are stacked along
the c-direction alternating with the Fe-squares
mentioned above to form Archimedean antiprisms.
Any substitution on the RE site will directly
influence the Fe-squares, subjecting them to
strains or distortions. If the Fe 3d-electrons
are mainly responsible for the superconductivity
of these compounds, clearly such strains may
adversely affect the superconducting properties.
The same argument works for the ternary rhodium
borides: There, the Rh atoms are clustered into
twodimensional sheets of tetrahedra [15,16]. The
RE site is spatially separated from the Rh
clusters. Substitutions on the RE site will
therefore cause changes in the Rh-Rh interactions which are expected to depend far less
critical on the size of the substituent atom
than is the case for the iron silicides.

In summary, the superconductors with $Sc_2Fe_3Si_5$-
type structure are extremely sensitive to the
effects of hydrostatic pressure and alloying.
The pressure coefficients of T_C are among the
highest values observed. $Y_2Fe_3Si_5$ shows a
dramatic initial increase of T_C which doubles
its value at 15 kbar before it decreases at
higher pressures. The T_C-variation in the series
of substitutional solid solutions is consistent
with the pressure results for small concentrations of the substituent, while at higher
concentrations factors other than volume change
become dominant.

REFERENCES

1   Braun, H.F., Physics Letters 75A (1980) 386

2   Braun, H.F., Acker, F., and Segre, C.U.,
    Bull. Am. Phys. Soc. 25 (1980) 232

3   Bodak, O.I., Kotur, B.Ya., Yarovets, V.I.,
    and Gladyshevskii, E.I., Sov. Phys.
    Crystallogr. 22 (1977) 217

4   Braun, H.F., in Shenoy, G.K., Dunlap, B.D.,
    and Fradin, F. (eds.), Ternary Superconduc-
    tors (Elsevier North Holland 1981) p. 225

5   Cashion, J.D., Shenoy, G.K., Niarchos, D.,
    Viccaro, P.J., and Falco, C.M., Physics
    Letters 79A (1980) 454

6   Cashion, J.D., Shenoy, G.K., Niarchos, D.,
    Viccaro, P.J., Aldred, A.T., and Falco, C.M.,
    J. Appl. Phys. 52 (1981) 2180

7   Braun, H.F., and Segre, C.U., Bull. Am. Phys.
    Soc. 26 (1981) 343

8   MacKay, H.B., Woolf, L.D., Maple, M.B., and
    Johnston, D.C., J. Low Temp. Phys. 41
    (1980) 639

9   Smith, T.F., Chu, C.W., and Maple, M.B.,
    Cryogenics 9 (1969) 53

10 Shelton, R.N., and Johnston, D.C., in Chu,
    C.W. and Woollam, J.A. (eds.) High Pressure
    and Low Temperature Physics (Plenum, New
    York, 1978) p. 409

11 Shelton, R.N., Moodenbaugh, A.R., Dernier,
    P.D., and Matthias, B.T., Mat. Res. Bull. 10
    (1975) 1111; Eiling, A., Ph.D. Thesis, Ruhr-
    Universität Bochum (1980)

12 Shelton, R.N., in Douglass, D.H. (ed.),
    Superconductivity in d- and f-Band Metals
    (Plenum, New-York, 1976) p. 137

13 Segre, C.U., unpublished results

14 Estimate based on Fe, Si, Fe_3Si : Landolt-
    Börnstein, New Series, VI/4 (Springer-Verlag
    Berlin, 1980)

15 Vandenberg, J.M., and Matthias, B.T., Proc.
    Natl. Acad. Sci. USA 74 (1977) 1336

16 Johnston, D.C., Solid State Commun. 24(1977)
    699

17 Shelton, R.N., Segre, C.U., and Johnston, D.
    C., Solid State Commun. (1981) in press

PHYSICS OF SOLIDS UNDER HIGH PRESSURE
J.S. Schilling, R.N. Shelton (editors)
© *North-Holland Publishing Company, 1981*

HIGH PRESSURE STUDIES OF THE SUPERCONDUCTIVITY OF LA-CHALCOGENIDES

A. Eiling, J.S. Schilling and H. Bach

Institut für Experimentalphysik IV
Ruhr Universität Bochum
D-4630 Bochum, Germany

Previous investigations by Shelton et al. of the pressure dependence of the superconducting transition temperature (T_c) of La_3S_4, La_3Se_4 and La_3Te_4 to 2 GPa are extended to 4.5 GPa hydrostatic pressure. La_3S_4 and La_3Se_4 exhibit a maximum in the $T_c(P)$-dependence at 2 GPa and 1 GPa, respectively. The initial volume dependence of T_c $(\partial \ln T_c/\partial \ln V = -22)$ is eight times larger for La_3S_4 and La_3Se_4 than for La-metal; it is, in fact, one of the largest known increases of any material under pressure.
To learn about the physical mechanism(s) responsible for this exceptional increase of T_c, we have measured on La_3Se_4 up to 1.5 GPa the pressure dependence of the Pauli-susceptibility χ_p, the temperature derivative of the upper critical field $\partial H_{c2}/\partial T$ and the residual resistivity ρ_{res}. Both χ_p and $\rho_{res}^{-1}(\partial H_{c2}/\partial T)$ yield information about the density of states at the Fermi energy. Our results indicate that the large enhancement of T_c at low pressures is due to a rapidly increasing density of states, while the decrease of T_c at higher pressures is caused by the stiffening of the lattice.

1. INTRODUCTION

Twenty-five years ago BCS theory (1) gave a complete explanation of the physical effect of superconductivity; nevertheless, theoretists and experimentalists are still actively engaged in unsolved problems concerning superconductivity. It is in general not yet possible to calculate reliable values of the superconducting transition temperature (T_c).(2) The object of the mutual stimulation between theory and experiments is not only to elucidate superconductivity as a physical effect, but also to learn how to produce new superconductors with better technical properties.(3) To explain the observed pressure dependences of superconducting elements and compounds there exist a number of theoretical approaches which calculate the variation of computable parameters under volume change. There is, however, a lack of experimental checks on the reliability of these calculations since measurements of microscopic quantities under pressure turn out to be very difficult.(4,5)

The almost universal reduction of T_c in single-phase s-p metals can be understood within a simple lattice model (5,6) where increasing pressure stiffens the lattice and shifts the phonon-spectra to higher energies. Since superconductivity is an effect arising from the interaction of electrons and phonons the stiffening of the lattice weakens the electron-phonon interaction and thus reduces the transition temperature.

Such a simple picture fails to describe the pressure dependence of T_c in a d-band metal or compound where $T_c(P)$ can be quite complicated, increasing or decreasing with pressure depending on the sample and pressure range.(7) Obviously in these materials the lattice stiffening alone cannot be responsible for the pressure dependence of T_c. Particularly the pressure dependence of superconductivity in lanthanum metal has received considerable attention through the years.(8,9) The T_c-value of fcc La-metal increases under pressure up to 17 GPa ($\equiv$170 kbar), an increase from 5.9K at ambient pressure to nearly 13K, making La the highest T_c element. Stimulated by this behavior the pressure dependence of La-compounds was investigated and the hypothesis was advanced that the La-La distance should be an important factor governing the pressure dependence of T_c.(10)

If one plots T_c of La versus the volume change, one recognizes that the volume dependence of T_c is not exceptionally large. Physically, the *volume* dependence is a more basic variation than the *pressure* dependence. The increase of the T_c of La with decreasing volume is comparable with the dependence of other early d-metals like V (see Figure 1), Zr (7) and Y (11). On the other hand, extraordinarily large increases of T_c with volume change were found by Shelton et al. (12) in the lanthanum compounds La_3S_4 and La_3Se_4, see Figures 1 and 2. The T_c-value of La_3Se_4 increases from initially 7.4K to 9.5K for 1 GPa pressure, decreasing at higher pressures. For La_3S_4 they found an enormous increase of T_c from 8.2K at ambient pressure to more than 11K at 2 GPa. The volume dependence of T_c for these La-chalcogenides is eight times larger than for La-metal. Previous quasihydrostatic measurements (12) on La_3S_4, La_3Se_4 and La_3Te_4 up to 17 GPa give for all three systems a monotonically decreasing $T_c(P)$ over the present pressure range (<4.5 GPa), indicating that $T_c(P)$ is sensitive to shear stresses and/or sample deformation.

A similar large increase of T_c has recently reportet by Segre and Braun (16) for $Y_2Fe_3Si_5$, the T_c-value increasing from 2.2K to 4.8K at 1.3 GPa pressure. The Chevrel-phase compund $CuMo_3S_4$ exhibits a similarly large enhancement of T_c from 10.5K to 11.7K at 0.6 GPa pressure.(17) The pressure increases of these two compound and the above mentioned La-compounds are the largest known for any substance. All the previous pressure measurements of $T_c(P)$ on these compounds are shown in Figure 2.

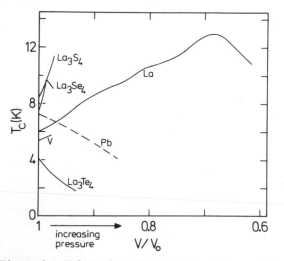

Figure 1.   Volume dependence of T_c of La-metal, $T_c(P)$ from (8,9), the $V(P)$-dependence above 12 GPa (13) was extrapolated by use of the Murnaghan formula. $T_c(V)$ of V from (7) and $T_c(V)$ of Pb from (14). Compared to these normal pressure dependences of T_c, the exceptional increase of T_c in La_3Se_4 and La_3Se_4 stands out. The $T_c(P)$-dependences of the three La-chalcogenides were measured by Shelton et al. (12), the $V(P)$-dependence was determined by bulk modulus data from ultrasonic measurements.(15) For La_3Te_4 we assumed a 30% increased compressibility compared to La_3S_4.

The above measurements on the lanthanum compounds raised the following questions: It would be interesting to know how much further T_c would increase at hydrostatic pressures *above* 2 GPa. Would it reach the temperature range of high temperature superconductors (∿20K)? Second, it would be of general interest to know what physical mechanism(s) is (are) responsible for the remarkably large increase of T_c with pressure in these compounds. Such a study would be pertinent to the more general question of how to construct systems with high T_c-values.

We are able to do pressure experiments under hydrostatic conditions up to 6 GPa (14,19), which extends considerably the pressure range of previous investigations. For P>2 GPa the $T_c(P)$ of La_3S_4 does not rise further, but passes through a maximum at T_c=12.5K for P≃2.1 GPa. Although it is disappointing that $T_c(P)$ didn't rise further,

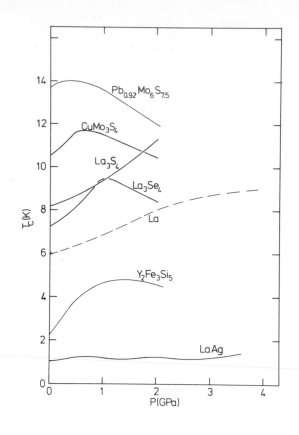

Figure 2.   Synopsis of the $T_c(P)$-dependence of those compounds which exhibit the largest known enhancements of T_c. For comparison we show La (8,9). Data of $Y_2Fe_3Si_5$ from Segre and Braun (16), La_3S_4 and La_3Se_4 from Shelton et al. (12), and $CuMo_3S_4$ and $Pb_{0.92}Mo_6S_{7.5}$ from Shelton et al. (17). LaAg (18) is shown as an example for a rather complicated pressure dependence, $T_c(P)$ having two minima and two maxima up to 3.6 GPa.

the systematics of $T_c(P)$ of the three La-chalcogenides La_3S_4, La_3Se_4 and La_3Te_4 is interesting in itself. To search for mechanisms responsible for the rapid T_c-increase, measurements of the electronic and phonon properties under pressure would be useful. One important quantity in superconductivity is the bare density of states at the Fermi energy $N(E_f)$ which is usually determined from the low temperature specific heat. Because of the relatively low accuracy of specific heat measurements under hydrostatic pressure (4), we chose two alternate approaches to determine the variation of $N(E_f)$ with pressure:
a) The initial temperature derivative of the upper critical field $\partial H_{c2}/\partial T$ and the residual resistivity ρ_{res} yield via the 'coherence length formula' the density of states.(20,21)
b) The Pauli-susceptibilty is also directly related to $N(E_f)$.

2. EXPERIMENTAL

Exact four-point measurements up to 2 GPa are carried out in a piston-cylinder apparatus.(22) To achieve higher pressures we use a metal-gasket technique.(19) In the metal gasket cell, shown in Figure 3, resistivity measurements can be carried out up to 10 GPa hydrostatic pressure at room-temperature and up to 6 GPa at helium temperatures. It is alternatively possible to measure the AC-susceptibility instead of the resistivity in the same pressure range.(14)

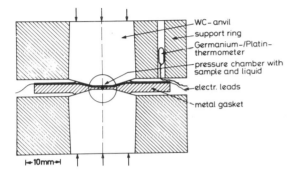

Figure 3. Hydrostatic pressure cell for four-point electrical resistivity measurements to 10 GPa. Opposing WC-anvils press into metal gasket with pressure chamber containing pressure fluid, sample, and Pb-manometer.

The magnetic susceptibility and the upper critical field $H_{c2}(T)$ are measured in a Faraday balance system (23), shown in Figure 4. In this apparatus a small pressure clamp (1.8cm dia., 5cm length, 100g total weight) is in a magnetic field of constant (H·dH/dz)-value i.e. constant force over the volume of the clamp. The balance system spans the temperature range 300-2.5K with 70kGauss (7 Tesla) maximum field at hydrostatic pressures up to 1.5 GPa.

The pressure is determined by use of a Pb-manometer. In the temperature range 300-40K the pressure is determined by the resistivity of the Pb-sample at the given temperature; at helium temperatures the pressure is determined by the value of the superconducting transition temperature $T_c(P)$ of Pb.(14) See Ref. 14 for a more detailed description.

All samples are single crystals produced in the same way. The respective chalcogen element (5N purity) and lanthanum sponge (4N purity) are mixed in stoichiometric quantities in a quartz ampule. This is slowly heated over a period of three days to 400 °C to complete the reaction. Then the temperature is slowly increased from 400 °C to 700 °C over a further two-day period to homogenize the compound. The resulting large-grained powder is compressed into tablets. These

are then placed inside a degased molybdenum crucible and sealed under a pressure of 250 mbar of argon. To homogenize the charge further, the crucible is heated for one hour at 2060 °C for La_3S_4, 1850 °C for La_3Se_4, and 1700 °C for La_3Te_4, in an induction furnace. To grow crystals by the Bridgman-Stockberger method, the crucible is lowered out through the temperature gradient at a speed of 1.5mm/h. The samples are analysed by x-rays and microprobe.

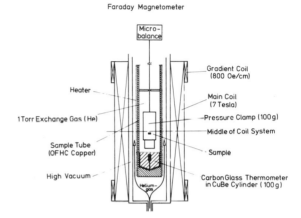

Figure 4. Faraday magnetometer for high pressure studies. A superconducting magnet generates both main and gradient fields.

3. RESULTS

3.1. $T_c(P)$-measurements

a) La_3S_4

As seen in Figure 5, by extending the previous pressure experiments (12) to pressures higher than 2 GPa, we find that $T_c(P)$ passes at approximately 2.2 GPa through a maximum and decreases monotonically at higher pressures. Thus, we confirm the result of Shelton et al. (12) that T_c of La_3S_4 is strongly enhanced by hydrostatic pressure, but, unfortunately, we find that T_c reaches its maximum value of 12.5K at moderate pressure. The T_c-value is limited to almost the same value La-metal reaches at 17 GPa.

Between the three measured La_3S_4 samples there exist minor differences in the T_c-value at ambient pressure (8.2K - 8.3K), in the slope of the increasing part of the $T_c(P)$-curve, and also in the maximum T_c-value under pressure. The steepest increase of T_c we find in a sample where T_c reaches 12.5K at 1.9 GPa, starting from 8.3K at ambient pressure. The maximum slope of this increase is $dT_c/dP = +3.2K/GPa$. At pressures above the respective maximum in the $T_c(P)$-dependence the $T_c(P)$-curve decreases with $dT_c/dP \simeq -0.8K/GPa$ for all samples (Figure 5).

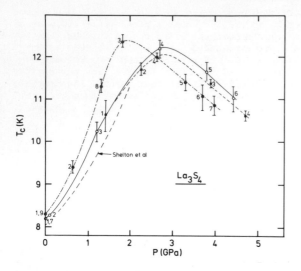

Figure 5. Pressure dependence of T_c for La_3S_4 (measurements on three single-crystal samples). The error bars correspond to the $\Delta_{90\%}^{10\%}$ width of the superconducting transitions. The full dots and squares (● , ■) are transitions measured resistively. The open circles (O) are from AC-susceptibility measurements. The solid (———) and dash-dot (—·—·—) lines are drawn for clarity. The numbers give the order of measurement. The dashed line (————) is the result of a previous measurement on a polycrystalline sample.(12)

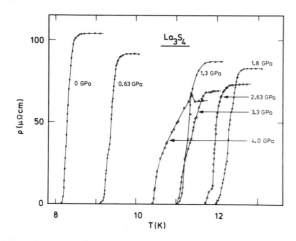

Figure 6. Resistive-measured superconducting transitions of La_3S_4 at different pressures.

Figure 6 shows the resistive-measured superconducting transitions of La_3S_4 at different pressures. With increasing pressure the transitions become broadened; this broadening arises

possibly from shear stresses in the solidified pressure transmitting fluid. The superconducting transitions of the Pb-sample stayed unbroadened up to the highest achieved pressures, indicating good pressure homogeneity. The measured pressure variation in the cell during cooling in the low temperature range, where the pressure transmitting fluid is solidified, supports this point of view.

b) La_3Se_4

In agreement with previous measurements (12) we find $T_c(P)$ for La_3Se_4 to pass through a maximum at approximately 1 GPa, as seen in Figure 7. We observe differences in the pressure dependence of samples with different T_c-values at ambient pressure (7.2K - 7.9K). We find that a higher T_c-value at ambient pressure is associated with a somewhat weaker pressure dependence (Figures 7 and 12).

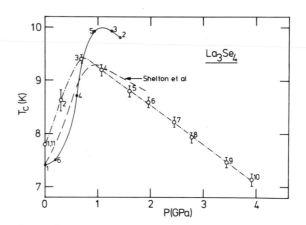

Figure 7. Pressure dependence of T_c for La_3Se_4. The open circles (O) are data from AC-susceptibility measurements, the squares (■) are from upper critical field data. Numbers give order of measurement. The solid (———) and dash-dot (—·—·—) lines are for clarity. The dashed line (————) is from previous measurements.(12)

Microprobe results indicate that La-atoms in La_3Se_4 can occupy intermediate lattice sites extending the stoichiometric portion of La-atoms. The La-atoms on the intermediate sites increase the T_c-value but apparently weaken the pressure dependence of T_c.

The sample with the largest pressure dependence of T_c exhibits a maximum slope of $dT_c/dP = +4.5K/GPa$, whereas in the decreasing part of the $T_c(P)$-curves, T_c decreases monotonically with $dT_c/dP \simeq -0.8K/GPa$ for all La_3Se_4 samples.

c) La_3Te_4

The $T_c(P)$-dependence of La_3Te_4 shows up to 4.5 GPa no notable structure (Figure 8), $T_c(P)$ falling monotonically with positive curvature. The initial slope of the $T_c(P)$-decrease is the same as we find for La_3S_4 and La_3Se_4 in the decreasing part of the respective $T_c(P)$-curves, namely $dT_c/dP \simeq -0.8K/GPa$. At 4 GPa the slope has diminished to $dT_c/dP \simeq -0.3K/GPa$.

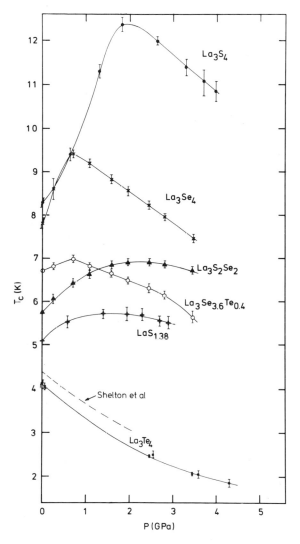

Figure 8. Synopsis of the pressure dependence of T_c for all measured La-chalcogenides. For clarity we plot only one run for La_3S_4 and La_3Se_4. The broken line represents a previous measurement on La_3Te_4.(12) All solid lines are drawn for clarity.

d) Other La_3X_4 compounds

In order to test if the La-La separation alone is an important parameter (10) for the value of T_c, we carried out measurements on $La_3X_{4-x}Y_x$ compounds, where chalcogen-atom Y is substituted for chalcogen-atom X. In this way the lattice constant can be varied linearly, since the lattice constant of these crystals with a mixture of chalcogen-atoms follows Vegards law. Such crystals also have sharp x-ray diffraction patterns and sharp superconducting transitions. Besides alloying it is possible to leave La-sites in the La-sublattice of these La-chalcogenides empty (24) and to vary in this way the La-La distance in a statistical manner. However, removing La-atoms also reduces the number of conduction-electrons.(25)

Such experiments revealed that the superconducting transition temperature is always lowered by alloying or doping with La-vacancies. The pressure dependence of $T_c(P)$ for all samples is smeared and broadened (Figure 8) compared to the pressure dependence of the stoichiometric La-chalcogenides. But for all alloyed and with vacancies doped samples a weak maximum in the $T_c(P)$-dependence is retained.

Because the pressure dependence of T_c of the mixed La-chalcogenides doesn't interpolate between the stoichiometric La-chalcogenides, the La-La separation is certainly not the sole parameter determining the value of T_c in these compounds!

The present determination of the pressure dependence of T_c indicates a systematic behavior in the three La-chalcogenides. La_3S_4 and La_3Se_4 exhibit a maximum in the $T_c(P)$-dependence at 2.1 and 1.0 GPa, respectively; the maximal value of $T_c(P)$ is of similar magnitude in La_3S_4, La_3Se_4, and La. In the decreasing part of the $T_c(P)$-curves we find the same slope dT_c/dP for La_3S_4 and La_3Se_4, and initially, La_3Te_4.

3.2. Structural phase transition and pressure dependence of resistivity.

The compounds La_3S_4 and La_3Se_4 exhibit at ambient pressure a structural phase transition from a cubic to a tetragonally distorted unit cell.(26) In the resistivity curve the phase transition is marked by a jump at the transition temperature T_m (La_3S_4: $T_m=100K$ and La_3Se_4: $T_m=65K$), see Figure 9. Note that the compound with the higher T_c-value exhibits the stronger negative curvature of the resistivity. A possible connection between the curvature and the T_c-value is found for a number of superconducting materials.(27)

Under pressure the resistivity decreases reversibly at all temperatures, as shown for La_3Se_4 in Figure 10. The decrease is relatively large and amounts to $d\rho/dP \simeq -20$ $\mu\Omega cm/GPa$ at 300K.

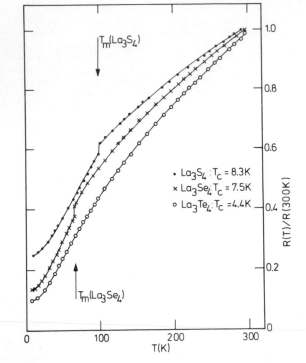

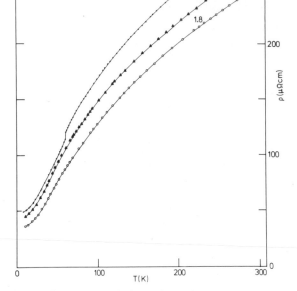

Figure 9.   Relative resistance of La$_3$S$_4$ (●),
La$_3$Se$_4$ (✗) and La$_3$Te$_4$ (○) versus temperature.
The solid lines are drawn for clarity. The structural phase transition of La$_3$S$_4$ and La$_3$Se$_4$ is
marked by an arrow. The resistivity values at
300K are: La$_3$S$_4$ 350μΩcm, La$_3$Se$_4$ 360μΩcm and
La$_3$Te$_4$ 250μΩcm.

Figure 10.   Resistivity of La$_3$Se$_4$ versus temperature at various pressures. The indicated pressure values are determined at 300K. Upon cooling
down to helium temperature, the pressure decreases by ∿0.3 GPa. The solid lines are drawn
for clarity.

The phase transition, which is clearly seen in
the resistivity at ambient pressure, disappears
in the resistivity even at moderate pressure.
For that reason it was not possible to determine
the pressure dependence of T_m. Since the structural phase transition appears to disappear in
a pressure range where T_c is far below the maximum T_c-value, there would appear to be no strong
correlation between T_c and T_m.

The pressure dependence of the resistivity itself
is smooth over the entire pressure range up to
the highest achieved pressure and shows no correlation to $T_c(P)$, see also Figure 11.

3.3. Measurement of the pressure dependence of
$N(E_f)$

a) Determination of $N(E_f)(P)$ from upper critical
field and residual resistivity.

We determined the pressure dependence of the
electron-phonon enhanced density of states
$N^*(E_f)=N(E_f)(1+\lambda)$ by measuring $(\partial H_{c2}/\partial T)\big|_{T_c}$ and
ρ_{res}. These two quantities are connected by the
'coherence length formula' which can be derived
from simple thermodynamic relations. The full
formula regards strong coupling corrections and
interpolates also between the 'clean' ($\ell > \xi_o$)
and the 'dirty limit' ($\ell < \xi_o$), where ℓ is the
mean free path of the superconducting electrons
and ξ_o is the BCS coherence length. This formula
is given by (21)

$$-\left(\frac{\partial H_{c2}}{\partial T}\right)_{T_c} (R(\lambda_{tr})) = (9.55\cdot10^{24}\ \gamma^2\ T_c\,(n^{2/3}\,(S/S_f))^{-2}$$

$$+5.26\cdot10^4\ \gamma\ \rho_{res})\,kG/K \quad (1)$$

where γ is the electronic specific heat coefficient in erg/cm^{3}K^2, n the number of conduction

electrons in cm^{-3}, ρ_{res} the residual resistivity in Ωcm, and S/S_f the ratio of the Fermi-surface area to the Fermi-surface of free electrons. $R(\lambda_{tr})$ is a function of the ration ℓ/ξ_o.(28)

The most reliable estimate for $N^*(E_f)$ is obtained in the 'dirty limit' ($\ell < \xi_o$); in this limit the Ginzburg-Landau parameter is dominated by ℓ.(20) In the 'dirty limit' the agreement between specific heat results and those determined by the 'coherence length formula' is reasonably good (21), the deviation being $\sim 10\%$. In the 'clean' and 'intermediate' ranges the estimate becomes somewhat less reliable as one has to know the effective Fermi-surface area. Corrections connected to this parameter S can be large, a factor of 2 or even greater.(21)

We chose to determine the variation of the density of states on a La_3Se_4 sample because the whole interesting range of this sample, including the maximum in $T_c(P)$, occurs within the pressure range of the clamp used ($P \leq 1.5$ GPa).

Before analysing the La_3Se_4 data, we first consider measurements on La_3S_4, as for this compound a full set of data (i.e. $\partial H_{c2}/\partial T$, residual resistivity and specific heat) exists at ambient pressure.(29) For $S/S_f=1$ we get identical results for the density of states measured both by specific heat and by $\partial H_{c2}/\partial T$ and ρ_{res}. Also setting $S/S_f=1$, we calculate the density of states from our data on La_3Se_4 at ambient pressure. Our result for the $N^*(E_f)$ of La_3Se_4 agrees within experimental error ($\sim 15\%$) with the literature value (30), determined by specific heat measurement.

The results of the measurement of the upper critical field derivative $(\partial H_{c2}/\partial T)|_{T_c}$ and of the residual resistivity are presented in Figure 11. The temperature derivative of the upper critical field increases under pressure rapidly, passing at 1 GPa through a maximum just as $T_c(P)$ does. The residual resistivity decreases under pressure and shows no exceptional behavior at any pressure.

We also measured a sample which had a large amount of impurities where the residual resistivity had the high value of 400$\mu\Omega$cm. Up to 1 GPa the T_c-value showed *no pressure dependence*. Measuring the temperature derivative of the upper critical field, it was found to behave similarly to $T_c(P)$ (Figure 12), as $\partial H_{c2}/\partial T$ decreases only slightly with pressure.

This establishes in a convincing way the connection between the pressure dependence of T_c and that of the critical field, which is in this case almost identical with the pressure dependence of the density of states.

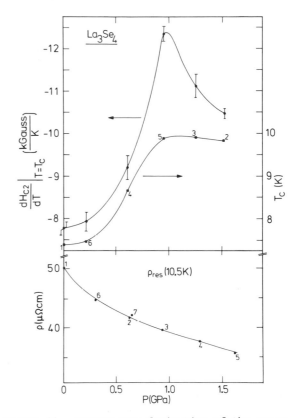

Figure 11. Temperature derivative of the upper critical field $\partial H_{c2}/\partial T|_{T_c}$, superconducting transition temperature T_c, and residual resistivity ρ_{res} of La_3Se_4 versus pressure. The numbers mark the measuring sequence, the solid lines are drawn for clarity. The absolute error in ρ_{res} is $\sim 30\%$ of the value, the relative error is $\pm 1\mu\Omega$cm. The error bars in $\partial H_{c2}/\partial T$ are due to a least square fit to the $H_{c2}(T)$-data. $T_c(P)$ from $T(H_{c2}-0)=T_c$.

b) Determination of $N(E_f)$ from DC-susceptibility data

An independent method to estimate the variation of the density of states under pressure is to measure the pressure dependence of the Pauli-susceptibility. Unfortunately, it is somewhat more complicated to determine $N(E_f)$ from susceptibility measurements than from the specific heat, because the susceptibility is a sum of three separate contributions

$$\chi_{exp} = \chi_{ce} + \chi_{dia} + \chi_{pi}. \qquad (2)$$

The total experimental measured susceptibility χ_{exp} includes the contributions due to the conduction electrons χ_{ce}, the ionic diamagnetism χ_{dia} and the susceptibility of the paramagnetic impurities χ_{pi}. To determine χ_{ce} one has to subtract off χ_{dia} and χ_{pi} from χ_{exp}.

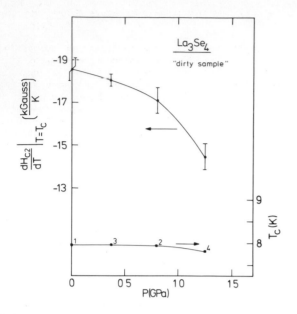

Figure 12. $(\partial H_{c2}/\partial T)|_{T_c}$ and T_c of a 'dirty sample' (residual resistivity 400μΩcm) versus pressure.

At low temperatures the effect of the paramagnetic impurities becomes larger, because it increases proportional to C/T, where C is the Curie-constant. We determined the Curie-constant in the low temperature range (6-10K) by assuming that the other contributions to the total susceptibility, namely χ_{ce} and χ_{dia}, are constant in this temperature range. By determining C, we obtain the temperature dependence of χ_{pi} (see Figure 13). The paramagnetic impurities are mainly rare earth ions contained in the lanthanum. From the low temperature increase of χ_{exp} we determine C=4.65·10⁻⁷emuK/g, which corresponds to ∿100ppm Ce.

The diamagnetic susceptibility was estimated from the values of La^{3+} and Se^{2-} (31), giving a temperature independent contribution to the total susceptibility of -3.42·10⁻⁷emu/g.

The conduction electron susceptibility itself consists of essentially three parts (32),

$$\chi_{ce} = \chi_P + \chi_{Ldia} + \chi_{orb}, \qquad (3)$$

the Pauli-susceptibility χ_P, the Landau diamagnetism of the conduction electrons (this is negligible in bands with large effective electron-masses), and orbital contributions χ_{orb}, which become important in narrow bands.

A susceptibility measurement at ambient pressure with the preceeding analysis is shown in Figure 13. The structural phase transition at 65K is marked by a decrease in the susceptibility at the transition temperature. The strong temperature dependence of the susceptibility is

indicative for the Fermi energy being positioned in a steep flank of a peak in the density of states curve, as χ_{dia} and χ_{orb} are essentially temperature independent.(33)

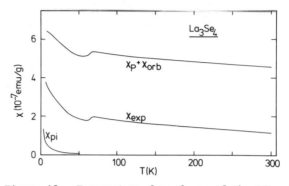

Figure 13. Temperature dependence of the DC-susceptibility of La_3Se_4. The drawing shows the different contributions to the total susceptibility. Subtraction off the diamagnetic susceptibility of the inner electrons χ_{dia}=-3.42·10⁻⁷emu/g and the susceptibility of the paramagnetic impurities χ_{pi} gives the electron-susceptibility $\chi_{ce}=\chi_{orb}+\chi_P$.

In order to extract the Pauli-susceptibility one has to subtract from χ_{ce} the orbital contributions. χ_{Ldia} can be regarded small because the high density of states at E_f indicates a narrow band, as shown by previous measurements.(30). Without any knowledge of χ_{orb}, we set it equal to a value, which adjusts χ_P to give the correct specific heat value of $N^*(E_f)$. At 300K we had to subtract 2.60·10⁻⁷emu/g or 57% of the conduction electron susceptibility (χ_{ce}).

Because of the influence of magnetic impurities in the pressure clamp, which causes a strong paramagnetic increase in the susceptibility at low temperatures (T<50K), we could measure the susceptibility of the sample with the necessary accuracy on ly at room temperature. These measurements at room temperature indicate only a slight increase in the sample susceptibility (Figure 14); the corresponding increase, calculated from

$$N(E_f) = \frac{\chi_P}{\mu_B^2} \qquad (4)$$

where μ_B is the Bohr magneton, is plotted in the same drawing. The shown increase of $N(E_f)$ (Figure 14) at 300K derived from susceptibility measurements is probably an upper bound of the increase of $N(E_f)$ at this temperature.

It is certainly possible that the pressure dependence of $N(E_f)$ is much larger at helium temperatures, as it is found for other compounds which show temperature-dependent susceptibility.(34) Therefore this result from the susceptibility is less relevant to the superconductivity than the measured variation of $N^*(E_f)$ by $(\partial H_{c2}/\partial T)$ and ρ_{res}, which determines $N^*(E_f)$ at low temperatures.

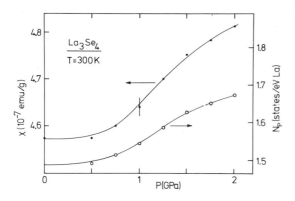

Figure 14. Pressure dependence of electron-susceptibility at 300K (left scale), the respective variation of the density of states (right scale). The indicated error corresponds to our measuring accuracy of ±20μg. The measurement was done on a single-crystal sample of 380mg total weight.

4. ANALYSIS AND DISCUSSION

4.1. Present theory and experimental results

Our analysis of measured quantities and their pressure dependence relative to superconductivity will be done in the framework of the McMillan theory.(35) The pressure (volume) dependence of T_C will be discussed using the McMillan equation

$$T_C = \frac{\Theta_D}{1.45} \exp\left(-\frac{1.04(1+\lambda)}{\lambda - \mu^*(1+0.62\lambda)}\right), \quad (5)$$

where Θ_D is the Debye temperature, μ^* is the effective electron-electron interaction, and λ is the electron-phonon coupling constant, which can be written

$$\lambda = \frac{N(E_f)\langle I^2\rangle}{M\langle\omega^2\rangle} = \frac{\eta}{M\langle\omega^2\rangle}, \quad (6)$$

where η is the McMillan-Hopfield parameter, which is a purely electronic quantity, whereas the denominator is mainly a phonon term.(36) $\langle I^2\rangle$ is the average of the squared electron-phonon matrix-element, M is the ionic mass, and $\langle\omega^2\rangle$ an averaged phonon frequency which can be approximated by (7)

$$\langle\omega^2\rangle = (0.8\Theta_D)^2. \quad (7)$$

To increase T_C, λ has to increase; this is the case if either $N(E_f)$ or $\langle I^2\rangle$ increase, or if $\langle\omega^2\rangle$ decreases.(37)

$N(E_f)$ is directly measured in our experiment. We approximate the pressure dependence of $\langle\omega^2\rangle$ by the relation (38)

$$\partial\ln\langle\omega^2\rangle/\partial\ln V = -2\gamma_{Gr}, \quad (8)$$

where γ_{Gr} is the high-temperature Grüneisen constant.

We now attempt to make a quantitative comparison between the measured pressure dependence of $N^*(E_f)$ and $\langle\omega^2\rangle$ with that of T_C. The pressure dependece of $N^*(E_f)$ is plotted in Figure 15. The actual value of $N^*(E_f)$ is calculated by use of the data from Figure 11 and equation 1. Because of the weak pressure dependence of the residual resistivity, $N^*(E_f)$ follows essentially the behavior of $\partial H_{c2}/\partial T$. Like $\partial H_{c2}/\partial T$, the density of states has a pronounced maximum at the same pressure as $T_C(P)$.

The absolute error in $N^*(E_f)$ is ~15% and is mainly due to the uncertainty in ρ_{res} at ambient pressure. The relative accuracy of the pressure dependence of ρ_{res} is much better so that the shape of the $N^*(E_f)$-curve is accurately determined.

The obvious similarity of $T_C(P)$ and $N^*(E_f)(P)$ under pressure is a clear indication that the initial variation of $T_C(P)$ is determined principally by the density of states. This conclusion is also supported by our results on the 'dirty sample' which is certainly in the 'dirty limit'; for this sample both $T_C(P)$ and $N^*(E_f)$ (or $\partial H_{c2}/\partial T$) had very little pressure dependence (see Figure 12).

The pressure dependence of $1/\langle\omega^2\rangle$ (Figure 15) is calculated using equation 8. Although the Grüneisen parameter was only measured for La_3S_4 (39), it seems reasonable to use the same value for La_3Se_4 because of the similarity of the elastic constants of La_3S_4 (15) and La_3Se_4 (40) (γ_{Gr}= +2 ±0.02). Using this Grüneisen parameter we obtain the pressure dependence of $1/\langle\omega^2\rangle$ shown in Figure 15.

Using equation 7 to estimate $\Theta_D(P)$ and assuming μ^*=0.1 and pressure independent, one can calculate from the McMillan formula (equation 5) the pressure dependence of the electron-phonon coupling constant $\lambda(P)$, also shown in Figure 15.

From the electron-phonon coupling constant $\lambda(P)$ and $1/\langle\omega^2\rangle(P)$ one can determine the value of the McMillan-Hopfield parameter by

$$\eta = \lambda M\langle\omega^2\rangle = N(E_f)\langle I^2\rangle. \quad (9)$$

The $\eta(P)$-curve (see Figure 16), calculated by equation 9, increases rapidly up to that pressure, where $T_C(P)$ passes through the maximum value, and then levels off, seeming to saturate. While the increase in η corresponds to the increase of λ the flat dependence at higher pressures means physically that the $T_C(P)$ dependence in this region is almost completely determined by the lattice stiffening. Any pressure dependence of $\langle I^2\rangle$ would also be mirrored in $\eta(P)$. Our results indicate that the initial increase of $N^*(E_f)$ is large enough to account alone for the required initial increase in η:

$\partial \ln \eta / \partial \ln V = -17.5$ and $\partial \ln N(E_f)/\partial \ln V = -23 \pm 5$. Therefore within the present experimental accuracy, $<I^2>$ appears to be pressure independent over the pressure range studied.

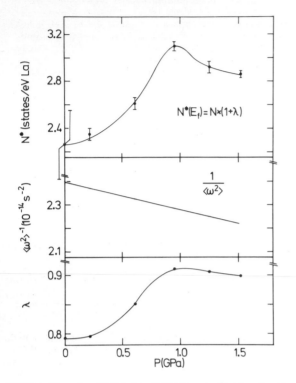

Figure 15. 1. Pressure dependence of the density of states at E_f, $N^* = N^*(E_f) = N(E_f)(1+\lambda)$, as calculated from the data in Figure 11 using equation 1. The error bar at zero pressure indicates the absolute error in $N^*(E_f)$ due to the uncertainty in ρ_{res}; the other error bars give the relative errors.
2. Pressure dependence of $1/<\omega^2>$ determined by equation 8 ($\theta_D = 195K$ Debye temperature at zero pressure (30))
3. Pressure dependence of λ calculated from $T_C(P)$-data of Figure 11 and pressure dependence of $1/<\omega^2>$ (see above) using the McMillan formula (equation 5).

4.2. Simple Model

We now discuss a simple model to account for the observed pressure dependence of T_C in the lanthanum chalcogenides. As there exists at present no bandstructure calculations on these compounds, we start with a density of states curve indicated by previous experiments (25), as seen in Figure 17.

For La_3S_4 and La_3Se_4 the Fermi energy is placed at the beginning of the steep increase of the d-band. The Fermi energy of La_3Te_4 is placed below the d-band which is consistent with the lower density of states $N^*(E_f)$ for La_3Te_4, derived

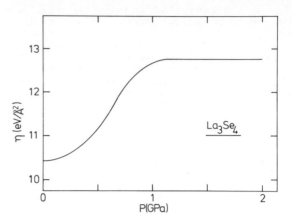

Figure 16. McMillan-Hopfield parameter of La_3Se_4 versus pressure. The $\eta(P)$-curve (solid line) was calculated from equation 9 using the $T_C(P)$-values of Figure 11 and the $1/<\omega^2>$ values from Figure 15.

from specific heat measurements.(30) The larger Te-atoms expand the lattice; this negative pressure to the La-atoms should shift the bottom of the s-band to lower energies more rapidly than the bottom of the d-band (41), thus pulling E_f out of the d-peak. This negative pressure effect would be somewhat canceled by the decreasing electro-negativity from S to Te. The decrease in electro-negativity should leave more electrons in the La-bands.

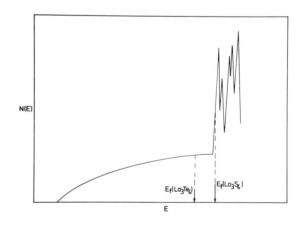

Figure 17. Hypothetical density of states curve of the La-chalcogenides. For La_3S_4 and La_3Te_4 the assumed position of the Fermi energy is indicated. The Fermi energy of La_3Se_4 should be located at almost the same position as that of La_3S_4.

On the other hand, *positive* pressure should cause the reverse effect, namely, that electrons are transferred from the s- to the d-band.(41) For La_3S_4 and La_3Se_4 this would mean that E_f

moves up the steep flank in $N(E)$, leading to the measured rapid initial increase of both $N(E_f)$ and T_c.

At pressures above the respective maximum in T_c, the increase of the density of states saturates, possibly due to the actual shape of the density of states curve or because the bands broaden under pressure. Once $N(E_f)(P)$ has saturated, the pressure dependence of T_c is apparently determined predominantly by the lattice stiffening.

La-vacancies or the substitution of other chalcogen atoms cause potential fluctuations and thus a broadening of the density of states curve. This would account for the lower T_c-values of the mixed compounds at ambient pressure and their more washed-out pressure dependence.

From our simple model we would predict that the T_c-value of La_3Te_4 should run through a minimum under pressure. As long as E_f is located in the s-band, which exhibits only a weak pressure dependence of $N(E_f)$, the phonons determine the pressure dependence of T_c; when E_f reaches the d-band edge the density of states increases and T_c increases. The measured decrease in the slope of dT_c/dP (Figure 8) indicates a minimum might be coming; indeed, Shelton et al. (12) observed a minimum in the T_c of La_3Te_4 at high quasihydrostatic pressures ($\sim$10 GPa).

In conclusion, these experiments give strong evidence for the importance of the value of $N(E_f)$ for superconductivity in La-chalcogenides. One can speculate that the observed large pressure effects in $CuMo_3S_4$ (17) and in $Y_2Fe_3Si_5$ (16) are caused by a similar mechanism.

ACKNOWLEDGEMENT

We would like to thank B.T. Matthias, S. Methfessel, R.N. Shelton and K. Westerholt for stimulating and very useful discussions, and P. Stauche and S. Erdt for their technical assistance in sample preparation and sample analysis. This work was supported in part by the Deutsche Forschungsgemeinschaft.

REFERENCES

(1) Bardeen, J., Cooper, L.N., and Schrieffer, J.R., Phys. Rev. 108 (1957) 1175.

(2) Glötzel, D., Rainer, D., and Schober, H.R., Zeitschr. Phys. B 35 (1979) 317.

(3) Applied Superconductivity, in Newhouse, V.L. (ed.), Vol. 2, (Academic Press, New York, 1975).

(4) Ho, J.C., Phillips, N.E., Smith, T.F., Phys. Rev. Lett. 17 (1966) 694.

(5) Zavaritskii, N.V., Itskevich, E.S., and Voronovskii, A.N., Soviet Phys. JETP 33 (1971) 762.

(6) Trofimenkoff, P.N., and Carbotte, J.P., Phys. Rev. B 1 (1970) 1136.

(7) Garland, J.W., and Bennemann, K.H., in Douglas, D.H. (ed.), Superconductivity in d- and f-Band Metals (American Institute of Physics, New York, 1972), p. 255.

(8) Balster, H., and Wittig, J., J. Low Temp. Phys. 21 (1975) 377.

(9) Wittig, J., Phys. Rev. Lett. 46 (1981) 1431.

(10) Smith, T.F., and Luo, H.L., J. Phys. Chem. Solids 28 (1967) 569.

(11) Wittig, J., this conf.

(12) Shelton, R.N., Moodenbaugh, A.R., Dernier, P.D., and Matthias, B.T., Mat. Res. Bull. 10 (1975) 1111.

(13) Syassen, K., and Holzapfel, W.B., Solid State Commun. 16 (1975) 533.

(14) Eiling, A., and Schilling, J.S., J. Phys. F 11 (1981) 623.

(15) Ford, P.J., Lambson, W.A., Miller, G.J., Saunders, G.A., Bach, H., and Methfessel, S., J. Phys. C Solid State Phys. 13 (1980) L697.

(16) Segre, C.U., and Braun, H.F., this conf.

(17) Shelton R.N., Lawson, A.C., Johnston, D.C., Mat. Res. Bull. 10 (1975) 297.

(18) Schilling, J.S., Methfessel, S., and Shelton, R.N., Solid State Commun. 24 (1977) 659.

(19) Fasol, G., and Schilling, J.S., Rev. Sci. Instrum. 49 (1978) 1722.

(20) Wiesmann, H., Gurvitch, M., Gosh, A.K., Lutz, H., Kammerer, O.F., and Strongin, M., Phys. Rev. B 17 (1978) 122.

(21) Orlando, T.F., McNiff, E.J., Foner, S., and Beasley, M.R., Phys. Rev. B 19 (1979) 4545.

(22) Peukert, H., Diplom Thesis, Universität Bochum (March 1980).

(23) Gerhardt, W., Diplom Thesis, Universität Bochum (March 1981).

(24) Methfessel, S., and Mattis, D.C., in Flügge, S. (ed.), Handbuch der Physik XVIII/1 (Springer, Berlin, 1968), p. 389.

(25) Westerholt, K., Bach, H., Wendemuth, R., and Methfessel, S., J. Phys. F 10 (1980) 2459.

(26) Dernier, P.D., Bucher, E., and Longinotti, L.D., J. Solid State Chem. 15 (1975) 203.

(27) Fisk, Z., Lawson, A.C., Solid State Commun. 13 (1973) 277.

(28) Werthamer, N.R., in Parks, R.D. (ed.), Superconductivity (Marcel Dekker, New York, 1969), p.321.

(29) Westerholt, K., Bach, H., Wendemuth, R., and Methfessel, S., Solid State Commun. 31 (1979) 961.

(30) Bucher, E., Andres, K., diSalvo, E.J., Maita, J.P., Gossard, A.C., Cooper, A.S., and Hull, G.W., Phys. Rev. B 11 (1975) 500.

(31) Landolt-Börnstein, Eucken, A. (ed.), Ionen und Atome 1 (Springer, Berlin, 1950), p.396.

(32) Clogston, A.M., Jaccarino, V., and Yafet, Y., Phys. Rev. 134 (1964) A650.

(33) Shimizu, M., Rep. on Progress in Physics 44 (1981) 329.

(34) Schilling, J.S., this conf.

(35) McMillan, W.L., Phys. Rev. 167 (1968) 331.

(36) Allen, P.B., in Horton, G.K., and Maradudin, A.A. (eds.), (North Holland, Amsterdam, 1980), p.95.

(37) Smith, T.F., and Finnlayson, T.R., Contemp.
     Phys. 21 (1980) 265.
(38) Barron, T.H.K., Collins, J.C., and White,
     G.K., Advances in Physics 29 (1980) 609.
(39) Pott, R., Güntherodt, G., and Bach, H.,
     Solid State Commun. (1981), to be published.

(40) Bucher, E., Maita, J.P., Hull, G.W.,
     Longinotti, L.D., Lüthi, B., and Wang, P.S.,
     Zeitschr. Phys. B 25 (1976) 41.
(41) Pettifor, W.E., J. Phys. F 7 (1977) 613.

PHYSICS OF SOLIDS UNDER HIGH PRESSURE
J.S. Schilling, R.N. Shelton (editors)
© North-Holland Publishing Company, 1981

Superconducting Transition under Pressure in Some Eutectic Alloys*

F.S. Razavi and J.S. Schilling

Experimentalphysik IV, Universität Bochum, 4630 Bochum, W. Germany

The superconducting transition temperatures T_c of $Ir_{.79}Y_{.21}$ and $Ir_{.8}Eu_{.2}$ eutectic alloys were measured under pressure up to 3.5 GPa hydrostatically. The results show negative slopes of -20 mK/GPa for the T_c. These results are compared with the theoretical model of Suhl et al. We also measured the T_c of $Ir_{.79}Y_{.21}$ quasihydrostatically up to 13 GPa. Several explanations of the experimental results are discussed.

In a short paper in 1980, Matthias et al. show that adding only ~1% of yttrium (Y) to Iridium (Ir) enhances the superconducting transition temperature (T_c) by a factor of more than 30, from .11 K to 3.34 K [1]. This result is not new since in 1965 Geballe et al. [2] showed that an excess of few percent of such elements as scandium, yttrium, lanthanum, cerium and lutetium similarly enhanced the T_c of Ir. Recent results [3] also show that an excess of europium (Eu) in Ir causes an enhancement of T_c, that is, Eu must be here in the valence 3 configuration. The striking feature in these alloys as, for example, in Ir-Y alloy is that Y is not superconducting, at all, at atmospheric pressure, and it only becomes superconducting above 65 mK under high pressures greater than 5 GPa [4].

Using X-ray and transmission electromicroscopy [1] it is found that far less than 1% of Y is soluble in Ir; any excess Y goes into a second crystalline phase Ir_2Y which is fcc with the structure of C15 compound. It has been shown [5] that Ir_2Y is not superconducting above 20 mK. Thus the question arises what actually causes the large enhancement of T_c in Ir-Y alloys. Matthias et al. [1] attribute the enhancement of T_c to the formation of a eutectic which is caused by the microscope mixture of Ir with small amount of Ir_2Y. The specific heat results [1] of Ir-Y eutectic alloys as a function of Y concentration show that the Debye temperature (θ_D) decreases by a factor of two from 420 K for pure Ir to 175 K at exactly eutectic composition, yet the electronic specific heat remains essentially unchanged. Therefore, Matthias et al. [1] attributed the enhancement of T_c to lattice softening at the eutectic composition. Later on, Suhl et al. [6] proposed a simple model which described this softening as a result of unrelieved long-range coherency strains extending throughout the approximately 1000Å thick lamellae of the eutectic. Based on the studies of Cline et al. [7] on NiAl-Cr eutectic alloys, Suhl [6] argues that this coherency strain is the result of a 2% lattice mismatch between successive layers of Ir

and Ir_2Y which causes a stretching of the Ir_2Y lattice along the lamellae. Suhl et al. [6] pointed out that for a 6-12 Lennard-Jones potential, a 2% strain along one particular crystal axis of a simple cube would cause a reduction of the Debye temperature by a factor of two. Two distinct explanations for the enhancement of T_c in Ir-Y eutectic alloys can be obtained within Suhl's model [6]. The enhancement of T_c is directly related either, firstly, to the lattice mismatch between Ir and Ir_2Y or, secondly, to the change in the lattice parameter of Ir_2Y which is due to the lattice mismatch. One way to check these models is to change the stretching effect of the lattice constant mismatch by the application of hydrostatic pressure. The application of pressure should then cause an increase or decrease of T_c depending on whether the mismatch is enhanced or reduced. On the other hand, assuming that the softening is solely a function of the Ir_2Y lattice parameter a, then the drastic drop in θ_D corresponds to an enormous Grüneisen constant to

$$\gamma = \frac{-\partial \ln \theta_D}{3 \partial \ln a} \approx \frac{0.5}{3 \times 0.02} \approx +10$$

which is about 4x larger than that normally observed. Therefore, the application of hydrostatic pressure should cause a rapid lowering of T_c as the eutectic quickly relinquishes its softness.

We have measured T_c as a function of both hydrostatic and quasihydrostatic pressure. The hydrostatic pressure cell [8] consists of a pressure chamber with a 2 mm diameter bore in a Cu-Be gasket filled with a 4:1 mixture of methanol-ethanol. Two opposed WC anvils are pressed into the Cu-Be gasket, thereby reducing the cell volume and generating hydrostatic pressure in the fluid. The quasihydrostatic pressure cell consists of a pyrophyllite gasket with steatite disks as the pressure transmitting medium [9]. In both techniques the T_c of Pb is used as an internal manometer. In the hydrostatic pressure cell an a.c. inductance technique is employed to measure T_c; a four point d.c. resistivity technique is used to measure T_c in the quasihydrostatic pressure cell.

We also measured the room-temperature compressi-

*This work has been supported by the Minister für Wissenschaft und Forschung des Landes Nordrhein-Westfalen.

bility of the two eutectic components Ir an Ir_2Y in a $Ir_{.79}Y_{.21}$ sample at the eutectic composition. The compressibility was measured in a diamond-anvil cell [10] with a powdered specimen mixed with NaCl, the latter serving as internal manometer. Our preliminary results show compressibility of Ir agrees well with previous results [11] $K_{Ir} = 2.76 \times 10^{-3}$ GPa^{-1}; and that of Ir_2Y is $K_{Ir_2Y} \sim 3.39 \times 10^{-3}$ GPa^{-1}.

The pressure dependence of T_C of the eutectic alloys $Ir_{.79}Y_{.21}$ and $Ir_{.80}Eu_{.20}$ is shown in Fig. 1. Within the experimental errors both alloys show a negative slope of $\partial T_C/\partial P \approx -20$ mK/GPa. In an independent measurement on $Ir_{.9}Y_{.1}$ of the static magnetic susceptibility as a function of pressure up to 1.2 GPa, we obtained the same values of $\partial T_C/\partial P$, as for $Ir_{.79}Y_{.21}$ given above. The above results thus indicate that the pressure dependence of T_C in Ir-Y is relatively independent of both the nature and concentration of the Y-component.

Another striking result is the effect of quasihydrostatic pressure in the $Ir_{.79}Y_{.21}$ compound (Fig. 2) which shows that T_C drops sharply as a function of pressure with an initial slope of -200 mK/GPa before seemingly saturating at $T_C \approx 2.8$ K for $P \gtrsim 8$ GPa. However, upon releasing the pressure, $T_C(P)$ shows a rather large hysteresis ($\Delta T_C \approx 0.5$ K). One possible explanation for this result is that quasihydrostatic pressure induces internal strains in the $Ir_{.79}Y_{.21}$ eutectic which reduces the T_C of the compound, similar to what is observed in A15 and C15 compounds [12]. In fact, we observed that prolonged

annealing at 1000°C of the $Ir_{.79}Y_{.21}$ eutectic enhanced the T_C value to 3.96 K.

In the McMillan equation [13] T_C is given by

$$T_C = \frac{\theta_D}{1.45} \exp \left[\frac{-1.04(1+\lambda)}{\lambda - \mu^*(1+.62\lambda)} \right]$$

where λ is the electron-phonon coupling constant and μ^* is the Coulomb pseudopotential. When considering Suhl's model [6] and McMillan's equation, one expects a direct relation between the relative lattice mismatch and the T_C of eutectic $Ir_{.79}Y_{.21}$. The linear compressibility can be defined by

$$K_{lin} = \frac{1}{a}\frac{\partial a}{\partial P} = \frac{1}{3V}\frac{\partial V}{\partial P} = K/3.$$

Using the measured values for K_{Ir} and K_{Ir_2Y}, we obtained the relative mismatch at 3.5 GPa;

$$\left(\frac{\delta a}{a_{Ir}}\right)_{P=3.5} = \left(\frac{2a_{Ir} - a_{Ir_2Y}}{2a_{Ir}}\right)_{P=3.5} = 0.245$$

whereas at zero pressure, we have $(\delta a/2a_{Ir})_{P=0} = 0.232$. The relative lattice mismatch between the Ir and Ir_2Y components thus <u>increases</u> under pressure. Assuming T_C is a linear function of the degree of mismatch, one would expect T_C to <u>increase</u> with pressure at the rate $\partial T_C/\partial P \approx +43\underline{mK}/GPa$. This is not in agreement with our results which show a negative slope $\partial T_C/\partial P = -20mK/GPa$. On the other hand, if we consider that in Suhl's model [6] T_C varies linearly with the change in the lattice parameter of Ir_2Y, i.e. with strain,

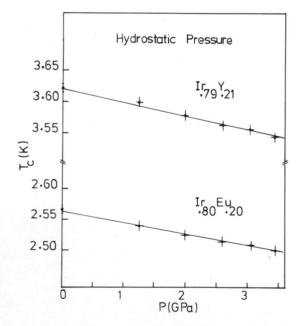

Fig. 1. The variation of T_C with hydrostatic pressure for $Ir_{.79}Y_{.21}$ and $Ir_{.8}Eu_{.2}$.

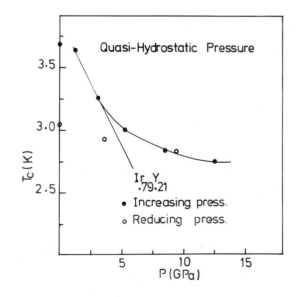

Fig. 2. The variation of T_C with quasihydrostatic pressure for $Ir_{.79}Y_{.21}$.

then,

$$\frac{\partial T_c}{\partial P} = K_{lin} \frac{\partial T_c}{\partial \ln a} = -170 \pm 20 \text{ mK/GPa},$$

which is a far more rapid decrease of T_c than what we observed experimentally. Using the McMillan equation, we calculated from T_c and θ_D for different concentrations of Y in Ir, assuming $\mu^* = .13$, the values of which are given in Table 1.

	T_c	θ_D	λ
Ir	0.11	420	0.33
Ir$_{.99}$Y$_{.01}$	3.34	375	0.55
Ir$_{.79}$Y$_{.21}$	3.67	175	0.68

As can be seen from Table 1, adding 1% Y to Ir causes a jump in the value of λ from 0.33 to 0.55; however, the Debye temperature θ_D changes only slightly. Yet between 1% Y and the eutectic composition one sees that λ changes only slightly ($\sim$20%) whereas the Debye temperature decreases by a factor of 2. Viewing these data, we believe the enhancement of T_c observed by Matthias et al. [1] is not directly related to the θ_D drop. λ is defined by [13]

$$\lambda = \frac{N(0) \langle J^2 \rangle}{M \langle \omega^2 \rangle},$$

where $N(0)$ denotes the density of state at Fermi energy, $\langle J^2 \rangle$ the sum over all electron-lattice matrix elements, M the ionic mass and $\langle \omega^2 \rangle$ is a measure of the average of the square of the phonon frequencies.

For 1% of Y in Ir one observes only a slight change in $N(0)$ [1]. Assuming changes in the electronic structure are negligible, the enhancement of λ might arise either from a change in the phonon frequency spectrum or from a change in $\langle J^2 \rangle$. We can argue that λ consists of two parts λ_1 and λ_2. λ_1 is related to the lower part of the phonon frequency spectrum which can be observed in low-temperature specific heat measurements and may be sensitive to the different concentrations of Y. λ_2, on the other hand, is the part which corresponds to the upper part of the phonon frequency spectrum which cannot contribute to specific heat measurements, but might be responsible for the T_c enhancement. Therefore, as shown in Table 1 the drop in the Debye temperature, or in other words, the phonon softening as a function of concentration, contributes only slightly to the enhancement of the T_c which would be sensitive to λ_2 from the upper phonon frequency spectrum. This part of the phonon frequency spectrum might be sensitive to external strain as can be seen from the quasihydrostatic pressure dependence of T_c. On the other hand, the electronic structure at the interface between the two phases in the eutectic Ir$_{.79}$Y$_{.21}$ compound might be also changed from the bulk values, leading to an enhanced value of λ and T_c.

In conclusion, our experimental results do not support Suhl's model [6]. If the value of T_c in the Ir-X eutectics is at all related to the lattice softening, then this model is too simple to account for this softening. For a better understanding of the T_c-enhancement in Ir-X based eutectic alloys, one should study alloys at low X-concentrations up to 1% where a drastic T_c-enhancement is already observed. Also, a complete phonon frequency spectrum obtained by neutron scattering would help in understanding the nature of the phonons in these alloys. We also suggest that the T_c-enhancement may arise from drastic changes in the electronic structure at the interfaces of the eutectic components; such changes might be too slight to be observed in the electronic specific heat of the bulk sample.

The authors would like to acknowledge the encouragement of the late B.T. Matthias to carry out these experiments and to S. Methfessel for his support. We are also grateful to H. Barz and M. Abd-Elmeguid for preparing the eutectic samples and to J. Masuch for his assistance with the X-ray studies. Special thanks are due to Prof. J. Carbotte for his discussions with F.S.R.

References

[1] Matthias, B.T., Stewart, G.R., Giorgi, A.L., Smith, J.L., Fisk, Z. and Barz, H., Science 208 (1980) 401.

[2] Geballe, T.H., Matthias, B.T., Compton, V.B., Corenzwit, E., Hull, G.W. Jr. and Longinotti, L.D., Phys. Rev. A137 (1965) 119.

[3] Probst, C. and Wittig, J., Superconductivity, in Gschneider, K.A. and Eyring, L. (eds.) Handbook on the Physics and Chemistry of Rare Earths, Vol. I., p. 749 (North-Holland, Amsterdam, 1978); see also paper by J. Wittig in the present volume.

[4] Matthias, B.T., Fisk, Z. and Smith, J.L., Phys. Lett. A72 (1979) 257.

[5] Smith, J.L., unpublished results.

[6] Suhl, H., Matthias, B.T., Hecker, S. and Smith, J.L., Phys. Rev. Lett. 45 (1980) 1707.

[7] Cline, H.E., Walter, J.L., Koch, E.F. and Osika, L.M., Acta Metall. 19 (1971) 405.

[8] Fasol, G. and Schilling, J.S., Rev. Sci. Instrum. 49 (1978) 1722; see also paper by A. Eiling et al. in the present volume.

[9] Eichler, A. and Wittig, J., Z. Angew. Phys. 25 (1968) 319.

[10] Lynch, R.W., J. Chem. Phys. 47 (1967) 5180.

[11] Gschneider, K.A. Jr., Solid State Physics 16 (1964) 275.

[12] Matthias, B.T., Superconductivity in d- and f-band Metals, Douglas, D.H. (ed.) (AIP, N.Y., 1972).

[13] McMillan, W.L., Phys. Rev. 167 (1968) 331.

PHYSICS OF SOLIDS UNDER HIGH PRESSURE
J.S. Schilling, R.N. Shelton (editors)
© *North-Holland Publishing Company, 1981*

ELECTRICAL BEHAVIOR OF CA, SR, BA, AND EU
AT VERY HIGH PRESSURES AND LOW TEMPERATURES

F. P. Bundy and K. J. Dunn

General Electric Company, Corporate Research and Development
Schenectady, New York 12301, USA

Compression of Ca and Sr initially causes an increase in resistivity, probably because of uncrossing of conduction and valence bands. Then at about 180 kbar for Ca and about 35 kbar for Sr the resistivity drops quite abruptly, following which the resistivity again increases with additional pressure, similar to the behavior of Ba starting at room pressure. The high pressure forms of Ba have already been reported to be superconducting, and our experiments confirm this. We find superconductivity appearing in Sr at about 350 kbar and developing strongly at higher pressures. In our 440 kbar experiment on Ca a resistance drop started at the lower threshold of our temperature capability, 2.1K, suggesting that Ca, too, becomes superconducting at sufficiently high pressures.

The high pressure form of Eu above 125 kbar was tested for superconductivity down to 2.2K with negative results.

In crystalline solids for which the electronic band gap is small (positive or negative) and in which the overlap may be changed by varying parameters such as pressure, there is the possibility of abrupt changes in the conductivity as the band gap approaches zero due to the formation of electron hole pairs, as suggested by Mott[1] (metal-insulator), by Knox[2] (insulator-metal via "excitons"), and by others. Jerome, et al.[3] and Halperin and Rice[4] suggested Ca, Sr, Yb, I, and other less likely semimetals in group V(a) elements as materials which might exhibit excitonic phase phenomena. McWhan, et al.[5] investigated Sr and Yb from 2 to 300K at pressures up to 50 kbar and found no such anomalies. Dunn and Bundy[6] studied molecular iodine at pressures in the region of band overlap without finding any resistance anomalies down to 2.2K. In the present work we did experimental studies of Ca, Sr and Ba in the temperature range of 2.2-300K and at pressures up to over 400 kbar looking for resistance anomalies, including superconductivity.

The apparatus and procedures used in this investigation have been described in detail elsewhere[7]. Briefly, the procedure consisted of pressurizing the specimen at room temperature, cooling it (under pressure) to about 2K in a cryostat; then observing the resistance as the temperature increased very slowly to room temperature.

CALCIUM:

The observed resistance behavior of Ca when compressed and decompressed at room temperature is shown in Fig. 1. Allowing for the establishment of good electrode contact with the specimen in the beginning of the compression (dashed line) it is seen that the specimen resistance increases with pressure to a maximum at about 180 kbar, then rapidly drops to a new low value, and finally increases to another maximum in two steps. The

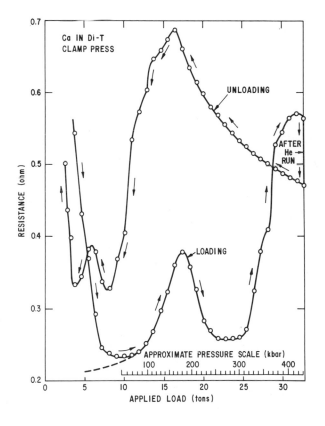

Figure 1. Loading and unloading resistance versus pressure curves for Ca at room temperature in the clamp press. The approximate cell-pressure scale applies to the loading curve only.

unloading curve shows the same features with the usual "hysteresis" observed in most pressure apparatus of this type. Thus we believe the resistance behavior shown is real, and that it indicates a number of phase transitions which are reversible.

When held at various pressures, temperature scans were made from about 2 to 300K with results as shown in Fig. 2. It is seen that dR/dT is positive and rather small at pressures of 62, 80 and 98 kbar, that it becomes negative in the 200-300K region at 150-180 kbar, and that at higher pressures it becomes positive and larger over the full temperature span. These and other features of the behavior of Ca with P and T may be more obvious when plotted as "synthesized" R(P) isotherms as shown in Fig. 3, by cross-plotting the data from Fig. 2.

The curves in Fig. 3 show: (i) that the first resistance peak at 180 kbar, room T, shifts toward lower pressure as the temperature is lowered, and (ii) that there is a possibility, noting the change in dR/dP of the curves near 310 kbar, that the up-jump of resistance which occurred at 360 kbar at room T (Fig. 1) might take place at a lower pressure for lower temperatures.

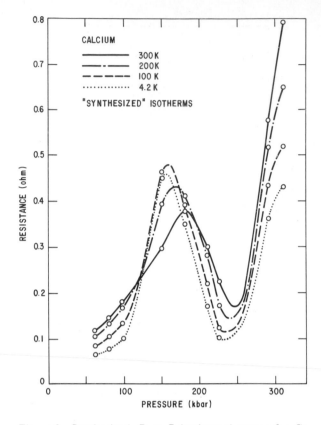

Figure 3. Synthesized R vs P isothermal curves for Ca data from Fig. 2.

Taking into account the above observations and the results of Jayaraman, et al.[8] for lower pressures and higher temperatures a tentative partial P,T phase diagram for Ca may be constructed as shown in Fig. 4. This diagram suggests that the 150-180 kbar phenomenon is a transition from the starting fcc phase to a phase III which in the 300 kbar region transforms to a phase IV, and this phase IV then goes on to a phase V in the 400 kbar region. A more detailed discussion of this diagram, and of the lack of experimental evidence for an excitonic state in Ca, is given in our current publication[9].

Our evidence for possible superconductivity in Ca is shown in Fig. 5 which presents our data for R(T) at 440 kbar. The insert which presents the lower end of the curve enlarged, shows how the resistance starts to drop down abruptly at the bottom edge of our attainable temperature range.

STRONTIUM:

The resistance behavior of Sr when loaded at room T in our apparatus is shown in Fig. 6. The 35 kbar resistance maximum followed by a drop (as observed by Bridgman and others in lower pressure apparatus) is obscured in our case by the changing electrode contact resistance during the tightening up of the cell and gasket. The Ba-like

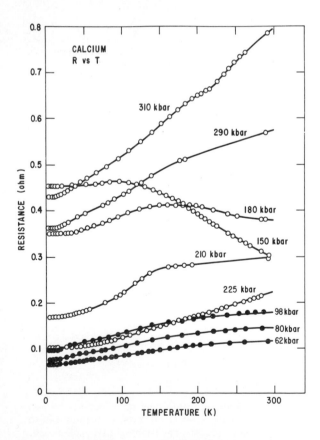

Figure 2. Resistance versus temperature curves for Ca at various fixed cell pressures.

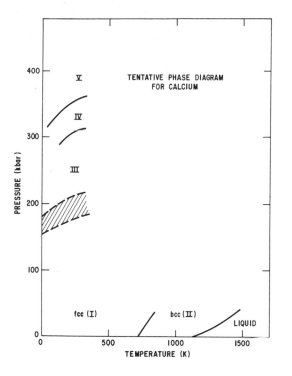

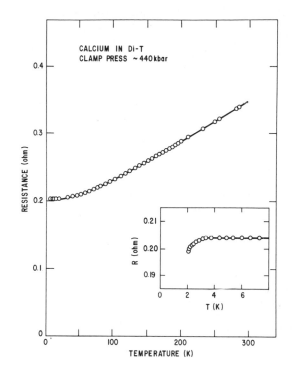

Figure 4. Tentative phase diagram for Ca.

Figure 5. R vs T curve for Ca at about 440 kbar. The insert shows the R vs T behavior in the low temperature region.

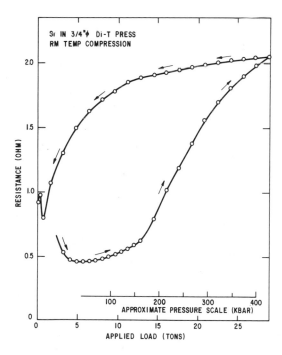

Figure 6. Resistance vs loading curve for Sr.

up-surge of resistance suggestive of band shifts or phase changes is evident at about 180 kbar. The R(T) behavior of Sr is shown in Fig. 7 for pressures ranging from 250 to 500 kbar. These curves all exhibit rather normal metallic behavior except the one for 350 kbar, which is the pressure at which superconductivity first appeared.

Fig. 8 illustrates the superconductivity behavior that we observed in Sr at various pressures above 350 kbar. The insert shows the R(T) curves at the pressures for which superconductivity occurred, and the main graph presents the T_c plotted against pressure. the open-headed arrow at 150 kbar, 1.5K is from Wittig[10] indicating no superconductivity was found. We found none down to 2.1K at pressures of 250 and 300 kbar, then it appears at about 3K at 350 kbar. As the pressure is increased the T_c seems to definitely increase.

BARIUM:

Our experimental data on superconductivity in Ba, and those of other observers, are presented in Fig. 9.

Although there are some discrepancies in the results of different observers in the 70 to 110 kbar region the gen-

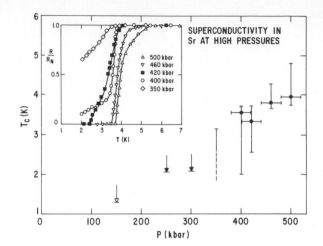

Figure 8. Superconductivity in Sr at high pressures.

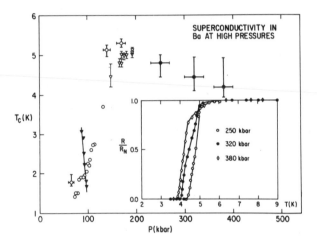

Figure 9. Superconductivity in Ba as a function of pressure.
o - Wittig and Matthias (1969). ∇ — Il'ina and Itskevitch (1970). ∇ — Il'ina Itskevich and Dizhur (1972). o - Moodenbaugh and Wittig (1972).

eral indication is that the first high pressure phase has a T_c around 2K, and the second phase (above about 120 kbar) has T_c much higher, around 5K. Our runs made at 250, 320 and 380 kbar indicate that the T_c decreases with increase of pressure above 150 kbar.

When our most recent results for Ca, Sr and Ba are put together with other reported T_c data as presented in Fig. 10, a possible trend for this family of elements emerges. As suggested by Probst and Wittig[11] application of high pressure converts Ba, through a series of crystallographic phase changes, into a 5d transition metal, the d character resulting from s-d hybridization. Beyond about 200 kbar additional pressure lowers the T_c in the normal manner. Our data on Sr suggests that it should be tested

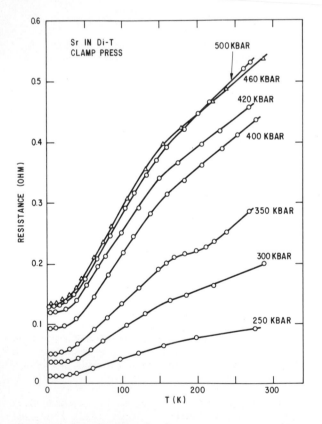

Figure 7. R vs T curves for Sr at pressures from 250 to 500 kbar. Note the anomaly at 350 kbar, the pressure at which superconductivity first appears.

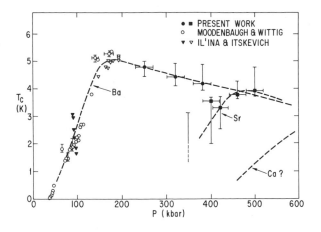

Figure 10. Summary of T_c vs pressure data for Ba, Sr and Ca.

in the 250-350 kbar range for very low temperature superconductivity behavior similar to that of Ba in the 35-90 kbar range. Similar very low temperature tests are indicated for Ca in the 400-600 kbar range.

EUROPIUM:

The lanthanide rare-earth series of elements is characterized by a gradual filling of the 4f shell which is generally located deep within the atom. A consequence is that the lanthanides are very similar to each other in their chemical and physical properties. Because of the tendency for the 4f shell to be either half-filled, or filled, two of the elements, Eu and Yb at normal conditions are divalent while the others are trivalent. However, the 4f and 5d electronic levels are very close and the actual state which the atom assumes is very sensitive to the density, or pressure. Some theoretical calculations[12] suggest that Eu and Yb would become trivalent like other lanthanides at pressures in the range of 150 kbar, and that in this new state there would be some possibility of superconductivity at very low temperatures. The occurrence of a high pressure phase was shown experimentally in 1964 by Stager and Drickamer[13] in electrical resistance experiments on several rare-earth elements up to pressures of about 250 kbar (modern 1975) scale) and temperatures ranging from 77 to 300K.

Having in place our ultra high pressure apparatus with cryogenic capability we were encouraged by the late B.T. Matthias to test the high pressure phase, or phases, for superconductivity. Altogether we carried out 15 resistance vs. temperature scans at pressures ranging from 55 to over 400 kbar. In general our results were in agreement with those reported by earlier observers, including the resistance up-jump phase transition of Stager and Drickamer[13], and the magnetic ordering transition around 90K reported by McWhan, et al.[14], and by Cohen, et al.[15] in the low pressure phase.

Typical R(T) curves at pressures below and above the phase transition are shown in Fig. 11. In the 125 kbar curve the sharp bend at about 90K marks the magnetic ordering transition. In the 260 kbar curve a similar sharp bend occurs at about 150K, and it occurs at this temperature in all the higher pressure runs. It is possible that this, too, is a magnetic ordering phenomenon, but we have not had the experimental means available to characterize it.

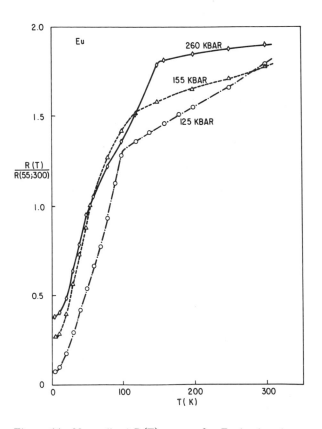

Figure 11. Normalized R(T) curves for Eu in the phase transition region 125-260 kbar.

Based on our R(T) data at various pressures we derive a P,T phase diagram for Eu as shown in Fig. 12.

The low pressure phase is known to be bcc[16], and the 90K magnetic ordering transition has been well studied and characterized by others. X-ray diffraction and magnetic studies would have to be done to characterize the structures and the phenomena which apply to the high pressure phase.

In our experiments no evidence of superconductivity could be detected down to about 2.1K. The semi-theoretical treatment of the lanthanides by Johansson and Rosengren[17] indicates that if the high pressure phase of Eu can become superconducting it would be at a fairly low temperature well below 2K.

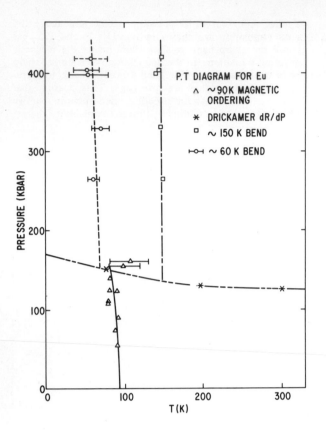

Figure 12. Proposed P,T phase diagram for Eu.

REFERENCES:

[1] N.F. Mott, Philos. Mag. *6* 287 (1961).

[2] R.S. Knox, Theory of Excitons of Solid State Physics Suppl. 5 of (Academic Press, New York 1963) p. 100.

[3] D. Jerome, T.M. Rice, and W. Kohn, Phys. Rev. *158* 462 (1967).

[4] B.I. Halperin and T.M. Rice, in Solid State Physics edited by H. Ehrenreich, F. Seitz, and D. Turnbull, (Academic Press, New York 1968) Vol 21, p. 115.

[5] D.B. McWhan, T.M. Rice, and P.H. Schmidt, Phys. Rev. *177* 1063 (1969).

[6] K.J. Dunn and F.P. Bundy, J. Chem. Phys. *72* , 2936, (1980)

[7] F.P. Bundy and K.J. Dunn, Rev, Sci. Instrum. *51* , 753 (1980).

[8] A. Jayaraman, W. Klement, Jr., and G.C. Kennedy, Phys. Rev. *132*, 1620 (1963).

[9] K.J. Dunn and F.P. Bundy, Phys. Rev. B (scheduled for Vol *24*, (Aug. 1981).

[10] J. Wittig, Phys. Rev. Lett. *24*, 812 (1970).

[11] C. Probst and J. Wittig, Phys. Rev. Lett. 39, 1161 (1977).

[12] B. Johansson and A. Rosengren, Phys. Rev. *B11*, 2836, (1975).

[13] R.A. Stager and H.G. Drickamer, Phys. Rev. *133A*, B30 (1964).

[14] D.B. McWhan, P.C. Souers and G. Jura, Phys. Rev. *143*, 385 (1966).

[15] R.L. Cohen, S. Hufner and K.W. West, Phys. Rev. *184*, 263 (1969).

[16] A. Jayaraman, Phys. Rev. *135A*, 1056 (1964).

[17] B. Johansson and A. Rosengren, Phys. Rev. *B11,* 2836 (1975).

PHYSICS OF SOLIDS UNDER HIGH PRESSURE
J.S. Schilling, R.N. Shelton (editors)
© *North-Holland Publishing Company, 1981*

ELECTRICAL AND MAGNETIC PROPERTIES OF PRESSURE QUENCHED CdS

C. G. Homan and D. Kendall

U. S. Army Armament Research and Development Command
Large Caliber Weapon Systems Laboratory
Benet Weapons Laboratory
Watervliet, New York 12189

and

R. K. MacCrone
Department of Materials Science
Rensselaer Polytechnic Institute
Troy, New York 12181

The anomalous diamagnetic and paramagnetic properties of pressure quenched CdS at 77 K are described.
The electrical behavior at the dia- to para- magnetic transition are also suggestive of a high temperature superconducting state.

INTRODUCTION

Pressure quenched CdS has exhibited very large diamagnetism at temperatures above 77 K.(1) Here we briefly review the previous work and present some new electrical measurements during the diamagnetic-paramagnetic transition.

EXPERIMENTAL METHODS

The CdS is pressurized in a conventional anvil configuration. The samples were prepared by the technique of "pressure quenching": the sudden release of pressure at rates of greater than 10^6 bar/sec.(2) This technique is analogous to the well-known technique of temperature quenching by which many metastable states are realized. The electrical resistance of the CdS specimens is monitored during pressure loading to provide an accurate measure of the state immediately prior to quenching. At about 30 kbar, the electrical resistance drops by about five orders of magnitude, signaling the well-known wurtzite to NaCl high-pressure phase transition.(3) Typical resistance versus pressure traces are shown in Fig. 1.

It has been discovered that the source from which the starting material is derived is important. Material of a particular lot from the Alpha-Inorganic produces the strongly diamagnetic and paramagnetic samples described here, while material from Eagle-Picher and a private source, similarly pressure quenched, does not. As can be seen, the pressure resistance traces are very different for specimens fabricated from the different starting materials.

Only specimens quenched from above 40 kbar show large diamagnetism and/or paramagnetism and

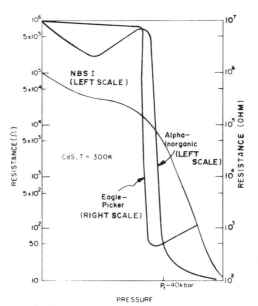

Figure 1: Resistance versus pressure traces of specimens of different source material. Eagle Picher < .01% Cl, α-Inorganic ∼ 1% Cl N.B.S. #1 ∼ 6% Cl.

electrical behaviour of concern here. After pressure quenching, the resistance at room temperature is more than four orders of magnitude less than the starting material and the specimens have a black sheen. Specimens which are slowly unpressurized revert to their original resistance and are a dark orange color.

The magnitude used in these *static* susceptibility measurements is a standard P.A.R., vibrating-sample magnetometer (VSM). Where magnetic measurements only were being made, a sample holder, rod, and enclosing jacket carefully designed to ensure no frictional contact near the pickup coils were used. The specimen space was evacuated with a rotary pump and nitrogen trap.

The magnetometer itself has been subject to an exhaustive series of tests to ensure the confidence of the results and was calibrated with both a $CuSO_4 \cdot 5H_2O$ sample and a nickel sphere sample for internal consistency.

Since the specimens are metastable at room temperature and below, it is not possible to perform magnetization and electrical measurements sequentially and make meaningful comparisons; the condition of the specimen will simply not be the same. To obtain incontrovertible results, it is necessary to observe changes in magnetization and at the same time changes in the conductivity in one and the same specimen.

A special specimen holder for a vibrating specimen magnetometer was thus constructed so that dc and ac resistance measurements could be made while the magnetic moment was being measured. Essentially the device consisted of a silica tube carrying the lower copper electrode, through which a second concentric silica tube passed and which carried the upper copper electrode. In this way, the electrode faces were maintained in alignment and the miniature coaxial shielded cable brought to the outisde world through a series of judiciously placed holes and slots. The specimen is simply slipped between the electrodes and maintained under light pressure from small springs far removed from the pick up coils. A cryostat provides controllable temperatures from 20 to 300 K (Air Products Helitran).

dc Resistance measurements were made using a current amplifier in a simple series circuit, with 1 to 6 volts supplied from a battery. ac Capacitance and conductive (or loss) measurements were made at 1 kJz using a G. R. 1615 capacitance and conductance off balance (which may be achieved by using two lock-in amplifiers) were simultaneously displayed on X-Y recorders, the abscissa of each being driven by the same time-base generator. In this way it is possible to obtain a complete one-to-one correspondence between any changes taking place amongst these quantities. The balance of the capacitance bridge was adjusted as required during the experiment, and similarly the current amplifier gain and/or offset.

For electrical contacts we relied solely on the mechanical pressure of the copper disks previously described. The metastability of the samples precludes the painting or evaporization of elecrodes. Gentle tapping and the rotation of the upper and lower contacts revealed that the contacts were stable. To test for surface effects, numerous I versus V curves at 77 K have been taken which show ohmic behavior, reversibility, and no polatization effects. During capacitance measurements, dc bias up to ~ 200 V/cms was applied from time to time which resulted in a (reversible) capacitance change of only ~ 1%. At room temperature, where this sample showed no anomalous flux exclusion, the relative dielectric constant ε of CdS was determined from the capacitance measurements to be 5.3 ± 1.0, which compares favorably with literature value of 5.2.[4] This agreement is taken as evidence that any contribution from the surface capacitance of a Shottky barrier is essentially absent. In this two terminal measurement, the stray components of capacitance are not magnetic field dependent.

The sample material used in this study, Optronic grade CdS powders from Alpha Inorganic stock no. 20130, was from the same lot used in the previous studies.[1,2,5] Preliminary chemical analysis of the starting powder yielded total metallic impurities in the 20 ppm range.[1] An x-ray flourescence spectroscopic analysis was performed by NBS to give a qualitative chemical analysis of all impurities (Z > 11). In addition both the starting materials and the pressure quenched samples have been characterized by x-ray measurements, differential scanning calorimetry, optical microscopy and metallography, results which have been reported elsewhere.[2,6] Electron and ion microprobe analysis have been performed on both starting materials and pressure quenched, magnetically active samples.[7] These techniques have provided a quantitative chemical analysis of all elements $Z \geq 1$ in our material and indicate that the starting materials used in these studies were heavily contaminated or doped with Cl and Si to levels in the 1 wt% range.

It is known that the concentration of Cl effects the fraction of NaCl structure retained on reducing the pressure.[8] The chlorine in the material which produces the anomalous behavior is probably complexed. Cote et al[6] has shown that CdS prepared by precipitation from an aqueous solution of $CdCl_2$ exhibits D.S.C. and x-ray spectra after pressure quenching similar to the active α-Inorganic material. On the other hand simple mixtures of CdS and $ClCl_2$ with the same Cl contant, are distinctly different, and in particular do not show unusual behavior.

The presence of some retained NaCl structure is necessary for the anomalous effects. The metastability of the anomalous effects suggest that they may be related to the metastability of the NaCl phase. The quenched samples had the morphology of a large fraction of powder compact matrix in which small lenticular platelets were embedded. Microhardness tests indicated that the platelets were harder than the matrix material, and chemical etching studies were in agreement with this conclusion. X-ray diffraction studies failed to reveal any new phase which could be identified with the lenticular platelets except possibly very broad diffuse lines from small crystallites or an amorphous phase.

Phase transformations are known to depend strongly on impurity levels as well as the conventional parameters, such as, quench rates, temperature, time, etc. Our present knowledge and understanding of this system in this regard is not yet sufficient to obtain complete control and consequent reproducability. We are presently systematically varying the chemistry and physical treatments and investigating the resulting structure and physical properties.

RESULTS

Figure 2 shows the diamagnetic response of a pressure quenched sample. These results were

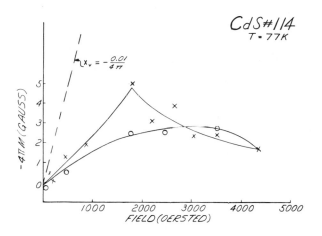

Figure 2: The diamagnetic moment of pressure quenched CdS No. 114 at 77 K as a function of magnetic field. Crosses, increasing field; circles, decreasing field.

obtained on a sample which had been mounted in the magnetometer at zero field at room temperature and then cooled to liquid-nitrogen temperatures at a rate of 100 K/min. This entire process was accomplished within six hours after the room-temperature pressure quench. After slowly warming this sample *in situ* to room temperature and recooling it was found that the magnitude of the susceptibility had decreased below the sensitivity of our magnetometer ($|\chi_V| < 10^{-5}$ cgs units). All susceptibility values are calculated using the entire sample volume of $3 \times 10^{-3} \mathrm{cm}^3$.

Diamagnetism and ac Conductivity as a Function of H

A pressure quenched CdS specimen was cooled to 77 K and the magnetization and ac conductivity measured as a function of magnetic field. In this case the dissipation, D, was measured rather than the conductance, G, which in this specimen was small. The results are shown in Figure 3.

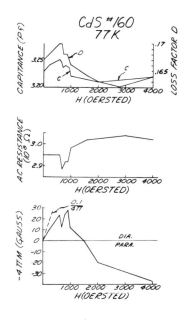

Figure 3: Lower: Magnetic moment of pressure quenched CdS as a function of applied magnetic field at 77 K.

Upper: Capacitance and loss observed at 1 kHz during the measurements shown.

Center: Calculated resistance.

At low fields the specimen exhibits a dismagnetic susceptibility amounting to 6% of $-1/4\pi$, the Meissner value. At about 600 Oe the specimen shows the onset of some instability accompanied by a capacitance and loss decrease. The flux exclusion is 4% at this point. Between 825 and 1500 Oe, the diamagnetism decreases to zero, while the capacitance and loss show a sharper drop which is almost complete at 1000 Oe. Above 1.5 kOe the specimen is in the positive magnetic state(5) in which the capacitance and loss show only a smooth behavior with no sign of phase transitions.

Diamagnetism and dc Conductivity as a Function of H

A pressure quenched CdS specimen was cooled to 77 K and the magnetization and dc conductivity measured. The results are shown in Figure 4.

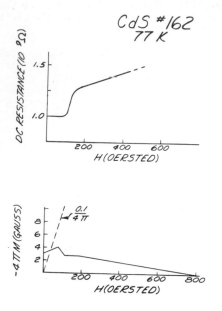

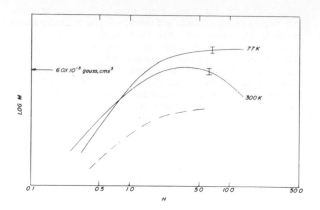

Figure 5: Magnetic moment M versus applied field H for a pressure quenched sample. The applied field H is shown in kOe.

Figure 4: Lower: Magnetic moment of pressure quenched CdS as a function of applied magnetic field at 77 K.

        Upper: dc Resistance observed during the magnetic measurements.

As the magnetic field is increased beyond 100 Oe, the dc conductivity decreases as the diamagnetism drops and flux enters the sample. [This specimen, showing constant diamagnetism up to 60 Oe, had been previously subjected to several exposures of fields of the order of 10 kOe while ac measurements were being made.]

As can be seen in the above Figure 3, the diamagnetic state can revert to a paramagnetic state. The paramagnetic state persists for much longer that the diamagnetic state. A paramagnetic response of a pressure quenched sample is shown in Figure 5 at two temperatures.

The behaviour is remarkably Langevin-like, as can be seen from the Langevin function which is also shown. By using the technique of v.d. Giessen(9), the number of moments n, and size of moment involved μ^* can be estimated. The values are shown in the Table.

Table

T (K)	μ^* (number of Bohr magnetons) $\times 10^{-3}$	n (cm^{-3}) $\times 10^{-17}$	N (cm^{-3}) $\times 10^{-21}$
273	9.4	2.5	2.35
77	1.33	25.9	3.44

Not only are the values physically reasonable, but the constancy of the product $n\mu^*$, as well as its value, is to be noted.

Discussion

The magnetic properties of the pressure quenched CdS at 77 K are very similar to the magnetic properties of well-known type-II superconductors observed at much lower temperatures.

A variety of mechanisms have been suggested for superconductivity at high temperatures. These include excitonic mechanisms, [10-13] proximity effects, [14] dislocations, [15] etc.

We note that many of the ingredients for superconductivity as proposed are present in our specimens: dislocations from the pressure treatment, two or more phases in intimate contact, and of course excitons. Of further interest, the metastability of the resultant crystallographic phases suggests that soft-mode instabilities may also be present.

The observed relation between the magnetic behavior and the electrical transport of these specimens is in general agreement with that expected of an inhomogeneous material containing superconducting regions below some percolation limit. The fact that a resistance increase is observed on an already large resistance implies that the normal state is highly resistive rather than metallic of low resistance. This is consistent with the observed decrease in the capacitance with the disappearance of the diamagnetism.

Rigorous analysis of the behavior expected of homogenous material containing superconducting elements which become normal confirm the intuitive expectations. Computer calculations of general random resistive networks always show a drop in conductivity as the value of a resistive element is increased, corresponding to a superconducting region becoming normal.(16) As pointed our by Landauer, the correspondence of J and E to D implies a corresponding decrease in capacitance as a capacitance element is added (again corresponding to a region becoming normal).

Thus the electrical behavior observed at the collapse of the diamagnetism in Figs. 1 and 2 is fully consistent with the electrical behavior expected of an inhomogeneous semiconductor containing superconducting inclusions. This conclusion is very general and is not dependent on ad hoc assumptions or specific morphology.

Some simple analysis is of interest. Consider, as a very crude model of the system, that the superconducting regions consist of thin sheets lying parallel to the plane of the disk. Then if d_s is the total thickness of superconductor, d the the total thickness of specimen, we find:

$$\frac{\Delta \sigma}{\sigma} = \frac{\Delta c}{c} \approx \frac{d_s}{d} \ .$$

The electrical results of Fig. 1 imply, for these ac measurements, that $\frac{d_s}{d} \approx 0.02$ with values ranging up to 0.09 in other specimens. The data of Figure 2 imply a ratio $\frac{d_s}{d} \sim 0.16$. These differences are within the uncertainties of this crude model, particularly as far as the ac response is concerned which does not address the frequency dependence. The order of magnitude agreement we consider satisfactory. Furthermore, the correlation between the estimate of this volume fraction of superconductor based on electrical behavior above and the volume fraction of platelets based on direct metallographic observation, $\sim 15\%$(2) typical of this material, is remarkable. Occurrence of superconductivity in the platelets produces a natural explanation for both the less than Miessner flux exclusion and the lack of zero

resistance in these specimens. We note that the magnetic flux exclusion decays as the electrical conductivity.

The existance of a paramagnetic state at high fields (and temperatures) is entirely consistent with free energy considerations.

It is interesting to note that the paramagnetic state displays rather bland electrical properties.

CONCLUSIONS

Pressure quenched CdS displays remarkable magnetic and electrical properties. One phase is diamagnetic, and upon transition to the paramagnetic state, show behavior very suggestive of superconductivity at high temperatures, 77 K. The paramagnetic state, although less dramatic, is also of considerable interest.

A major problem is to determine precisely the chemical and physical requirements for the effects.

ACKNOWLEDGEMENTS

We would like to thank P. J. Cote for helpful discussions, P. Au and W. Yaiser for technical assistance and expertise, S. Bloch and G. P. Piermarini (N.B.S.) for powders and chemical analysis.

The financial support of Benet Laboratory, ARRADCOM under contract DAAA-22-80-C-0256, the Air Force Office of Scientific Research under contract AFOSR-79-0216, and the Office of Naval Research under contract N00014-80-C-0828 is gratefully acknowledged.

BIBLIOGRAPHY

1) E. Brown, C.G. Homan and R.K. MacCrone, Phys. Rev. Lett, 45, 478 (1980).

2) C.G. Homan and D.P. Kendall, Bull. Am. Phys. Soc. 24, 316 (1979).
Details are presented in Benet Weapons, Lab Report No. ARLCB-TR-79-004, available from NTIS, Springfield, VA., AD No. A-69-609.

3) G.A. Samara and H.G. Drickamer, J. Phys. Chem. Solids 23, 457 (1962).

4) J.C. Phillips, Phy. Rev. Lett. 20, 550 (1968).

5) C.G. Homan, and R.K. MacCrone, J. Non Crystal Sol. 40, 369 (1980).

6) P. Cote, G.P. Capsimalis and C.G. Homan, Appl. Physics Lett. 38, 927 (1981).

7) Private Communication - U.S. National Bureau of Standards, Report No. 553-33-Y-81, 1980 (unpublished).

8) J.A. Corll, J. Appl. Phys. 35, 3032 (1964).

9) A.A. v.d. Giessen, J. Phys. Chem Solids 28, 343 (1966).

10) B.I. Halperin and T.M. Rice, Rev. Mod. Phys. 40, 755 (1968).

11) A.A. Abrikosov, Pis'ma Zh. Eksp. Teor. Fiz. 27, 235 (1978) [JETP Lett. 27, 219 (1978)].

12) V.L. Ginzburg, Pis'ma Zh. Eksp. Teor. Fiz. 14, 572 (1971) [JETP Lett. 14, 396 (1971)].

13) T.H. Collins, A.B. Kinz, and R.S. Weidman, in Recent Advances in Quantum Theory of Polymers, edited by J. Ehlers et al., Lecture Notes in Physics, Vol. 113 (Springer-Verlag, New York, 1979), p. 240.

14) D. Allender, J. Bray, and J. Bardeen, Phys. Rev. B 7, 1020 (1973).

15) C. Elbaum, Phys. Rev. Lett. 32, 376 (1974).

16) R. Landauer, "Electrical Conductivity in Inhomogenous Media", in A.I.P. Conference Proceedings, No. 40, Eds. J.C. Garland and D.B. Tanner, 1978, "Electrical Transport and Optical Properties of Inhomogenous Media", Ohio State University, 1977, p. 2.

PHYSICS OF SOLIDS UNDER HIGH PRESSURE
J.S. Schilling, R.N. Shelton (editors)
© *North-Holland Publishing Company, 1981*

CONCLUDING REMARKS

The final session of the conference, which was called the "Jam Session", was so successful that the editors felt it worthwhile to ask all contributors to this session to submit a written summary of their remarks. Most contributors did respond so that the following represents a reasonable summary of the actual session.

N. W. Ashcroft:

One rather obvious development that I have noticed in the high pressure scene over the nearly 20 years that I have been an observer is the degree to which pressure P has become an accepted thermodynamic parameter in the exploratory sense. It may not yet rival T, but as the papers have shown most clearly in this conference, it has become almost routine to subject condensed matter to experimental tests under pressure for a variety of both equilibrium and transport properties. And this is surely important since if our aim is the understanding of the physical properties of materials in the universe, then it is well to keep in mind that very little matter is actually under conditions of low pressure.

It is both stimulating and encouraging to hear that in the laboratory we can now generate pressures in the range of megabars both statically and dynamically, and in either case with impressive precision. Though this presents a considerable advance over the last few years, it is still somewhat salutory to realize that the atomic unit of pressure ($e^2/2a_0^4$) is 147 Mbars. If we take this as a guide to the regime in which to seek possible new orderings in condensed matter (less dramatic rearrangement may, of course, take place at lower pressures), then we might draw two tentative conclusions: (i) that in theoretical developments, precision of calculation is paramount since energies small on the atomic scale ($e^2/2a_0^4$) can still translate into large pressures, (ii) that perhaps it is not inappropriate to enquire about the ultimate pressures expected from current devices such as diamond cells, and to think about the next generation of devices. On point (i) we see already some truly impressive advances in the computational aspects of band theory and the associated density of state and thermodynamic functions of the materials being studied. Doubtless these advances will continue in the area of non-equilibrium properties and in the systematic inclusion of many-particle corrections. On (ii), perhaps we need to understand the phase diagram of diamond a little better, and some data on its band closing as a function of P would assist here. Is diamond the ultimate superhard material under conditions of high compression? The traditional answer could be challenged on many grounds. A lurking instability may well be a determining factor in reaching the pressures required for the metallization of hydrogen, now thought to be in the vicinity of 2 Mbar. The understanding of the way in which diatomic order in this system is progressively destroyed by the application of pressure might also be aided by further studies on iodine and, as Drickamer suggests, bromine. So also might further measurements on the band gap in hydrogen, if not masked (in a diamond cell) by the gap in diamond itself. If hydrogen can be metallized, then, as noted earlier, the resulting system can be expected to exhibit interesting quantum behavior.

To a lesser extend these characteristics might also be expected in other light elements possessing extended valence structure (boron and lithium, for instance). For these systems, for liquids, for glassy and amorphous materials, and as we have seen, for ordered states, such as magnetism, pressure may yet have much to say.

F. P. Bundy:

Having been an active worker in the field of ultra high pressure science for about 30 years, I am naturally interested in the prospects for doing meaningful experimentation at pressures still higher than those currently attainable, and for doing more significant experiments at the pressures which are now realizable. For the production of the highest static pressures we all use the hardest, strongest material known: diamond. I know of no theoretical reason which indicates that there is any possible substance which is harder or stronger. Even with perfection of the stress fields in diamond pressure apparatus there is an ultimate *natural* limit set by the pressure at which diamond collapses to a metallic tin-like state. This is probably in the vicinity of 2 megabars at room temperature.

While it is recognized that extremely high pressures can be generated at the very small contact areas between "points" and "planes" of very hard strong materials, we must remember that a *specimen* has to be in the pressurized zone and to learn anything about its behavior under such conditions we have to be able to monitor it reasonably accurately via electrical, optical, or other methods. So I urge potential apparatus designers to keep this practical requirement in mind.

The much higher pressures attainable by shock compression methods inevitably have very high temperatures associated with them, especially for the more compressible materials. As has been pointed out already here by others the temperature rise can be made much smaller by arranging for slower isentropic compressions, and by starting the shock compression with the specimen at a high static pressure (density) and low temperature.

It will be most interesting to see how far these ideas can be developed in the future.

W. B. Daniels:

Neil Ashcroft has commented on rather fundamental experimental problems which face those who wish to transform hydrogen to its metallic phase, such as the possibility that diamond would transform at a lower pressure than the hydrogen. I would like to suggest a rather crude but simple and inexpensive technique which might be useful in doing highly preliminary tests on such questions. It proceeds as follows: Partially fill the sample cavity of a diamond anvil cell with a diamond "powder" having an appropriate particle size and charge the cell in the usual way with hydrogen or other material. Given the proper filling fraction of diamond powder, compression of the cell will generate a predetermined fluid pressure before the diamond anvil faces compact the powder. At this point, each diamond particle in contact with a neighbor will constitute a crude "Ruoff Diamond-Indenter" loaded to high pressure with sample gas, with the corresponding possibility of generating at the contacts pressures substantially higher than in the conventional anvil cell. Further, in as much as a high ambient pressure enhances the strength of the individual diamond chips, there is a potential increase in performance over an indenter not imbedded in a high pressure matrix. Observations of conditions at the contact points would at least include direct microscope observations for black metallic spots. Just think of it, $\sim 10^6$ indenters per experiment!

H. G. Drickamer:

The problem of metallic hydrogen has been under investigation for years, frequently with more regard for the spectacular than for scientific understanding. At present, theorists appear on sounder ground than experimentalists. A number of analyses indicate that hydrogen will become a diatomic metal by band overlap in the range 2-3 megabars and a monatomic metal at high pressures (3-3.5 megabars). It seems doubtful to me that these pressures will be attainable in any static apparatus now in sight, and, in particular, that the variety of measurements necessary to show what is actually occurring can be made.

We have available one example of a diatomic molecule which is a model for the anticipated behavior of H_2. Twenty years ago it was shown that iodine becomes a metal in the range 130-170 kbar, using optical absorption edge shift, resistance, and the temperature coefficient of resistance. More recently, x-ray studies have shown that it is a diatomic conductor in the range 170 to about 210 kbar, at which point it becomes a monatomic metal. Another example or so would solidify our

understanding of H_2. With all the apparatus available which apparently can reach 0.5 to 1.0 megabar or higher in almost effortless fashion, it is difficult to understand why a thorough investigation of bromine has not been undertaken. The handling problem is surely not insurmountable. Organic chemists use it daily. One could, it would seem, measure the absorption edge shift, the resistivity and its temperature coefficient, and the x-ray structure. The P-V data from the x-rays on the molecular phase would aid theorists in analysis.

A follow up would be the study of solid chlorine. It is possible that at least the absorption edge could be shifted into the visible region of the spectrum and P-V data for the molecular solid obtained.

Until these investigations are undertaken, it is difficult for me to take seriously experimentalists' interest in the hydrogen problem.

R. D. Etters:

I have two comments. First of all, the application of high pressures has rapidly moved us into a regime where the static and dynamic properties associated with intramolecular behavior is strongly coupled to intermolecular behavior. This complicates the analysis of properties that could previously be understood by decoupling the degrees of freedom associated with these two parts of the system. Two examples dramatize this point. Strong coupling occurs near the molecular to atomic phase transition, where the distinction between lattice and internal modes actually becomes ambiguous, as appears to be the case for iodine near 21 GPa. It also occurs in solid H_2, even at low pressures where the pressure dependence of the intramolecular bond length and the protonic zero motion are important components in the analysis of bulk behavior, such as the equation of state. With the added complexity also comes a richness of information not previously appreciated.

Secondly, it is gratifying that evidence of good agreement between shock compression data and data determined from other methods, such as diamond cell technology, is growing. This knowledge is particularly timely since we may be reaching the limits of pressure attainable by diamond anvils, leaving shock compression as an important alternative for probing higher pressures.

J. M. M. Franse:

Comment on Wohlfarth's paper--Connected with the difficulties that arise in describing homogeneous and heterogeneous magnetic systems are the problems of evaluating the magnetic parameters, especially the Curie temperature, from magnetization measurements. The determination of the Curie temperature and its pressure derivative requires a detailed knowledge of the model that is appropriate for the system under consideration.

It may happen that by a closer inspection of experimental data some of the materials that belong to Table A are found to have a more complicated relation between $\partial T_c/\partial P$ and T_c and should be classified under Table B.

Comment on Glötzel's paper--The experimental value of -0.29 Mbar^{-1} for the logarithmic pressure derivative of the spontaneous magnetization of nickel at low temperature is probably taken from the work of Kondorski and Sedov (1960). The error in this number is $\pm$0.05 Mbar^{-1} which means that for nickel the calculated result is in better agreement with experiment than you suggested in your talk. Recently (Hölscher and Franse, 1980) a more accurate value (-0.23$\pm$0.01 Mbar^{-1}) became available from volume magnetostriction measurements which makes the agreement between calculations and experiment almost perfect for nickel. For iron and cobalt the experimental data for the logarithmic pressure derivative of the spontaneous magnetization as obtained from volume magnetostriction measurements amount to -0.29 $\pm$ 0.01 Mbar^{-1} and -0.26 $\pm$ 0.01 Mbar^{-1}, respectively.

Comment on Smith's paper--The study of the thermal expansion of weakly magnetic metallic systems provides us, also in my opinion, with information on the magnetic interactions that is of comparable quality with the information obtained from specific heat measurements. In some cases a magnetic transition is even better observable in anomalies in the thermal expansion than in the specific heat. The transition from ferromagnetism to paramagnetism upon changing the nickel content in the disordered Ni-Pt alloys is a good example as is Ni-Cr, apparently.

The present study on spin-glass systems reminds me of thermal expansion measurements on PdMn alloys in the ferromagnetic and spin-glass region (Brommer et al., 1981). Just as for CuMn and AgMn there is no clear anomaly visible at the spin-freezing temperature of Pd + 10 at% Mn, for instance. By combination of the magnetic contributions to thermal expansion and specific heat a temperature independent Grüneisen parameter was determined for Pd + 10 at% Mn. For Pd + 2 at% Mn, a ferromagnetic material below 5.8K, the corresponding Grüneisen parameter is strongly temperature dependent and changes even sign below 5K. By separating the magnetic contribution into a ferromagnetic and antiferromagnetic cluster part two temperature independent Grüneisenparameters could be determined. For the PdMn alloys, however, it was possible to go beyond the Grüneisen parameters by interpreting them in terms of a long-range polarization of the matrix by the ferromagnetically ordered Mn atoms and a short-range polarization of the matrix connected with the breaking up at higher temperatures of the antiferromagnetic pairs of Mn atoms.

W. A. Harrison:
One basic change in direction for high-pressure research, which seems inevitable from the papers at this meeting, is in the use of high pressures to learn about the electronic structure of solids at normal pressures. High pressures have been used as a tool to probe the theoretical predictions of electronic structure and its relation to solid-state properties. Note that very little work has been described here concerning simple metals, semiconductors, nor simple ionic solids. I believe that the reason is that those systems are already very well understood and almost any property of a well-characterized system can be calculated - under pressure or not. There is little motivation to do an experiment for which the result can be calculated. The "well-characterized" condition is an important one. One cannot calculate properties of a system for which the composition is uncertain or for stress distributions that are uncertain or complex, and even under the best of circumstances the prediction of structural changes is uncertain. However, the theory of changes of Fermi surfaces, band gaps, dielectric properties, and Grüneisen constants is in reasonably good shape. Simple theories, such as I described in my paper, can give reasonable estimates and full calculations, such as those described by Johansson and others, can give quite accurate estimates.

Thus much of the discussion in this category has concerned the most difficult systems, cerium and americium, for example, and I would judge these also to be rather well understood now. We have nearly run out of systems and the future does not lie, it seems to me, either in finding more and more systems nor in obtaining more and more accurate results for the systems which have already been studied. Certainly such activities will continue but they seem not to be nearly as exciting as the studies which have brought us this far. I think that the real frontier may be in exploiting the knowledge which we have gained, applying it in novel circumstances to make interesting systems with useful properties. In just this way the study of semiconductors has turned from learning about the bands of perfect crystals to the construction of composite systems and heterojunctions with remarkable, though predictable, properties. This frontier may be endless and will require inspiration and insight comparable to that which was required to bring us to where we could move on to this new phase.

B. Johansson:
To me it has been extremely satisfying to learn about the beginning of a world-wide use of synchrotron radiation within the field of high pressure research and also that x-ray absorption studies are now becoming feasible under pressure. The latter technique was nicely used to demonstrate the pressure-induced valence transitions in SmS (Ingalls et al.) and Ytterbium metal (Syassen et al.).

In the sixties quite a lot of experimental studies were performed on the pure elements. For example, we all know about the work made by Jayaraman on the melting temperature as a function of pressure. This type of work seems to have declined during the seventies and comparatively few structural investigations have been performed, despite the fact that very many phase transformations are known to occur under pressure and have not been structurally classified. However, with the new developments one may hope that this will change during the present decade. The time seems especially ripe for this now since first principle theoretical calculations of equation-of-states have rather recently become possible and hopefully the accuracy will turn out to be sufficient for a study of structural changes. Indeed, the present state of the art is such that people now have started to seriously consider the influence of lattice vibrations on the equation-of-state (Glötzel and McMahan in their Cs work, and Ross).

High pressure work has shown that elements which are not superconductors at normal conditions might become superconducting under compression. This has for example been demonstrated for Cs, Ba, Y, Sc, and Lu (Wittig). At this conference we learned about that Sr and (most likely) Ca also become superconductors at high pressure (Bundy and Dunn). From this a pattern emerges and superconductivity should now also be expected for the alkali metals Rb and K at attainable pressures. A similar pattern concerning the crystal structures also seems to emerge and we just heard that yttrium under compression attains the same crystal structure as La under ambient conditions (dhcp) (Vohra et al., Paderborn-preprint). This type of interrelationship between the crystal structures of elements belonging to the same group in the Periodic System shows that high pressure work can provide the missing link in the occurrence of various structures in the solid state and thus may reveal a beautiful periodicity of the crystal structures of the Elements.

S.H. Liu:
Since high pressure affects directly the lattice of a solid, it is very important to study the lattice dynamics of solids under pressure, especially those solids which undergo phase transitions, because any anomaly in the phonon spectrum will shed light on the basic mechanism of the phase transition.

For those materials which undergo second order phase transitions, it may be worthwhile to study the effects of pressure on the critical exponents. This can tell us whether there is a change in the effective dimensionsality or the effective degree of freedom (e.g. Heisenberg spin vs. Ising spin) under pressure. Information of this kind is useful in explaining any deviation from the normal high pressure behavior.

D. B. McWhan:
There are striking contrasts in at least two areas of research at high pressure between this conference and the one in 1965. The general trend of non-metals becoming metals at high pressure had been established experimentally by 1965, and the occurrence of superconductivity in almost all of these new metallic phases had been found. Many of the trends in crystal structure in the periodic table had also been found. By 1980 band structure calculations had progressed to the point where many of these trends are understood using the idea of a competition between a transfer of electrons from one band to another versus a band to localized transition within a band. Another area of contrast is in the field of melting. In 1965 melting was being studied experimentally at high pressure in a number of laboratories. No experimental studies of melting were discussed at this conference even though there has been substantial theoretical progress in recent years, mainly stemming from studies of lower dimensional systems. Perhaps this is an area for future study. Finally synchrotron radiation is already being used at high pressure for EXAFS and structural studies, but the potential for doing high resolution and time dependent experiments at high pressure remains untapped.

D. Rainer:
I would like to remind you of a brief but important message on T_C in superconductors which we got during this conference. It was somewhat hidden in a talk on bandstructure calculations, such that perhaps many of us have overlooked it. A.J. Freeman mentioned in passing: While studying structural transformations in La under pressure, he incidentally also calculated T_C, and found 6.6K. Experiments have verified this remarkably accurate result (T_C^{exp}=6.0K!), as reported here by J. Wittig. This is a striking refutation of B.T. Matthias' opinion on theories of T_C: "They have been totally and entirely useless."

G. A. Samara:
I was pleased to see that at this conference a number of speakers emphasized the systematic trend in the pressure behavior of given classes of materials. This is very good and more attention should be devoted to the elucidation of such trends for it is through trends in behavior that we develop a better understanding of physical phenomena in real systems.

A number of attendees are under the impression that displacive (soft mode) transitions are always first order. This is certainly not the case. In fact, some of the best known and understood cases (eg. the 104K transition in $SrTiO_3$ and the 129K transition in K_2SeO_4) are second order. In such cases the soft mode frequency vanishes precisely at the transition temperature, T_c, and the order parameter (which, for example, is the angle of rotation of TiO_3 octahedra in $SrTiO_3$) evolves continuously below T_c before reaching a saturation value in the low temperature phase.

In his summary comments, Prof. Wohlfarth drew attention to the apparent generality of the parabolic dependence of T_c on pressure which I discussed in my paper on displacive transitions, namely, $T_c \propto |P-P_c|^{1/2}$, as $T_c \rightarrow 0K$. He noted that this relationship is also observed in some ferromagnetic and superconductive systems. I have recently learned that it appears to also hold for a broken symmetry phase transition in solid ortho-deuterium at $P \geq 278$ kbar and low temperature in which the D_2 molecules, which are spherically symmetric at low pressure, go into an orientationally ordered state above 278 kbar (see, I.F. Silvera and R.J. Wijngaarden, Phys. Rev. Lett. 47 (1981) 39). Thus, the above result (which is a mean field result) appears to be independent of the specific nature of the interactions.

I agree with Prof. Wohlfarth that it would be of much interest to investigate and understand deviations from this behavior. Indeed, such deviations are known to occur in the incommensurate transition in Cr, in some superconductors and in ceramic ferroelectrics. The effects of dimensionality on this behavior would also be of interest.

J. S. Schilling:

I would like to make two comments--(a) The rather rapid increase of the superconducting transition of La with pressure has sometimes been attributed to the presence of special 4f electron character at the Fermi energy. I would like to point out that the increase of T_c versus *volume change,* which is the relevant parameter, is *not* particularly rapid, being no larger than that found for vanadium and other d-band metals. The increase of T_c versus volume change for systems like $CuMo_3S_4$, La_3S_4, La_3Se_4, and $Y_2Fe_3Si_5$ is far more rapid than for La, as is pointed out in several papers in the present volume. In his paper, J. Wittig has shown that a pressure increase of 200 kbar is required to raise T_c in *both* La *and* Y by 8K. There is certainly no special f-electron character near E_f in Y. It is, therefore, reasonable to conclude that whatever 4f-character La might have at its Fermi energy, this 4f-character is neither important for the pressure dependence of T_c, nor for the rather high value of T_c (5-6K) in La at ambient pressure.

(b) The second comment is that it is perhaps no accident that $T_c(P)$ for La, La_3Se_4, and $CuMo_3S_4$ goes through a maximum at about $T_c \approx 10\text{-}13K$. Lest one say, "Well, almost all contain La" one should be reminded that La in La has *three* valence electrons, whereas La in La_3S_4 has only 1/3 valence electron. These experiments may be telling us something important about the reasons why the T_c-value is limited. In fact, only a very small

fraction of all superconductors has T_c above 10K. Careful experimental and theoretical studies of the superconductivity of systems where T_c "maximizes out" as a function of pressure could help clarify why some systems, like the A-15's, often have high T_c-values, while others don't. Such studies could also shed some light on the general question of the most effective way to "engineer" a high T_c-system, having the entire periodic system at one's disposal.

J. W. Shaner:

With the rapid development of diamond anvil technology, the pressure range of static compression has spread into the area previously accessible only to shock waves. At the same time ab initio theoretical calculations are now providing insight into the effects we have observed in dynamic pressure experiments. The three approaches now overlap in a very satisfying way.

As a result of these developments, there have been changes in emphasis for the shock compression of solids. One new emphasis is on the development of techniques for measuring stress tensors and elastic constants. These measurements are needed for more detailed comparisons with static experiments and theoretical predictions. Another emphasis is a result of the fact that the high temperature high pressure corner of the solid stability field is still accessible only by shock waves. Our new experiments are particularly appropriate for observing high-temperature solid-solid and, especially, melting transitions.

Although we would still like to approach isentropic conditions in dynamic compression to isolate effects of density changes, such experiments at multi-megabar pressures remain extremely inaccurate, difficult, and expensive. I am not optimistic that many questions about solids at very high compressions will be worth the expense of developing such experiments.

T. F. Smith:

Following Peter Wohlfarth's comments on the possible application of his simple mean field pressure law to a variety of solid state transitions, including superconductivity, I would draw attention to the close adherence of the superconducting transition temperature, T_c, to a linear dependence upon volume (T.F. Smith, High Pressure Science and Technology, 6th AIRAPT Conference, K.D. Timmerhaus and M.S. Barber, (eds.) (Plenum Press, N.Y., 1979) Vol. I, p. 309). The suppression of superconductivity under pressure has not been pursued with the same degree of enthusiasm as the promotion, but the nature of the volume dependence as $T_c \rightarrow 0K$ is an interesting question which could now be examined well beyond the present limit of $T_c/T_{co} \approx 0.07$ for aluminum (D.U. Gubser and A.W. Webb, Phys. Rev. Lett. 35 (1975) 104).

The report by Francis Bundy of superconductivity in strontium under pressure and the positive value for dT_c/dP fits in nicely with the general pattern for the early transition metals (T.F. Smith, Superconductivity in d- and f- Band Metals, AIP Conference Proc. 4, D.H. Douglas (ed.) (ATP, N.Y., 1972)p. 293). The band structure calculations for these elements, which indicate an increase in d-electron character under pressure, also provide theoretical support for the earlier suggestion that such an increase could be responsible for the superconducting behaviour under pressure.

However, I would comment that in the enthusiasm for describing the influence of pressure upon the electronic properties, the phonons have been virtually ignored at this meeting. Certainly, the phonons are important and in some cases the distinction between phonon and electronic contributions may be quite arbitrary.

J. Wittig:
Lanthanum Ought to be a 4f Band Metal--Under this title I have published a brief paper some years ago (J. Wittig, Comments on Solid State Physics 6 (1974) 13). In the present remark I wish to maintain my previous position against opposite opinions which have been discussed at this conference. What I always meant and still mean by the term "4f-band metal" is the following: I believe that La's metallic properties (crystal structure, melting point, superconducting T_c, etc. and also the variations of these quantities under pressure) are so significantly anomalous within the family of trivalent metals (Sc, Y, Lu and La) that I'm obliged to conclude that this circumstance must be inherently linked to a qualitatively different electronic band structure with rather strong 4f character at the Fermi surface in contrast to the pure sd bands of Sc, Y, and Lu. In the meantime, several band calculations for La revealed the anticipated share of itinerant atomic-like 4f electrons. Nonetheless, Pickett, Freeman, and Koelling think that this 4f charge is insignificant and that La is a typical $(sd)^3$ transition metal in terms of its electronic configurations (W.E. Pickett, A.J. Freeman, and D.D. Koelling, Phys. Rev. B 22 (1980) 2695).

Theoretical work by Duthie and Pettifor (Phys. Rev. Lett. 38 (1977) 564) revealed a structural sequence from hcp to Sm-type→dhcp→fcc for the trivalent rare earth metals (including the end-members La and Lu) as a function of increasing d-band occupancy with increasing pressure ("s→d transfer"). In their view, La occurs in the dhcp structure at ambient pressure since it happens to have about 0.6 5d electrons per atom more than Lu.

In our view, the Sm, dhcp and fcc structures are related to some 4f character at the Fermi surface in rare earth metals with open 4f shells. According to Duthie and Pettifor's prediction, the pure d-band metal Lu (and perhaps even Y and Sc)

should exhibit the above structural sequence. In our view, Lu should presumably behave differently.

Liu (J. Phys. Chem. Solids 36 (1975) 31) has in fact discovered a phase transition in Lu at 230 kbar and claims that it is from hcp to Sm-type, contrary to our expectation. We think these preliminary data must be judged with caution. An apparent drawback of the reported experiment is that is was carried out on a two-phase sample only with several diffraction lines of the remnant hcp phase coinciding with possible lines of the anticipated Sm-type structure. Most recently, Vohra et al. (preprint from Y.K. Vohra, H. Olijnik, W. Grosshans, and W.B. Holzapfel) claim that the high pressure phase YII (above ≃ 130 kbar) also has the Sm structure and is followed by a phase YIII at higher pressure which is believed to be dhcp. Since these structure determinations are based on five fairly broad diffraction lines only, some doubt arises, whether these experiments present "conclusive experimetal evidence" already for a pressure-induced hcp-Sm-dhcp sequence in the pure d band metal Y.

If the hcp-Sm-dhcp-fcc structural sequence should in fact turn out to be correct for trivalent pure d-band cases like Y and Lu, I will, as I said at the conference, publicly admit that my expectation was wrong. More structural research at high pressure will obviously be extremely fruitful in connection with subtle questions of electronic structure of rare earth metals on which I touched in this note. If the subject should be essentially settled at the next conference of this kind, I will provide a keg of beer for all those participants who have been active or interested in this problem, regardless of whether I'm right or wrong.

EDITORS' COMMENT:
Both editors, J.S. Schilling and R.N. Shelton, would like to emphasize that we are extremely interested in the problem discussed by J. Wittig.

E. P. Wohlfarth:
Two remarks--(a) As an example of the universality of physical laws it may be noted that the parabolic law we obtain in magnetism, using a mean field model, i.e. $T_c(P)=T_c(0)(1-P/P_c)^{1/2}$, should be applicable for other phenomena involving phase transitions. These include superconductivity, ferroelectricity and some structural phase transitions. We have attempted to discuss *deviations* from the parabolic behaviour in terms of metallurgical effects, such as concentration fluctuations. It may be interesting to have a similar approach where such deviations occur in the non-magnetic cases quoted.
(b) In an earlier paper (Phys. Lett. 75A (1979) 141) we have proposed that ε-Fe may be a high pressure, low temperature superconductor and have estimated $0K<T_c<0.25K$. We understand this proposal will be tested in the near future and wish to draw attention to this exciting possibility.

AUTHOR INDEX

ABE, K., 141
ANDERSEN, O.K., 245
AOKI, K., 141
ASHCROFT, N.W., 155

BACH, H., 385
BALOGH, J., 47
BARANOWSKI, B., 231
BATLOGG, B., 215
BATURIĆ-RUBČIĆ, J., 117
BAUBLITZ, JR., M.A., 81
BELL, P.M., 137
BLECKWEDEL, A., 323
BLOCH, D., 203
BOPPART, H., 301
BRAUN, H.F., 381
BUCHER, E., 295
BUNDY, F.P., 401
BURAS, B., 305

CHU, C.W., 357, 365
CLARKE, R., 223
CROFT, M., 335, 341
CROZIER, E.D., 67

DANIELS, W.B., 23
DE BOER, F.R., 181
DIATSCHENKO, V., 357
DRICKAMER, H.G., 3
DUNLAP, B.D., 271
DUNN, K.J., 401

EICHLER, A., 323
EILING, A., 385
ETTERS, R.D., 39

FELDHAUS, J., 319
FILIPEK, S., 231
FINGER, L.W., 137
FLEMING, R.M., 219
FRANK, K.H., 319
FRANSE, J.J.M., 181
FREEMAN, A.J., 193
FRINGS, P.H., 181
FRITSCH, G., 239

GAL, J., 271
GERWARD, L., 305
GLÖTZEL, D., 263
GREENBLATT, M., 215
GUPTA, L., 341

HARRISON, W.A., 57
HAZEN, R.M., 137
HELMY, A.A., 39
HOCHHEIMER, H.D., 295
HODGES, J.A., 341
HOMAN, C.G., 407
HOR, P.H., 365
HUANG, S.Z., 357

INGALLS, R., 67

JAMIESON, J.C., 47
JAYARAMAN, A., 295
JOHANSSON, B., 169, 245, 305

KAINDL, G., 319
KALVIUS, G.M., 271
KEMLY, E., 341
KENDALL, D., 407
KLUKOWSKI, M., 231
KOBAYASHI, T., 141
KOMMANDEUR, J., 203
KRISHNAN. A., 341

LAM, D.J., 271
LEBECH, B., 305
LEVINE, H.H., 335
LIN, T.H., 357, 365
LIU, S.H., 327
LÜSCHER, E., 239

MACCRONE, R.K., 407
MAINES, R.G., 215
MANGHNANI, M.H., 47
MAO, H.K., 137
MCMAHAN, A.K., 161, 169
MCWHAN, D.B., 219
MEDINA, F., 23
MENG, R.L., 357
MENONI, C.S., 73
MENOVSKY, A., 181
MIGNOT, J.M., 311
MING, L.C., 47
MINOMURA, S., 131
MOSER, J., 271
MURGAI, V., 341

NAKAI, J., 149
NEIFELD, R., 335
NICOL, M., 33
NOWIK, I., 271

OTTO, A., 125

PARKS, R., 341
POTZEL, W., 271

QADRI, S., 47, 73

RAZAVI, F.S., 397
ROSS, M., 161
RUBČIĆ, A., 117
RUOFF, A.L., 81

SAMARA, G.A., 91
SCHIFERL, D., 47
SCHILLING, J.S., 345, 385, 397
SCHIRBER, J.E., 207
SCHMIDT, J., 23
SEGRE, C.U., 381
SHANER, J.W., 99
SHASHIDHAR, R., 109
SHIMOMURA, O., 131
SHIRAKAWA, T., 149
SKELTON, E.F., 47, 73
SKRIVER, H.L., 169, 245, 279, 305
SMITH, T.F., 369
SPAIN, I.L., 73
SPIRLET, J.C., 271
STAUN OLSEN, J., 305
STEENSTRUP, S., 305
SYASSEN, K., 33, 125, 319

TAKEMURA, K., 125
TAKEMURA, K-I., 131
TEI, T., 141
THOMPSON, A.H., 365
TOMIZUKA, C.T., 1
TRANQUADA, J.M., 67
TUPS, H., 125

VETTIER, C., 203
VOIRON, J., 203

WACHTER, P., 301
WEBB, A.W., 73
WEGER, M., 199
WERNER, A., 295
WHITMORE, J.E., 67
WILDERMUTH, A., 239
WILLER, J., 239
WILLIAMS, F., 15
WITTIG, J., 283, 311
WOHLFARTH, E.P., 175
WORTMANN, G., 271, 319
WU, M.K., 357, 365

YAMAMOTO, K., 141

ZHANG, D.L., 365
ZOU, G.T., 137

419

AUTHOR INDEX

ABE, K., 141
ANDERSEN, O.K., 245
AOKI, K., 141
ASHCROFT, N.W., 155

BACH, H., 385
BALOGH, J., 47
BARANOWSKI, B., 231
BATLOGG, B., 215
BATURIĆ-RUBČIĆ, J., 117
BAUBLITZ, JR., M.A., 81
BELL, P.M., 137
BLECKWEDEL, A., 323
BLOCH, D., 203
BOPPART, H., 301
BRAUN, H.F., 381
BUCHER, E., 295
BUNDY, F.P., 401
BURAS, B., 305

CHU, C.W., 357, 365
CLARKE, R., 223
CROFT, M., 335, 341
CROZIER, E.D., 67

DANIELS, W.B., 23
DE BOER, F.R., 181
DIATSCHENKO, V., 357
DRICKAMER, H.G., 3
DUNLAP, B.D., 271
DUNN, K.J., 401

EICHLER, A., 323
EILING, A., 385
ETTERS, R.D., 39

FELDHAUS, J., 319
FILIPEK, S., 231
FINGER, L.W., 137
FLEMING, R.M., 219
FRANK, K.H., 319
FRANSE, J.J.M., 181
FREEMAN, A.J., 193
FRINGS, P.H., 181
FRITSCH, G., 239

GAL, J., 271
GERWARD, L., 305
GLÖTZEL, D., 263
GREENBLATT, M., 215
GUPTA, L., 341

HARRISON, W.A., 57
HAZEN, R.M., 137
HELMY, A.A., 39
HOCHHEIMER, H.D., 295
HODGES, J.A., 341
HOMAN, C.G., 407
HOR, P.H., 365
HUANG, S.Z., 357

INGALLS, R., 67

JAMIESON, J.C., 47
JAYARAMAN, A., 295
JOHANSSON, B., 169, 245, 305

KAINDL, G., 319
KALVIUS, G.M., 271
KEMLY, E., 341
KENDALL, D., 407
KLUKOWSKI, M., 231
KOBAYASHI, T., 141
KOMMANDEUR, J., 203
KRISHNAN, A., 341

LAM, D.J., 271
LEBECH, B., 305
LEVINE, H.H., 335
LIN, T.H., 357, 365
LIU, S.H., 327
LÜSCHER, E., 239

MACCRONE, R.K., 407
MAINES, R.G., 215
MANGHNANI, M.H., 47
MAO, H.K., 137
MCMAHAN, A.K., 161, 169
MCWHAN, D.B., 219
MEDINA, F., 23
MENG, R.L., 357
MENONI, C.S., 73
MENOVSKY, A., 181
MIGNOT, J.M., 311
MING, L.C., 47
MINOMURA, S., 131
MOSER, J., 271
MURGAI, V., 341

NAKAI, J., 149
NEIFELD, R., 335
NICOL, M., 33
NOWIK, I., 271

OTTO, A., 125

PARKS, R., 341
POTZEL, W., 271

QADRI, S., 47, 73

RAZAVI, F.S., 397
ROSS, M., 161
RUBČIĆ, A., 117
RUOFF, A.L., 81

SAMARA, G.A., 91
SCHIFERL, D., 47
SCHILLING, J.S., 345, 385, 397
SCHIRBER, J.E., 207
SCHMIDT, J., 23
SEGRE, C.U., 381
SHANER, J.W., 99
SHASHIDHAR, R., 109
SHIMOMURA, O., 131
SHIRAKAWA, T., 149
SKELTON, E.F., 47, 73
SKRIVER, H.L., 169, 245, 279, 305
SMITH, T.F., 369
SPAIN, I.L., 73
SPIRLET, J.C., 271
STAUN OLSEN, J., 305
STEENSTRUP, S., 305
SYASSEN, K., 33, 125, 319

TAKEMURA, K., 125
TAKEMURA, K-I., 131
TEI, T., 141
THOMPSON, A.H., 365
TOMIZUKA, C.T., 1
TRANQUADA, J.M., 67
TUPS, H., 125

VETTIER, C., 203
VOIRON, J., 203

WACHTER, P., 301
WEBB, A.W., 73
WEGER, M., 199
WERNER, A., 295
WHITMORE, J.E., 67
WILDERMUTH, A., 239
WILLER, J., 239
WILLIAMS, F., 15
WITTIG, J., 283, 311
WOHLFARTH, E.P., 175
WORTMANN, G., 271, 319
WU, M.K., 357, 365

YAMAMOTO, K., 141

ZHANG, D.L., 365
ZOU, G.T., 137